I0084005

Henry Hartshorne

Essentials of the Principles and Practice of Medicine

Henry Hartshorne

Essentials of the Principles and Practice of Medicine

ISBN/EAN: 9783337311803

Printed in Europe, USA, Canada, Australia, Japan

Cover: Foto ©berggeist007 / pixelio.de

More available books at **www.hansebooks.com**

ESSENTIALS

OF THE

PRINCIPLES AND PRACTICE

OF

MEDICINE.

A

HANDY-BOOK

FOR

STUDENTS AND PRACTITIONERS.

BY

HENRY HARTSHORNE, M. D.,

PROFESSOR OF HYGIENE IN THE UNIVERSITY OF PENNSYLVANIA, AUXILIARY FACULTY OF
MEDICINE; FORMERLY PROFESSOR OF PRACTICE OF MEDICINE IN THE MEDICAL
DEPARTMENT OF PENNSYLVANIA COLLEGE; LATELY PHYSICIAN TO THE
EPISCOPAL HOSPITAL OF PHILADELPHIA; FELLOW OF THE
COLLEGE OF PHYSICIANS OF PHILADELPHIA;
MEMBER OF THE AM. PHILOS. SOCIETY,
ETC. ETC.

PHILADELPHIA:
HENRY C. LEA.
1867.

Entered according to the Act of Congress, in the year 1867, by

HENRY C. LEA,

in the Office of the Clerk of the District Court of the United States in and for
the Eastern District of the State of Pennsylvania.

PHILADELPHIA:
COLLINS, PRINTER, 705 JAYNE STREET.

PREFACE.

THIS manual is an unambitious effort to make useful the experience of twenty years of private and hospital medical practice, with its attendant study and reflection. Whatever defects the book may have, the author does not concede that it is necessarily a *fault* that it is *small*. That science whose facts and laws may be stated in the fewest words, is the most advanced. Even in the complex and still very incomplete science of Medicine, something like the same idea may, at least, justify the briefest statement, compatible with clearness, of its most important and best established doctrines and precepts. While brevity has been aimed at upon all subjects, the most extended consideration has been given to those which especially require the attention of the student, from their difficulty, comparative novelty, or intrinsic importance. In the practical part, however, mere novelty has not been made a ground of selection of the means recommended for the treatment of disease. Rather than to enumerate a long list of remedies which have been proposed or

1*

employed in each case, the author has preferred to give, chiefly upon the basis of his own experience, those which he believes to be the best. A didactic, almost dogmatic method of expression has thus resulted, from the effort to condense within the smallest possible limits the essentials of Practical Medicine.

H. H.

PHILADELPHIA, July, 1867.

CONTENTS.

viii CONTENTS.

SECTION III.

GENERAL THERAPEUTICS.

SECTION IV.

NOSOLOGY.

PART II.

SPECIAL PATHOLOGY AND PRACTICE.

CONTENTS.

PRINCIPLES AND PRACTICE

OF

MEDICINE.

INTRODUCTION.

SYSTEMS OF MEDICINE.

BEFORE Lord Bacon, and before, in fact, all others whose writings have come down to us, Da Vinci, the architect, painter, and engineer, proclaimed in the first half of the sixteenth century, that in the study of natural truth we must consult *experience, experience* rather than reason. "Those," said he, "who in the study of the sciences do not consult nature, but authors, are not the children of nature, they are only her grandchildren." "Nature begins from the reason and ends in experience, but we must take the reverse course, begin from the experiment and try to discover the reason." "Theory is the general, but *experiments* are the *soldiers*."

Not that these were the first utterances in all time in favor of the value of observation and experiment in acquiring a knowledge of. nature; but only that now, for the first time, these began to be the *governing ideas* of science and philosophy. Aristotle was a naturalist, although still more emphatically a dialectician; Leucippus and Democritus founded a school whose dependence was almost exclusively on the evidence of the senses; and even Cicero, who paid little attention to natural science, wrote this wise sentence; "Præstat naturæ voce doceri, quam ingenio suo sapere." But it is especially interesting to us to recall the fact that most clearly, perhaps, of all the ancients, was this reliance upon nature enunciated, and most practically was it exemplified, by Hippocrates of Cos. He asserted again and again in his works that "nothing should be affirmed concerning the nature of man, until after having acquired a certainty of it by the aid of the senses." And, although this may seem very obvious indeed to us, yet it is a familiar fact that the great intellects of antiquity, from the sages of the Vedas and from Pythagoras and Plato, downward, had more confidence in the truth-compelling powers of their own reason; and even Hippocrates himself often forgot his own maxims, and became dogmatic beyond his knowledge.

2

It is not my purpose here to go into any historical discussion of philosophy; which would be inappropriate in this place. Nor will I attempt to crowd into a few pages the history of medicine itself. But it appears to me that I cannot better occupy space, in this introduction, than by endeavoring to place before the mind such a succinct view of the *most essential* phases and mutations of medical opinion, in times past and present, as will enable us to apprehend all that bears upon the aspects and prospects of the theory and practice of Medicine.

In the midst of the multitude of authors who have written upon medicine, in every age which has possessed a literature, the number of cardinal ideas, of distinctive methods, opinions, or principles, has not been great. Those who may be considered to have been original thinkers or leaders in medical philosophy have been few; or, if we cannot refuse to a larger number the credit of originality, yet that of actual novelty is not often theirs, as they have merely started anew an idea, a principle, a system, or theory, which had long ago had its propounders, its advocates, and its opposers, although it may have been again forgotten.

Yet, few as these essential ideas have been, it will be impossible to do more than mention them, as it were, in catalogue, at present.

A work, for example, might be, and more than once has been, written upon the doctrines of Hippocrates, and of the writings classed under his name alone. Suffice it for us to recollect, that the leading idea of this greatest of physicians was *reliance* upon, and *observation* and *imitation* of *nature*. Yet he theorized upon health and disease, upon the four elements and the four humors, and his system has, therefore, been styled *Dogmatic*. To him, also, is traced the principle of medication by contraries; τα ἐναντια των ἐναντιων ἰστιν ἰηματα. The greatest value of the Hippocratic writings undoubtedly consists in their numerous and admirable descriptions of the symptoms of disease, and of the relations of symptoms to prognosis. The study of hygienic laws and influences also received from his school much attention.

Contrasted with the Hippocratic reflective or dogmatic method of studying nature, was the more detailed and less systematic plan of the contemporaneous Cnidian school.

Later, with Philinus and Serapion of Alexandria, the distinctly *Empirical* method was promulgated; in which observation, and this alone, especially as to the use of remedies, was urged. No reasoning about *why* or *how*, but only *what*, engaged the minds of these industrious men; whose materials thus accumulated only too fast for their limited powers of classification. Their most elegant writer was Aretæus; who is not always credited to them; but whose descriptions of disease have seldom been equalled, even down to our own day.

It is less easy to characterize, in a few words, the school curiously called *Methodist;* which originated with the opinions of Cleophantus of Alexandria, and of Asclepiades the Bithynian, the friend of Cicero, and was established by Themison, their disciple, at Rome. Dismissing the expectant study of the course of diseases inculcated by Hippocrates—which they laughed at, as a " meditation upon death"—and denying his theory of " coction" and " crises," they dogmatized in a different way, upon the changes occurring under disease in the condition of the solid structures of the body, and in the movement of its

atomic components. Making but two essentially different pathological states, the "laxum" and the "strictum," they simplify the theory
of medicine very much. Chiefly, however, was Asclepiades distinguished for the moderation of his practice; rejecting complex, violent,
and perturbatory remedies, and aiming, as he said, to cure "tuto, cito,
et jucunde." A somewhat complicated course of alterative treatment
is, however, ascribed to his successors, by Cœlius Aurelianus, under
the name of the "metasyncritic circle."

The most judicious, as well as one of the most learned of physicians, was Aulus Cornelius Celsus. He selected, from the opinions
and practice of his predecessors and contemporaries, those of the
greatest soundness; so that, not having propounded any exclusive
dogma, nor yet being limited by the narrow results of observation
alone, he may be justly styled *eclectic;* or, as that term has been made
odious of late by the usurpation of a set of pretenders, *episynthetic,*
or *comprehensive,* might be a preferable title.

Galen, less carefully selective, although undoubtedly an admirable
man, excellent practitioner, and learned writer, renewed and added
further strength to the hypothetical as well as the practical views of
Hippocrates.

From his time, but little of original force appeared in medical
literature, until after the period of eight or nine centuries of mediæval
darkness had been broken in upon by the revival of learning and
intellectual activity, in the fifteenth and sixteenth centuries.

In this revival, it was natural that much recourse should be, at
first, had to the treasures of the ancients. Plato and Aristotle
divided the newly-revisited realm of philosophy; while Galen, as the
exponent of Hippocratic doctrine, almost monopolized that of medicine—until Da Vinci, Telesius, Cæsalpinus, Campanella, and *Bacon*
established the *inductive* method of observation and experiment,
most obviously necessary for advancement in the physical sciences,
of which medicine is one; one, too, which, as Lord Bacon expressed
it, had been previously "more labored than advanced."

Chemistry, which had already received much attention from the
Arabians, and which, under the fascination of alchemy, had reached
valuable discoveries—which, in fact, in the hands of Albertus Mag-
nus, Roger Bacon, Basil Valentin, Isaac Hollandus, and others, had
performed wonders; and, in the trumpetings of Paracelsus, the arch-
quack, had made still more extraordinary pretensions—chemistry
was now ripening into a great science. In the seventeenth century,
its influence upon physiological and pathological theory much increased; and the practice of medicine could not fail to be consequently affected. By Sylvius of Amsterdam and Leyden, and by
Thomas Willis especially, a school of *Chemiater,* iatro-chemists, or
chemical physicians was instituted.

Following the discovery by Harvey of the circulation of the blood,
in the investigations of Sanctorius and Borelli, of Pisa, *mechanics*
likewise found a place in the study of the functions of the body, in
health and disease. An *Iatro-mechanical* school may be thus said
to have existed, to which the distinguished Sénac, physician to
Louis XIV., among others, contributed, in a work of great ability.

Boerhaave, professor at Leyden, endeavored to combine these, the chemical and mechanical modes of studying the body and its disorders, into an ingenious but complex eclectic system of his own; which his influence, as a man of genius, and one of the first of modern clinical lecturers, enabled him to extend far and wide. It was rather a dogmatical than an empirical eclecticism.

The latter was admirably exemplified in the writings of Sydenham ; who has been well called the modern Hippocrates. Certainly there was a great resemblance between the methods of the Greek and the English fathers of medicine.

At the beginning of the seventeenth century, there grew up, in the University of Halle, two opposing theories : the *Animism* or psycho-vitalism of Stahl, and the *Solidism* and neuro-pathology of Hoffmann. Stahl's doctrine was, in brief, that the *soul* of man governs health and disease. An expectant or do-nothing practice naturally followed from such a view. Hoffmann taught a less simple scheme; but that part of it which seemed to the renowned and learned Cullen, the nosologist of Edinburgh, to be the most worthy of adoption, was his appreciation of the importance of the *nervous system* in the production of the phenomena of disease.

But the most brilliant of the meteors that have crossed the horizon of medical science, not disappearing, indeed, any of them, without leaving some solid precipitate of knowledge, was the *Sthenic* system of John Brown, of Edinburgh ; the pupil, friend, rival, and enemy of Cullen.

All life, according to this bold and able, although too reckless dogmatist, depends upon *stimulation;* all disease upon too much or too little excitement, causing direct or indirect debility. Ninety-seven cases of sickness out of a hundred, in his therapeutics, require stimulation for their relief or cure. Wonderfully simple, this ! Haller's doctrine of the irritability of organic tissues was, very possibly, its source ; but so nearly akin was it to the great idea of *vitalism*, dimly seen by Pythagoras, announced by Hippocrates, but lost for ages until revived and distorted by Van Helmont and Stahl, and afterwards rendered more positive by J. Hunter and Bichat,—so near was it to this imperishable idea, that Brown's theory, thus supported at once by ancient philosophy and modern discovery, had an unprecedented influence upon medicine. All theories and theorists, during and since his time, unless we except the discreet vitalism of Barthez, of Montpelier, have reflected or refracted, with various modifications, the Brunonian ray.

What have we had since, in fact ? Rasori, in Italy, adopted Brown's physiological basis, but considered that excitability and excitement were *multiple*, and *unequally distributed*, in disordered states, in different organs ; and, moreover, that over-excitement was much more frequent and demanded more attention in practice than Brown had supposed. Hence arose his sedative or "contro-stimulant" method, by large bleeding and tartar emetic ; so famous once, especially in the treatment of inflammatory affections of the chest.

Broussais, in France, proceeding upon the same original basis, saw in *local irritation* and *inflammation*, mostly of the alimentary canal, the seat and centre, the *fons et origo* of the dynamic or excitational error which caused all diseases. His practice varied from both Brown and Rasori; his whole object being to calm and allay the central irritation by diluents, demulcents, local depletion, and counter-irritation, avoiding all heroic treatment.

In this country, which can hardly be said to have had a system before Rush, that noble and independent mind was also influenced by the Brunonian radiation; although a still different view of pathology and therapeutics resulted from his reflections and observations. The "unity of disease" was, with him, a favorite idea. Although his strong good sense did not allow this to exclude from his appreciation remedies and modes of treatment not easily reconciled with such a scheme, Rush evidently leaned much toward the opinion, that all acute diseases were but different "states of fever;" for the mitigation of most of which the lancet was the most potent and indispensable remedy. In this he resembled Rasori rather than Brown or Broussais. Even earlier than the universal dissemination of the teachings of the latter, the distinguished successor of Rush, Dr. Nathaniel Chapman, of this city, claimed and afforded evidence that he had first taught the theory of the local origin of fever, in irritation of the alimentary canal; but he did not allow it to modify his practice in the same manner as Broussais.

Now,—let us look back. Who have been the ιατρο-προφῆται, the great leaders in medical speculation, the reformers and deformers of medical practice? The list is not a long one, although its scope of time reaches over two thousand years. Permit me to hazard their enumeration. Hippocrates, Serapion, Asclepiades, Celsus, Sylvius, Harvey, Borelli, Sydenham, Boerhaave, Stahl, Hoffmann, Haller, Cullen, Avenbrugger, Brown, Jenner, Hunter, Bell, Bichat, Barthez, Pinel, Rasori, Rush, Hosack, Laennec, Broussais, Louis, Liebig, Virchow.

And what have been their essential *ideas,* stripped of all their complexities and environments? *Naturalism, empiricism, eclecticism, humoralism, solidism, chemicism, mechanicism, neuro-pathology, stimulism, phlogisticism, pyrexism, vitalism,* and latest, of the present date, *cellular pathology.*[1] I leave out of the list the Thomsonian extravaganza of thermalism, and the Hahnemannic homœopathism, as, however serious may have been their detrimental effect upon the welfare of the public at large, they have scarcely influenced the progress or present status of medical science, either for good or evil.

By *naturalism* I mean dependence upon nature, and systematic imitation, in practice, of her spontaneous curative processes. We have already referred to this as the leading Hippocratic idea. It was rejected by the early Methodists, practically repudiated by Cullen, and systematically excluded by Rush. Quite recently it has been conspicuously illustrated and defended in the lucubrations of Sir John Forbes, following those of Dr. Bigelow, of Boston, upon "Nature and Art in Disease;" to the former of which the soubriquet of "Young Physic" has been applied.

Empiricism is strict adhesion to experience; the accumulation of

---

[1] We may add, now, the *ganglio-therapy* of Dr. John Chapman; built upon the vivisections of Bernard and Brown-Séquard, and an ephemeral (though now current) physiological speculation, in regard to the circulation.

2*

means of treatment simply by observation and experiment, independently of physio-pathological reasoning. The most favorable example of this among the ancients, was Arctæus; of distinguished moderns, Sydenham, Laennec, and Louis.

Eclecticism or *episynthetism* is, of course, the selection or combination of what is deemed best in several methods, as, in practice, of means or measures, some of which have been obtained by mere observation, and some from physiological reasoning or deduction. Celsus afforded the most beautiful early example of this; it has been exemplified, although at the same time, somewhat paradoxically, derided, in our own period, by Trousseau, of Paris.

All of the other systems which I have named are phases of *Rationalism*, which is the proper antithesis of Empiricism.

Solidism, first broached perhaps by the ancient school of Asclepiades, with its *laxum* and *strictum*, was urged to its farthest limit in the *mechanicism* of Borelli, and in the neuro-pathology of Hoffmann, Cullen, and Henle. It was taught in this city for twenty-five years by Professor Chapman.

Humoralism, the older view, which saw in changes of the fluids all that was essential in disease, pervaded the system of Galen, and the Galenists of the fifteenth and sixteenth centuries. The chemicists generally have had a natural leaning towards it. In this country it was represented at one time by Dr. Hosack, of New York. A very distinguished example of it has lately been known and respected in England, in the lamented Robert Todd, of London.

Chemicism was boldly inaugurated by Sylvius De Lebö, in the seventeenth century, but has received its ripest contributions in the two last decades; especially from Liebig, of Giessen, and the other chemical physiologists, Lehmann, Moleschott, &c.

Mechanicism, as an exclusive system of physiological or pathological reasoning, was never permanently established; its influence, as affording even a predominant bias, having been always confined to a few thinkers during a brief period.

Neuro-pathology has had a more important place; dividing with a modified humoralism the domain of medical theory, even down to the present hour. We can never dismiss the consideration of the nervous centres and their communicating nerves, from the study of the human functions, healthy or morbid. So that, although it is decidedly an error to say, as some do, that man is *all brain*, or that the "nervous mass" is the animal, yet the nervous system must be made prominent in all medical inquiries.

Enough has been said already to explain the nature and powerful influence of the Brunonian theory, of excitement, or of sthenia and asthenia, which I have named under the title of *stimulism*. It was one step toward the application to pathological and medical truth of that dynamic physiology, that study of the *forces* of the living body, in connection with the constantly acting forces of external nature, which is now, or soon, destined to rule supremely, not as excluding, but as *guiding* our investigations of the chemical and mechanical changes both of the solids and of the fluids. Life is *not merely* excitation; but normal excitation is one of the requisite *conditions* of the performance of all the functions of the body, not even excepting that of growth and development itself; since to this a certain degree of *heat* at least is essential.

Rasori was, moreover, right in saying that excitement is *not a unit* for the whole body, but may be unequal in its different parts; and, moreover, that *excess* of excitement of one or more organs or functions, is at least as frequently present in acute disease as the reverse.

So, too, Broussais made a just amendment of the same scheme, to a certain extent, in noticing the *sympathetic* and *secondary* effects of *local irritation;* although he, as well as our Chapman, undoubtedly exaggerated the relative importance of irritation of the stomach and intestinal canal.

We need not pause for a moment over the Stahlian theory of the organic soul or *autocrateia;* although very lately a view much like it has been again taken, by Laycock and Morell, under the cognomen of the "unconscious soul."

The last phase of revolution in the scientific basis of medical opinion, has been that which, in the language of its most eminent leader, Virchow, of Berlin, we may designate as *cellular pathology.* Associated in similar, although not quite identical views, have been Prof. Bennett, of Edinburgh, and the late Dr. Addison, of London.

It has been a favorite idea with the physiologists of our period, that in the general law of organic cell-genesis, in the fact that every living being, human, animal, and vegetable, springs from a globoid *germ-cell,* while most of the separate tissues also have the cell for their first starting form; that in this we have the great central radical fact of physio-pathology, out of which (as in physical science out of the Newtonian law of gravitation) all truth in the history of the animal organization, and thus in medicine, must grow. But Dr. Bennett, one of the most earnest students and teachers of the cell doctrine,[1] denies that to it can be awarded such a place or potency; as it is not a *universal* law, but has its manifest exceptions. Dr. Beale, also, a leading British authority in histology, insists upon some essential modification in Virchow's theory.

It does not belong to me to discuss this point here; but, as it bears largely on the theory of medicine, I will merely say, that if there *be* one fact or idea which more than any other is the gravitative centre of all truth in physiology, pathology, and medicine, it is that of the peculiarity and supremacy in the body of the *life-force,* and of its intimate relations with the other physical forces;[2] of its being, in fact, capable of *degrees* of *life temperature,* like those of heat temperature, in the body as a whole, and in its various parts and organs; of its possessing *attributes,* like the other forces or phases of impetus and molecular movement in nature; which must be much more patiently and thoroughly studied than they have yet been, before we can be said to understand the human economy, even so well as astronomers now do the solar and sidereal systems.

This brings us towards the conclusion of our inquiry. We have been examining, in this brief manner, several schemes of *rationalism.*

---

[1] Clinical Medicine, Introductory Lecture.
[2] See *Grove, Carpenter,* and others, on the Correlation of Physical and Vital Forces; *Inman,* Foundation for a New Theory and Practice of Medicine; *Chambers,* Renewal of Life; &c.

But, as the use of facts and ideas in the *practice of medicine* is our stand-point, we must now ask, Is rationalism *available* for the treatment of disease? Is physiology perfect? How much of it is positive? We are compelled to answer—Physiology, and with it necessarily pathology, is one of the least-matured, because one of the most complex of the sciences. What would be said, then, were a man to undertake to repair a watch, when he had never seen its *works in motion*, and had no *proven* knowledge of the mode of action of nearly all its machinery? If he should find, on *trial*, that hanging it up, or laying it down, or shaking it when it stopped, or keeping it warm or cold, promoted its good time-keeping—very well; let him do so. But if, in this state of uncertain knowledge, he should seize and alter, with fingers or forceps, the delicately arranged and complicated wheels and springs, would not the chances be, that he would do more mischief than good? Nor would reasoning about possible or probable watches, theories in chronometry, avail him much toward the medication of the particular timepiece in his hands. Yet this is our position, as physicians, regarding the present relation of physiology and pathology, to the actual treatment of disease. It seems, therefore, only a slight over-statement of Trousseau's, that "La rationalisme ne conduit en médecine qu'à des sottises." *Rationalism in medicine leads only to absurdities.*

We might easily confirm Trousseau by other authorities, early and late. Stahl spoke of the materia medica of his time as a "stable full of offal." Sydenham complained that practice was "pestered with too many eminent remedies." It is said, that when Sydenham was asked by Sir R. Blackmore, what book to begin his medical studies with, he replied, *Don Quixote.* Bichat denounced the vague theories of medication prevalent in his day, and declared that but little was really positive in our knowledge of the action of remedies. Pinel had so little confidence in therapeutics, that his only study of disease was for a naturalistic classification: "Given a certain malady—to find its place in the nosological system." Laennec considered physiology and pathology "vain amusements of the mind." Says Lebert, "We cannot yet, unhappily, construct therapeutics on the basis of scientific medicine; and with the best intentions in the world we can regard the greater part of its precepts but as the result of empiricism."

But, some may exclaim, this is treason! This would remove the practice of medicine from science altogether, and leave it at the mercy of Paracelsus, and Cagliostro, and the old women! Not so. We have only to turn back to the grand platform of Bacon, on which all modern science is built, to see that to found the practice of medicine on *observation* is to make it *eminently* scientific. What science can do without empirical observation? Can physics, or astronomy, or chemistry? None of them. How irrational, then, to attempt to reason out. *a priori,* therapeutics, or to place it upon any other principal basis than clinical observation! Blind, uninstructed, unsystematic empiricism is a bane to society, and a disgrace to the human intellect.[1]

---

[1] I advise no one to imitate the follies of Cato the Censor; who, while he forbade his son ever employing a physician, yet dosed his own wife to death; attempted to reduce dislocations by repeating magical words, and wrote a book, in which he recommended cabbage as a sovereign remedy for many diseases.

But *scientific* empiricism constitutes the most rational practice attainable, while physiology is imperfect. What is most wanted now, is more *positivism* in medicine; more *exact observation* of clinical and therapeutical facts. It is otherwise in most of the natural sciences. Agassiz, one of the great leaders in science, has remarked, that *thought* and *generalization* are now especially required amongst naturalists; who are in danger of being buried among their multitudinous detailed facts, as knights of old sometimes were borne down by the weight of their own armor. But it is not so in our science. *Medicine needs more fact and less theory.* I could sustain these positions by argument, by citation, and by example; but we have no room. The proposition must be barely stated, that the most complete knowledge possible of a disease will never *alone* inform us, what will be the effect upon it of any remedies, until *experience* has put them to the test. The two blades of the scissors of practical medicine are, diagnosis and clinical proof. Nor does our total ignorance of the *modus operandi* of any agent in the least interfere with its availability in the treatment of disease, when that treatment has been proved to be successful. We do not know—nor does the chemist require to know—*why* sugar is sweet, or sulphuric acid sour; or why the latter will redden litmus, while an alkali will render turmeric brown. It is no more *necessary*, although it would be interesting, to know how bitters improve the appetite, or iodide of potassium cures syphilitic rheumatism. We may use opium to produce sleep, or lull pain, although we know little more than Molière's doctor—"opium facit dormire, quia est in eo virtus dormitiva."

It was, in fact, as it was long ago observed, "only after men had found remedies, that they commenced to reason upon them." The most remarkable treasures of medicine have been discovered almost by accident, and have obtained their place in the materia medica often against the protests of the theorists. Opium is one of the oldest of drugs.[1] Iron is nearly as ancient. Mercury was a contribution of the alchemists. Arsenic and colchicum appeared first as secret remedies. Iodine (in burnt sponge) and sulphur were popular and domestic before they were professional medicines. And did not the French academy formally denounce tartar emetic? Did not all the schools disbelieve in cinchona, because it neither sweated, puked, nor purged? And Jenner, who drew the idea of vaccination, by a most sagacious induction, from a popular tradition of the country, against what strong theoretic opposition did his noble discovery have to establish itself!

Nolens volens, then, we have to acknowledge our indebtedness, in therapeutics, to empirical observation. But it is the vocation of the true physician to make it scientific. To know that two cases of disease are *really* alike, and not only apparently so, in order to the application to them of the same remedies; to make *accurate comparison* of the virtues of different modes of treatment, avoiding the "post hoc propter hoc" fallacy; and to appreciate the conditions and circumstances which *modify* the actions of medicines, as they do the course of diseases; these are tasks which enlist the highest faculties of analysis, as well as of observation.

Pereira.

Moreover, medicine is progressive. Even an incompleted physiology may suggest safe and proper experimentation. And for good *diagnosis* we need pathology; for pathology, physiology is indispensable. We do not admit, then, with Laennec, that these beautiful sciences are but "vain amusements." We look forward to the day, when the laborious and intelligent culture they are now receiving, will be repaid by a tenfold harvest, practical as well as abstract. The time *may* come, when the why and the how of therapeutics may be largely as well as accurately explained. But, practical medicine, having its crying necessities, cannot wait for such an era; let it use its facts, and not be misled by false expectations.[1]

Yet, we must remember, that it is the facts, not of the experience of an *individual*, which most of all is "experientia fallax," but of the *aggregate* experience of the *whole profession*, in *all time*, that constitute the body of therapeutical science; which should, as Dr. Todd has said of pathology, be reviewed and reconsidered from time to time; but which can never be abandoned or rejected. It is not well, then, to call the great physicians, our predecessors, as Dr. Bennett has done, "blind guides." Rather, may we, with the late Dr. Alison, believe that a disagreement between a newly-broached pathology and the practical experience of all time, is a much better reason for setting aside the new pathology than the old practice.

We may now sum up the substance of the foregoing remarks, by asserting that the therapeutical methods or principles upon which we may deal with the treatment of disease are essentially three;[2] the *natural*, hygienic, or expectant, the *specific* or purely empirical (including the tentative), and the analytical or *conditional* (including the perturbative or alterative). Of the first two quite enough has been said. Of the last it will suffice to say that it is the most of all open to suggestion from positive physiology and enlightened pathology. It comprises the rational treatment of diseased *conditions* of the body, for which no direct or *specific* remedy has been discovered; a part of medicine of very great importance, but in which the greatest variation has necessarily occurred in the past, and continues yet to exist. This is the debatable ground, upon which tournament upon tournament and crusade after crusade have been fought; the world at large looking on sometimes with more amusement than profit. The lesson of these petty wars, however,—pre-indicated clearly by the old classical writers upon medicine,—has been at last tolerably well learned : *Not to do harm when we are unable to do good;* the reversal of the old maxim, "melius anceps quam nullum remedium; because, in the restoration of a patient from disease, the physician is not the only nor even the stronger agent, nature being the principal, he only the accessory.[3] Some have given credit for this medical gospel to distinguished recent writers, as Dr. Bigelow and Sir John Forbes; but they are revivers of the doctrine only, not its discoverers. Hippocrates dis-

---

[1] "When there is no certain knowledge of a thing, a mere opinion about it cannot discover a sure remedy." "Medicine ought to be rational; but to draw its methods from the *evident* causes; all the obscure being removed, *not from the attention of the artist, but from the practice of the art.*"—CELSUS, Treatise on Medicine.

[2] Lordat. See Renouard's Hist. of Medicine.   [3] Chomel.

tinctly recognized the self-limitation of many diseases. Τὰ χρινόμενα καὶ τὰ κεκριμένα ἀρτίως μη κινέειν μηδὲ νεωτεροποιέειν, μήτε φαρμαχείησι, μήτ' ἄλλοισιν ἐρεθισμοισιν, ἀλλ' ἐάν.[1] So also did Asclepiades, notwithstanding the protest of his sect against Hippocratism; when he said, that the best cure for a fever is the fever itself.[2] So did Sydenham[3] and others who wrote long ago.

And now, although Dr. Bennett, of Edinburgh, predicts the "approaching downfall of empirical practice," yet his late co-laborer, Dr. Todd, of London, urged in his last words the importance of its support in its clinical research; and the philosophic medical historian, Renouard, seconding the efforts of Louis, the founder of the numerical method, foretells the coming *triumph* of *rational empiricism* or *inductive medicine*. We may well believe that this prophecy will yet be fulfilled.

[1] Aphorism 20, Section 1st. [2] Cælius Aurelianus.
[3] "To imagine that nature always needs the aid of art is an error, and an unlearned error too."

PART I.

PRINCIPLES OF MEDICINE.

SECTION I.

GENERAL PATHOLOGY.

Disease may be defined as a perversion either of the functions or of the structure of the body or of any of its parts. It is, in other words, a deviation from the normal physiological state or action of the organism, under the disturbing influence of **morbid causes**.

The **seat** of disease may be
In the **constitution**: *e. g.* secondary syphilis; tuberculosis.
In special **tissues**: *e. g.* mollities ossium.
In particular **apparatus**: *e. g.*, dyspepsia; hysteria.
In individual **organs**: *e. g.* pneumonia; cirrhosis; hydatids.
In the **blood**: *e. g.* anæmia; scorbutus; typhus.

Of course, disease may be, and generally is, not limited to what is to be regarded as its principal or original seat. For example, in cholera, while its cause, no doubt, acts first upon the blood, the *ganglionic* system is also affected, as well as the stomach and bowels, &c.

Morbid states of the **system**:—
Fever;
Toxæmia;
Anæmia;
Plethora;
Degeneration of organic force (Cachexia);
Depression;
Exhaustion.

Morbid states of **organs**:—
Hyperæmia;
Hypertrophy;
Irritation;
Inflammation;
Hyperæmæsthesia, or "chronic inflammation;"
Atony; Exhaustion;
Atrophy;
Degeneration.

Of the above, the *most important general or systemic* morbid states may be included under **fever, toxæmia,** and **cachexia**; constituting a sort of "tripod" of systemic disease.

3

A similar tripod of the most frequent and important *local* disorders may be established, of **irritation, inflammation,** and **atrophy.**

GENERAL PATHOLOGY OF AFFECTIONS OF THE SYSTEM.

FEVER.

In using the term **fever,** as applied to a morbid *state* of the system, we must remember that the same word is also used as a part of the designation of several *complex diseases :* as typhus fever, yellow fever, remittent fever, etc. This double use of the word is unfortunate, but cannot now be avoided.

Symptoms of Fever.—*Increased heat* of the whole body;
Dryness of the skin, mouth, etc.;
Diminution in *bulk* of the *excretions ;*
Muscular debility ;
Frequency of the pulse ;
Functional disturbance of stomach, brain, etc.
Heat is the most essential characteristic of the febrile state, having given name to it in all languages.

Notwithstanding the scantiness in *quantity* of the stools, urine, and perspiration in fever, it has been shown by Virchow, Vögel, Böcker, Parkes, Jenner, and Hammond, that the *actual amount of solid matter excreted,* especially by the kidneys, *is generally increased.* Although none of these observers have made chemical examination of the *expired air* during fever, we have, in the *heavy, offensive odor* of the breath, evidence that it, too, contains an excessive amount of decomposing organic material. It is highly probable, also, that *much excrementitious matter is,* during fever, *retained in the blood.* It has been observed that if a *local inflammation,* as pneumonia, occurs during the febrile attack, the *excess of excreted solids* (urea, etc.) *disappears* until the inflammation has passed.

This increase of the disintegration of the substance of the body (tissue-metamorphosis) is, at present, one of the most prominent and interesting phenomena connected with the pathology of fever. The whole subject, however, is surrounded by obscurity, notwithstanding the fact that the *symptoms* and aspects of the febrile state have been familiar ever since man became a prey to disease.

On the basis of the facts observed and scrutinized at the present time, I think we may venture to throw out a comprehensive **theory** of fever. Thus,—its essential phenomenon is increased **heat** of the body ; this being produced by **excessive tissue-metamorphosis,** under an abnormal "tension-condition" (Virchow) of the **ganglionic nerve-centres;** which abnormal condition is the result of (Addison) either 1, **corpuscular** toxæmia, or 2, **plasmic** toxæmia, or 3, (Campbell & Müller) sympathetic **irritation** from local inflammation.

A pathological classification of fevers, convenient for some purposes, is, into **irritative, reactive,** and **toxæmic** fevers.

TOXÆMIA.

Toxæmia, more properly *toxicohœmia*, (from τοξιχον, poison, and αἱμα, blood), is a term used to indicate *poisoning of the blood*.

After all the long and reiterated disputes between the advocates of the exclusive *solidist* and *humoral* pathologies, it has now become a matter of general recognition that *both* the fluids and the solids are involved in almost every disease,—their mutual interdependence making the contrary impossible.

Certain diseases, however, more than others, are believed, upon the strongest evidence, to depend upon a chemical and dynamic *change in the blood*, to which the name of toxæmia is applied.

Toxæmia may originate :—1. By the introduction into the blood of morbid poisons *from without;* as in syphilis, smallpox, remittent fever, &c. 2. By morbid alteration from processes occurring *in the blood itself.* 3. By *absorption* of poisonous material, by the vessels, from diseased parts of the body; as in purulent infection after wounds, &c. 4. By the non-excretion, and consequent accumulation in the blood, of post-organic or excrementitious substances, which, by their own properties, or by the chemical changes they undergo, prove injurious to the system. Obstructive jaundice, and uræmia, afford the best examples of this last occurrence.

1. All of the *zymotic* diseases (*e. g.*, exanthemata, yellow fever, diphtheria) have their origin explained by the first of the above modes of blood-poisoning.

Yet, our knowledge of the very existence of several of these "morbid poisons" is inferential only. Our idea of their nature is conjectural; and our reasonings upon their mode of action upon the blood and system at large are entirely speculative.

Some facts, however (see *Simon's Lectures on Pathology*), are well deserving of notice.

1. The effects of these poisons, when introduced into the body, are, both *local* and *constitutional* symptoms. The *constitutional* symptoms, which begin the attack, are *nearly alike for them all;*—the *local* symptoms are *peculiar for each one.*

2. The smallpox virus is the most readily studied of all of these causes. This material is evidently **volatile,** as it acts often through *considerable distances;* and it is **soluble,** because it infects, sometimes, the fœtus in utero, which has no communication of fluids with the mother, except by placental *endosmosis.* The poison of **primary** *syphilis* is not thus transmissible, although it is directly **contagious** by *inoculation;* that of **secondary** syphilis is *not*[1] contagious, but *is* **transmissible** by descent.

3. **One** attack of smallpox, scarlatina, measles, hooping-cough, usually gives **immunity** for the rest of a lifetime. It may, from this, be argued, that besides the *materies morbi* or *causative* matter, *another* material must exist *in the blood* of the susceptible person, which *combines* with the former (thus producing *the disease*), and which is *ex-*

[1] Recent experiments have occasioned some doubt as to the entire correctness of this commonly accepted statement.

haustible.[1] (Illustration: the saturation of a carbonated alkali by an acid; after a certain portion of the latter has been added, it will cease *to effervesce* with any subsequent addition of the same.)

Vaccination can be best explained upon this view. Just as more than one acid will neutralize potassa or soda, etc., so that after it has been saturated with sulphuric acid it will not react with nitric—so the virus of the vaccine disease appears capable of *saturating* and *exhausting that material in the body, the presence of which constitutes the susceptibility to variola.*

No such immunity after a single attack is found to exist in the case of the miasmatic fevers (remittent, intermittent). The element in the blood for which *their* morbid poison has affinity is, therefore, not ex-. haustible. Several reasons exist for conjecturing this element to be the *red corpuscles* themselves.

Of the different hypotheses propounded in regard to the *modus operandi* of **zymotic** (epidemic, endemic, infectious, or contagious) causes upon the blood, that which has best stood its ground is that of **catalysis**, or continuous molecular action.

Liebig first urged this theory, upon the analogy between the action of yeast in producing fermentation, and that of a virus, as of smallpox, in producing its effects upon the system, through the blood. It is true, that the blood does not *ferment;* the action is therefore not *similar,* but **analogous.** Chemical action, of a certain kind, going on in the particles of the yeast, or of a virus, is, by their contact with another substance, communicated to or instigated among the particles of the latter. A *mechanical* analogy to illustrate this, is the setting in motion of one cog-wheel by another. A physical illustration, less remote, is that of the extension of fire from a burning body to other combustibles near it.

2. Toxæmia from *spontaneous changes in the blood itself,* under causes or conditions which do not affect the solid structures of the body, if it occur, must be rare. *Heat-stroke* may be an example; that is, the dangerous or fatal effect of extreme heat, with exhaustion, away from the direct influence of the rays of the sun. Here the blood probably undergoes a chemical change which renders it unfit to neutralize the nerve-centres and other organs.

Perhaps **pyæmia, ichoræmia,** or **septæmia** (pus-forming blood,—contaminated blood,—blood-decomposition), as nearly the same affection is called by different authors, may be supposed to occur sometimes spontaneously. More often, however, such an affection is ascribed to the next mode of causation.

3. Absorption of deleterious material, by the bloodvessels or lymphatics, from parts of the body at the time in a state of disease, may cause *purulent infection,* or **pyæmia.**

Absorption of pus (containing pus-cells) cannot occur without a solution of continuity in the vessels; the pus-*cells* cannot go through their *walls.* But in *arteritis* or *phlebitis,* suppuration of the artery or the vein may introduce pus into the blood.

---

[1] Reflection will show that there is no real incompatibility between this theory and that of *zymosis*, to be mentioned presently. We have not space, however, for the farther discussion of so speculative a subject.

In the greater number of cases, it is not pus, but an unhealthy material of a less definite nature, which contaminates the blood by its absorption. This may take place after wounds or surgical operations,—from the womb in the puerperal state, &c. Pyæmia is attended by great prostration, rapid pulse, copious perspirations, low delirium, and the depositing of pus and formation of abscesses in different parts of the body. It often begins with a chill.

4. Toxæmia from **non-elimination** of the secretions may follow, of course, upon prolonged constipation, obstruction of the biliary duct, inaction of the liver, or suppression of the action of the kidneys.

Effort is made (according to the adaptations of nature), when one emunctory fails to act, to carry out its *excreta* by other channels. Especially the skin and kidneys act thus *vicariously* for each other.

When the blood is in no way rid of those effete particles which should make the *solids of the urine*, the resulting condition is called **uræmia.** Its symptoms are, pain in the head, dulness of sight and hearing, vertigo, nausea and vomiting ; ending, unless relieved, in convulsions, coma, and death. Pregnant women sometimes have *uræmic convulsions (C. Braun)*, from fœtal pressure obstructing the renal circulation.

The term **cholæmia** is less used, though quite as justifiable as uræmia. It means, retention in the blood of the excrementitious matter of the bile, from inaction of the liver. **Cholesteræmia** is a term preferred by Dr. A. Flint, Jr., who asserts cholesterin to be the excretory ingredient of bile.

Jaundice is well known to be of two origins : 1. **obstruction** of the biliary ducts, with reabsorption of bile into the blood ; 2. **suppression** of the secretion of the liver. (A third is possible,—perhaps present in the jaundice of infancy ; viz., **excessive formation** of yellow pigment in the blood, and its deposit in the skin, &c., without disorder of the liver.)

In jaundice from obstruction and reabsorption, the symptoms are milder and the state less dangerous than in that from suppression of the action of the liver. Severe, and even fatal disease of the liver may occur, however, without jaundice.

Dr. Harley has shown that the diagnosis between these two forms of jaundice may be made, on analysis of the urine, by finding the *coloring* matter of the bile always in the urine in *both*, but the biliary *acids* only in the *obstructive* form.

Slight and transient cholæmia is, no doubt, common. Although the term "biliousness" is much abused, it is not always quite a misnomer. As signs of the condition mentioned, we find nausea, bitter taste in the mouth, constipation and dizziness, with yellowness of the tongue, conjunctiva and skin.

ANÆMIA.

Anæmia (*spanæmia, hydræmia*) is the common term indicating poverty of the blood. The density of that liquid is diminished, and there is a deficiency in the number of the red corpuscles. Exhausting hemorrhages or discharges, severe attacks or long continuance of disease, insufficiency of food, &c., may cause this condition.

It is shown by paleness (sometimes with occasional flushes) of the face, even of the lips, and tongue, as well as of the hands; debility; feebleness and excitability of the pulse; frequently, palpitation of the heart, and a *bellows murmur*, audible especially near the base of the heart, to be carefully distinguished from valvular murmurs of organic disease. *Nervousness*, and neuralgic pains, are also very common in the anæmic.

Chlorosis, although by some separated from anæmia, is generally regarded as a variety of it, occurring in young females. The name is given on account of the peculiar sallowness of the complexion. *Perverted appetite*, as for charcoal, slate-pencils, &c., is one of its occasional symptoms.

Morbus Addisonii, Addison's disease, is a rare constitutional malady, in which anæmia coincides with bronzing of the skin, disease of the supra-renal capsules, and progressive debility, usually ending in death. The capsular disease is not shown by any definite local symptoms during life; and its connection with the *cachexia* has not been explained.

Leucocythæmia (leukæmia) has been, after Virchow and Bennett, recognized as a condition in which the number of colorless corpuscles in the blood is *increased;* sometimes numbering one to four, three or two, instead of one to fifty (normal) of the red corpuscles. This, of course, can be only ascertained by means of the microscope.

Enlargement of the liver, and still more of the spleen, and disease of these organs as well as of the thyroid and lymphatic glands, are found to attend this order. It most frequently affects men. Its symptoms are, pallor, emaciation, diarrhœa, epistaxis (bleeding from the nose), or other hemorrhages, and dropsy.

Melanæmia is the name given (Frerichs) to a state of the blood most common in severe malarial fevers, in which the coloring matter (pigment) escapes from the corpuscles, and is deposited in the liver and other organs.

Plethora involves an *excess* in the density of the blood, and in the number of its red corpuscles; the opposite to anæmia. It is shown by a high color, distension of the bloodvessels, a full, strong, but rather slow pulse, and general roundness of the figure. It may exist without actual deviation from health; but the plethoric are especially liable to acute inflammations, active congestions, and hemorrhages.

CACHEXIA.

Cachexia (from κακος, bad, and ἑξις, habit), is usually understood to mean a *depraved habit of system;* an error of development and nutrition, affecting the general state of the blood and organs with perversion.

There is, at the same time, no good reason why we should not speak of *local* as well as general cachexiæ; although this has not been usual.

Addison's disease, and *leucocythæmia*, may be regarded as cachexiæ. Much more frequent, and therefore important,—the most common and destructive of all cachectic affections, is tuberculosis.

While diverse opinions exist as to the essential nature of tubercle, and its origin, there is a general agreement among pathologists and clinical observers, upon most of the following points.

1. **Tuberculosis** and **Scrofulosis** are **identical.**[1] The term *scrofula* is generally applied to certain slow inflammations, abscesses, ulcerations, and other disorders of the skin, mucous membrane, glands, and bones, which occur especially in young persons, and are characterized by the *moderate* degree of vascular excitement attending them, with the great *obstinacy* or *chronicity* of their career. In many cases, also, of external scrofula, particularly in the glands, a deposit of curd-like or cheesy material is found, *not distinguishable from tubercle.*

2. Of the **causes** apparently connected with the production of the tubercular or scrofulous diathesis (to which the general name *tuberculosis* may be applied), the most obvious and constant is *hereditary* **predisposition.**

3. This diathesis may, however, undoubtedly be **acquired** without inheritance. *Change of* **climate,** from a warm to a *cold and damp* locality, will often induce it. Other **depressing** influences promote it, such as want of *food, light,* or *warmth,* sedentary habits, etc. But all of these often fail to generate any form of tubercular disease.

4. Tuberculosis may be pathologically defined as a constitutional tendency to the formation of blood, the plasma of which is defective in organizable capacity; so that, in nutrition, instead of healthy tissue, it forms, in one, or very often in many, of the organs, **aborted blastema,**[2] which accumulates as a deposit. This deposit is called tubercle; the process, **tuberculization.**

5. But the tubercular- **diathesis** (tuberculosis) may exist *without* *tuberculization.* Its influence is then shown, especially, in **modifying** *inflammatory* or *other morbid processes;* giving them a *lower, slower,* and more persistent or intractable type. Thus, many cases of what is called *tubercular meningitis* in children occur, with fatal result, in which (Bouchut, Hughes Wilshire, etc.) *no tubercular* **deposit** *is found;* yet the disease is modified by the diathesis.

6. Tubercle is distributed either in **regularly-formed** masses (miliary tubercles, etc.), or **irregularly,** through the tissue of organs. The most amorphous (shapeless) and homogeneous examples of it are called **infiltrated** tubercle. The *size* of the masses of tubercle varies from that of a pin's head to that of a hen's egg.

7. The two essential varieties of tubercle are the **semitransparent, gray, granular,** and the **yellow, opaque, caseous** tubercle.

8. **Neither** of these forms ever undergoes **organization.** They are never *vascular.* They are deposited **outside** of the bloodvessels only, and not in *non-vascular* tissues, such as cartilage, etc.

9. The **gray** tubercle, when *alone,* is subject (Rokitansky) to one change only, **cornification;** *i. e.* drying into a horny substance as

---

[1] Dr. C. West is among those who deny this. *Diseases of Children*, Phil. ed., p. 583.

[2] From βλαστανω, I *bud;* used to mean *tissue-forming* material. This view of tubercle has been denied of late, it appears to me upon insufficient grounds, by Dr. T. K. Chambers and others. Some even assert a *specific* character in tubercle. Lebert, of Breslau, declares that he has propagated it in animals by inoculation.

hard as a shot. When *with the yellow* tubercle, the gray may undergo softening.

10. **Yellow** tubercle *usually* **softens**; sometimes it **cretifies**; *i. e.* becomes chalk-like, by degeneration.

11. The softening of tubercle is **spontaneous**; not depending upon any agency of surrounding parts. In regularly formed tubercles it commences at the **centre**; in the irregular, at *any* part. *Tuberculous* **softening** must not be confounded with **suppuration** of inflamed tissue; although they are often mingled.

12. Examined with the microscope, tubercle is found to consist essentially of: 1. An **amorphous**, *granular* material, containing *irregular solid* corpuscles (tubercle-corpuscles), considered (Virchow) to be **shrivelled nuclei**. 2. Elements of **disintegrated tissue** of the part involved; as epithelial cells, fibres, etc. 3. Results of **degeneration**; *e. g.* oil-globules, pigment, calcareous particles, etc. 4. Results of **inflammation** of surrounding parts; lymph, pus, exudation-corpuscles. 5. Extravasated **blood-corpuscles**, from hemorrhage, the effect of obstruction or ulceration of vascular trunks.

13. Tubercle contains, then, **no specific**, heterologous **form**. All that it holds is the consequence of abortion and degeneration.

14. The process of **tuberculization** or deposit of tubercle in the organ may occur (Rokitansky)—

a. **Insensibly,** in the course of ordinary nutrition.

b. With **hyperæmia**, or local determination of blood.

c. With **inflammation**; *i. e.* as a *product* or concomitant of the inflammatory process.

15. The effects of the tubercular deposit *upon the part* are: 1. **Obstruction**, and arrest or impairment of function. 2. **Inflammation**; *e. g.* in phthisis pulmonalis (consumption), which has, in its usual form, been designated (Condie) *tubercular pneumonia*, from the common occurrence of inflammation of the lungs with the deposit of tubercle. 3. **Ulcerative destruction** of the tissue by the repeated new formation and softening of tuberculous matter, producing **cavities**.

16. Tubercle, once thrown out, is never (as a whole) *absorbed*. It can only be **eliminated, cretified, or cornified.** Elimination is the most common. After this has happened, sometimes *callous* cavities are formed by a process of **cicatrization**.

17. The *order of frequency* with which different organs are affected with tubercle is (Rokitansky) as follows :—

Lungs ;   Spleen ;
Intestines ;   Kidneys ;
Lymph glands ;   Liver ;
Larynx ;   Bones ;
Serous membranes ;   Uterus ;
Brain ;   Testicles.

Except in the case of *children*, in whom the *lymph-glands* and the spleen stand first on the list.

18. But the organs most frequently *first invaded* by tubercle are, at all times of life, the **lungs** and **lymph-glands**.

19. The *parts especially preferred* by tubercle for its deposit are, in the lungs, the apex ; in the pia mater, about the base of the brain ; in the brain, the gray substance ; in bones, the cancellated structure ;

in the bowels, the lowest part of the ileum; in the testicle, the epididymis; in the female generative apparatus, the Fallopian tubes and fundus of the uterus.

20. The immense experience of Rokitansky gives origin to the statement that tubercle has certain general incompatibilities; the most important of which are, with cancer, with typhus, with ague, and with goitre (bronchocele, enlargement of the thyroid gland). These incompatibilities are, however, *general*, not *universal;* as, for instance, a considerable number of cases have been observed, in which cancer and tubercle were undeniably present in the same patient.

21. The only possible cure of tubercular disease (*e. g.* of the lungs in phthisis) after the deposit has occurred, consists in the total elimination (or absolute quiescence by cornification or cretification) of the tuberculous matter, and *improvement in the general* hæmatosis (*i. e.* blood-formation), so that *no new tubercle is formed.* The two great indications, therefore, in the treatment of pulmonary consumption (Radclyffe Hall) are, to gain time and tone: *time*, by allaying or preventing pulmonary or bronchial inflammation and irritation ; and *tone*, by strengthening the patient's system by all possible hygienic and therapeutic measures.

GENERAL PATHOLOGY OF AFFECTIONS OF ORGANS.

HYPERTROPHY.

Hypertrophy is, strictly, *overgrowth;* an increase of the size and weight of a part without change of tissue. It is only in recent times that this has been clearly distinguished from *enlargement with alteration of tissue;* which is really, in many cases, a *degenerative* change, and therefore akin rather to *atrophy* than to hypertrophy.

Hypertrophy is often, *per se*, physiological or natural; although depending on a morbid or pathological cause. When the bladder, for instance, becomes hypertrophied in consequence of obstruction by an enlarged prostate, although the *latter* is morbid, the increase in the strength and thickness of the muscular coat of the bladder is as normal as is that of the uterus during gestation ; in due proportion to the necessities of its use.

A constant law of the animal economy is, that, within certain limits, the growth of an organ is in proportion to its exercise; provided that this exercise is not too violent, and is alternated with sufficient periods of repose.

The three causes of hypertrophy, then, are (see *Paget's Surgical Pathology*):—

1. Increased *exercise* of a part in its healthy functions.
2. Increased accumulation in the blood of the particular *materials* which a part appropriates in its nutrition or secretion.
3. Increased *afflux* of healthy *blood* to the part.

We may illustrate the first of these modes of causation by the blacksmith's arm, the legs of the *danseuse*, the cuticle of the laborer's hands, *the heart* in cases of *valvular obstruction*, etc.

An example illustrative of the second is found in the enlargement of a healthy kidney, when the opposite one fails, from disease, to remove from the blood its due share of urea, etc.

The third is exemplified in the large growth of hairs around an inflamed ulcer or osseous fracture; by the growth of the bones of the limbs when their nutrition is increased by exercise; by hypertrophy of a bone, a portion of which has been subject to disease with vascular excitement; and by Hunter's interesting experiment of the transplantation of the spur of a cock to its *comb.*   ·

Adaptive hypertrophy is remarkably seen in the changes undergone by the skull in proportion to its contents. The cranium is subject to—1, eccentric, and 2, concentric hypertrophy. The first occurs in cases of *hydrocephalus,* the second in cerebral *atrophy;* the bony case in the one instance *expanding* with its contents, in the other *thickening* so as to fill up the abnormal void.

Corns illustrate hypertrophy extremely well.

Intermittent pressure, or attrition, causes *hypertrophy; constant* pressure, *atrophy* or absorption.

The formation of corns upon the foot illustrates the former of these laws; the wasting away of vertebræ under the incessant pressure of an aortic aneurism affords an example of the latter. *The Chinese woman's foot, with corns,* exemplifying *both.*

ATROPHY.

Atrophy requires but a few words in this place. Simple atrophy is exactly the reverse of simple hypertrophy; viz., wasting and diminution of a part, without change of structure. But most pathologists include also under the same term such defects of nutrition as result in **degenerative** changes; constituting the two classes of—1, **quantitative,** and 2, **qualitative** atrophy. The latter (*e. g.* fatty degeneration) is frequently attended by *increase* instead of diminution of *bulk* in the parts affected.

The **causes** of atrophy are:—   ·

1. Deficient *exercise* of a part;
2. Deficiency in the *supply of blood ;*
3. Defective supply of *nervous influence ;*
4. *Disease* (inflammation, etc.) in the part.

Of the first of these, the atrophy of the mamma of the old maid may afford an example.

Of the second, softening of a portion of the brain from the obstruction of one of its arteries by a coagulum.

Of the third and first together, the muscles of a paralyzed limb.

Of the last, there are many instances familiar to the pathologist, although obscurity often attends their individual history; as, the *gouty kidney,* etc.

Quantitative or qualitative atrophy may affect the heart, arteries, brain, muscles, bones, liver, kidneys, pancreas, testicles, etc., and *morbid products, e. g.* inflammatory exudations, cancer, etc.

Qualitative atrophy, or **degeneration,** will again be alluded to presently.

IRRITATION.

Irritation and inflammation are at once the most familiar in their phenomena, and the most obscure in their nature, of all pathological processes or occurrences. I shall confine myself to a broad statement of what I believe to be the most important truths concerning them; although a somewhat argumentative tone may be unavoidable, upon topics which are subject to so much controversy.

Stimulation and **irritation** are often inconveniently confounded. It would be desirable to confine the *former* term to excitation *within physiological* (*i. e.* healthy or normal) *limits;* applying the *latter*, irritation, only to such an *excessive* action upon a part as produces *morbid* effects.

With this, which seems to me a necessary postulate—I would define **irritation** as an *arrest of vital movement in a part*. This could only be elucidated, by an extended allusion, inappropriate here, to the *correlation of physical and vital forces; life* being considered as *a molecular motion.*

In regard to the *circulation*, to the old and accepted maxim—
    Ubi *stimulus*, ibi *affluxus*—
may be added a second—
    Ubi *irritatio*, ibi *stasis.*
And, anticipating the account about to be given of *inflammation*, a third—
    Ubi *phlogosis*, ibi *effusio.*

The **stasis** of **irritation** may be either *partial* or *complete; limited* to a very small surface, or widely *extended;* and *transient*, or *continued* for a considerable time.

If *complete, extended*, and *continued* in a tissue at all vascular, in-**flammation** *follows.*

If the influence of the irritant be very *limited* and *transient.* a temporary stasis and functional and sensational disturbance only follows.

If it be *extended* and *continued*, or *repeated*, and yet of power enough to produce a *partial stasis only*, a condition may result to which the name of **chronic inflammation** has (improperly) been commonly given; of which more will be said hereafter.

The *effects*, or **symptoms**, of irritation differ according to the tissue or organ affected. When a nervous expansion or centre is involved, *pain* is the most familiar result. Functional disorder of the part innervated also occurs. Irritation of muscular tissue causes tonic *spasm.*

INFLAMMATION.

Inflammation must be considered, briefly, as to its *symptoms, minute phenomena, products, terminations* or *effects*, and *post-mortem* appearances.

Its recognized **symptoms** or signs, in a part open to inspection, are *redness, heat, swelling, and pain.*

In **internal** organs inflammation is detected chiefly by *pain, increased by pressure* or *motion; obstruction* or *alteration* of the *functional action* of the organ; and *general* (sympathetic) *vascular excitement.* Certain *physical signs* also aid in the diagnosis of inflammation of particular organs (see *Semeiology*).

The **minute phenomena** of inflammation, as seen under the microscope, have been variously construed by different observers. The use of the *term* itself has been, of late, distorted (Virchow) from its old meaning; and attempts have been made by some (Andral, Eisenmann, Bennett) to do away with it entirely; attempts which fail, because, in proposing other terms, a *part* only is substituted for the *whole*. It is curious, that, of the three terms proposed by three leading pathological writers, **hyperæmia**, **stasis**, and **exudation**, to take the place of the old word inflammation, *each* expresses a *single* part or element of · the process, which *can only be defined by including them all;* while cell-multiplication, made by Virchow almost the whole of the process, is only an incidental attendant upon it.

The essential minute phenomena of inflammation are, as regards the circulation,—

Central stasis;
Concentric hyperæmia;
Exudation.

Other changes, affecting the red and white corpuscles, etc., occur, but are of secondary consequence.

The *nature* and *cause* of these phenomena require, for their comprehension, a close consideration of the laws which govern nutrition, the capillary and arterial circulation, and innervation, in their mutual relations, under the influence of *normal stimuli* and of *morbid irritants*.

What are the actual causes of inflammation?

Not section of the nerves; nor division of the arteries (*per se*); nor division of the veins; nor ligation of arteries nor of veins; nor (*per se*) of lymphatics. Only such causes as modify the **molecular state** of the tissue, and arrest, for the time, the usual interchange of material between the tissue and the blood, can induce a true inflammation.

Let us, then, revert to our maxims. *Ubi stimulus, ibi affluxus. Stimulation* causes *active hyperæmia. The arteries thus exhibit* **reflex action**; a fact which, in spite of the teachings of Unzer, Hunter, and C. Bell, has been denied or misunderstood by nearly every other physiologist and pathologist down to the present day.[1]

---

[1] Much of the accepted pathology of to-day, and some current notions in therapeutics, are founded upon an erroneous view of the physiology of the circulation, especially in regard to the mode of action of the arteries. The error is, the statement (based upon experiments whose results were only *morbid*, not normal) that the *normal active* contraction of the arteries always *diminishes* the supply of blood through them; as Virchow expresses it *"the more active the vessel, the less the supply of blood."* Another generation will attain to the correction of this; and, with it, a revolution in the pathology of inflammation must occur. See the author's *Prize Essay on the Arterial Circulation.* Trans. Amer. Med. Association, 1856. More recently (1858) *Lister* has asserted reflex action as occurring in the vessels in inflammation; as well as the central arrest of nutrition.

Next, *ubi irritatio, ibi stasis.* Stimulation, carried to *morbid excess,* interrupts, by the molecular disturbance it induces, the *normal life-movement* of the part, and checks the interchange of particles going on between the capillaries and the tissue. Thus *the circulation in the capillaries of the part is arrested;* **stagnation** ensues.

Both of these results, active arterial hyperæmia and capillary stasis, follow from the *same* or *similar* causes acting in *different degree.* They may and do exist *together;* the one (capillary stasis) at the very point of irritation, the other (active hyperæmia) in the vessels surrounding it.

What follows? Hydraulics may answer this question. A quantity of fluid, in (minutely) porous vessels, being forced upon a centre whose condition allows little or none of it to be transmitted, an **effusion** must result, through the more or less distended coats of the vessels.

This is expressed by our third maxim: *Ubi phlogosis* (inflammation), *ibi effusio.* This phenomenon, the " exudation," has attracted almost all the attention of many recent pathologists, to the exclusion of other occurrences, which precede and accompany it; an exclusion which has had detrimental results (J. Hughes Bennett) as regards the practical and therapeutical deductions made therefrom.

Virchow, of Berlin, has still another theory of inflammation, forming a part of his "Cellular Pathology." He identifies (confounds) **stimulation,** which is physiological or healthy, with **irritation,** always abnormal or pathological in kind or degree. All irritation, in vascular or non-vascular parts, is, in effect, either *functional, nutritive,* or *formative.* Exudation, or transudation of fluid into the substance of an inflamed part, is admitted only of the more vascular, soft, and superficial tissues. In others (parenchymatous inflammation) the essential effect of irritation of a high degree is said to be, the taking, by their own action or attraction, of more fluid into the *cells* of the organ or tissue. Thus they swell, and become clouded in aspect under the microscope. Next the *nucleus* divides; and afterwards the cells themselves multiply by division, or *proliferate.* The origination of pus or other cells from entirely liquid lymph, as asserted by Paget and others, Virchow denies, in accordance with his maxim, "omnis cellula e cellulâ." Either the epithelial cells or those of the connective tissue (common germ-stock of all tissues) must give rise, by change, to pus-cells. At a certain stage, cell-enlargement and proliferation become destructive of function; the parts then degenerate. But Virchow does not with any distinctness at all state the relation between this **degeneration** and that *nutritive* or *formative* **action** which he considers the one effect of "irritation;" nor does he allow to the condition of the blood-vessels any importance in what, in any tissue, he calls inflammation.

This eminent pathologist has *added* to previous knowledge, that of the changes going on in the cells of an organ, a-part of which is inflamed. These are important. But, he omits, in his account of the process, much; and makes, on the whole, the least satisfactory theory of it now held by any authority.

To return to our account of it; an example of the three stages or processes of stimulation, irritation, and inflammation, may be well studied in the action of a common mustard plaster applied to the skin.

4

Its first effect (the only one if the mustard be diluted) is **stimulant** merely; the skin grows warmer, and redder, and its sensibility is moderately heightened. Next (if it be strong and allowed to remain), **irritation** is produced; shown by *pain*, tenderness on pressure, and a deeper or more purple redness. If the irritating matter be now withdrawn, all of these may subside without going further. But if the irritation be continued up to a certain point of duration and intensity, **inflammation** occurs. Then we have redness, heat, pain, and swelling, with effusion of lymph, which, after a sinapism or cantharidal plaster, raises the cuticle in a blister.

I express then what I hold to be a correct theory of the nature of the inflammatory process, in this definition : *Inflammation is a local lesion of nutrition, with concentric vascular excitement;* resulting either in *exudation* or *cell-distension* and *proliferation;* this last being *destructive* at the centre of the inflamed part, but often *formative around* and at some distance from it.

The **products** of inflammation (by exudation) are (see *Paget's Surgical Pathology*): 1. **Serum.** 2. **Blood.** 3. **Mucus.** 4. **Lymph.** The inflammatory effusion of non-fibrinous **serum** is *rare.* The term is often applied, however, clinically, to a serosity which contains a small proportion of fibrin; as the effusion which follows pleurisy, etc.

Blood is exuded *occasionally* only ; e. g. in dysentery, in nephritis, and (dissolved) in pneumonia.

Mucus, a certain portion of which constantly moistens the surface of mucous membranes in health, is altered both in character and in amount by inflammation. The general statement is, that when a mucous membrane is inflamed (e. g. in bronchitis), its secretion of mucus is *at first arrested,* then *increased,* and lastly *perverted* in character.

Coagulable **lymph** is, however, the•characteristic ingredient of inflammatory exudations.

Inflammatory lymph is divided by Paget into—1, **fibrinous,** and 2, **corpuscular** lymph ; with the assertion that, as a general fact, the more fibrin a specimen of lymph contains (provided it be *healthy* fibrin), the greater the probability of its being organized into tissue ; while the larger its. proportion of *corpuscles,* the greater is the likelihood of suppuration or some other degenerative process, and the more tardy its development into any kind of tissue. (Note an apparent *exception to this* in the case of *diphtheritic* exudation ; explained by the fibrin of the latter *not being healthy.* See *Rokitansky's Pathological Anatomy.*)

Fibrinous (coagulable or plastic) lymph is very well seen in the autopsy of any case of acute pleurisy, peritonitis, meningitis, &c. It is a whitish or yellowish-gray substance, opaque or semitranslucent after coagulation, arranged in fibrous bands, meshes, or layers, and causing adhesions between contiguous portions of the tissues affected.

Corpuscular lymph may be studied in the fluid of the vesicles of herpes, or of an ordinary blister; especially if the surface of the latter have been exposed to the air for a short time.

The lymph- or exudation-**corpuscles** which it is found (under the microscope) to contain, are about $\frac{1}{3500}$ of an inch in diameter, "round or oval, pellucid, but appearing, as if through irregularities of surface, dimly nebulous or wrinkled." Examined after a few hours, under the action of water, a round and pellucid nucleus is observed within and attached to the cell-wall. It is, however, impossible, in a given instance, to make a *positive* microscopical diagnosis between these corpuscles of inflammatory lymph, and the normal lymph or chyle corpuscles, colorless corpuscles of the blood, and pus corpuscles.

The "biography" of the lymph of exudation consists in its **resorption**, or its **development** into connective, fibrous, elastic, osseous, cartilaginous, or vascular tissue, or into epithelium, etc. (rarely into muscular or nervous tissue); or its **degeneration** into *pus*, or *granule-cells, exudation granules*, etc.

The rapid **resorption** of a moderate amount of exuded lymph constitutes the **resolution** of an inflammation.

Its **development** is also a form of resolution, but with modification of the condition, dimensions, etc. of the part. This is, in some instances, merely restorative.

The **degeneration** of the exudation results in its being *thrown off*, as pus, or *finally absorbed*, in the form of molecular exudation-granules.

Whether immediate absorption, development, or suppurative or granular degeneration shall occur in any particular case of inflammation, will depend—

1. On the *state* of the **blood**;
2. On the **seat** of the inflammation;
3. On the **degree** of inflammation.

(See Paget's[1] experiments as to the influence of the state of the **blood** on the lymph of vesication.)

As to the **seat** of the attack, *generally, serous* and *synovial* tissues (pleural, peritoneal, arachnoid, articular) are most subject to *adhesive* inflammation, *i. e.* with the exudation of *fibrinous* lymph. *Mucous* tissues seldom exhibit this, being more prone to *suppurative* inflammation. (Exceptions in *croup, diphtheria*, etc.) *Parenchymatous* tissues, as those of the lungs, liver, etc., when inflamed, may suppurate, or the lymph exuded may degenerate into exudation granules, and be finally absorbed.

The **degree** of the inflammation exercises an important influence. The greater its intensity or severity (*i. e.* the more decided and extended the *local lesion of nutrition* and *concentric hyperæmia*), the farther will the lymph exuded be removed, in its primary character, from that transuded in the natural state of the part, and the more will its subsequent changes differ from those of normal nutrition and development.

Degeneration may affect both the *fibrinous* and the *corpuscular* portion of inflammatory lymph.

The **fibrinous** part is subject to—

1. Drying into *horny concretions* (as on the valves of the heart, from endocarditis).
2. *Fatty softening.*
3. *Liquefactive* degeneration.

Both of these last contribute, no doubt, to the process of **suppura-**

---

[1] Surgical Pathology, Phil. ed., p. 220.

tion. Calcareous and pigmental degeneration are also described as occurring occasionally, but they are less important.

The **corpuscular** portion of lymph may also undergo—

1. *Withering* and *drying* (as in scrofulous inflammation of glands).

2. Conversion into *granule-cells* (inflammatory globules of Gluge), by *fatty degeneration.*

3. *Calcareous,*

4. *Pigmental* degeneration.

5. *Most commonly,* degeneration of the lymph-cells into **pus-cells**; the whole of the lymph being transformed into **pus.**

Pus is a greenish-yellow, creamy fluid, consisting, under the micro-. scope, of the *liquor puris* and *pus-cells* or corpuscles. The latter are definite cell forms, larger than blood or lymph corpuscles, somewhat more irregular, and often containing several nuclei. Their characters, however, are *not invariably distinctive;* as might be anticipated, from their being merely *transformed or degenerated* lymph or epithelial corpuscles; or, in a wound or ulcer, cells of granulation. Chemically, pus may be approximatively tested by its solubility in *liquor potassæ.*

Suppuration is either **circumscribed** (as in abscess), **diffusive** (in erysipelas), or **superficial** (in leucorrhœa, etc.)

The **effects** of inflammation *upon the part or organ involved* are—

Enlargement ;   *Degeneration ;*
Induration ;   *Ulceration ;*
Softening ;   *Mortification.*

We thus see that very different or even *opposite* results may follow. from different *degrees* or *kinds* of inflammatory action.

· **Specific** inflammations require merely to be mentioned here. They are, chiefly, **scrofulous, erysipelatous, rheumatic, gouty, exanthematous, syphilitic, gonorrhœal.** These are distinguished from ordinary inflammation and from each other in that—1, each exhibits a peculiar *plan* of morbid process ; 2, each depends upon a peculiar *cause ;* 3, the effects of the said cause are irrespective of its *quantity* or *extent* of *application ;* 4, they are especially *diffusible* from one part of the body to another ; 5, they sometimes exhibit definite *stages* of the morbid process (*e. g. primary* and *secondary* syphilis) ; 6, they are *nearly all,* in a more strict sense than other inflammations, *self-limited ;* the morbid process *dying out* after a certain time. (This *last* statement applies especially, if not only, to *exanthematous, rheumatic, gouty,* and *gonorrhœal* inflammations ; hardly to the scrofulous, erysipelatous, and syphilitic.)

The **post-mortem appearances** of inflammation are important. They can be *generalized,* so as to avoid, to a great extent, the necessity of their reiteration in connection with the description of particular diseases. It is, at the same time, necessary for the student to *familiarize* himself with them, in their local manifestations, by availing himself of every opportunity for autopsic study.

A part which has been inflamed will exhibit after death some, or perhaps all, of the following signs :—

Redness;
Enlargement of
bloodvessels;
Tumefaction;

Coagulable lymph;
Pus;
Softening;
Induration.

The **redness** of inflammation must be distinguished with care from —1, *hypostatic* injection, or cadaveric settling of blood in the lowest parts, by gravitation; and 2, *physiological* redness, as of the stomach during digestion, the ovaries during menstruation, etc. Inflammatory redness is usually more *unequal* than either of the above, and is *stellated*, or in *streaks* and *patches*.

Enlargement *of the* **bloodvessels** of a part may occur as the result of a *chronic* affection, different from acute inflammation. This sign, therefore, is to be interpreted with great caution. The same is true of *tumefaction*.

Softening, if not *cadaveric* (as when the body has been long defunct), may have been produced by *chemical action*, as in poisoning by corrosive sublimate, etc., by *acute and rapid inflammation*, or by *slow, non-inflammatory degeneration*.

Induration may also follow either acute inflammation or slow, atrophic degeneration.

The presence of bands or membranes of coagulable **lymph** is indisputable evidence of inflammation having occurred in the part. But it is not easy, in *all* cases, to determine with certainty whether such formations are *old* or *new*.

The existence of **pus** is a still stronger sign of the recent existence of inflammation; but, occasionally, instances occur in which pus, produced by inflammation in *one* part, is conveyed (as in phlebitis) by the veins, etc., **and** *deposited in another*. This, although a rare event, is *possible* at least.

Clearly, therefore, *no one* of the above post-mortem signs of inflammation is sufficient *alone*. *Several of them together* will make the diagnosis certain. *Redness* and *enlargement of bloodvessels*, with *lymph* or *pus*, and softening or slight induration of tissue, will leave little or no doubt in any case.

The variations in the appearance of different organs and portions of the body, in fatal cases of inflammatory disease, are not such as to interfere with the correctness and availability of this general description.

CHRONIC INFLAMMATION.

The term "**chronic inflammation**," *as commonly* applied, is a *misnomer*. Although the cases so designated exhibit more or less redness, heat, swelling, and pain, yet they are wanting in *exudation;* without which, pathologically, there is *no inflammation*. There is also, in the same cases, only a *partial stasis* or *none;* and the *hyperæmia* is less intense and less strictly concentric than in acute inflammation.

The characteristics of this state, for which a new term is wanted, are—

1. **Enlargement** of the **bloodvessels** of a part (chronic hyperæmia), with the flow of a large amount of blood **through** it.

4*

2. Exaggeration of the **sensibility** of the part (*hyperæsthesia*) and morbid **irritability**.

3. **Deficient** or irregular *functional* **power**.

4. Unusual **proneness** to acute or subacute *attacks* of **actual inflammation.**

For this familiar combination of pathological elements I propose the name **hyperæmæsthesia.**

It has, lately, been usual to designate it by the term **"irritable,"** in connection with the name of the particular part affected, thus—

Irritable uterus;
Spinal irritation;

to which I would add—

Irritable eye (chronic ophthalmia);
Irritable stomach (chronic gastritis);
Irritable brain, etc.

DEGENERATION.

Degeneration has been already defined as **qualitative atrophy**; *i. e.* a substitution, under decline of the organic force incessantly active in nutrition and repair, of *abnormal* for *normal* structure and material. The forms under which this occurs are—

Fatty degeneration;
Calcification;
Pigmental degeneration;
Fibroid, colloid, or amyloid degeneration;[1]
Liquefactive and corpuscular degeneration.

In regard to all of these except the last, it may be stated (see *Paget's Surgical Pathology*) that

1. They are changes such as may be observed **naturally occurring**, in one or more parts of the body, at the **approach** of the natural **termination of life.**

2. The new material is of **lower chemical composition** than that normal to the part; *i. e.* it is less removed from the inorganic state: as fat, gelatin, calcareous matter, etc.

3. In **structure**, it is **less developed**; being crystalline, granular, simply globular, etc.

4. In **function**, it is **less powerful.**

5. In **nutrition**, it is **less active** and capacious.

6. *Generally*, although *not always*, **constitutional** debilitation precedes, and (we may infer) institutes the local alteration of structure.

7. *Inflammation* or other *local* **disease** may, by *impairing* the *nutrition* of a part, **cause** it to degenerate.

The form of degenerative disease which has received the most attention from pathologists is **fatty degeneration.** This has been carefully studied, as it occurs in the heart, arteries, brain, muscles,

---

[1] *Amyloid* degeneration has been described by Virchow and others, as occurring in the brain, spleen, liver, etc. It consists in the conversion of tissue into a substance having physical and chemical properties resembling those of starch or cellulose.

bones, liver, kidneys, and morbid products. It must be distinguished carefully from mere fatty *accumulation* or adiposity.

Our knowledge of the facts concerning degenerative disease, and of the share which it claims in the domain of structural pathology, once almost entirely usurped by inflammation, is among the most important of the acquisitions of the medical science of the last quarter of a century.

Tumors, and **morbid growths,** benign and malignant, which may be best classified as forms of structural degeneration, or vitiated nutrition, I leave, at present,—except some brief consideration of the pathology of cancer.

Cancer falls under the notice and care of the medical practitioner, when it attacks parts or organs within any of the great cavities of the body.

There is no essential impropriety in classing, pathologically, *all* **malignant** growths[1] together as cancerous; their *subdivisions* being clinical or surgical. (By *malignant,* we mean, prone to unlimited increase: disastrous in effect or result; and difficult or impossible of arrest or cure.)

Cancers may, then, be divided simply into

Scirrhus, or hard cancer (fibro-carcinoma);

Colloid, or gelatiniform (alveolar) cancer;

Encephaloid, or brain-like (medullary) cancer.[2]

Each of these contains, as its anatomical elements, **fibres, fluid,** or semi-fluid jelly, and **cells.**

Scirrhus is composed mainly of a fibrous or filamentous tissue, with little fluid, and comparatively few cells. It never becomes encephaloid, nor does encephaloid cancer ever become scirrhus.

Colloid cancer has a variable amount of fibrous tissue, arrayed as a *matrix* (compared often to the structure of an orange), containing a *jelly-like* substance; cells may be also found in it, but in less proportion.

Encephaloid cancer is (so to speak) the *highest development* of carcinomatous formation. It consists of a fibrous matrix, containing an abundance of abnormal, multiform cells, and a peculiar fluid.

When a cancer, of either type originally, is based upon and includes bony structure, it constitutes *osteo-sarcoma.* If it develop itself upon the skin, or other epithelial tissue, or, wherever occurring, display similar structure, it may be called *epithelioma.* If its location involve especial *vascularity* and hemorrhage, it assumes the form and name of *fungus hæmatodes.* The *pain* of cancer (which is not always present) appears to depend upon the extension of the disease to a tissue well endowed with nerves.

The cells, fibres and fluid[3] of cancer are all *abnormal.*

---

[1] Tubercle, it will be remembered, is not a *growth,* but an *abortion* of tissue.
[2] Other names are frequently used, as *epithelial, melanoid, osteoid, hæmatoid,* and *villous* cancer.
[3] The milky or creamy "cancer-juice," which emulsifies with water, is considered highly characteristic by many observers. The *malignity,* or proneness to increase, and extend destructively, of a tumor, is generally in proportion to its succulence or juiciness.

Cancer-cells are of *various* shapes, resembling gland-cells, but larger, averaging about $\frac{1}{1000}$ of an inch in diameter.

But, are these forms *heterologous; i. e.* different from anything normal or natural to the body? Are they, so to speak, *implantations*, or **distortions**? I believe, fully, that the latter is the correct view. The "cancer-cells" are no longer held by micrologists to be pathognomonically distinctive; they are homologous with other cells found in the body. Yet, they are such forms as *do not normally belong to the part;* being produced by morbid alteration or perversion of its natural elements.

The most rational theory of cancer, then, is *dynamic.* The disease consists in a morbid **tendency**; a tendency to enormous and unhealthy growth of a formation which is, at the same time, vascular and sensitive, showing subserviency, although under *perversion*, to the physiological laws of the organization.

The origin of cancer, in many cases, is **constitutional**; it is not unfrequently *hereditary.*

Genuine cancer may always be expected to return after removal— although exceptions occur, and it has occasionally been known to undergo spontaneous degeneration.

The order of choice which cancer exhibits, as to the parts it attacks, is (Rokitansky) as follows:—

Uterus;	Liver;	Testicle;
Mamma;	Bones;	Ovary;
Stomach;	Skin;	Tongue;
Rectum;	Brain;	Œsophagus.
Lymph-glands;	Eye;	

Colloid cancer, in particular, prefers the
    Stomach, rectum, peritoneum.
Scirrhus, the
    Mamma, stomach, intestines.
Encephaloid may occur in any organ: it *alone* attacks the
    Liver, kidney, lung, testicle, eye, lymph-glands.[1]

NEUROPATHOLOGY.

The pathology of the **nervous system** is, itself, an extensive field, of which the merest *coup d'œil* is possible here.

For the purposes of pathological study, we must remember that the *anatomical* elements of the nervous apparatus are—1, gray, vesicular; and 2, white, tubular nervous substance; the former being arranged in *ganglia*, the latter in *nerves* and *commissures.*

Physiologically, the functions of the *ganglia* (nerve-*centres*, and, probably, impressible *peripheral* ganglionic *expansions* also) are, to *receive, reflect, accumulate* (generate?), and *distribute* nerve-force. The sole function of *nerves* and *commissures* is, to *transmit* or conduct it.

As a whole, we may state the offices of the nervous apparatus to be as follows:—

---

[1] Dr. DaCosta has recently noticed an unusually low temperature of the body in cancerous patients.

Excito-motor;
Excito-secretory;
Sensory;
Voluntary motor;

Internuncial. *i. e.*
Sympathetic and
Synergic (co-ordinative) ;
Psychical, *i. e.*
Intellectual ;
Emotional.

The .primary disorders to which this apparatus is liable, are (see *Simon's Lectures on Pathology*) :—

1. **Anæsthesia**; *i. e.* that condition in which the patient remains without cognizance of impressions made on a surface which is normally sentient. This may result, *a*, from disease of the nervous expansion at the **surface**; *b*, from disease or injury of the **conducting nerve**, somewhere on its track ; *c*, from disease of the **cerebral subcentre** of sensation (sensorium). The *thalami* (miscalled *optic*) are believed by physiologists generally to be the *aggregative* centres of sensation ; and local lesion (apoplectic clots, tumors, softening, etc.) in or near .them is frequently associated with hemiplegia, etc. The paralysis is commonly observed (from *decussation* of the nerve fibres in the medulla oblongata) on the side *opposite* to that on which the lesion has occurred.

2. **Subjective** impressions and sensations; *i. e.* those which affect the consciousness of the individual *without the action of any external or peripheral cause*. These subjective impressions may be divided into—*a*, those which are **central** in their origin, as when disease of the optic thalamus causes neuralgia of the fifth pair of nerves ; and *b*, those whose origin is **intermediate**; as, when inflammation of the *sheath* of a nerve, or disease of the *spinal axis*, gives rise to pain referred by the patient to the *termination* of the nerve.

Subjective **hyperæsthesia,** or perversion of sensibility or psychical impressibility, may be, in its causation (as regards the nervous apparatus), either **functional**[1] or **organic;** and the difference between these is often practically important.

3. **Muscular paralysis**; or that condition in which a central volition (or the excitation equivalent to it) fails to produce its normal effect of muscular contraction. Of this defect, also, the pathological origin may be, as to its seat, either **peripheral, intermediate, subcentral** (in the corpora striata or cerebellum), or **central** (in the convolutions of the cerebrum). Muscular as well as sensational paralysis, dependent on an affection of the brain, occurs on the *opposite side* to that of the encephalic lesion. **Hemiplegia** and **paraplegia** have already been defined (see *Semeiology*). Scarcely ever are either of these varieties of palsy confined *exclusively* to sensation or to voluntary motion—although the *proportion* of impairment of the two functions may vary considerably in particular cases. Both kinds are occasionally *reflex* (Brown-Séquard).

4. Involuntary contraction of voluntary muscles, or **convulsion**. Only very *local*, and usually *transitory* spasmodic affections are *peripheral* in their origin. Usually, convulsive affections are accounted for by excessive functional excitement of the (spinal) **motor centres ;**

---

[1] *Functional* nervous disorder results generally (Todd) from an abnormal state of the *blood*.

the causation of which is made up of three elements, in variable proportion, viz.: *a*, morbid irritability of the spinal *excito-motor apparatus itself; b*, imperfect *control* over the *subordinate nervous centres* by the brain, from an abnormal condition of the latter; *c*, the disturbing influence of a peripheral irritant—as, the tension of the gums in teething, worms in the bowels, undigested food in the stomach, etc.

The three forms of spasmodic disturbance to which the muscles are liable under a morbid alteration of innervation, viz.: the tonic, choreic, and clonic, are illustrated respectively in tetanus, chorea, and epilepsy.

5. Excito-secretory action (Longet, Campbell) becomes morbid under conditions often like those which produce convulsion; for example, the *diarrhœa* of infants, so common at the time of dentition. A subject of great interest, almost neglected until within the last ten years, is that of the effects of various agencies, *through* the nerve-centres, upon the bloodvessels. But while the *vaso-motor nerves* are now recognized, and their special relation to the *ganglionic* or sympathetic system is beginning to be appreciated, a remarkable confusion on this subject pervades the medical literature of the present time.

The therapeutic system (neuro-pathy—ganglio-therapy) associated with the name of Dr. John Chapman, will receive attention under *general therapeutics.*

A further important pathological subdivision exists as to the method of origination of those functional disturbances of the nervous system to which we have been alluding.

The source of any of the above forms of nervous disorder, hyperæsthesia, anæsthesia, muscular paralysis, or convulsion, may be (when not purely local) either

    1. Central organic disease;

    2. Blood-perversion, or defective nutrition;

    3. Purely sympathetic disturbance.

It is far from easy, in many cases, to mark the diagnosis between these different modes of causation of nervous symptoms; but, when the decision has been made, in any instance, the *prognosis* is *most* favorable in the *last* case; less so in the *second;* and most unfavorable in the *first, i. e.* when the symptoms have their origin in an actual organic lesion of an important nerve-centre.

Other subjects (hemorrhage, dropsy, &c.) which might be considered as belonging to general pathology, will be taken up in Part II. of this book.

MODES OF DEATH.

Death may occur

1st. By asthenia; the dynamic force of the system being exhausted or destroyed, so that the heart ceases to beat; as in lightning-stroke, poisoning by prussic acid, &c. Syncope (fainting) simulates or threatens this.

2d. By anæmia; the blood being rendered insufficient for life; as from hemorrhage after labor, surgical injuries, bursting of aneurisms, &c.

3d. By **apnœa**, or **asphyxia**; that is, arrest of respiration, either from disease of the lungs, obstruction of the air-passages, deficiency or impurity of the air.

4th. By **coma**; the brain and medulla being made incapable of sustaining innervation ; as in apoplexy, opium poisoning, &c.

Sudden death may occur from
Apoplexy;
Valvular heart-disease (especially mitral) ;
Rupture of heart (or syncope) in fatty degeneration ;
Bursting of an aneurism, or abscess, within the thorax or abdomen ;
Suffocation ;
Violent mental shock or alarm.

SECTION II.

SEMEIOLOGY.

I. RATIONAL SYMPTOMATOLOGY.
II. PHYSICAL DIAGNOSIS.

Rational symptoms and **physical signs** are distinguished (somewhat arbitrarily) thus : a rational symptom is a sign of disease which is obvious to the patient himself or to the practitioner without close inspection. A physical sign is one determined by examination into the properties and material conditions of the organs of the body ; as by palpation, auscultation, percussion, etc. Symptoms guide us, generally, by *physiological inference;* physical signs, by *anatomical necessity.*

Symptoms and physical signs together contribute to **diagnosis;** *i. e.* the knowledge of the character of the morbid process or state in given cases ; the answer to the question, " what is the matter ?"

Prognosis is the anticipation of the *progress* and *results* or *terminations* of disease. The essential elements of prognosis are, a knowledge of the cause or causes of disease present ; of the *condition of the organs;* and of the *general* vital state, or *degree of vital force* of the system. Prognosis depends therefore upon diagnosis ; but is governed, in a majority of cases, chiefly by those rational symptoms which indicate the organic energy of the patient, and the kind and rate of change that his system is undergoing.

SYMPTOMATOLOGY.

Symptoms, or rational signs, are—
Local, or constitutional ;
Idiopathic (primary), or secondary ;
Premonitory (prodromata) ;
Critical;
Pathognomonic (characteristic).

We examine the symptoms of disease as connected with the digestive, circulatory, respiratory, tegumentary, secretory, motor, sensory, and psychical apparatus.

SYMPTOMS CONNECTED WITH THE DIGESTIVE.SYSTEM.

The **tongue** may be natural, pale, cold, red, furred, brown, black, cracked, or fissured.

It is *pale*, in anæmia.

Cold, in *collapse*, as of cholera, etc.

Red, in scarlatina, stomatitis, sometimes in gastritis.

Furred, in indigestion, gastro-hepatic catarrh, fever, etc.

Brown or *black, cracked* or *fissured*, in low fevers : as typhus or typhoid.

Protruded with difficulty, in low fevers, and in apoplexy; *to one side*, in paralysis.

The *manner of cleaning* of the tongue during convalescence should also be noticed, as affording prognostic indications.

The **teeth** are *covered with sordes* in low febrile states.

They are *loosened* by severe salivation.

Their *rapid decay* shows impairment of constitution ; but this is unfortunately very common.

The **gums** are swollen, soft, and spongy, and prone to bleed, in scurvy.

A *blue line along the gums* is observed in lead poisoning.

A *red* line along their edge is sometimes noticed in phthisis.

Swelling and soreness of the gums, with tenderness of the teeth, and a coppery taste, occur in *salivation*.

Increased flow of saliva gives name to this effect of mercury on the mouth.

Deficiency and thickness or viscidity of the saliva occurs generally during fever; and often also in chronic diseases, especially of the throat and stomach.

The **taste** is morbidly

Bitter, in hepatic derangements, dyspepsia, etc.;

Sour, in gastric indigestion;

Saltish, in phthisis pulmonalis, hæmoptysis, etc.;

Putrid, in gangrene of the lungs.

Appetite is generally *deficient* (anorexia) in disease, especially of an acute character.

Excessive appetite (bulimia) is not often important ; sometimes it occurs in nervous affections, in *diabetes*, and in persons having *worms* in the alimentary canal.

Perverted appetite is one of the symptoms of chlorosis, hysteria, etc.

Thirst is *excessive* in *two very opposite* conditions : *high fever* and *low collapse*.

Difficulty of Swallowing (*dysphagia*) may result from—

Inflammation of the fauces, tonsils, or pharynx ;

Spasmodic constriction of the throat ;

Stricture of the pharynx or œsophagus ;

Obstruction by a foreign body, tumor, etc. ;

Retro-pharyngeal abscess ;

General debility, as in the moribund state.

Nausea and vomiting may occur from—
Indigestion : egesta,[1] partly digested food, mucus, etc. ;
Colic : eg., ditto, bile, etc. ;
Pregnancy : eg., mucus, food, etc. ;
Gastritis : eg., abundant and altered mucus, etc. ;
Hysteria : eg., gastric and biliary secretions, more or less altered ;
Cholera morbus : eg., gastric and biliary secretions, diluted ;
Cholera maligna : eg., copious watery fluid (rice-water) ;
Bilious fever : eg., altered mucus, bile, &c. ;
Yellow fever : eg. (advanced stage) black vomit ;
Ulcer of stomach : eg., mucus, lymph, blood ;
Cancer of stomach : eg., ditto, with cancer-cells, fibres, etc. ;
Disease of the brain : eg., not peculiar in character.
Bright's disease of kidney : eg., not peculiar ;
Strangulated hernia : eg., stercoraceous (fæcal) ;
Poisoning : as by tartar emetic, arsenic, &c.
Sarcinæ, or microscopic, wool-sack like vegetable parasites, are occasionally found in matters vomited, in cases of disease of the stomach. Epithelial cells, starch granules, torulæ (also vegetable) and vibriones (animalcular) are often discovered by the microscope.

SYMPTOMS CONNECTED WITH THE CIRCULATORY SYSTEM.

Palpitation or disturbed action of the heart may depend upon—
Pericarditis or *endocarditis ;*
Hypertrophy of the heart ;
Chronic valvular disease ;
Anæmia ;
Nervous irritability (nervousness) ;
Disorder of the brain ;
Dyspepsia.
The pulse should be examined when the mind and body of the patient are as tranquil as possible. It is most rapid in the standing posture, less so when sitting, slowest in the recumbent position. Dr. Guy asserts it to be most rapid in the morning. It is increased in force and frequency by exercise, food, and emotional excitement. The pulse of the female is slightly more rapid, as a rule, than that of the male sex. It diminishes in rapidity from birth to old age; but in very aged people it again becomes somewhat accelerated.
In obscure cases we should examine the pulsation of other arteries besides those at the wrist; and should especially observe the character of the *impulse of the* **heart**.
In adults, the average number of beats in health is, for the male, 70 ; for the female, 75.

At birth, 120 to 140.	Middle life, 75 to 65.
Infancy, 120 to 100.	Old age, 70 to 60.
Childhood, 100 to 90.	Decrepit age, 75 to 80.
Youth, 90 to 75.	

We judge by the pulse (inferentially) of the force of the **heart's**

---

[1] *Egesta,* matters thrown out.

5

action, of the force of the **arterial** impulse, of the **excitability** of the nervous system, of the **fulness** of the bloodvessels, and of the tone and physical **condition** of the arteries.

The pulse in disease may be *natural*, or *strong, weak, firm, yielding, full, small, bounding, compressible, rapid, slow, quick, jerking, hard, soft, tense, gaseous, corded, wiry, thready, imperceptible, regular, irregular, intermittent, dicrotous.*

Not considering it necessary here to define each of these terms, it may be remarked that an important difference exists between a *rapid* pulse and a *quick* pulse, and between one that is merely *full* and *large* and one that is *strong.*

The pulse of **fever** is characterized by *moderate* **acceleration**, with variable increase of force in the beat.

The pulse of **inflammation** (with constitutional excitement) is not only *accelerated*, but **hard** or *tense*, and, commonly, full. Whatever may be said to the contrary, this character of the pulse is, in acute inflammations, of great consequence as an indication of treatment; although, of course, it must not be depended on *alone.*

The pulse of **nervous irritation** is usually **quick**, and **variable** in rapidity and force, under excitement or repose.

A *jerking*, abrupt pulse is associated (Stokes) with *deficiency of the aortic valve.*

The pulse of *extreme* **debility** is nearly always (as in the dying state) very **rapid** and very **small**, or "thready."

Irregularity of the pulse is occasionally congenital; sometimes it comes on with old age. It may be a transient symptom, accidental, as it were, during the progress of an acute malady; or at the commencement of convalescence, as from remittent fever. It is directly related to the nature of the disease, in certain cases of *disease of the heart*, and in *meningitis* (inflammation of the membranes of the brain) during the stage of effusion.

The **dicrotous** or double pulse is observed especially during *continued fevers*, either typhous or typhoid. It is explained by a loss of muscular tone in the arteries, so that the arterial impulse is separated from that of the ventricles by a perceptible (though slight) interval.

The state of the **capillary and venous circulation** often affords signs of disease. Torpor of the circulation is marked by slowness in the return of the blood after it has been displaced by pressure; for instance, upon the cheek or the back of the hand. The *veins* of the hand or arm may be similarly examined with advantage; as in *cholera, pernicious intermittent, low continued fever*, etc. The venous circulation is affected not unfrequently in *heart-disease: e. g.* pulsation of the jugular veins, from valvular disease involving the right side of the heart; *cyanosis*, or *blueness*, from imperfect separation of the arterial from the venous blood, etc.

Pulsation of the veins does not, however (notwithstanding the dictum of authorities), *always* depend upon disorder of the heart. The author has seen three cases in which jugular pulsation was evidently the result of *local irritation*, exaggerating the muscular activity resident in the organic muscle-fibres of the vein.

The **blood** itself is perhaps the most important of all subjects of inquiry in connection with disease. Little, however, as yet, is known of its morbid changes. The principal facts are, that—

In *anæmia*, there is a deficiency of the red corpuscles;
In *plethora*, an excess of the red corpuscles;
In *leucocythemia*, an excess of the colorless corpuscles;
In *inflammation*, and in *chlorosis*, excess of fibrin;
In *gout*, excess of uric acid;
In *rheumatism* (perhaps), excess of lactic acid;
In *melanæmia*, excess of free pigment;
In *jaundice*, excess of biliary matter;
In *Bright's disease*, excess of urea, etc. (uræmia);
In *diabetes*, excess of sugar;
In *malignant cholera*, deficiency of water and salts.

These peculiarities require minute inspection, with the aid of the microscope or of chemical reagents. To the eye, differences sometimes exist which may be instructive: *e. g.* as to the bright red or very dark color of the blood; as to the magnitude, form, and firmness of the clot, and the rapidity of coagulation, etc.

In cases of lingering prostration, clots may form in the heart or large arteries before death. After very rapid malignant diseases, the blood is sometimes found uncoagulable.

Hemorrhage from different parts of the body is often important as a symptom, but requires to be interpreted with care. Its consequence varies much with its *quantity*, and the *source* of the blood thrown out.

Thus, in **epistaxis**, or bleeding at the nose, the flow may result from—

Mechanical injury;
Congestion of the Schneiderian membrane;
Congestion of the brain;
Typhoid fever;
Hemorrhagic diathesis;
Suppressed menstruation.

This variety of hemorrhage is, however, *most* frequent during *childhood* and early adolescence.

In **hæmoptysis**, or spitting of blood, the source of the hemorrhage may be the—

Gums;
Posterior nares;
Throat (e. g. ulcerations, etc.);
Bronchial mucous membrane;
Lungs;
Stomach.

In the *last* case, being *vomited* into the mouth, it is properly called **hæmatemesis**. Sometimes it requires care to determine *what is* the source of blood coming from the mouth. We must notice what are the *symptoms preceding* the hemorrhage; and the *manner of its ejection*, whether by *coughing* or *vomiting*, etc., as well as the appearance of the blood, whether mixed with food, gastric fluid, etc.

True **pulmonary hæmoptysis** may arise from—

Active congestion of the lungs;
Passive congestion, from heart disease;
Tubercular phthisis;

Hemorrhagic diathesis;
Vicarious monthly flow, in the female;
Mechanical injury, as fractured rib, etc.;
Rupture of aortic aneurism.

Hæmatemesis, or vomiting of blood, may be—
Hysterical;
Ulcerative;
Cancerous;
Vicarious; etc.

Uterine hemorrhage, other than the normal menses, may be—
Congestive;
Ulcerative;
Cancerous; as well as, in the *pregnant* female, *placental,* technically called " unavoidable hemorrhage ;" that of *abortion;*
or, after parturition.

Hemorrhage from the bowels may be connected with—
Hemorrhoids, or piles;
Dysentery;
Ulceration of the bowel;
Intussusception;
Cancer of rectum, etc.;
Rupture of aneurism;
Hemorrhagic diathesis;
Typhoid, or *yellow fever;*
Vicarious menstruation.

Hæmaturia, or bloody urine, may result from—
Mechanical injury of the bladder, prostate gland, or urethra;
Renal inflammation;
Calculus;
Hemorrhagic diathesis;
Passive senile congestion of the kidneys;
Scarlatina.

SYMPTOMS CONNECTED WITH THE RESPIRATORY ORGANS.

The normal, average rate of breathing in the adult, while at rest, is sixteen or eighteen respirations in the minute. In fever it is much accelerated. In *extreme narcotism* it becomes slower than natural. In some cases of *fatty degeneration of the heart* it is sighing and interrupted.

Dyspnœa, or difficulty of breathing, when great, is called **orthopnœa,** from the erect posture required by the patient. **Cervical** respiration occurs in cases of great exhaustion, or of obstruction of the respiratory function by disease.

Dyspnœa may be caused by—
Chlorine or other irrespirable gases in the air;
Morbid change of the blood, as in cholera;
Laryngeal or tracheal obstruction, as in croup, etc.;
Bronchial spasmodic constriction, as in asthma;
Bronchitis; pneumonia; pleurisy; phthisis;
Heart disease; aneurism of thoracic aorta;
Cancer within the chest; hydrothorax; ascites.

Coughing may depend upon a variety of causes, the nature of which may often be concluded upon from its character. Thus, usually, Cough is *dry and hollow*, or *hacking*, when nervous or sympathetic;

Dry and tight, in early bronchitis;

Soft, deep, and loose, in advanced bronchitis;

Hacking, in incipient phthisis pulmonalis;

Deep and distressing, in confirmed consumption;

Short and sharp, in pneumonia;

Barking and hoarse, in early or spasmodic croup;

Whistling, in advanced membranous croup;

Paroxysmal, and whooping, in pertussis.

Expectoration is—

Mucous, in catarrh, and early bronchitis;

Purulent, in severe and protracted bronchitis;

Rusty, in the early and middle stages of pneumonia;

Bloody and muco-purulent, in phthisis;

Nummular and heavy, etc., in advanced phthisis;[1]

Putrid, in gangrene of the lung.

The **temperature** *of the breath* is *increased* during the *febrile state*. It is *lowered*, sensibly, only in aggravated prostration; as in the collapse of cholera. Coldness of the breath is an almost certain prognostic of dissolution.

The **odor** of the breath is rarely perfectly agreeable except in the healthy infant or child. It is very heavy at the commencement of fever; sour during indigestion; offensive, often, from decayed teeth; rotten, in gangrene of the lung.

Hiccough (singultus) is produced by a spasm of the diaphragm. It may depend upon *indigestion, nervous disorder,* or *exhaustion.* It is serious in prognosis only when the latter is present or is anticipated.

Stertorous *respiration*, from relaxation of the *velum palati*, results from *cerebral oppression;* the cause of which may be *apoplexy, fracture of the skull, dead drunkenness,* or *narcotism* by opium, etc.

SYMPTOMS CONNECTED WITH THE TEGUMENTARY APPARATUS.

The skin is **hot** and **dry** during the presence of **fever.**

Moisture is almost always a *favorable* sign.

The exceptions are, the profuse *colliquative* sweats of phthisis, etc., and the *cold and clammy* perspiration of extreme prostration. Coldness of the skin, or inequality of temperature, is always more or less unfavorable.

Emaciation is often an important sign. It generally occurs in severe chronic diseases, but is sometimes rapidly brought on in acute affections; *e. g.* diarrhœa or dysentery. The changes which occur in the adipose tissue, and in the plumpness and roundness, or flabbiness and shrunken appearance of the surface of the body, are often *extremely rapid in children.*

---

[1] Microscopic examination has discovered (Schrœder von der Kolk) portions of disintegrated *lung-tissue* in the expectoration of phthisical patients; arched and anastomosing fibrils of pulmonary and bronchial elastic tissue.

The color of the skin varies much in disease. Thus, the face is
Pale, in anæmia, syncope, etc. ;
Flushed, in fever, congestion of brain, etc.
Cheeks brightly flushed, in hectic fever ;
Forehead and eyes flushed, in early stage of yellow fever ;
Purple or livid, in low continued fever ;
Yellow, in jaundice, bilious fever, yellow fever ;
Sallow, in chlorosis, dyspepsia, cancer ;
Blue, in the collapse of cholera, and in cyanosis ;
Black, almost, in asphyxia.
Eruptions upon the skin are characteristic of certain diseases.
Their description belongs to Special Pathology.

SYMPTOMS CONNECTED WITH THE SECRETIONS.

These must always be considered along with *other* explanatory
symptoms ; and the character of the *discharges* should never be
overlooked. Thus,
Constipation may denote—
Torpor of the muscular coat of the bowels ;
Deficient *secretion* of the liver, or intestinal glands ;
Defective *innervation*, from spinal or encephalic disease ;
Stricture of rectum or *colon, pregnancy*, or *cancer ;*
Intussusception, strangulated hernia, etc.;
Sympathetic disturbance from fever, etc.
Diarrhœa and dysentery will be considered under another depart-
ment. It may be mentioned, however, that in *dysentery* the discharges
contain blood, mucus, lymph (in small quantity), and, when ulceration
has occurred, pus. In *diarrhœa* they are either fæcal, mucous, bilious,
or serous—the latter being of importance especially in the diagnosis
of *cholera*.

Symptoms Connected with Urination.

Dysuria, or difficult urination (strangury).
Ischuria, retention of urine.
Enuresis, incontinence.
Diuresis (diabetes), excessive discharge of urine.
Morbid character of the urine itself.
The average quantity of urine passed by a healthy adult in twenty-
four hours, is from thirty to forty ounces—greatest in the winter.
In reaction to test-paper, the urine is normally *acid ;* reddening
litmus, or restoring to turmeric its yellow after it has been made
brownish red by an alkali.
The *color* of healthy urine is that of *amber*.
The average *specific gravity* of human urine (water being 1000) is
1017–20 ; containing about twenty grains of solid matter to the ounce.
Deviation, to a certain extent, from any or all of the above stand-
ards as to quantity, reaction, color, and weight, is quite compatible
with ordinary health ; but a very decided and persistent deviation is a
proof of disease.

Retention of urine may be caused by—
Spasmodic constriction of the vesico-urethral muscular fibres;
True *stricture of the urethra ;*
Enlargement of the prostate gland ;
Calculus in the bladder or urethra.
Percussion and palpation, as well as catheterism, are sometimes necessary to determine the fact of retention of urine.

Suppression of urine, from inaction of the kidneys, is a most serious symptom under all circumstances. If long continued, it becomes fatal by *uræmic poisoning*—coma, and often convulsions, preceding death. *Partial* suppression of urine occurs, sometimes transiently, in cholera, scarlet fever, etc.

Excessive urination is frequently present in hysterical cases—the water being pellucid, and of low specific gravity (diabetes insipidus). The influence of cold and of diuretic medicines produces a similar watery excess, with little increase in the solids of the urine.

Diabetes mellitus is, however, a more important affection; in which the urine is not only excessive in quantity, but *heavy*, and *loaded with sugar.*

For the accurate estimation of the changes occurring in the urine in disease, some scientific skill is requisite. To pursue *original investigations* upon the subject, considerable practical knowledge of *analytical chemistry*, and of the use of the *microscope*, is indispensable. But for the *application* of the conclusions of pathological chemists and micrologists to *diagnosis*, a much more moderate amount of skill will suffice. There is wisdom in the remark of Dr. Todd (*Clin. Lect. on Urinary Organs*, etc., p. 73), that, "while it is clearly a duty not to neglect any means of observation and investigation, it is desirable that you should be as little as possible dependent on means which are not always at hand, and which it does not fall to the lot of every eye and hand to use with equal readiness and skill."

I shall state, on this principle, only the *most important* and *available* points in urinary pathology and diagnosis.

Allowance must always be made, or correction obtained, for the *variation* the urine undergoes in the course of the same day. It is divided technically into the *urina* **sanguinis**, *urina* **chyli**, and *urina* **potus**; the first being that after a night's rest, the second that after dinner, the third after a very light meal with fluid, as tea. All of these should in each case be examined and compared.

The questions in regard to any given specimen of urine are (see Bowman's *Medical Chemistry*), as to its **general appearance, specific gravity, acidity** or alkalinity, the chemical or microscopical character of its **sediments**, and the **effect of reagents** upon the clear fluid.

General appearance. If *clear*, after standing a few hours, note the *color*. *Deep-colored* transparent urine, of high specific gravity, indicates excessive metamorphosis of tissue. In *jaundice*, the urine is generally very yellow, and sometimes as dark as porter.

If the urine be *opaque*, it is either *white* or *dark*. *White* opaque urine contains either *mucus*, or *pus*, or undissolved *earthy salts*, or all of these together. Mucus floats more distinctly in a separate cloud than pus; purulent urine is generally opaque throughout, and of a creamy yellow color at bottom. Pus can, however, be more readily diffused by agitation than mucus. Purulent urine is oftenest *acid;* mucous urine, generally *alkaline.* Pus contains *albumen,* as shown by testing; mucus does not. Acetic acid coagulates mucus, not pus. *Dark-colored opaque* urine is most frequently tinged with *blood,* giving it a pinkish or brownish hue. The latter color prevails especially in cases of passive hemorrhage from the kidney—the former, in fresh hemorrhage from the bladder, or acute renal hemorrhage. Urine may also be dark from the presence of *bile* (as in jaundice), or of *purpurine.*

For biliary coloring matter (biliphæin, cholepyrrhin) a good **test** is (Gmelin's) the addition of nitric acid, drop by drop, to a little of the urine on a white dish. It will become pale green, violet, pink, and· yellow, in succession. Or (Heller's) shake a little solution of albumen (white of egg) with the urine, and then add a slight excess of nitric acid; if bile be present, the coagulum will be dull green or bluish. Or (Cunisset), add half its bulk of chloroform to the urine; the yellow coloring matter will be carried down.

Pettenkofer's test for bile consists in the addition (after separating albumen, if it be present in the urine, by coagulation and filtration) to the fluid of a grain or two of white sugar, and then, drop by drop, two-thirds of the volume of strong sulphuric acid. If bile be present, a very distinct and characteristic violet-red color will be produced, which is intensified by heat.

Purpurine is, probably, a morbid modification of the *coloring matter* of urine; derived, originally, from that of the *blood.* Some pathologists believe it to be one of the indications of disease of the liver. It frequently accompanies deposits of urate of ammonia. Urine containing purpurine is pink or purple; not unlike bloody urine in appearance.

As tests,—liquor potassæ makes purpurine *greenish brown;* carbonate of potassa, *yellow.* Alcohol will dissolve purpurine from an evaporated extract of urine, receiving and retaining its color. Hydrochloric acid, added to urine containing purpurine, will, if heat be applied, give it a lilac or decidedly purple tinge.

The **specific gravity** of urine is easily ascertained by means of the *urinometer;* a small glass instrument so marked that, when floated in the urine at 60° Fahr., it will show, in thousandths, the excess of density of it above that of water.

Excessive weight of the urine is caused by its containing an unusual quantity of *salts,* or of *urea;* or by *sugar.* The quantity passed in twenty-four hours must always be considered in connection with its specific gravity, so as to judge of the actual quantity of solids passed, as well as their degree of dilution.

The *heaviest* urine is that of *diabetes mellitus* (glycosuria); sometimes reaching 1060 or 1070. The *lightest* is observed in hysteria, and in Bright's disease; running down sometimes even to 1003.

The degree of **acidity** of urine may be approximately estimated by the more or less decided redness given by it to litmus paper. If it be *alkaline,* it will make turmeric brown, and restore the blue to litmus reddened by an acid.

Alkalinity of the urine is *uncommon;* unless (Bence Jones) immediately after a meal. If it does occur at other times, it depends upon either fixed (potassa, soda) or volatile alkali (ammonia). If the former, it is usually associated with *nervous debility* or general depression of vital power; except when accounted for by the medicinal use of potassa, soda, or lithia. Excess of the phosphatic salts, and of oxalate of lime (oxaluria), often accompanies alkalinity of the urine. The importance of the presence of oxalate of lime has probably been overrated.

Carbonate of *ammonia*, when present in the urine, causes it to effervesce on the addition of an acid, from the escape of carbonic acid gas. The change of color produced by ammonia in turmeric paper will, also, *disappear when it is heated.*

Ammoniated urine becomes so by the decomposition of urea, and its conversion into carbonate of ammonia. When the bladder is inflamed, and contains unhealthy mucus, this decomposition occurs, either in the bladder, or in the urine shortly after it is passed; making it alkaline in reaction, and effervescent when acid is applied. In cases of *much less frequency*, urine will effervesce with acid from the presence of *carbonate of lime.*

Sediments occur in the urine, either when first passed or after standing, from its containing substances (1) insoluble in it, or (2) precipitated upon its cooling, or (3) from chemical changes rapidly taking place. Such sediments may be examined both chemically and microscopically.

A *fawn-colored* deposit, not crystalline, which is *redissolved when the urine is heated*, consists of urates of *ammonia* and soda. Urate of ammonia is also immediately dissolved by solution of ammonia or of potassa.

A much more *rare* deposit, of cystine, has a similar color; but it is not soluble by heat, and is but slowly dissolved by alkalies. Cystine, under the microscope, shows rosette-like or hexagonal crystals, sometimes like those of chloride of sodium; the latter, however, is much the most soluble in water. The crystals of "triple phosphate" are known from those of cystine by being freely soluble in *dilute acids.*

Heavy red sand, at the bottom of the vessel, insoluble in hydrochloric acid, but *dissolved by nitric acid*, and also by *alkalies* (as liquor potassæ), is uric (lithic) acid. When *strong* nitric acid is added to deep-red urine containing urate of ammonia in excess with purpurine,—solution occurs, with effervescence; and a brownish deposit falls, of uric acid chiefly. When evaporated to dryness, the addition of ammonia to the deposit will produce the purple *murexide.* Similar reactions occur with uric acid itself, acted upon by strong nitric acid.

Blood-corpuscles sometimes fall to the bottom of urine as a colored sediment. They are not soluble in acids or alkalies, and may be distinguished by aid of the microscope.

A *whitish* deposit, *not at all dissolved by heat*, but dissolved by nitric acid, consists of earthy salts, phosphatic or oxalic. If *oxalate* of lime, it will *not* be dissolved by *acetic* acid; if *phosphates*, that acid will render the liquid clear. Phosphatic deposits occur (even if not excessive in amount) when the urine is alkaline. Nevertheless, excess of the phosphates is indicated, by the urine becoming turbid when heated, and clearing up when acetic acid is added.

A *creamy-white* flocculent and *ropy* deposit, not dissolved on agitating the liquid, is probably **mucus**. A greenish-yellow settling, diffused when shaken, and which is dissolved by and forms a *jelly* with liquor potassæ, may be concluded to be **pus.**

The **microscope** may detect, even in urine scarcely opaque, or in its residue after evaporation,—

Blood-corpuscles, disk-like, or jagged and out of shape;
Mucus-corpuscles, mingled with epithelial scales or cells;
Pus-corpuscles, granular, containing several nuclei;
Epithelial cells or scales, from the kidney or bladder;
Tubular casts from the kidney (desquamative nephritis);
Spermatozoa ; thread-like, with one end ovate and expanded;
Uric acid crystals, variously shaped, as lozenges and square prisms;
Triple phosphate of magnesia and ammonia, in three-sided prisms with bevelled ends ;
Phosphate of lime, granular, or in long needle-shaped crystals;
Oxalate of lime, in transparent octohedral or dumb-bell crystals;
Oil-globules (rare) with dark, smooth and well-defined outline ;
Chyle-corpuscles, found in urine of a milky appearance.

Urine **free from deposit** should, in suspected cases, be tested for **albumen** and for **sugar.**

The test for albumen is the *successive* addition to the urine of *heat* and *nitric acid*. If it become and continue turbid under their combined influence, it is albuminous; but neither alone will suffice. Another test (Millon's) is the acid nitrate of mercury ; which causes with albuminous urine a pink precipitate. Fibrin and casein have this reaction also, but they will scarcely ever be found in urine, in the absence of albumen. Other mineral salts (ferro-cyanide of potassium, bichloride of mercury, &c., will precipitate albumen ; but the first-mentioned test is the most available.

We must remember, however, that **albuminuria** is no longer synonymous with Bright's disease. Albumen occurs, *transiently*, in the urine in many acute affections, as scarlatina, diphtheria, and renal congestion from cold and wet. It is only when *persistent* as a symptom that this becomes pathognomonic of degeneration of the kidney. In rare instances, moreover, this degeneration has been found (post-mortem) to exist, *without* albuminuria.

The principal tests for **diabetic sugar**[1] (glucose) are *Moore's*, *Trommer's, Maumené's, Böttger's*, and *fermentation*.

Moore's : Boil the liquid with half its bulk of liquor potassæ. If saccharine, it will become first yellow and then brown, and ruby red by transmitted light.

Trommer's : Add a few drops of strong solution of sulphate of copper to the urine, in a test tube, and then pour in liquor potassæ, to about half the bulk of the urine. On the careful application of heat, a yellowish or reddish-brown precipitate (sub-oxide of copper) is thrown down.

Maumené's : Dip into the liquid a strip of flannel (not linen or mus-

---

[1] Bence Jones has found a very small quantity of grape sugar in healthy urine.

lin) saturated with a solution of bichloride of tin in twice its weight of water. The strip, on being heated over a fire or lamp to near 300° Fahr., will at once become brownish-black, like caramel.

Böttger's: Add a few drops of dilute solution of nitrate of bismuth in nitric acid ; make the liquid alkaline with carbonate of soda, and boil for a few minutes. When sugar is present, it becomes dark, and will gradually throw down a grayish black deposit. If the urine were healthy a *white* deposit would fall.

Fermentation, on the addition of yeast at 80° Fahr., will only occur in saccharine, not in ordinary urine. During this process, the white scum which forms is found, under the microscope, to contain the oval vesicles of *torula* which characterizes vinous fermentation.

Trommer's test is the one most generally employed. Occasionally, a substance called *alkapton* may be present, which likewise reduces the oxide of copper; but it will not ferment, nor cause a dark deposit with bismuth. Its possible existence does not interfere with the practical value of the test for sugar.

Coloring matters taken as medicine or food may sometimes occur in the urine; as rhubarb, senna, logwood, coffee, &c. Mineral acids will change the color of rhubarb or senna to a bright yellow. Aqua ammoniæ will turn the orange hue of rhubarb to crimson. Indigo coloring matter (*indican*) is said to have been found in or educed from normal urine.

The **quantitative** analysis of urine, to determine the amount and proportion of its different ingredients, requires considerable chemical proficiency.

The following statement of the **normal average** amount of the constituents of healthy urine, passed in twenty-four hours, is from Thudichum :—

Solids, altogether	850 to 1020	grains.
Urea	463 to 617	"
Uric acid	7.5	"
Creatine	4.5	"
Creatinine	7.0	"
Hippuric acid	7.5	"
Chloride of sodium	154 to 200	"
Sulphuric acid	23 to 38	"
Phosphoric acid	56	"
Earthy phosphates	19	"
Ammonia	10	"

Besides *sarkine, uræmatine, uroxantine, potassa, soda, lime, magnesia, iron, trimethylamine, carbonic* acid, *phenylic* acid, and *damaluric* acid, in undetermined amounts.

Heavy and **dark-colored** urine (diabetic urine is straw or amber colored), with a strong odor, may be inferred to contain an excess of solids from waste of tissue; among which **urea** is the most important.

When excess of urea is present, the addition of a few drops of strong colorless nitric acid to the urine on a watch-glass will throw down a number of crystals of *nitrate of urea* (delicate, rhomboidal, like those of saltpetre).

Excess of **phosphates** *is generally associated* with disintegration of brain and nerve tissue.

Chloride of sodium has been found (Redtenbacher) to *disappear* from the urine in the height of an attack of *pneumonia*, and (Beale) to appear at the same time in excess in the *sputa*. This may be tested by the addition of a few drops of nitric acid, followed by solution of *nitrate of silver*—a white precipitate of chloride of silver indicating the presence of the chloride of sodium.[1]

Fatty matter in the urine is detected by the microscope, and by the use of ether, which will dissolve the oily particles from the extract.

Kyestein is a greasy pellicle found on the surface of the urine (after a day or two standing) of pregnant women, or (Kane) in those whose mammary glands are excited by sympathy with uterine irritation. *Urostealith* is a solid adipose concretion (Roberts), now and then making part of a calculus.

In a **low state of vitality** (Inman) the urine, after being passed, *undergoes decomposition* more rapidly than during health.

For chemical analysis of urine by the student or practitioner, simple apparatus will answer. There are needed half a dozen test-tubes, a urinometer, a spirit-lamp, litmus and turmeric paper, and a small amount of each of the following reagents : nitric, hydrochloric, sulphuric and acetic acids, liquor potassæ, aqua ammonia, nitrate of silver solution ; and, sometimes, alcohol, solution of chloride of barium (to test for sulphuric acid or sulphates), and ether.

Microscopical examinations may be made satisfactorily for ordinary purposes with an instrument of moderate cost; such as Woodward's student's microscope.[2] (Guidance should be sought for in the works of Beale, Carpenter, Hogg, or Wythes on the microscope.)

Urinary Calculi.

Gravel, formed in the kidneys and passed, sometimes with much pain, along the ureters to the bladder, and thence out through the urethra, consists usually of *urates* of ammonia and soda, and uric acid.

Calculi, of larger size, are in a majority of cases, composed of **uric acid.** Such are smooth or but slightly rough outside, formed in concentric layers of different thickness. They will dissolve in a dilute solution of potash ; or in strong nitric acid, with effervescence. The microscope will show the uric acid crystals.

Next in frequency is the calculus formed of a mixture of **phosphate of lime** with *triple phosphate* (of magnesia and ammonia), called *fusible* calculus ; because, before the blowpipe, it melts readily without combustion.

Calculus of *phosphate of lime* is generally smooth and polished on the outside ; it *chars* before the blowpipe. So does the rather rare calculus of *triple phosphate* alone.

Oxalate of lime forms the so-called *mulberry* calculus ; irregular and rugged in structure throughout.

---

[1] Chloride of ammonium will produce the same reaction ; but this salt is rare in the urine or sputa.
[2] Made by J. W. Queen, Philada.

Cystine is occasionally found forming the substance of rather soft, brownish or greenish-yellow calculi. Uro-stealith has been already mentioned.

Gall-Stones.

These are concretions of biliary matter, formed in the gall-bladder, and frequently causing great pain in their passage along the cystic and common biliary ducts. They are of various sizes, averaging about that of a pea. **Cholesterin** forms the greater part of their substance, mixed with the biliary resinous and acid constituents (cholic and cholcic acids, taurine, &c.) and coloring matter.

From a solution of the gall-stone in boiling alcohol, cholesterin will crystallize, on cooling, in *fine scaly crystals*. Though allied to the fatty bodies, it differs from them in not dissolving in a solution of potash. The observations of Dr. A. Flint, Jr., make it probable that cholesterin is converted normally into *stercorin*, in which form it is excreted.

Other Secretions.

The **milk** of a mother may be affected in quality as well as quantity by the physical or even mental state of the individual; so as to become innutritious, or even injurious to her offspring.

In the **parturient** state, the *sudden* arrest of the formation of milk in the mammary gland, with the cessation of the uterine *lochial* discharge, is an alarming symptom,—threatening child-bed fever.

Menstruation is not exactly a secretion; rather a periodical discharge of somewhat altered blood, along with monthly *ovulation*, or escape of a germ from the ovary. Its occurrence, however, is necessary to the health of the female, from 15 to 45 years, about, and its variations and deviations are important signs of disease.

Abnormality in menstruation is, principally, either **amenorrhœa, dysmenorrhœa, or menorrhagia.** The first, amenorrhœa, is either (1) non-appearance, (2) suppression, or (3) retention of the menses. The last of these (retention) is rare.

Suppression or irregularity of menstruation, apart from pregnancy, may result from uterine or ovarian disease, or from *constitutional* conditions affecting the uterus or ovaries functionally. The latter is more common in general practice. The amenorrhœal woman is generally, though not always, *anœmic*.

Dysmenorrhœa (painful menstruation) may be either *obstructive, spasmodic*, or *neuralgic*. The *first* may occur from congenital smallness of the orifice of the neck and mouth of the womb; or, from retroversion, anteversion, or obliquity of that organ; or pressure by a tumor; or occlusion of the os by inflammatory bands or adhesions of lymph. The diagnosis of such affections from *spasmodic* or *neuralgic* dysmenorrhœa belongs to *obstetric* surgery or medicine. So does the consideration of *retention* of the menses.

Perspiration.—Changes affecting the secretion of the *skin* have been already alluded to, in connection with the signs of disease be-

6

longing to the tegument. Strong *odor* of the perspiration indicates vicarious excretion by the sweat-glands, and commonly accompanies insufficient action of the bowels. *Acidity* of the perspiration is sometimes dependent on the presence of an excess of *uric acid;* which, in gout, in the form of urate of soda, is occasionally concreted palpably upon the surface of the body. *Sudoric* acid is said by Favre to take the place of uric acid, normally. The perspiration contains also chloride of sodium, urea, lactic acid, ammonia, &c. The *odor* of the perspiration is *peculiar* in smallpox, typhus, gout, albuminuria, &c.

SYMPTOMS CONNECTED WITH THE MOTOR APPARATUS.

The **decubitus,** or mode of lying down, of a patient, should be noticed. *Inability to rise* may depend upon *general debility, paralysis* of the extremities, rheumatic or gouty *inflammation* of the joints, etc., or *injuries,* such as fractures or dislocations.

Inability to lie down is most frequently the result of dyspnœa (orthopnœa)—the respiratory muscles having the freest scope in the erect position.

In *colic,* the patient generally prefers to lie upon the belly.

In *peritonitis,* the characteristic position is upon the back, with the *knees drawn up,* to relax the abdominal muscles.

Lying *upon one side* is often significant in disease. In the early stage of pleurisy, the patient prefers to lie upon the *healthy* side; when effusion has taken place, this is reversed. In irritative disorder of the liver, with enlargement, the patient will often lie most comfortably upon the right side. When the heart is enlarged or violent in its action, the sufferer *generally* cannot lie upon the left side. The exceptions are most frequent in cases of long duration.

In *aneurism of the aorta,* the *prone* or semi-prone position (as leaning forward over a bed) is sometimes preferred.

Muscular debility may be the result of acute disease, as fever, or of actual exhaustion and prostration. *Total want of exercise* will enfeeble the muscles; as, when a limb is long confined in splints on account of a fracture or other injury.

Spasm is of three kinds: *tonic, clonic,* and *choreic.*

Tonic spasm is *fixed rigidity;* such as *emprosthotonos* (arching of the body forwards) or *opisthotonos* (arching backwards) in *tetanus.* *Clonic* spasm is ordinary *convulsion:* i. e. *successive* contractions of the muscles at short intervals. *Choreic* spasm is a term suggested to indicate the jerking, irregular movement of the muscles, not controllable by the will, in cases of chorea.

Subsultus tendinum, or jerking of the tendons at the wrist, is one of the symptoms of low states of continued fever.

Paralysis will be alluded to presently.

SYMPTOMS CONNECTED WITH THE SENSORY APPARATUS.

Of these, the most important is **pain.** Pain may be—
Acute, sharp, cutting, as in pleurisy;
Shooting, darting, as in neuralgia;
Lancinating, in cancer;
Gnawing, tearing, in rheumatism;
Dull, heavy, aching, in pneumonia;
Griping, twisting, in dysentery;
Bearing down, in second stage of labor;
Pulsating, in the formation of an abscess;
Burning, smarting, in erysipelas;
Stinging, nettling, in urticaria;
Constant or *intermittent;* fixed, or wandering.

Tenderness *on pressure* is generally associated with *inflammation;* although some affections designated as neuralgic also present it—possibly from inflammation of the sheaths of the nerves. *Exhausted muscles* also have it, with pain (*myalgia,* Inman).

Sometimes pain is *relieved* by pressure; as in many cases of *colic* and *dysmenorrhœa.* This is a sign, usually, of the *absence* of inflammation.

Pain is *not always at the seat of disease.* Thus, in disease of the hip joint (morbus coxarius), the pain is felt chiefly at the knee; in calculus of the bladder, at the glans penis; in ovarian disease, sometimes, along the limbs; in disorder of the liver, often, under the scapula; in dyspepsia, frequently, about the sternum; and in irritation of the uterus, on the top of the head.

Total **loss** of sensation, local or general, is called **anæsthesia. Acinesia** (a term seldom used) is loss of muscular power.

Paralysis of one side only, of the body, *e. g.* the right arm and leg, is *hemiplegia.* Paralysis of both lower extremities, *paraplegia.* These terms are commonly applied either to loss of power, loss of sensibility, or the more usual combination of both.

The **eye** affords many indications of disease. A prominent and turgid condition of both eyes occurs in acute ophthalmia, and in congestion of the brain. If *one eye* alone becomes prominent, local disease, *e. g.* a tumor behind the orbit, may be suspected. The eyes are sunken, in phthisis, and in other wasting maladies. Sinking of *one* eye indicates local atrophic disease.

The **movements** of the eyes should be noticed, especially in children. Rolling of the eyeballs from side to side is a commom symptom of nervous restlessness or cerebral irritation in infants. *Squinting,* occurring as a symptom in disease, is of unfavorable import.

The **color** of the eyes varies in disease. In *conjunctivitis,* the bloodvessels are generally enlarged, and the membrane reddened. In *sclerotitis,* the enlarged vessels are seen *converging toward the margin of the cornea.* In *iritis,* discoloration, *irregularity,* and sometimes fixedness of the pupil occur.

The cornea in old people occasionally exhibits the **arcus senilis**— a sign of fatty degeneration. It is an opacity around the circumference of the cornea.

The *lustre* of the eye is lessened generally in depressing acute diseases, and especially just before death.

The eyes are often remarkably bright during the progress of phthisis. They have a *glare* in some cases of inflammation of the brain, and of mania.

The **pupil** is generally *contracted* in—
Inflammation of the retina;
Inflammation of the brain;
Narcotism by opium, or the Calabar bean.

It is *dilated*, usually, in—

Apoplexy;　　　　　　　　　　Amaurosis;
Hydrocephalus;　　　　　　　Cataract;
Narcotism by belladonna or stramonium.

An *immovable* state of the pupil, or a *difference between the two eyes* under the same light, gives rise to suspicion of ophthalmic or cerebral disorder.

Photophobia is a dread of or shrinking from the light, such as occurs in ophthalmia, and in meningitis or cerebritis. Other symptoms connected with the eye are—

Photopsia, flashes of light passing before the eyes.
Muscæ volitantes, moving spots, or spectra.
Amblyopia, dimness of vision.
Diplopia, double vision.
Hemiopia, half-sight; *i. e.* seeing but one-half of an object at a time.

Tinnitus aurium, or ringing in the ears, may attend *congestion of the brain; nervous debility;* or *quininization.* **Deafness** may proceed from *coryza.* (a cold in the head); *wax* in the ears; *quininization; typhus* or *typhoid fever; disease of the ear; cerebral* softening.

Pain in the head (cephalalgia) may be specially alluded to as depending upon—
Neuralgia;
Rheumatism of the scalp;
Congestion of the brain;
Toxæmia (*e. g.* by narcotics, alcohol, etc.);
Fever (remittent, yellow, typhoid, etc.);
Chronic disease of the brain;
Uterine irritation, etc.

The distinction between these different forms of headache is by no means always easily made out. As a general statement, it may be said that *neuralgic* headache is mostly on *one side* (hemicrania), and extends more or less to the *face;* it is usually accompanied, also, with sensitiveness of the scalp, and is *shooting* or *darting* in its character. *Rheumatism* of the head is attended by *stiffness of the muscles* which move the head from side to side. *Congestive, febrile,* and *toxœmic* headaches are accompanied by *heat* of the head, and are *throbbing* or *pulsating.* That of *uterine irritation* is on the *top* of the head. The pain of *chronic cerebral disease* (tumors. etc.) is commonly *constant* or *periodic, in one spot,* and is attended by some functional disorder.

SYMPTOMS CONNECTED WITH THE PSYCHICAL APPARATUS.

The **expression of the countenance** is usually altered by disease, especially of an acute kind. The change from anxiety or distress to serenity is always a favorable prognostic, except where *gangrene*, or *paralytic anæsthesia* accounts for it.

Great anxiety of expression is seen especially in organic disease of the heart, and in acute disorders of the abdominal viscera. In hypochondriasis, a *sad and desponding* expression prevails.

Terror is shown, by the countenance, in delirium tremens.

Rage, in hydrophobia, and sometimes in acute mania.

Insanity and *imbecility*, although not characterized by any *special* cast of countenance, yet modify its expression so as to enable the mental state to be detected by one accustomed to the observation of deranged persons.

The *facies Hippocratica* is the countenance of extreme exhaustion or of the moribund state.

Delirium is described as being either *active* or *passive*. Active delirium is present in cases of acute meningitis; passive or low muttering delirium, in typhus fever, etc.

Coma presents itself in practice chiefly in four forms: Alcoholic stupefaction; Opium poisoning; Apoplexy; Typhus; also, fracture of the skull with compression of the brain.

Typhous stupor is generally easy of recognition; the others may give some trouble in the diagnosis. Between narcotism by opium and dead-drunkenness, we have the distinctions, that in opiate poisoning the pupil is almost always *firmly contracted*, and that the breath smells of alcohol (or aldehyde) in the intoxicated subject.

GENERAL VITAL CONDITION.

Lyons (Hospital Practice) remarks as follows: "The highest skill in physical diagnosis, and the most profound knowledge of pathological anatomy, will leave you but very imperfect and unsafe practitioners, incapable of clear judgment and self-reliance in difficult cases in which you have to rest on your own responsibility, if you do not from the first endeavor to master and acquire for yourselves that unwritten and indescribable knowledge which constitutes the consummate skill of the experienced medical man. It consists in a faculty of appreciating the vital state of your patient; of forming a rapid but complete and accurate estimate of the nervous and muscular force which he possesses; or, in general terms, of the powers of life which remain to him—his *viability*, so to speak, or the power which his system retains of resisting the morbid or fatal influences of injury or disease."

PHYSICAL DIAGNOSIS.

The *idea* of physical exploration for the purpose of diagnosis has been well defined by Piorry, in the word "**Organography**:" *i. e.* the determination of the actual and relative *position*, material *condition*, and functional *action* of the organs contained within the body. The methods in use for this purpose are modern, dating from Auenbrugger, of Vienna, the inventor of diagnostic percussion, in 1761, and Laennec, the great originator of auscultation, about 1818.

The modes of examination of the chest, abdomen, etc., are— Inspection; Mensuration; Palpation; Succussion; Spirometry; Percussion; Auscultation.

By **inspection**, we estimate, with the eye, the *form*, size, and *movements* of the chest, etc.

By **mensuration**, we obtain a more *accurate* knowledge especially of *deviations* and *alterations* of size and form.

Palpation aids in the determination of the character of surfaces and of subjacent parts, and, in the chest, detects changes in the degree or extent of the *movements* of respiration and of the heart, and in the *vibrations* connected with the voice, cough, and breathing.

Succussion, or *shaking* the chest suddenly, is of use occasionally, in establishing the presence of *fluid* in the thoracic cavities.

Spirometry is the measurement of the capacity of the lungs for air.

By **percussion** we learn much of the physical condition of the lungs, heart, and abdominal viscera, through the variations of *resonance* and *resistance* when the walls of the thorax or abdomen are lightly struck.

Auscultation is equally important, but somewhat more difficult in its application, on account of the complexity of the signs afforded by it. It consists in direct *listening* to the sounds produced within the cavities of the body, by placing the ear, with or without the stethoscope, upon the surfaces thereof.

The **Regions** of the *Chest*, for the purpose of physical exploration, may be most conveniently divided into the following:—

ANTERIOR.	POSTERIOR.
Upper and lower sternal;	Interscapular;
Right and left clavicular;	Dorsal;
Right and left subclavian;	Lower dorsal;
Right and left mammary;	Right and left acromial;
Right and left infra-mammary.	Right and left scapular;
	Right and left infra-scapular.

LATERAL.

Right and left axillary;
Right and left lateral;
Right and left lower lateral.

The most important peculiarities of these different regions, in the normal state, are connected with *percussion-resonance*. The clearest and fullest sound, on percussion, is given over the *subclavian* and *lateral* regions; the dullest and smallest, over the *acromial*, the *right infra-mammary* (hepatic), and the *left mammary* or præcordial region.

MENSURATION AND PALPATION.

For **mensuration,** various stethometers or chest-measurers have been devised; but, with care and judgment, the common tape-measure will suffice.

The dimensions to be compared are the—

Circular: around the chest opposite the base of the ensiform cartilage. This averages thirty-three inches. The *right half* of the thorax is nearly always half an inch to an inch **larger** in circumference than the *left*.

Transverse: from the nipple to the middle of the sternum.

Vertical: from the clavicle to the lower margin of the ribs.

Antero-posterior: from the clavicle anteriorly to a corresponding point in the scapular region.

General expansion and local bulging of the chest, and general retraction and local depression, are the signs most frequently determined by inspection and mensuration.

General expansion or **local bulging** of the chest, usually upon one side only, may be caused by—

> *Pleuritic effusion;*
> *Pneumothorax;*
> *Emphysema* of the lung;
> *Aneurism, cancer,* etc.; or, more rarely, by
> *Hydrothorax;*
> Pneumonia;
> Incipient tuberculization.

Retraction or **local depression** of the thoracic walls may result from—

> *Absorption* of pleuritic effusion;
> *Tuberculization;*
> Pneumonia;
> Pleuro-pneumonia;
> Infiltrated cancer of the lung.

By **palpation,** we observe **diminution** of the **expansion** *and* **elevation** *of the* **ribs** *in breathing,* in—

Pleurisy;	Emphysema;
Pneumonia;	Intercostal rheumatism;
Tuberculization;	Paralysis;
Pneumothorax;	Hydrothorax.

Increased expansion and elevation of the ribs in breathing occurs in—

> *Asthma;*
> *Croup;*
> *Spasm of the glottis;*
> Foreign bodies in air-passages.

Increased vibration *of the walls of the chest* with the **voice** and **cough** is noticed in—
Tuberculization;   Pulmonary apoplexy;
Pneumonia;   Dilatation of bronchi.
Diminished vocal and tussive vibration occurs in—
Pleuritic effusion;   *Emphysema;*
Pneumothorax;   Cancer of the lung.
Rhonchal vibration, occasionally, in bronchitis.
Rubbing, or to-and-fro vibration, in—
Pleurisy;   ·   *Pericarditis.*
Pulsatile vibration in—
Aneurism of aorta;   ·
Cancer of lung or pleura;
Pneumonia.
Fluctuation in—
Large pleuritic effusion.
Purring vibration (frémissement cataire) in—
Aneurism of aorta;
Valvular heart disease;
Anæmia.

SPIROMETRY.

For **Spirometry,** Hutchinson's, Pereira's, Coxeter's, and Mitchell's[1] spirometers have been used.

Dr. Hutchinson made elaborate investigations into the comparative breathing power of individuals, by which he proposed to conclude upon their *vital capacity.* A man 5 feet 8 inches in height, and of 155 pounds weight, was found, on the average, to expire, after a full inspiration, 230 cubic inches.

For every inch of height above this, a definite increase in the quantity breathed was observed. The proportion was *less constant* with *weight* and with *age.* After fifty-five there was a decrease.

In the first stage of *consumption,* the average (for the adult of ordinary height) was found to be 154 cubic inches; second stage, 131; late stage, 108, etc.

In practice, however, spirometry is not extensively used.

PERCUSSION.

Percussion is either *mediate* or *immediate.* In immediate percussion, we tap with the ends of the fingers at once upon the body; in mediate percussion, a pleximeter (*stroke-measurer*) is used. The latter is almost universal; but a difference exists as to the kind of pleximeter employed. Louis and Walshe prefer one made of caoutchouc; Piorry and Skoda, one of ivory; Wünderlich uses an ivory disk, upon which he strikes with a small steel hammer, the head of which is covered with caoutchouc.

A majority of practitioners, however, are satisfied (with good reason) with the use of the *middle finger of the left hand* as a pleximeter.

[1] Consisting of a small *gas-meter,* with a mouth-piece.

(Percuss by movement of the hand on the *wrist*; not by sledge-hammer motion from the *shoulder*).

In using percussion as a means of physical diagnosis, we note—

1. The **clearness** or **dulness** of the resonance produced.
2. The **duration** of the resonance.
3. Its **special character.**
4. The **degree of resistance** felt.

Certain terms are in common use to describe particular characters of resonance: as, *wooden* sound, *thigh* sound, *stomach* sound, *tympanitic* or drum-like resonance, *amphoric* or pitcher-like sound, *bruit de pôt fêlé* or *cracked-pot sound*, etc.

It is indispensable, in commencing the study of percussion (or other modes of physical diagnosis), to become familiar with the *normal and natural* sounds observed in *health*. To be anything more than a *routine* diagnostician, moreover, it is necessary to understand the *principle* of the exploration, and as far as possible, the *reason* of the meaning of every sign.

Two or three very simple facts explain the use of percussion in diagnosis.

When any solid body is struck, the sound elicited varies according to its *material, form, size*, and, *if hollow*, the *condition of its walls*, and that of its *contents*.

The human thorax (or abdomen) having a certain general form, size, condition of its walls, and proportion of air, blood, and solid structure in its contents, will give forth a certain degree and kind of resonance.

Whatever alters either the *state of its walls* or the *proportion of air, fluid, and solid* contained within them, gives rise to an alteration of percussion-resonance.

Alteration of the state of the *walls* of the thorax seldom occurs in disease in such a way as to modify percussion-resonance. Changes in the proportion of **solid, liquid,** and **air,** in the lungs and pleural cavities, as well as in the similar relations of the heart and pericardium, aorta, etc., are frequent. The *more air*, and the *less liquid or solid* contained within the part of the chest which is percussed, the clearer and fuller the resonance, and, as a general rule, the less the resistance to the finger. Any *increase* in the relative proportion of *liquid* (as in pleuritic effusion), or of *solid* (as in tuberculization), must cause a duller or lesser degree of resonance, and, other things being equal, a greater degree of resistance.

Thus, **diminution** *of* **clearness** *and* **duration** of the percussion-sound, with **increased resistance** of the walls of the chest, occurs in—

Pneumonia;   Pulmonary apoplexy;
Pleurisy;   Hydrothorax;
Tuberculization;   Cancer, etc.;
            Collapse of lung.

The extent over which dulness on percussion is observed sometimes *varies* with the *position of the patient*. This is practically important in the diagnosis of *pleuritic effusion, empyema, hydrothorax*, and *hydropneumothorax*.

Increased clearness and **duration** of resonance, with **decrease** of **resistance**, occurs in—

Pneumothorax;   Emphysema;
Atrophy of lung;   Anæmia;
Hypertrophy of lung;   Emaciation.

Increased clearness of sound with **increase** of **resistance** is observed when there is a *tubercular cavity near the surface of the chest, with its outer wall thin, hard,* and *adherent to the pleura.*

Tympanitic resonance of the chest is present in—
Pneumothorax;
Emphysema;
Pulmonary atrophy, etc.

Amphoric resonance, when there is a large tubercular *cavity,* with *solid* and *tense* walls, near the surface of the chest.

The **cracked-pot** sound indicates an *anfractuous* cavity, *i. e.* one whose walls are broken or incomplete, communicating with the bronchial tubes. It may be imitated by clasping the hands loosely and then striking the back of one of them upon the knee.

Skoda's classification of percussion-sounds has the merit of great simplicity. He distinguishes them as

Full; empty; (*large* and *small* resonance);
Clear; dull;
Tympanitic; non-tympanitic;
High; low; (pitch).

A sound may be at the same time *full* and *dull,* or *clear* and *empty* (small).

Skoda does not value very highly the information obtained from differences in the *pitch* of percussion sounds. Other authorities differ from him, however, upon this point; and with good reason.

AUSCULTATION.

In **auscultation**, as well as in percussion and other modes of physical exploration, a comparison is made not only with the normal standard, but *between the two sides* of the chest.

The *stethoscope* is, in auscultation, generally speaking, a superfluous instrument. If any be used, a simple wooden tube with one end slightly expanded is the best. Camman's *double* stethoscope is approved by Dr. Flint; but it requires a good deal of practice to use it well.

The beginner must familiarize himself with the *natural breathing sound,* as heard when the ear is placed over any part of the *lungs,* and with that heard in the *sternal* and *inter-scapular* regions. The latter is **bronchial,** the former is the **vesicular murmur.** The tubular, blowing character of the respiration, as heard in the bronchi, and its soft, breezy nature when the ear is placed over the lungs, are essential elements in diagnosis by auscultation.

The pulmonary vesicular murmur is always *louder in infants and children.* **Puerile** respiration is, therefore, the name given to **exaggerated** breathing-sound in the adult.

In a healthy state of the lungs, the *expiratory* murmur is very faintly heard. A *prolongation,* and increase in loudness, of the sound of expiration, is sometimes a sign of disease (tuberculization).

The sounds detected by auscultation of the chest are divided into **respiratory** and **secretory** sounds, **friction** sounds, and modifications of **vocal resonance.**

Respiratory Sounds.

Normal vesicular murmur;
Puerile respiration;
Prolonged expiratory sound;
Harsh, tubular, blowing,
Bronchial, and cavernous respiration;
Amphoric respiration.

Secretory Sounds.

Dry.

Sibilant rhonchus (hissing or whistling);
Sonorous rhonchus;
Dry crackle.

Moist.

Fine crepitation or crepitant râle;
Coarse crepitant râle (mucous râle);
Humid crackle or gurgling;
Metallic tinkling or dropping sound.

Friction-sounds are peculiar to pleurisy and pericarditis, at the stage of adhesion, or, at least, of effusion of plastic lymph.

Modifications of **vocal resonance** *are—*
Bronchophony;
Pectoriloquy;
Ægophony.

The above is. essentially, the classification commonly adopted by auscultators. That of Skoda is, however, still more simple. He divides *respiratory* sounds into

Vesicular;   Amphoric;
Bronchial;   Indeterminate.

Skoda denies, also, the validity of the distinction between *pectoriloquy* and *bronchophony;* and shows that *ægophony* cannot have the precision of meaning supposed by Laennec and others to belong to it.

No description of the sounds of auscultation can do more than *guide* and *assist* their actual clinical study. For this purpose the simplest and clearest terms are, of course, the best.

The **normal respiratory murmur** as heard in the *lungs*, is well illustrated by Skoda as resembling (during inspiration) the sound caused by narrowing the opening of the mouth, and then drawing in the air. The *consonant* of this murmur is *f* or *p*. The *expiratory* murmur may be represented as somewhere between *f* and *h*. That of the *larynx, trachea*, and *bronchi*, by the guttural *ch*, or between that and *h*. The vesicular murmur is slightly louder (Flint) and lower in pitch, on the left side.

By **bronchial respiration** as a sign of disease in the lung, we mean

a breathing-sound heard while listening over the lung, like that nor
mally heard when auscultating the middle regions of the chest, over
the bronchial tube. It occurs when the lung is *solidified* or *condensed*.
(The *explanation* of bronchial respiration is by *conduction* or by *consonance*. The latter theory, that of Skoda, is *preferred*.)

Cavernous respiration is that which is inferred to occur during the
passage of air into or out of a *cavity* in the lung (as in tubercular disease). Notwithstanding the truth of the statement urged by Skoda,
that it is *often* impossible to draw a certain demarcation between
bronchial and *cavernous* respiration, yet, in a *sufficient number of cases*
this *can* be done, and the term cavernous, therefore, should be retained.

Of the **secretory** sounds, the **sibilant** and **sonorous rhonchi** are
the results of narrowing and obstruction, by congestion, mucus, etc.
of the bronchial ramules; the *smallest*, in the case of the sibilant or
whistling rhonchus; those *somewhat larger*, so that the air passes
through in irregular and varying bubbles, in the sonorous (snoring or
roaring) rhonchus. Both of these sounds are characteristic of bronchitis. The use of the term *dry* sounds, as applied to them, is not
strictly correct; but it is convenient, as designating the impression
which they convey to the *ear* as compared with those technically
called moist sounds.

The **dry crackle** is associated with *incipient* or infiltrated *tubercle*.
Among the **moist** sounds, the finest or most delicate is the **crepitant rale**, or fine crepitation, of pneumonia. It is very well imitated
by rubbing a few hairs of one's head between the thumb and finger,
near the ear. Its *cause* is, probably, the penetration of the air into
the air-cells of the lung at a time when their walls are rendered
slightly *adhesive* by effusion of coagulable lymph. The gentle *forcing
apart* of these adherent walls, or of portions of the viscid lymph
itself, produces the fine crackling sound, as a modification of the
natural vesicular murmur. It is only heard during *inspiration*.

Coarse crepitant or "mucous" râles are heard whenever any fluid
exists in the lungs in quantity sufficient to modify respiration without
arresting it, whether that fluid be mucus, pus, blood, or serous effusion.

The **humid crackle** or *gurgling* is pathognomonic of advanced
tuberculization. It is heard during the later stages of nearly all cases
of consumption.

Friction or *to-and-fro* sounds are produced by the rubbing of two
surfaces, as of the pleura or pericardium, when made adherent or
slightly roughened by inflammatory lymph. It sometimes requires
an acute and practised ear to discriminate these from other sounds.
They are heard both with inspiration and expiration.

Bronchophony, or bronchial vocal resonance, corresponds in its
history with *bronchial respiration*. It is simply a resonance of the
voice, to the ear of the auscultator placed over the *lung* of the patient
while he speaks, loud, near, and clear, as it is normally when the ear
is placed over the *bronchial tube*. The same reasonings will apply
to the explanation of this sign by the two theories of *conduction* and
consonance, which have been urged in regard to bronchial respiration.
The *latter* theory, as in that instance, I prefer; but, practically, all
agree as to the *circumstances* under which the sign occurs (solidifica-

tion of the lung). Normally, the vocal resonance is loudest under the *right* scapula. Modifications of the sound of the *whispered* voice are spoken of by Dr. Flint as exaggerated bronchial, amphoric, and cavernous whisper.

Pectoriloquy (chest speaking) is merely a *yet nearer* and *louder* resonance of the voice, heard on auscultation, than that called bronchophony; the sound seeming to be *vocalized* in that part of the lung which is *immediately under the ear*. Skoda objects that this cannot be definitely distinguished from loud bronchophony. But, although this is *generally* true, a certain number of cases occur in which it may be so distinguished, as indicative of a very different pathological state of the lungs, viz., a large cavity.

Aegophony, bleating, or goat-like resonance of the voice, has been, since Laennec, supposed to be an almost certain sign of the existence of *pleuritic effusion* or *hydrothorax*. Skoda's observations, and those of others also, show that it is occasionally heard in pneumonia, in phthisis, and even in the healthy state of the thoracic organs. It is, therefore, not *pathognomonic* of the presence of fluid within the pleura; but it is among the signs which render that diagnosis probable.

Amphoric resonance is heard especially in connection with the sound produced by *coughing*. A tense condition of the walls of a large cavity will explain it, as well as the phenomenon called *metallic echo* of the voice or cough.

Metallic tinkling is usually accounted for by the *dropping of fluid* in a large cavity (as in hydropneumothorax, with collapsed lung) with tense walls.

Resuming the consideration of respiratory sounds, **puerile** or *exaggerated respiration* occurs in the *healthy* lung, or part of the lung, when the *other* lung or portion of the same is *obstructed*, as by a foreign body, or by bronchitis; *condensed*, as by

Pneumonia; Pleuritic effusion;
Tuberculization; Tumor.

A lung, a *portion* of which is permanently expanded by *emphysema*, or *hypertrophied*, may also give an exaggerated vesicular murmur; and, transiently, this is observed in a lung *just released* from the paroxysmal obstruction of *asthma*.

Feeble respiratory murmur is heard in one or both lungs in cases of

Croup; Collapse of lung;
Foreign bodies in air-passages; Pulmonary apoplexy;
Bronchitis; Emphysema;
Pneumonia; Pneumothorax;
Pleurisy; Hydrothorax;
Asthma; Intercostal rheumatism;
Infiltrated tubercle; Paralysis;
Cancer; or other tumor.

Harsh respiratory murmur, passing by gradations into **blowing** and **bronchial,** in

Dry bronchitis; *Pneumonia;*
Incipient tubercle; Pulmonary apoplexy;
Pleurisy (condensing lung); Bronchial dilatation.

7

Cavernous respiration, in case of
Tubercular cavity;
Excavation from
   Abscess of lung;   Softening of cancer;
   Gangrene;   Large bronchial dilatation.

Amphoric respiratory sound is particularly associated with the existence of a fistulous opening between the pleural cavity and one of the bronchial tubes; the cause of which fistula may be either tuberculous softening, or abscess, etc.

The **sibilant** and **sonorous rhonchi** occur nearly always in *bronchitis;* occasionally in pulmonary *emphysema,* and when the bronchi are pressed upon by tumors, etc.

Dry crackling indicates the existence of hard tubercle in the lungs.

Humid crackling or **gurgling,** tubercle in the softened state.

The **crepitant rale** has already been explained as peculiar to pneumonia.

The **coarse** crepitant[1] *rale,* or **mucous** râle, is observed frequently in
Capillary bronchitis;   Pulmonary *hemorrhage;*
Bronchorrhea;   Pulmonary *œdema;*
Last stage of *pneumonia;*   Pulmonary *abscess.*

Friction sounds have been before alluded to as connected with *pleurisy* and *pericarditis.*

Resonance of the voice is **feeble** in
   Emphysema;   Atrophy of the lung;   Pneumothorax.

Bronchophony occurs in case of
Tubercle;
Hepatization (pneumonia);
Pleurisy (condensing lung); *Cancer;*
Dilatation of bronchi.

Aegophony, in
Pleurisy; Hydrothorax; Pneumonia.

Pectoriloquy, in case of
Tubercular cavity;
Dilatation of bronchi;
Excavation from
   Abscess; Cancer; Gangrene.

Metallic tinkling and **echo,** in
Pneumohydrothorax;
Large tubercular cavity.

The *sounds of the* **heart** *are heard at an unusual distance from the heart itself,* in some cases of
   Pneumonia;   Pleurisy;
   Tubercle;   Cancer, etc.

(This affords some argument for the *conduction* theory of Laennec.)

Displacement of the heart, diaphragm, liver, spleen, stomach, sometimes occurs from pleuritic effusion or empyema, cancer of the lung, etc. In *rare* instances, such a displacement may be *congenital;* as of the heart on the right side side.

---

[1] The term *sub*-crepitant râle or rhonchus, appears to the author to be very objectionable, as leading to confusion.

Diseases of the Heart.

The physical diagnosis of **diseases of the heart** is conducted upon exactly the same principles as that of affections of the lungs and pleura.

By **inspection** we can detect bulging or distortion in the præcordial region, and, in some cases, judge of the extent, force, and character of the heart's *impulse*. By **mensuration**, changes in the thoracic dimensions consequent upon diseases of the heart can be more accurately determined.

By **palpation**, the *impulse* of the heart may always be examined and estimated. This is very important, not only in actual *diseases of the heart*, but also in *fevers*, etc.; in the course of which the movements of the heart, as the centre of the circulation, are often seriously affected.

In *hypertrophy*, this impulse is *increased in force*; in *dilatation*, it is *extended*; in *atrophy*, it is *diminished*.

Percussion aids us in detecting some very important pathological changes in the heart; as *hypertrophy, dilatation, pericarditic effusion*. The percussion resonance is *unusually clear* in *atrophy* of the heart.

In **auscultation** of the heart there is often a convenience, although no necessity, in the use of the stethoscope. The learner must in the first place make himself familiar with the *natural sounds* of the heart.

The **first** sound is the longest and loudest; the succession being imitated by the syllables *lubb, dup*. If the time from the commencement of one pulsation to that of another be divided into five equal parts, *two* of them will be occupied by the first sound, *one* by the second, and *two* by the interval of repose.

The first sound accompanies the *systole* or contraction of the *ventricles*; the *impulse* of the heart occurs at the same moment. The second sound is *diastolic* as regards the ventricles.

The *causes* of the first sound are believed to be the contraction of the powerful ventricular muscles, the *tension of the closed auriculo-ventricular valves*, the rush of blood into the great vessels, and the impulse of the heart against the walls of the chest.

The cause of the **second** sound has been proved to be the *flapping together*, during the diastole or dilatation of the ventricles, of the pocket-like, *semilunar valves* of the aorta and pulmonary artery.

The essential points in the "medical anatomy" of the heart are as follows :—

The *semilunar valves of the* **pulmonary artery** lie behind the junction of the cartilage of the third rib with the sternum.

The *semilunar valves of the* **aorta** are just *below* these, between the cartilages of the third and fourth ribs.

The **tricuspid** or *right auriculo-ventricular valve* is behind the sternum, on a level with its articulation with the fourth rib.

The **mitral** or *left auriculo-ventricular valve* lies behind the cartilage of the fourth rib, a little to the left of the sternum.

The *heart's* **apex** *strikes*, during the impulse, at a point just below and outside of the left nipple. The point of greatest dulness on percussion is slightly within the left nipple. The diameter of the *normal* region of dulness does not exceed two inches.

Using terms of *convenience* merely, the valves of the heart may be said to be of two kinds; *cavity* valves and *vascular* valves. The cavity valves are both nearer to the *middle* and *apex* than to the base of the heart; the vascular valves (aortic and pulmonary arterial) nearer to its *base, i. e.* as the heart is situated in the chest, its *upper* part.

By auscultation of the heart we may detect **valvular** *murmurs,* **anæmic** *murmurs,* and **friction** *sounds.* Details in regard to these can be best given in connection with the *special pathology* of the heart. A few main points only require mention here.

The valves of the **left** or *systemic* portion of the heart are much more often affected by disease than those of the *right.* Practically, in most cases, those of the right side may be left out of the question of diagnosis.

The following is Harvey's statement of the comparative frequency of the different valvular affections :—

1. *Aortic obstructive.*
2. *Mitral regurgitant.*
3 *Aortic regurgitant.*
4. Aortic obstructive and mitral regurgitant together.
5. Aortic obstructive and regurgitant together.

If a murmur (not anæmic) is **systolic**, *i. e.* is heard with the *first sound* of the heart, and is loudest at the **base** of the heart, it may be inferred to be **aortic obstructive.**

If **systolic**, and loudest at the **apex, mitral regurgitant.**

If **diastolic**, *i. e.* with the *second* sound, and loudest at the **base** of the heart, **aortic regurgitant.**

If **diastolic**, and loudest at the **apex, mitral obstructive.**

The *rationale* of these inferences is explained by the physiology of the heart's action, in connection with the position of the several valves.

Much complexity attaches necessarily to the *exact* diagnosis of affections of the heart; but we have the excellent authority of Dr. Stokes for the principle, that the *important practical* questions in each case are—do the abnormal sounds present have origin in **organic** disease or lesion, or not ? and, how far is the **functional action** and *capacity* of the heart interfered with or impaired ?

Upon these, as upon all other questions in diagnosis, it is proper *never* to *confine our attention to physical or immediate signs alone.* To rest either upon symptomatology or physical exploration *exclusively,* would be like hopping constantly on one foot, instead of walking upon two.

Friction *sounds,* in the region of the heart, are connected with *pericarditis.* Their *narrow limits,* and *association, in time,* with the *sounds* of the *heart,* serve usually to contrast them with *pleuritic* sounds. It is sometimes difficult to distinguish them from *valvular murmurs.*

The signs of **aneurism** *of the thoracic* **aorta** may be alluded to in another place. They are, chiefly: 1. A *second impulse* (often with a *thrill*), apart from that of the heart. 2. Dulness on percussion. 3. Bulging. 4. *Symptoms of dyspnœa, cough,* and *dysphagia,* from pressure upon the trachea, œsophagus, etc.

Abdominal Diseases.

The physical diagnosis of **abdominal** affections comprises inspection, mensuration, palpation, percussion, and auscultation. The belly is divided, externally, into the *epigastric, umbilical, hypogastric,* two *hypochondriac,* two *lumbar,* and two *iliac* regions.

By abdominal **inspection** we can observe the alteration in size and shape caused by *pregnancy, hernia, tympanites, ascites,* or *ovarian dropsy.* By **mensuration,** we can ascertain the *exact changes* which may occur from time to time in dropsical accumulations, etc.

By **palpation** of the abdomen, we may develop the symptom of local *tenderness on pressure:* as in *gastritis, hepatitis, peritonitis, cystitis,* etc. By the same method of examination, more forcibly employed, we detect *enlargement* of the *liver* or *spleen, ovarian tumor, mesenteric disease, cancer, aneurism, fœcal accumulation,* etc.; and, with the aid of both hands, prove the presence of *fluid* (ascites, etc.) by the sign of *fluctuation.*

Percussion assists materially in the diagnosis of diseases of the abdominal viscera. The ordinary percussion-resonance, in health, is clear, full, and *slightly* tympanitic, all over the abdomen. It becomes more drum-like in distension of the intestines with gas (meteorism), or in tympanitic distension of the peritoneum. *Dulness* of resonance occurs, with limits and peculiar characters, in *enlargement of the liver or spleen, ascites, ovarian dropsy, pregnancy, cancer, aneurism, retention of urine, fœcal accumulation,* etc.

Auscultation of the abdomen is especially useful in the diagnosis of *pregnancy,* by detecting the sounds of the fœtal heart, and the placental *soufflet.*

Morbid sounds are occasionally appreciable in abdominal affections, as in tape-worm, etc. ; but they are subject to so much uncertainty as to be hardly available for practical purposes. *Friction-sounds* in *peritonitis* have recently been studied by Dr. Seidel.

The Laryngoscope.

Instrumental aid in examining the interior of the larynx was first thought of by Levret, in 1743. Laryngoscopic mirrors were devised by Dr. Babington, of London, in 1829. Türck, of Vienna, applied them anew to diagnosis in 1857; but shortly afterwards Czermak pursued laryngoscopy with so much acuteness and energy as to have associated his name with it pre-eminently.

The apparatus required, (Semeleder) is a laryngeal mirror, an illuminating mirror, and a tongue-depressor. Glass or polished metal may do for the mirrors.

The laryngeal mirror may be round or square, preferably the former; and about an inch or less in diameter. It should be attached at an obtuse angle (120° to 125°) to a stem, which may be fastened into a slender handle so as to be drawn out or pushed in.

The illuminating mirror is larger (from 3 to 12 inches in diameter) and concave, to concentrate reflected light. It may be held by a handle in the operator's mouth, or fixed by a band to his forehead, or, best, as used by Semeleder, perforated in the middle and fastened to

the bridge of a pair of spectacles (with or without the glasses) so
as to rest before-one of the eyes and be looked *through*.

The laryngeal mirror is introduced (after being *warmed* to prevent
condensation of moisture) so that its back pushes the uvula upwards
and backwards, its lower edge presses upon the posterior wall of the
pharynx, and its stem rests in the angle of the mouth.

Sunlight, horizontal (morning or evening), is the best for laryngoscopy, but artificial light, as of a good lamp, may suffice.

The *difficulty* of the operation is produced by the *irritability* of the
fauces and larynx.   Few can allow of a successful examination on the
first attempt; practice makes tolerance.   To hasten this, bromide of
potassium has been given by some.   The frequent insertion, and retention for a while, of the finger of the patient, or of an instrument
in the fauces, accustoms the parts to pressure.   Holding ice in the
throat, just before the examination, also lulls sensibility.

By laryngoscopy, tumors, ulcerations, inflammatory changes, etc.,
in the larynx may be inspected, topical applications, as of nitrate of
silver (H. Green) made, and surgical operations performed, with a
precision not otherwise possible.

Rhinoscopy is the examination, in a similar manner, of the *posterior nares.*   It requires merely a *smaller* mirror (less than three-
fourths of an inch diameter) than for laryngoscopy, and at about a
right angle to its handle.

The **ophthalmoscope** (for the eye), **otoscope** (for the ear), **endoscope** (for the urethra), **uterine speculum**, etc., are instruments for
surgical and obstetric diagnosis, not demanding description here.

TEMPERATURE IN DISEASE.

The thermometer (De Haen, 1754, Wunderlich, J. Davy) is a useful aid in diagnosis and prognosis; making exact that information
which every physician constantly obtains by the touch.   It is especially valuable in the chemical study of febrile disorders; as, since
Galen, fever is essentially defined by the words "preternatural heat."

The **axilla** is the best part for examination of temperature.   The
instrument should be kept there from three to five minutes at a time.
Normally, in the armpit, the temperature averages 98.4° Fahr.; with
a range in health (Davy) from 99° to 97.92°.   It is about 1° higher
in tropical than in temperate climates.   In the temperate it is *highest*
on waking in the early morning; lowest at midnight.   In tropical regions, it is *lowest* in the early morning, and highest during the day.[1]
It is one or two degrees higher in children than in adults.

A rise of temperature, in disease, of 1° Fahr. corresponds, as a rule,
with an increase of the pulse, of eight to ten beats per minute.   The
thermometer in the axilla may, in some febrile cases, mark 106°, 108°,
even 112°.   It has been found highest in scarlet fever, yellow fever
(Dowler) and tetanus.   Dr. H. C. Wood, Jr., found it 109° in the
axilla of a man dying with heat-stroke; and 110¾ in his abdomen after
death.

---

[1] See Aitken's Science and Practice of Medicine, 4th ed., vol. i. p. 39.

In intermittent fever, during the paroxysm, even when the patient shivers and feels cold to himself, his heat by the thermometer is always above the natural degree.

[1] "When the temperature is increased beyond 98.5° it merely shows that the individual is ill; when it is raised as high as 101°–105°, the febrile phenomena are severe; if above 105°, the patient is in imminent danger; with 108° or 109°, a fatal issue may without doubt be expected in a comparatively short time.

A person, yesterday healthy, who exhibits this morning a temperature above 104° Fahr., is almost certainly the subject of an attack of ephemeral fever or of ague; should the temperature rise to or beyond 106.3°, the case will certainly turn out one of some form of malarious fever. It cannot be typhoid fever.

A patient whose temperature rises during the first day of illness up to 105° or 106° Fahr., certainly does not suffer from typhus or typhoid fever. In a patient who exhibits the general typical signs of pneumonia, but whose temperature never reaches 101.7° Fahr., it may be concluded that no soft infiltrating exudation is present in the lung.

If a patient suffer from measles, and retains a high temperature after the eruption has faded, it may be concluded that some complicating disturbance is present.

In typhoid fever a temperature which does not exceed any evening 103.5° indicates a probably mild course of the fever. 105° in the evening, or 104° in the morning, shows danger, in the third week. In pneumonia, a temperature of 104° and upwards indicates a severe attack. In acute rheumatism a temperature of 104° is always an alarming symptom, foreboding danger, or some complication such as pericardial inflammation. In jaundice, otherwise mild, a rise of temperature indicates a pernicious turn. In a puerperal female an increase of temperature shows approaching pelvic inflammation. In tuberculosis an increase of temperature shows that the disease is advancing, or that untoward complications are setting in.

A fever temperature of 104° to 105° Fahr., in any disease, indicates that its progress is not checked, and complications may still occur."

Certain diseases have been found to have typical ranges or daily fluctuations of temperature throughout their course; so that their "differential diagnosis" may be thus assisted materially. This has now been determined, especially, in malarious fever, typhus, typhoid, smallpox, scarlatina, measles, rheumatism, pyæmia, pneumonia, and acute tuberculosis. Dr. Da Costa has observed that in some cases at least, cancer is attended by a lowering of temperature.

In continued fevers the temperature is generally less high in the morning than in the evening. Stability of temperature from morning to evening is a good sign; on the other hand, if a high temperature remains stable from evening till the morning, it is a sign that the patient is getting or will get worse.

When the temperature begins to fall from the evening to the morning, it is a sure sign of improvement; but a rise of temperature from the evening till the morning is a sign of his getting worse.

_____

[1] See Aitken, op. citat., vol. i. p. 44.

Convalescence from disease does not begin until the normal temperature of the body returns, and maintains itself unchanged through all periods of the day and night.

INSPECTION OF THE BODY AFTER DEATH.

In conducting post-mortem examinations, with a view either to pathological study or medico-legal investigation, *order* and *method* are of great importance.

The three great cavities—the **head**, the **chest**, and the **abdomen**—should always be examined, whether suspicion of disease in them exist or not. First, however (the autopsy being made from twelve to thirty-six hours after death), we should note the **external appearance** of the body; its *size, weight, conformation, color of the skin,* etc. (In cases of suspected violence even abrasions should be minutely described.)

To examine the **Head,** an incision should be made through the scalp, across the top of the head, from ear to ear—the two flaps thus formed should be reflected, the one over the forehead, the other over the occiput. The nature of the attachment of the occipito-frontalis muscle to the bone beneath is such as to allow, very easily, the loosening of the scalp. The cranium (calvaria) is now to be removed by means of a small saw.

For the purpose of holding the head firmly during the use of the saw, Dr. Demmé has furnished, as a substitute for the craniotome of Mr. Lund, of London, a *cranium-holder*, which enables the operator to make a section of the skull in any direction. It consists simply of a bar of iron curved like the letter U, at each extremity of which two drill screws are placed, which, when forced down upon the bone, hold the bar firmly in situ, and enable the examiner to control the head. The legs of the instrument, for use, are placed upon the lateral portions of the skull, over the squamous portions of the temporal bones.

The section of the cranium with the saw should be made through its outer table, completely around the head—from *before, backward,* from below the frontal protuberances to the squamous portion of the temporal bone, and from *behind, forward,* from the occipital protuberance to the squamous portion of the temporal bone, meeting the line just described. The shape of the piece thus cut out enables it to be maintained in its proper position when the parts are readjusted. It is removed by the aid of an elevator or chisel, and hammer, fracturing the inner table of the skull by strokes so applied as not to pierce the brain.

The dura mater is next to be cut through, on each side of the superior longitudinal sinus; after which, dividing the *falx cerebri,* the brain may be raised carefully with the hand placed under its anterior portion. The internal carotid artery, and cranial nerves, etc., are now to be severed by the knife, and, finally, the vertebral arteries and spinal cord. The brain itself may then be taken out, and inspected, by slicing it, from the upper part downward, in successive horizontal layers.

To examine the **spinal column,** an incision should be made from the occipital protuberance to the extremity of the os coccygis. The deep muscles of the back should then be loosened from their attachments, so as to expose the laminæ and spinous processes of all the vertebræ. With the chisel and mallet, or saw, we must cut through the arches of the vertebræ on each side, close to their articular processes. After thus opening the spinal canal, the cord is to be exposed by dividing the dura mater through its whole length.

To examine the **Neck,** an incision should be made through the skin, extending from above the hyoid bone to the upper part of the sternum. Avoiding penetration of the large veins of the neck, the parts to be examined may be carefully dissected, and, if desirable, removed from the body. The thyroid gland, larynx and its appendages, tongue, pharynx, œsophagus, bloodvessels, and nerves of the neck, may be thus viewed.

To examine the **Chest,** two incisions are desirable: the one from the root of the neck, in front, to the extremity of the ensiform cartilage; the other at right angles to this, across the middle of the thorax. The cartilages of the ribs are to be cut through at the line of junction with the ribs. The ensiform cartilage, being drawn outward, is to be detached from the soft parts, the knife being held *close to the sternum.* The sterno-clavicular articulation may now be opened, and the sternum with the costal cartilages raised from its position—a cautious use of the knife being made to remove the adherent soft parts.

The thoracic viscera are now exposed, and may be drawn out with care, and inspected in detail.

To examine the **Abdomen,** make a crucial incision: the one branch extending from the sternum to the pubes, passing to the left of the umbilicus; the other, transversely across the middle of the abdomen. Care must be taken, in making these incisions, not to injure the subjacent viscera.

Before removing the stomach or any portion of the intestines, ligatures should be placed above and below the part that is to be separated.

When—as is always desirable if possible—both of the large cavities of the trunk are to be opened, a single incision, extending from the top of the sternum to the symphysis pubis, may be made.

In every case incisions through the skin should be made, as far as practicable, only in those parts which are usually covered by the clothes of the deceased. It is generally advisable, when the abdomen or thorax has been opened, to fill the cavities with bran or sawdust. After the examination has been completed, the edges of the divided integument should be brought together, and retained in apposition by the common suture.

MEDICO-LEGAL EXAMINATIONS.

In cases of suspected *poisoning,* the following practical directions are given by Professor Reese, of the University of Pennsylvania; to be observed by those who have charge of the *post-mortem* examinations :—

1. Ascertain whether the individual has labored under any previous illness; and how long a time had elapsed between the first suspicious

symptoms and his death; also, the time that had elapsed after death before the inspection is made.

2. Note all the circumstances leading to a suspicion of murder or suicide—such as the position and general appearance of the body, and the presence of bottles or papers containing poison about his person, or in the room.

3. Collect any vomited matters, especially those *first* ejected, and preserve them in a clean glass jar, carefully stoppered and labelled. The vessel in which the vomited matters have been contained should be carefully inspected for any *solid* (mineral) matters which may have sunk to the bottom, or adhered to the sides. If no vomited matters be procurable, and vomiting has taken place on the dress, bed-clothes, furniture, etc., then portions of these must be carefully preserved for future examination.

4. Before removing the stomach, apply *two* ligatures beyond each extremity, dividing between each pair, so as to prevent the loss of any of the contents.

5. If the stomach is opened for inspection, this should be performed in a perfectly clean dish, and the contents collected carefully in a graduated vessel, so as to properly estimate their quantity. [Note here, also, the presence of blood, mucus, bile, or undigested food.] These contents should be preserved in a perfectly clean glass jar, securely stoppered, covered over with bladder, and sealed. The contents of the *duodenum* should be collected and preserved separately.

6. Carefully inspect the state of the *throat, œsophagus*, and *windpipe* for the presence of foreign substances, and for marks of inflammation and corrosion.

7. Observe the condition of the *large intestines*—especially the *rectum:* the presence of hardened feces would indicate that purging had not very recently taken place.

8. Note any morbid changes in the *lungs*, as congestion, inflammation, or effusion;—in the *heart*, as contraction, flaccidity, presence of a clot, and condition of the contained blood.

9. Examine the state of the *brain* and *spinal marrow :* and, in the female, the condition of the uterus, ovaries, and genital organs. [Poisons have sometimes been introduced into the vagina.]

10. Along with the contents of the stomach and duodenum, the viscera that are to be reserved for chemical analysis are the stomach and duodenum (to be kept separate from the others); the liver and gall-bladder, spleen, kidney, rectum, and urinary bladder with its contents. Sometimes, also, a portion of the *blood* may be required for the examination.

11. As the legal authorities will rigorously insist upon proof of the *identity* of the matters alleged to be poisonous, it is of the greatest importance to preserve such matters from all possible contamination by incautious contact with a surface or vessel *which is not absolutely clean.* Avoid the use of colored calico or paper for wrapping up the specimens. When once the suspected articles are deposited in the hands of a medical man, he must preserve them strictly under lock and key, and confide them only to a trusty agent for transportation. Many cases are on record where the chemical evidence failed, simply

from a want of power clearly to establish the *identity* of the matters analyzed.

Actual testing for poisons in cases of suspected criminality ought to be undertaken only by those whose chemical knowledge and skill are considerable. (See Taylor's Medical Jurisprudence.)

SECTION III.

GENERAL THERAPEUTICS.

REMEDIES have been classified, for the study of *Materia Medica*, in a manner (see *Wood's Therapeutics and Pharmacology*) which is perfectly well adapted to the present state of that science.

I propose the following classification, from the standpoint of the *practitioner; i. e.* according to the **indications** of **treatment**, or *objects proposed.*

Thus regarded, remedies may be studied as—

Anodyne and **calmative**: *e. g.* opium; ether; chloroform; aconite; hydrocyanic acid.

Protective: *e. g.* demulcents; surgical dressings.

Balancive: *e. g.* cold to an over-vascular part; pediluvia; blood-letting.

Economic: rest; astringents; retarders of tissue-metamorphosis.

Eliminative: *e. g.* colchicum in gout; purgatives; iodide of potassium, etc.

Antidotive: *e. g.* hydr. ox. of iron for arsenical poisoning; antacids; cinchonization in intermittent.

Alterative: *e. g.* nitrate of silver in scarlatinal sore throat; arsenic in skin diseases; electricity in cancer.

Recuperative: stimulants; tonics; chalybeates; oleum morrhuæ; travelling.

An elaborate work might, of course, be written upon the topics just enumerated. It is appropriate to our purpose, only to state them; dwelling, presently, upon another yet more brief classification, of the modes of treatment *most frequently called for*, in the management especially of acute and subacute affections.

First, a few words upon **balancive** measures. These constitute a very large part of therapeutics; one of the most constant elements of disease, and of diseases, being, a disturbance of the *proportion* of circulation, nutrition, innovation and action in different parts.

For example: when one "takes cold," what has occurred? Chilling the surface, as by damp air, has *checked* perspiration, *contracted* the superficial bloodvessels, causing *congestion* of interior organs, and partial contamination of the blood, from *retained excretory matter.* What, then, is the "indication" or pointing of nature?

Clearly, it is, to **restore** the lost balance; by *warmth* to bring on perspiration (unless *fever* occurring demand another method), purgatives and diuretics, with plenty of water, to *relieve* the blood of its morbid excess of excreta.

Again, in flatulent colic, unequal distension and spasmodic contraction of a bowel occur, from gaseous accumulation or the presence of irritating ingesta. Aromatics, such as ginger; stimulants, as hot water or brandy; or anodynes, as camphor or opium, by a diffusive action on the whole surface of the affected intestine, and upon its innervation, when they are absorbed and reach the ganglia, will renew a **proportionate** contraction (peristaltic) of the muscular coat, and remove the pain. Very often gentle friction, pressure or *kneading* the abdomen, or external warmth all over it, will have a similar balancive effect.

Laxatives for deficient movement of the bowels, astringents for excess of the same, cold to a too hot head, and mustard and hot water to cold feet, are all balancive means. So is the familiar, and always safe use of a *mustard-plaster* to the skin, over any part of the body which suffers pain. Pain denotes a morbid innervation, from some cause. Apply something which, like mustard, causes a strong impression in a different place, not too remote, and the "error loci" of nerve-tension (or *debilitation*,[1] as the case may be), is done away with—the balance is restored.

Notice may be here taken, briefly, of a new "method" in therapeutics, extending the balancive principle systematically,—called the "neuropathy" (*ganglio-therapy*) of Dr. John Chapman. The origin of this is really to be credited to the vivisections of Bernard and Brown-Séquard,[2] and conclusions based upon them, especially by the latter. The experiment of most importance in this connection has been the section of the sympathetic nerve in the neck of a rabbit; which was found to be followed by dilatation of the blood-vessels of its ear. It was concluded from this and other facts similarly obtained (all **traumatic** or **pathological,** as, indeed, are *all* facts of vivisection), that to *increase* the amount of blood and sensibility in any part of the body, you must *paralyze*, partially or temporarily, its sympathetic ganglion. To *diminish* its vascularity and sensibility,— you should *excite* its vaso-motor nerve-centre.

Dr. John Chapman systematizes the application of these principles, by the application of elastic bags, containing ice, or hot water, along the spine; so as to act upon the ganglia located thereupon, and, through them, to affect the viscera, both palliatively and curatively, in disease.

If the *physiological theory* of Brown-Séquard, above mentioned, be true, the *therapeutical process* of Chapman, deduced from it, may be justified. I deny the truth of the one, and, *a priori*, have no belief in the validity of the other. Clinical experience, however, must decide the claims of the practice.

It must be remembered, that a *practice* may prove beneficial, whether the *theory* which suggested it be correct or not. Hot and cold applications to the spine must make (especially when *alternated*,

---

[1] *Radcliffe* "On Epilepsy, Pain and Paralysis;" *Inman*, op. citat.
[2] See his *Lectures on the Nervous System*, p. 205, &c. See, also, a discussion of the experiment mentioned, in the author's *Essay on the Arterial Circulation.*

as Brown-Séquard proposed for bed-sores) a strong impression on the whole system; this *may* prove a **rapidly alterative** impression in certain cases of disease. *Should* this prove so, the use of such means should be adopted, just as, and *so far* as, experience shows it to be useful. No such utility of applications whose explanation may be reached in many different ways, can make the *theory* above cited seem, to the present writer, other than erroneous.

The *modes of treatment most frequently called for*, in ordinary medical practice, may be designated as—

The **antiphlogistic**;   The **supporting**;
The **febrifuge**;   The **antidotive**;
The **alterative** treatment.

Under the first head, the **antiphlogistic** (*i. e.* the treatment of *inflammation*), we place—

Rest;—position;	Tartar emetic;	Digitalis;
Cold applications;	Nitrate of potassa;	Ergot;
Venesection;	Ipecacuanha;	Mercury;
Local depletion;	Veratrum viride;	Opium;
Purgation; Diet;	Aconite;	Counter-irritation.

The necessity of **rest** during *active inflammation* of any organ is a rule without exception.

A choice of **position** is often dictated by the sensations of the patient. When one of the *extremities* is inflamed its *elevation* is advised in order to allow the blood to return from the overloaded vessels.

Cold applications are very important in the treatment not only of *inflammation*, but of *active hyperæmia* or congestion (*e. g.* "determination of blood" to the head). The two precautions most necessary in their use are, that the cold be not *excessive*, and that it be not *ill timed*, so as to arrest desirable *perspiration*.

Blood-letting, by *venesection, leeching*, and *cupping*, is one of the oldest, and has been one of the most universal of remedies for inflammation. Although "αἱμορφοβοι," or "blood-fearers," have occasionally appeared in all ages and nations, yet the aggregate testimony of the profession, from Hippocrates down to the present time, has been in favor of the use of the lancet and of local blood-letting in the treatment of violent inflammation and congestions.

Now, however, it must be admitted that blood-letting has more opponents and fewer defenders than at any previous period in medical history. Why is this? By reason of—

1. *Reaction* from previously existing *abuse* of the remedy.
2. A *change in the average human constitution*, occurring under the artificial habits of *civilized life*.
3. *False construction and misapplication of recent science*.
4. *Leadership* and *fashion*.

I must briefly remark, that the *reaction* alluded to has proceeded *too far*, going from one *extreme* to another.

The *change* occurring *especially in large cities*, in the average human constitution, affords good reason for *limiting* the use of the lancet to a smaller number of cases than was once thought necessary; and for

8

using especial caution as to the *amount of blood abstracted ;* but not for abandoning the remedy altogether.

The improved condition of the sciences of *semeiology* and *pathology* gives us the power to discriminate more narrowly in our use of blood-letting, as well as of other remedies. But, we should not, for this, throw aside as useless all the experience of our predecessors; as if every new fact was necessarily the heir of some dead old one. *All* facts, old and new, should be retained.

In the *physiological* and *pathological* science which bears upon the question, I hold that false construction and misapplication of observed facts have been operative. An important threefold error has been committed ; viz.—

1. In physiology, the denial or depreciation of the *active* part taken by the *arteries* in the circulation; and of the great fact, without recognition of which no theory of inflammation can stand, that the arteries are subject to *reflex* excitement,—the most normal form of which constitutes active hyperæmia, the most abnormal and excep-tional, tonic constriction of the vessels.

2. The error of Prof. J. H. Bennett, of giving attention, in regard to the pathology of the inflammatory process, to the *exudation* alone.

3. That of Prof. Virchow, in considering that no important differ-ence in *kind* exists between morbid lesions of nutrition in vascular and in non-vascular tissues ; and that *stimulation, irritation,* and *inflammation* are, *essentially* and *practically,* as well as causatively, only degrees of the same vital impression.

The theory of inflammation which has been already laid down,[1] as entirely consistent with the *observations* (whatever may have been the reasonings) of the most accurate pathologists (*e.g.* Rokitansky, Paget, Wharton Jones), inculcates, that the determination of blood toward an inflamed part *conspires with* the central stasis in *causing* the exu-dation ; and that a constant *proportion* exists between the *degree* of this active turgescence and the *amount* of the exudation, and the *character of the changes* which it subsequently undergoes.

Now, of the cardinal elements of the inflammatory process, the local *arrest of nutrition* and *capillary stasis* cannot be *directly* affected by treatment. Nor, *when the exudation has occurred,* can any but palliative or expectant measures be applied to the management of its changes. But, the active concentric determination of blood—the *ar-terial excitement*—cannot *this* be essentially *modified* by *treatment ?* Yes.

By abstraction of blood, we lessen (for a time at least)—

1. The **fulness** of the vessels;
2. The number of **red corpuscles;**
3. The force of the **heart's** impulse;
4. The force of the **arterial** impulse;
5. The excitement of nerve-centres.

And by each and all of these influences, we diminish the vascular excitement connected with an inflammation ; and thus (I repeat) *lessen the amount* of the resultant exudation, and (Paget) render its

---

[1] See *General Pathology,* Section III.

" biography" more normal, its changes less degenerative and destructive.[1]

If this be true, it is altogether an erroneous assumption, of Professor Bennett, and others, that inflammation is a " self-limited process which cannot be cut short nor interfered with to advantage." If there be anything positive in medical experience, I believe the contrary of this to have been established.

Thus much, perhaps, must be allowed to the influence of recent ratiocinations and experimentations in medical practice without the lancet; that *local* blood-letting may be admitted, in almost every case, to have all the advantages which can be claimed for venesection, except *convenience;* and that, in doubtful cases, the smaller quantity abstracted ought always to be an argument in favor of local rather than general depletion. This admission may be made without surrendering, in the least degree, the principle of therapeutics upon which blood-letting is scientifically justified, and according to which, if we are to interfere at all with disease, it is often one of the mildest, most beneficent, and least hazardous of remedies. Dr. J. H. Bennett admits that *relief* of pain, dyspnœa, &c., sometimes follows bleeding; and this concession carries a good deal with it.

How, then, it is proper to ask, do we *define* or *classify* the remedial action of blood-letting?

It is **balancive.** What do we mean by *reducing treatment?* The answer to this question is important.

I do not know of a single case of any kind of disease, in which the indication or object of medical treatment is to reduce the strength, or lower the vital power of the patient's system.

What we aim to **reduce** is, *disproportionate vascular* **excitement,**

---

[1] It may be hoped that the time has gone by when any question in therapeutics can be decided by *leadership.* But the "blood-letting controversy" has shown, that the medical mind is not yet *absolutely* free from its influence. As to authorities, old and new, it may interest the student to remember, that, of ancient opponents to the lancet, Chrysippus and Erasistratus were the most noted; of the modern European schools, Van Helmont, Dietl, and Skoda, in Germany; Grisolle, in France; Bennett and Todd, in Great Britain. Exemplification of Sangrado's practice, on the contrary, has been especially accredited to Cullen in England, Rasori in Italy, Chomel and Bouillaud in France, and Rush (the father of American medicine) in this country. We should place in the class of *moderate* bleeders, of antiquity, Hippocrates, Asclepiades, Celsus, Galen, Avicenna, and " οἱ πολλοί;" of earlier English and French teachers, Sydenham, Huxham, Gregory, Laennec, etc. ; of the present date, the recently deceased Professor Alison, with Watson, Christison, Copland, Lawrence, Chambers, Parkes, C. West, W. T. Gairdner, Symonds, F. Winslow, Aitken, and others in Great Britain ; Wunderlich in Germany ; and, in this country, as a representative of American medical conservatism, G. B. Wood. As *statistics* have been especially appealed to by the opponents of blood-letting, it may be proper to quote here the conclusion of an able analysis of all the evidence of this kind made public (*Brit. and For. Medico-Chirurg. Rev.,* July, 1858). It is as follows:—

" While the non-bleeding plan has a demonstrable advantage over that of indiscriminate and repeated bleedings, we maintain that the discriminating practice of moderate and early bleeding, general or local, in cases of more or less sthenic inflammation, and of refraining from it altogether in asthenic cases, whether as regards the character of the disease or the constitution of the patient, is *pressed upon us both by experience and science.*"

or **congestion**; to restore the **balance** of the circulation. It is a mere *imagination* that abstraction of a small quantity of blood *must* *always* lower the patient's strength. Under some circumstances, it actually increases it. At the same time, there are many persons who *will never bear* bleeding, from an actual deficiency or defective quality of blood.

Taking these propositions as established, we may draw blood, locally or generally, for
1. High, **sthenic** inflammation:
2. *Active* **congestion, threatening** inflammation or hemorrhage;
3. *General* **plethora**, ditto;
4. Sthenic *spasm;*
5. Sudden passive congestion (not toxæmic) in robust persons.

It must be remembered that, at the present time, *no one* thinks of bleeding *for fever, as such.*

Repeated venesections are seldom now resorted to; the time for the lancet, if at all, is always in the **early** stage of a phlegmasia.

It would be instructive, if compatible with our plan, to allude farther, especially, to the use of blood-letting in **certain cases** of **pregnancy**, and of convulsions; and to the **caution** necessary in its application to the treatment of **senile** apoplexy. Old persons rarely bear bleeding well.

We might also, if space allowed, illustrate the principles above laid down, by examples; as, of 1, erysipelas; 2, pneumonia; and 3, meningitis. Why is bleeding seldom called for in the first, more frequently in the second, and quite often in the last?

Our answer is,—that it depends chiefly upon the **anatomical relations** of the tissue involved. The skin is unlimited in its opportunity of hyperæmic expansion, and escape of exudation. The lungs are partially confined and limited, by the pleura and walls of the chest. The brain and its membranes are shut entirely within the closed skull. Therefore the influence of **vascular pressure** (which is most affected by blood-letting) is most marked and most important in congestion or inflammation of the brain, next so in that of the lungs, and least of all in that of the skin, as in erysipelas.

In **uræmia**, when the patient will bear it, venesection may do good, by taking out the excreta with a portion of the blood, and favoring secretion by general relaxation.

As to the quantity of blood taken by venesection, twelve fluidounces may be stated as a full, though not large, bleeding for an adult man; ten fluidounces for a woman. For infants and children, one ounce under one year, two ounces under three years, three ounces under five years, four ounces under ten years, would be a full average. Bleeding from the jugular vein is sometimes preferred in young children. The practitioner should judge for himself of the *effect* upon the pulse, &c. It is remarkable how small an amount will sometimes do a great deal of good.

Cut cups and **leeches** act alike as to the abstraction of blood; but the former have a more revulsive or counter-irritant effect.

Leeching, being somewhat less violent, is more applicable than cupping to parts which are very *tender;* as, the side in acute pleurisy; the abdomen in peritonitis; a much inflamed joint, &c.

Leeches are usually, for the same reason, applied as near as possible to the part inflamed; cups, sometimes, at a short distance from it. In *bronchitis*, it is ordinarily best to apply leeches or cups to the upper *sternal* region.

In *pneumonia*, they may be preferably applied *between the shoulders*, as a general rule, thus leaving room for counter-irritation in front.

In pleurisy, it is desirable to use *leeches* immediately over the inflamed part.

Purgation, especially by *saline* cathartics, is a frequently useful part of antiphlogistic treatment.

Cathartics are to be **avoided** in **enteritis** and **peritonitis**; for obvious reasons.

Diet, in cases of **sthenic** inflammation, should be *non-stimulant;* but it may be sufficiently *nourishing* (vegetable, farinaceous) at the same time. *Starving* patients is not now thought of, unless they are fearfully *plethoric*. In the **later** *stages* of inflammatory disorders—in fact, *as soon as the exudation has all been thrown out*, **generous** diet is usually required. Some patients will never bear a purely vegetable diet under any circumstances; and *some* cases, even of inflammation, · require stimulation from the first.

The idea of the association, always, of *low diet* with *inflammation*, has been too absolute in common practice. When, in acute disease, the stomach refuses to digest food, it is vain to force it upon it. But, it will often digest **liquid** food when it cannot solids. And, as some degree of debility is constant in disease, alimentation being necessary, *concentrated* liquid food, *e. g.*, beef-tea, will frequently be appropriate, when no solid substance at all can be taken.

I believe the *principal* requisites of diet in illness to be, liquidity and facility of digestion and assimilation. In an irritable, febrile state of the system, the presence of a solid body, as meat, or bread, in the stomach, when nó digestive fluid is secreted to act upon it, has the effect of a foreign substance,—namely, irritation; sympathy with which may disturb or increase existing disturbance of the whole economy.

Practically, I have seen, in a person not robust, suffering from catarrhal fever, the drinking of a wineglassful of beef-tea followed by a copious perspiration and cooling of the skin. Still, a young and previously healthy patient will often do best in the early stage of inflammatory disease upon small or moderate amounts of what are called articles of sick diet; as, oatmeal gruel, toast-water, panada, arrow-root, &c. In disorders not affecting the bowels, fruits, especially white grapes and oranges, need seldom be withheld. They are often refrigerant and useful; and such is very frequently the effect of lemonade; which acts as a good diuretic and diaphoretic.

The most powerful of antiphlogistic (arterial sedative) **medicines** is **tartar emetic.**

The "contro-stimulant" plan, of giving *very large* doses of this drug in pneumonia, pleurisy, etc., has been abandoned as excessive and injurious. We need never give more than $\frac{1}{4}$ of a grain of tartar emetic at a dose to an adult—oftener $\frac{1}{8}$, $\frac{1}{16}$, etc. *Children* require *especial* caution in its use, on account of the sensitiveness of their alimentary canal. I have known severe vomiting to be induced in an infant by $\frac{1}{8}$ of a grain. No other medicine as yet discovered, however, is so useful in violent inflammations of the pleura, lungs, bronchial tubes, etc.

Tartar emetic (of course) must **never** be given in **gastritis** or **enteritis**.

Nitrate of potassa is a very valuable adjunct to, or in some cases substitute for, the antimonial tartrate. It is often given in *too small* doses; ten grains may be a minimum for an adult, if the stomach is in an ordinary state.

Ipecacuanha is especially valuable in *bronchial, tracheal,* and *laryngeal* inflammations, and in *dysentery*.

Veratrum viride is, lately, assuming an important place as a cardiac and arterial sedative, and promoter of the secretions. It is a very certain reducer of the pulse, but requires caution in its use.

Aconite is, likewise, a favorite medicine with some practitioners, in the management of pleurisy, pneumonia, etc.

The power **digitalis** has been supposed to possess, to reduce the rate of action of the heart,[1] has induced the expectation that it would prove a reliable antiphlogistic remedy; but this expectation has been *disappointed.* It is, however, occasionally useful in bronchitis, etc.

Ergot has been employed with the same view, rather as a direct sedative to the *smaller arteries.* It is quite possible that its powers have not as yet been sufficiently tested.

The place of **mercury** had appeared, until within a few years, to be settled. Twenty-five years ago, nothing was more common than intentional mercurial *salivation* in the treatment of almost all violent acute and even chronic diseases. In the management of inflammation, in addition to its powerful *alterative* influence, tending to displace, by its own impression, morbid actions and conditions, it was *believed* to exert a peculiar control over the *blood, lessening the tendency to the effusion of coagulable lymph.*

In recent times, the " salivating" practice has been almost entirely abandoned, as disproportionately violent, as well as uncertain. A reaction, somewhat similar to that occurring in the case of blood-letting, has even shaken the confidence of many practitioners in the value of mercury as an antiphlogistic.

My own opinion, very decidedly, is this. That experience fully warrants the inference that mercury is *a general stimulant to all those functions of organic life which are performed under innervation from the ganglia of the* (so-called) *sympathetic* system. It is probable that its action is directly *upon* these ganglia. Thus, mercury tends to *diffuse* and *equalize* secretion,[2] and the circulation of the blood, aiding, in this way, to break up local congestions and inflammations.

I believe that calomel and blue mass, etc., *have been shown* to be useful in the treatment of several of the acute phlegmasiæ. I do not think that a due regard for the principles of evidence in *therapeutical*

---

[1] Facts have been urged of late which lend force to the opinion that digitalis is, primarily, rather a *tonic* than a sedative to the heart.

[2] Very few points in practice are, for instance, so well sustained by experience, as the familiar use of small doses of *blue pill*, in the treatment of indigestion with torpidity of the liver and bowels, etc.

science can allow us to put aside the proof of this, deduced from actual experience.

Moderate doses, at the same time, are capable of doing all that we can *safely* aim to effect with the use of mercurials. I do not know of any variety or form of disease in which I should, at the present moment, feel justified in *intentionally* causing *full salivation* as a means of medical treatment.

Mercury is especially *contra-indicated* in the presence of the *tubercular diathesis.*

Opium, always the most reliable and potent of *anodyne* medicines, has, in latter times, assumed a more important position as a *remedy* in the treatment of inflammatory diseases.

Experience has warranted this, while certain theoretical considerations have also been urged in regard to it.

1. The influence of the *nervous centres* upon *inflammation* (as upon normal nutrition, circulation, etc.), and the intimate *interconnection* of the two portions, organic and sensori-motor, of the nervous apparatus, are now more fully recognized than formerly.

2. Opium (morphia) is believed by some to act *directly, not only* upon the cerebro-spinal, but also upon the *ganglionic* nerve-cells as a peculiar stimulant, thus affecting the circulation, nutrition, etc., otherwise than by mere sympathy.

Yet, in estimating the adaptation of preparations of morphia or opium to the treatment of inflammations of important organs, in different stages, we must remember, that

Opium is an *arterial stimulant*, and is, therefore (as a *general rule*), inappropriate in the *early stage* of an active sthenic phlegmasia.

Opium *first excites*, and then *oppresses* the *brain;* in a word, promotes determination of blood to the head, and is, therefore, contra-indicated by an already existing tendency to *cerebral congestion.*

Opium also *constipates the bowels*—a fact of less importance than either of the two preceding, as the constipating tendency can be counteracted, if desirable, by other medicines; while, in certain cases, it aids in the treatment (as in dysentery).

In *peritonitis*, where the extent and visceral connections of the tissue affected induce more rapid *prostration*, and more serious *nervous irritation* than in any other phlegmasia, *opium has become the* **main dependence** with very many practitioners, even from the *beginning* of the attack. The same reasoning will apply, to a somewhat less extent, to its use in *severe* cases of *pleurisy* and *pericarditis.*

Counter-irritation is a measure of treatment often of great service, especially in the *later* stages of inflammation (after local or general depletion, etc.), or in cases unattended with much vascular excitement. In the very *incipiency*, or rather *incubation*, of an inflammatory attack, *i. e.*, in the stage of mere *irritation* or *congestion*, counter-irritation (*e. g.* by a *sinapism*) may *prevent* the further progress of the inflammatory process. But, if the stasis and concentric hyperæmia be already developed, all *powerful* counter-irritants should be avoided (lest they prove *co*-irritants) until the vascular disturbance has subsided.

Counter-irritation is, usually, the *most important* part of the treatment of **hyperæmæsthesia**, or "chronic inflammation."

To recapitulate the **order** of time, in which long recognized experience has prescribed the use of the different means now included under the term " antiphlogistic" treatment:—

Supposing all[1] of the main remedies of this class to be called for in a given case, we would resort *first* to *venesection* ; or, if this be undesirable, to *cupping* or *leeching* ; next, to *saline purgation* ; then, to tartar emetic, nitrate of potassa, ipecacuanha, veratrum viride or aconite ; mercury at the same time, or immediately following these sedatives ; opium sometimes with it or them—oftener, a little later ; counter-irritation by blisters,[2] etc., last. The *subsequent debility,* especially in cases of *suppurative* inflammation, may call for *tonics* or even stimulants, with generous diet, etc. ; while certain cases will even require such treatment from the *first.*

The treatment of *subacute* or *chronic* inflammation, in external or accessible parts, by *astringents* or *stimulants* (e. g., *nitrate of silver*), does not require, in this place, extended discussion ; as it usually comes under the domain of Surgery. One example, however, of its *medical* utility, may be named—viz., the administration of nitrate of silver (gr. ¼—¾ *ter die*) in *chronic gastritis.* The change which it undergoes in the intestines, when given by the mouth, explains the fact that the same medicine fails to exhibit a similar beneficial influence in chronic *enteritis.* In prolonged *dysentery,* however, *enemata* containing this or some analogous mineral salt, as sulphate of zinc, sulphate of copper, or acetate of lead, are often very valuable remedies.

We cannot leave the subject of the management of inflammatory disease without reminding the student of the important practical difference between **sthenic** and **asthenic** inflammations.[3]

This difference is constituted—

1. By the *state of* **system** of the patient affected ;
2. By the nature of the *producing* **cause.**

One whose constitution has been prostrated by previous disease, or recent excess, will have, when exposed to the ordinary causes of inflammation, an *asthenic* attack ; i. e. one in which, with all the local symptoms of phlogosis, *the general organic functions are sympathetically affected rather with* **depression** *than with excitement.*

Again, certain **morbid poisons** induce, with toxæmia, local inflammation ; and blood-disease (dyscrasia), arising from various causes, may have local inflammation as a *secondary effect.* In these cases the type of the inflammation is generally *asthenic,* and the *treatment* must be modified accordingly—depletion being avoided, or used with the greatest caution, and full diet and even stimulation being not unfrequently called for.

As examples of inflammations which may be either sthenic or

---

[1] Of course, this supposition, of the successive use of all of the remedies named in this paragraph, does not, in very many cases, need to be realized.

[2] Dr. Inman, of Liverpool, suggests that the so-called counter-irritants really act, by absorption, as direct *stimulants,* to parts enfeebled by disease. It is quite possible that, in some cases at least, this may be true.

[3] Granting that *all* disease is *debilitating* (Inman), the distinction is still valid and important, as to the different *degree* of depression produced by its different forms or types.

asthenic, I may mention *erysipelas, dysentery, peritonitis, pneumonia, gout.*

The first three[1] of these are at times epidemic; and *then* it is that the greatest number of asthenic cases is observed. The following maxim may be considered as fully established:—

Whenever any local affection, as dysentery, peritonitis, catarrh or pneumonia, occurs sometimes **sporadically** (*i. e.* in altogether separate or independent cases) *and sometimes endemically or* **epidemically** (*i. e.* a number of cases at the same place and time, under a common local or temporary cause), *the* **latter** *cases exhibit, as a rule, the greatest tendency to* **depression** *in their symptoms, the largest mortality, the least tolerance of depletory treatment, and the most frequent need of stimulation or support.*

By **febrifuge** treatment I mean, that which is proper during the existence of the febrile state. It comprises no violent measures of any kind.

Remembering that the essential phenomena of fever are, increased *heat*, especially of the exterior of the body, *dryness* of its surfaces, *scantiness* of *fluid* in all the discharges, with actual *increase* in their *solids*, from accelerated tissue-metamorphosis—our therapeutics must be adapted to these conditions. Apart from the necessity of removing or antagonizing, if possible, the *cause* of the febrile disturbance, the indications are, to **allay** the **heat** and **dryness** of the surfaces of the body, tegumentary and mucous, and, to *favor* the **removal** of **excreta**, accumulated in unusual amount in the blood and organs.

For these purposes, we may use

<table>
<tr><td>Moderate laxatives;</td><td>Cold drinks;</td></tr>
<tr><td>Saline diaphoretics;</td><td>Tepid ablutions.</td></tr>
</table>

Of these measures, I have no doubt of the propriety of the designation of **water** as the *heroic* remedy, to which the others are merely adjuvants. Diaphoretics will scarcely act at all without free imbibition of water, and the operation of laxatives is much promoted by it. Water *alone* is diaphoretic, diuretic, and laxative; but it may be *aided*, to an important degree, in alleviating the symptoms of fever, by the addition to it of citrate of potassa, acetate of ammonia, etc.

I have already laid emphasis upon the statement, that no one now thinks of *bleeding* for fever, as such. In a much more strict sense, pathologically speaking, than inflammation, the febrile *nisus* is *self-limited*, although variable in its duration according to the cause inducing it. The object of the physician is *not* to *cut it short* (*jugulare*), but to **conduct** it safely to a critical termination. In an equally important practical manner, this principle applies, not only to the management of a brief or ephemeral exacerbation or paroxysm of irritative or reactive fever, but also to those of longer duration, under toxæmic (zymotic) causation: as *exanthematous* (rubeolar, scarlatinal, variolous) or *continued* (typhus, typhoid) fevers. An exception is believed by many to exist, in the case of autumnal, miasmatic, periodical fevers: *i. e. intermittent, remittent,* and *pernicious* (con-

---

[1] Pneumonia is also sometimes endemic or epidemic, in the form of *typhoid* pneumonia.

gestive) ; in which, interference, by the *antidotal* remedy, cinchona or its alkaloids, is considered *safe at all times*. and sometimes necessary before the subsidence of fever. But I believe this exception to be only partial, since the most careful recorded experience has given rise to the conclusion[1] that quinine is *never necessary* during the *height* of the exacerbation of either type of miasmatic fever, and that in large doses at that period it *may do harm*. It is, I consider, the best practice always, in the treatment of autumnal remittent (bilious) fever, to wait, *until the febrile stage has passed its climax*, and its symptoms have *begun* to decline—the urgency of the case, and all its circumstances, then, guiding the practitioner as to *how soon*, as well as how largely, the special remedy, whose interference is called for, must be introduced.

It is a matter of general remark, that patients scarcely ever die during the *hot stage* of any kind of fever. In the most intense form of miasmatic poisoning, called *pernicious fever*, the danger exists in the extreme depression of the *cold stage ;* if fever comes on, the patient is comparatively safe for the time.

The **supporting** treatment is that adapted to states of *prostration* or *debility*.

General weakness of the body (when not a *congenital* defect) occurs under three forms :—

> **Exhaustion ;**
> **Depression ;**
> **Oppression.**

We are familiar with the first, **exhaustion,** as the effect of overexertion, loss of sleep, deficiency of food, excessive discharges, etc., and as *following* acute, or constituting a *part* of chronic disease.

The second, **depression,** is to be discriminated from exhaustion, as resulting, not from expenditure or waste of the material or forces of the body, but from *interference* with their normal activity by some disturbing cause. To use a mechanical illustration, *exhaustion* is the *running down* of the clock ; *depression*, the *arrest* of the *impelling movement* of the weights, by which its wheels are kept in motion.

Oppression, then, may be compared to the *obstruction* of the *machinery* by some foreign body, or by some mechanical disarrangement among the wheels, which clogs their action until it is removed or corrected.

Exhaustion and *depression* have their chief seat in the nerve-central sources of dynamic force ; *oppression*, in the circulation of the blood, or in some subordinate organs or functions.

This distinction, however recondite in *theory* it may seem to be, is of high **practical** importance. This will be seen on consideration of the remedies used and required in the different forms of debility.

Supporting measures may be classified as—

> 1. **Stimulant;**
> 2. **Analeptic** (recuperative, restorative).

Under the first head we rank the preparations of *ammonia* and *alcohol*, etc., as usually employed.

---

[1] See Medical Statistics of United States Army, 1839–54.

Under the second are included *generous diet, tonics, chalybeates, cod-liver oil, change of air,* etc.

Now the *first* of these (**stimulants**) are adapted especially to *acute prostration* or *depression;* the second class (**analeptics**), to *chronic prostration* or *exhaustion. Oppression* or *counterfeit* debility, generally requires *neither,* being benefited by very different treatment. A *mingling* or *blending* of these states is of course possible; and then a union of measures is right, to meet the conjoined indications.

Oppression (simulating depression) is every day illustrated by the condition of a patient in the early stage of any of even the mildest acute disorders; as, catarrh or bronchitis, indigestion, tonsillitis, measles, etc. In all of these cases, especially where *fever* is developing, the patient is very *weak;* not only as to his muscular apparatus, but in the performance of all the animal and organic functions. But *stimulation,* for such a condition, in persons of ordinary constitutional vigor and soundness, would be generally *inappropriate,* often injurious, sometimes dangerous.

A more serious degree of oppression occurs in some cases of visceral congestion, particularly of the lungs or brain; and in violent spasmodic affections of the alimentary canal, with constipation of the bowels. *Uræmia,* from inaction of the kidneys, presents another case of oppression, in which even a fatal result may occur.

Counterfeit debility or oppression, then, to recapitulate, may occur in—

> The *first stage* of all acute diseases;
> The *febrile state;*
> · *Indigestion* or *dyspepsia;*
> *Congestion of the brain, lungs,* etc.;
> *Obstruction of the bowels;*
> *Uræmia.*

The *first* of these instances is to be treated usually by measures which promote reaction in the mildest manner. More doubt exists, however, if the *cold stage* itself be intense or profound—as in pernicious intermittent—constituting a *depression,* under **toxæmic** *influence.* Of this, a word or two presently.

The *febrile* oppression is to be managed simply by those palliative measures mentioned already under the head of *febrifuge treatment.*

That of *indigestion* is usually *temporary* or *occasional* only; and gives way under the use of antacids, carminatives, blue pill, etc.

Violent *congestion* of the *brain* or *lungs,* occurring in a person of previously good constitutional strength (although it may produce the most absolute debility, which, especially in the case of *pulmonary* congestion, masks the cause of the disorder), calls, if the diagnosis be clear, for counter-irritation and the local or general abstraction of blood. In doubtful cases a **tentative** plan may be pursued: abstracting but a minimum quantity at first, being encouraged to repeat and enlarge the depletion only if the result be favorable.

Constipation, producing oppressive debility, is of course to be met by agents calculated to unload the bowels; anti-spasmodics, anodynes, etc., being also indicated, if colic exist, and be not relieved by laxatives alone. In absolute (mechanical) *obstruction* of the bowels, causing or endangering enteritis or strangulation, the treatment now generally adopted is, to depend upon *opium* and *rest, avoiding* purgatives.

Urœmia demands all the means within our power to restore the action of the kidneys; and to aid them in their eliminating duty by favoring the *cutaneous* transpiration and secretion.

I have already said that **mixed** cases of oppression and depression occur, in which the indications of treatment are, to a certain extent, obscure and doubtful. Such are, the cold stage of pernicious (congestive) intermittent, the incipiency of the algid or collapsed state of epidemic cholera, etc.

It is clear that **reaction** is here to be brought about, if possible ; and that *external* stimulation, by powerful rubefacients, frictions, etc.; is altogether appropriate ; but, however authoritatively rules may have been laid down, it is not so certain, in *every* case, whether *alcoholic stimulation* or *venesection* would afford the better result, or whether some cases might not be benefited by *both combined*. The *incompatibility supposed* to exist between blood-letting and stimulation is in fact *not intrinsic*, but *circumstantial*. Holding distinctly in our minds the principle that the object of blood-letting is to **balance** the quantity, consistency, and distribution of the blood, and *not* to reduce the strength of the patient, it is far from impossible that the balancive action especially of **local** blood-letting may be called for in a case in which the forces require to be sustained at the same time by "**supporting** treatment."

Personal experience, however, is indispensable to the application of these, or any analogous principles, to cases, in regard to the management of which the profession has been, but we may hope will not always be, divided. The present tendency of medical practice is quite too much in the direction of *over-stimulation*.

Depression is exemplified in the state produced by—

Severe injuries ; e. g. railroad accidents ; extensive burns, etc.

Mental shocks ; e. g. terror or great grief.

Withdrawal of accustomed stimulation ; e. g. delirium cum tremore.[1]

Intense toxæmia ; e. g. cholera collapse, etc. (see above).

Gouty spasm, of the heart or stomach, etc.

Stimulation by alcohol, ammonia, ether, opium, camphor, turpentine, capsicum, etc. is needed, with greater or less urgency, and in larger or smaller doses, in all of these conditions ; always bearing in mind the probability of *reaction*, and avoiding, as far as possible, the exaggeration of this reaction into fever.

The prostration of *typhus fever*, in a *majority* of cases (*not in all*), and that of *typhoid* fever, in a *minority* of cases, requires, especially after the height of the fever has passed, alcoholic stimulation, as well as support by concentrated liquid nutriment (beef-tea, milk) at short intervals.

The instances of these fevers afford us a sort of intermediate gradation between what I have called **acute** and **chronic** debility.

In regard to the *latter* (the debility of convalescence, chronic dis-

---

[1] It is not intended, of course, to indicate that *all* cases of delirium tremens are referable to this cause.

ease, etc.), certain principles are agreed upon by all physicians, on the ground of experience, confirmed by the deductions of physiological science. We shall first briefly consider some of these, and then one or two debatable points akin to those already alluded to.

The two most important and familiar results of clinical experience in the treatment of debility, are, the superiority of the pure vegetable **bitters** in **stomachic** and *digestive* weakness, and of **iron** in **anæmia** (spanæmia). The influence of **quinia** and cinchonia in **nervous** debility is almost as assured. The confidence of many physicians is strong in the utility of the **mineral** tonics (zinc, copper, arsenic, and silver salts)¹ in debility with **nervous symptoms**; *e. g.* chorea, hysteria, etc. I believe this confidence to be deserved, to a considerable extent; but, some of the diseases in which these medicines are given (*e. g.* epilepsy) will, in many instances, baffle all treatment. The use of **strychnia** in certain cases of **paralysis** is also well established; although requiring much care and *discrimination.*

Cod-liver oil holds, at the present time, a very high place in the list of analeptics. All medical observers are not of one opinion in regard to its value; but some of them believe it (on the basis of experience in practice) to be the best and most reliable (where it is tolerated) of all recuperative medicines; not only in consumption, but in all other wasting diseases.

The theory of the *mode of action* of cod-liver oil as an analeptic is an interesting subject. Dr Bennett's view of phthisis is, that the error of hæmatosis, from which tubercle results, consists in an *excess of* **albumen** in the blood, with a *deficiency of* oil; so that, in the process of cell-formation, the first step of which is believed (Ascherson) to be the investment of *oil globules* in *albuminous envelopes,* an imperfection exists, fatal to the subsequent development of the cell and obliging it to abort. But, the debilitation of the *digestive and assimilative* functions in phthisis renders it impossible, by ordinary food, to supply the desiderated oleaginous matter to the blood. *Cod-liver oil* is fatty matter which, by the assimilating action of the liver, following the process of digestion, is **prepared** *for immediate absorption* and **appropriation** by the blood, for purposes of nutrition. This rationale of its influence is, although not demonstrable, much more probably correct than that which refers it to the presence of *iodine, phosphorus,* or any other special ingredients which it may contain. Allusion to the more newly introduced dogfish oil, shark oil, etc., would be more proper in connection with the subject of *materia medica* than here.

The **phosphates and hypophosphites** have attracted a great deal of attention. I do not consider the question at all settled, as yet, as to their value; and do not feel competent to pronounce an opinion upon them. My *impression,* however, is, that the phosphate of *iron* is the best of them all, and that they will be found secondary and inferior to cod-liver oil.

What is the proper place of **alcoholic** beverages or preparations in the treatment of **chronic** debility, such as that of phthisis, etc. ?

¹ Bromide of potassium promises to assume a somewhat important place in this list.

9

This important question opens a discussion, only the main elements of which can be noticed in this work.

In the first place, the theory of the action of agents called stimulants is almost universally misstated in authoritative treatises. It is commonly laid down that "one of the laws of all stimulation, whatever may be its degree, is, that it is followed by a depression proportionate, at least approximately, to the previous exaltation of the function or functions excited."

The true law is this: that all stimulation which is **excessive** is followed by a depression *corresponding to the excess*; while all that merely excites any function *up to par* (to use a familiar expression), *i. e.* to or toward its **normal** activity, does, so far, only good, with no *resulting* debilitation, however it may fail, from want of other conditions, to sustain the organ or system at the point desired. To deny this, would be to ignore some of the most obvious of physiological facts. Heat is a stimulant to life force; oxygen to all the active functions; blood is an excitant as well as food to all the tissues it reaches; and all those impressions upon the exterior of the body which give rise to instinctive and automatic actions are stimulant, without any necessary ulterior depression. Nor do I see how the use of stimulants in any supposable case of disease could be rationally justified, if we *practically* admitted the force of the law as commonly stated; since if, after every dose of an excitant, the patient should sink as far *below* the condition for which he was treated, as the intended remedy raised him for the moment *above* it, of course a mere *oscillation*, and no advantage, must be the result.

This, however, is theory; which has not governed practice on this subject. Another interesting physiological question—"does alcohol contribute to the **material,** or to the **force** of the economy, or only excite some of its organs to exhaustive **action?**"—has been the topic of able and learned disquisitions. I venture merely the opinion that it may do *either* of the three, or *neither*, according to the **circumstances** and the **quantity** of its administration. When there is *scarcity* of *food*, or *difficulty* of *digestion*, alcohol may contribute to the needed material; its carbon, hydrogen, and oxygen going to repair the *adipose* tissue at least. When there is *excessive exertion*, alcohol may sustain the flagging forces of the system. When given in *mere excess*, as with the intemperate, it excites to exhaustive action, organic if not motor; even when the bloated body shows increase in *quantity* of material, its *quality* being more or less degenerate.

In a word, then, the phrase "**accessory** food" is a happy one. When *unnecessary,* as in full health, alcohol is **injurious** precisely in proportion to the quantity used; and the same is true in disease, when the quantity given is **disproportionate.**

This is the important practical precept. Alcoholic stimulus should never be taken in quantities which produce circulatory or cerebronervous *disturbance* or *super*-excitation. If this rule be observed, not only will it be a valuable *supporting* agent in phthisis and other complaints, but no *dipsomania* (methomania) or morbid thirst for it will arise; *that* terrible disease always growing out of **excess.** Upon this principle, in the use of alcoholic beverages in cases of ordinary debility, the common table doses are, medically speaking, *too large.*

Alcohol, in advanced or advancing consumption, in low fever, and in other analogous cases, when used in due proportion, is useful—

1. By its **direct** excitant supporting power.

2. By aiding the enfeebled **stomach** to digest a larger supply of food.

3. By tending to **retard tissue-metamorphosis.**

This last action is one which alcohol has been shown (Böcker. Hammond) to have, in common with other agents, used as medicines or luxuries; coffee, tea, morphia, quinia, etc. I have alluded to it in our classification of remedies, under the head "economic medicines."

It is not supposable, however, that the retardation of the change of tissue in the body is always beneficial. It may, especially in febrile disease, when accumulation of effete matter in the blood and organs is a present evil, be injurious.

For this and other reasons, but most of all from clinical observation, we are prepared to condemn without hesitation or qualification the practice introduced by the late Dr. Todd, of London (foreshadowed by that of the famous Dr. Brown of the last century), of giving alcohol as the remedy or proper medicine "for all acute diseases." Enough for our present purposes to cite two impartial testimonies as to the *results* of that practice.

1. The physician whom Dr. Todd entrusted with the task of analyzing his own records of hospital practice[1] asserts, that the mortality from fever in the hospital attended by Dr Todd was in a marked degree greater than that of any other fever hospital in Great Britain.

2. Statistics of the London Hospital,[2] recently published, show a large increase, since 1858, in the use of stimulants in that hospital, and, with it, a closely coincident ratio of increase in mortality.

Stimulism, as we may call the theory and practice of Dr. Todd, as now followed by many others, confounds three distinct propositions: 1. That all disease is debility; 2. That all debility should be treated by the use of stimulants; 3. That alcohol is always the best stimulant. Granting, with some qualification, the first of these, we emphatically deny the truth of the second and third. It is a practice which, like many other *specialisms,* will *have its day.*

The following classification of the " genuine effects of stimulation," when properly used as to time and *dosage* (remembering the *often opposite* effects of *small* and *large* doses)—is from Anstie.[3]

" I. Relief of pain. II. Removal of muscular spasm, tremor, or convulsion. III. Reduction of undue frequency of the circulation. IV. Reduction of excessive secretion. V. Removal of general debility, or of special fatigue of muscles, brain, or digestive organs. VI. Removal of delirium or maniacal excitement, and production of healthy sleep. VII. Support of the organism in the absence of ordinary food. VIII. Local increase of nutrition where this is deficient."

From the same writer comes also the following terse summary of the stimulating agencies most available therapeutically.

" 1. Quickly digested and nutritious *food.* 2. *Opium* in doses of one

[1] British and Foreign Medico-Chirurg. Review, October, 1860, p. 331.
[2] British Med. Journal, Dec. 9, 1865.
[3] On Stimulants and Narcotics; pp. 112, 113.

or two grains; or *morphia* (sulphate, &c.) a quarter to half a grain. 3. *Carbonate* and *muriate* of *ammonia*, in doses of five and ten grains respectively. 4. *Alcohol*, in doses just too *small* to produce flushing of the face or sweating of the brow. 5. Chloroform, inhaled (in the proportion of about two per cent. to the bulk of atmospheric air) for a short time; or taken internally, in doses of a few drops. 6. Certain fœtid gum resins. 7. Many aromatic volatile oils. 8. The bitters, pure and aromatic. 9. Counter-irritation, as it is called; stimulation, as it should be termed, through the skin."

The subject of the treatment of debility, acute and chronic, must not be dismissed without one further remark, upon the importance of rest in cases of exhaustion from over-exertion. The popular truism, that *exercise is beneficial to health*, has been often *abused* by applying it almost *universally* to invalids or valetudinarians.

The one remedy for the immediate effects of **over-exertion** *is absolute and prolonged* **repose.**

The time required for recuperation, after *cerebral* over-fatigue, may be counted rather by *months* than weeks or days; and it is quite possible for *irreparable* mischief to be done to the brain or spinal marrow by neglecting too long the demand of nature for rest. With many others, the author must acknowledge indebtedness to Professor Jackson, of the University of Pennsylvania, for the judicious emphasis of his teaching upon this point.

It is an important hygienic and therapeutic law, that exercise, to be beneficial, must be proportioned to the strength of the individual; and must never be carried to the extent of actual fatigue or temporary exhaustion.

Antidotal treatment is a topic of great interest. Its *idea* is probably the oldest in medicine. *Specifics* always have been looked upon as the *magna bona* of therapeutical science. Unfortunately, however, their number, instead of increasing, has *diminished* under the inexorable scrutiny of modern investigation. Yet, there is room for hope that they may again *positively* increase, with the diligent application of the same means of observation and discovery.

In the largest extension of the term, antidotive remedies may be classified thus :—

Positive antidotes :

Chemical palliatives;	*Chemical antidotes*;
Antacids;	Antitoxics.
Antilithics.	Parasiticides.

Constructive antidotes :

Anti-periodics;	*Anti-scorbutics.*
Anti-syphilitics;	

Tentative antidotal remedies :

Anti-arthritics;	*Antiseptics;*
Antirheumatics;	*Anti-diphtheritics;*
Anti-zymotics.	

The familiar use of **antacids** as palliatives in dyspepsia, etc., needs no remark.

Nor have we occasion to dwell, here, upon **anti-lithics**; *i.e.* solvents for urinary solids, prescribed on chemical principles ; as, alkalies for excess of uric acid or the urates, mineral acids for excess of phosphates or oxalates.

The subject of **chemical** *antidotes* for poisons belongs to *Toxicology*. **Anthelmintics** are best treated of in the department of *Practice of Medicine.* (Part II.)

Anti-psorics, or specific remedies for scabies (itch), are represented generally by sulphur ; which, although not at all the *only* agent capable of destroying the morbific *acarus*, is the most convenient. Other cutaneous parasites (nosophyta) are also destroyed, but with less certainty, by preparations of mercury, etc., called **parasiticides.**

Of "**constructive** antidotes," the most important are the alkaloids of **cinchona,** applied to the treatment of **miasmatic** affections (antiperiodics). Medical men are divided upon the question whether quinia arrests intermittent fever, etc., by antagonizing (chemically) the miasmatic poison itself in the system, or (physiologically) by causing such an opposite impression upon the nervous centres as is capable of subverting the condition on which the periodical or paroxysmal affection depends. The last is the prevailing view. But, in either aspect, the cure of autumnal fevers and allied affections occurring under miasmatic influence (neuralgias, etc.) by *cinchonization,* is properly called **specific** treatment ; as—

1. No other remedies (yet discovered) have the same power.

2. These remedies have no such control over any *other* diseases (*e. g.* typhus and typhoid fever, yellow fever).

The second proposition is asserted with positiveness, notwithstanding the experimental use of quinine in full doses, by a few practitioners (Dundas), in typhus and typhoid fever, and its frequent administration in yellow fever.

In stating that no other medicines, yet discovered. have the *same* power, I mean, to a degree or with a certainty at all comparable to that of the cinchonic alkaloids. The nearest approach to this is afforded by *arsenic.*

It is, however, a remarkable and important fact, that, when the recurrence of the paroxysms of intermittent fever has been allowed for a long period (**chronic** intermittent), and the system of the patient has become debilitated and **anæmic,** quinine will only **interrupt,** but *will not cure* the disease. **Iron** *is, then, the remedy.*

Opinion is divided as to the value or necessity of **mercury** as an **anti-syphilitic.** In the primary disease, I am a full believer in its importance ; against which its frequent *abuse* furnishes no argument. In secondary syphilitic affections, *especially* syphilitic **rheumatism, iodide of potassium** exhibits decidedly specific powers.

Anti-scorbutics are most valuable as *preventives* of scurvy ; but will promptly relieve it, also, when it has occurred. All **fresh vegetables** belong prominently to this class ; certain plants not so used, as the cactus opuntia, are included in it ; the juice of **lemons,** limes, etc., is of service for the same end, and the neutral salts of potassa have been largely employed, with variable results.

Tentative antidotal treatment—for diseases in which there is evidently (as a *part* at least, if not the primary part of the malady) *humoral* disorder, such as gout, rheumatism, the exanthemata, etc.—

9*

affords a large field for study and ratiocination. The positive *facts*, so far, are few ; the *hypotheses*, legion.

In **gout**, **colchicum** has long held, deservedly, the first place, as either an *eliminative* or *antidotal* remedy. Most observers have given it the first title ;[1] Dr. Garrod's experiments induce him to prefer the *idea*, if not the phrase of the latter. **Alkaline salts** *of organic acids,* as bicarbonate of potassa, soda, or lithia (Garrod), or tartrate of potassa and soda, and the *alkaline earth, magnesia,* have also a large share of confidence in the treatment of gout. Experience satisfies me that this confidence is well founded. After all, however, so incomplete is any curative plan as yet devised, that a large margin is left for patience and opium.

The same is true of **rheumatism**; especially in its distinctive form, of *acute articular rheumatism,* or **rheumatic fever.** Colchicum is here also much given; but *in the absence of the gouty diathesis,*[2] hereditary or acquired, it will *generally disappoint.* **Alkaline salts** are, at the present moment, the favorite tentative anti-rheumatics. Lemon-juice has been freely employed by some practitioners. Phosphate of ammonia was for a brief time in vogue. Calomel and opium are still the reliance of many. Certain enfeebled cases, with free perspiration, will recover speedily under *quininization.* But in all these modes of treatment there is no.*specific* **certainty.** The attack will last from one or two to six or eight weeks with all. In **chronic** rheumatism we resort, with the same *hope,*·to guaiacum, spirits of turpentine, iodide of potassium, cod-liver oil; but often our hope is much lengthened out. Of *propylamin,* a remedy for rheumatism imported not long since, I have had some experience, and have no basis for a favorable opinion. The search for and trial of such remedies is, however, certainly altogether legitimate.

In the management of the **zymotic** affections, the only great triumph of medical art has been one of **prevention.** **Vaccination** affords an instance of effectual control over one of the most destructive and loathsome of pestilences, by the interference of the physician. As to the **treatment,** even of smallpox[3] itself, when it has occurred, and of scarlet fever, measles, chicken-pox, hooping-cough, and mumps, we are forced to confess our powerlessness, except to conduct the case, by the aid of **palliative** measures, to its natural and spontaneous termination.

[1] Colchicum has been shown, by Krahmer and Hammond (*Proceedings of Biological Department of Acad. of Nat. Sciences of Philadelphia,* Nov. 1st, 1857), to *increase the amount of the solids* of the urine, more decidedly than any other vegetable diuretic.

[2] Garrod insists on the diagnostic importance of the *uric acid test* for gout. It is easily applied, as follows : Take about f℥jss of the serum from a blister, or from the blood drawn by venesection or cupping, and place it on a flat dish or watch-glass. Add to this fifteen drops of acetic acid, and place in it two or three threads of cotton. Allow the glass to stand in a warm room for one or two days, to evaporate. If the cotton fibres be then removed and examined microscopically with an inch object-glass, they will be found, if the serum contained uric acid, to be covered with its crystals, arranged somewhat as the crystals of sugar-candy form on a string.

[3] *Sarracenia* has proved valueless upon fair trial.

This is equally true of **yellow fever.** There is **no specific** yet known for this terrible disease. Quinine, mercury, etc., have failed in the hour of need too often to be relied upon. It is to be **palliated,** as it *cannot be cut short.*

Nor have we any specific for epidemic **cholera.** Anti-spasmodics, at very short intervals of administration, and ice, with free external stimulation, will conduct many cases to a successful close; but this is not **antidotal** treatment. I am fully satisfied that *calomel,* in true epidemic cholera, is altogether *useless.*

In the medication of zymotic affections having, as a local symptom, inflammation of the mucous membranes, with unusual tendency to (septic) decomposition or **disorganization,**—*e. g. scarlatina* and *diphtheria,*—**chlorate of potassa** and other preparations of chlorine, as tincture of **chloride of iron,** have achieved recently a very widespread reputation.

The last-named of these medicines, the tincture of the chloride of iron, appears also to have an excellent effect (although we can hardly call it antidotal) in *asthenic* **erysipelas.**

Anti-diphtheritic power has of late been strongly asserted of *lime-water* locally applied.

Professor Polli, of Milan, has, not long since, proposed the internal use of the *sulphites* of soda, lime, and magnesia, in toxæmic diseases, as antizymotics or antiseptics. The chemical rationale of their action is very plausible. Success is already asserted[1] for them in pyæmia, etc., and in glanders in the horse. As a tentative practice it is worthy of careful trial.

With the view of encouraging such trial, the following account of the treatment is given:—

Dr. Giovanni Polli, of Milan, in 1864, communicated to the Royal Institution of Lombardy his researches and experiments upon the influence of materials taken into the body upon its morbid and destructive changes. Believing that, in all zymotic diseases, two morbid elements must exist, the "ferment" and the fermentable material, he endeavored to ascertain how, if we cannot destroy the morbid poison, we may at least arrest or retard the decomposition it produces. Many agents which check fermentation and putrefaction are not safe as medicines. Sulphites and hyposulphites of alkalies and alkaline earths prove, upon trial, to be so.

First, Dr. Polli experimented upon dogs; saturating their bodies, through the stomach, with sulphites, and then killing them. They were exceedingly slow in undergoing decomposition. Then, he inoculated several with mucus of glanders, putrid blood, and unhealthy pus, and dosed some of them with sulphites, others not. Those so dosed suffered much the least severely.

So encouraged, he extended his experiments to human beings. Without detail, it may be said that he satisfied himself and fellow practitioners of the safety of sulphites and hyposulphites of magnesia, soda, and lime, as medicines; and of their exhibiting a hopeful degree of power in mitigating and even sometimes arresting

[1] See Amer. Journal of Med. Sciences, Oct., 1863; and later numbers of the same Journal.

septic diseases. Such experiments are well worthy of extended, careful and systematic trial. Thus it is that science should *suggest* remedies for experience to *prove;* empiricism may thus be made rational, and rationalism in medicine become practical. Even if disappointment attend a certain set of experiments, or the trial of a particular remedy, such a trial is fully justifiable in its principle.

Justice to Dr. Polli's ideas and inquiries calls for a clear recognition of what he does assert as matter for experimental practice. He claims, for instance, that intermittent fever is cured by sulphite of magnesia, an ounce or an ounce and a half in 24 hours, in divided doses; that the attack is not cut short as by quinine, but that the symptoms are gradually diminished; but, that it cures the disease more completely, and in a greater number of cases, than quinine does. It is asserted that mitigation of symptoms is observed in pyæmia, scarlet fever, diphtheria, and other acute febrile disorders, under the sulphites; and that in cattle-plague, one third attacked recover under their use.

The sulphites appear in the urine about twenty minutes after they are taken; also in the sputa and saliva; but they are gradually changed in the system into sulphates. M. Carey Lea, of Philadelphia, in a paper published in 1865,[1] reports a series of careful experiments, in which he found evidence that when a small quantity of sulphite or bi-sulphite of soda is taken, less than a hundred grains, it disappears by oxidation in the system; but if large amounts be ingested, a considerable portion passes unchanged in the urine, and sulphurous acid may even be detected in the breath.

Polli asserts that the sulphites are not usually decomposed in the stomach. If they are, sulphurous acid gas is evolved. A little magnesia should then be added to each dose to correct this effect. The tolerance of the sulphites and hyposulphites is greatest when they are freely diluted. They are decomposed by all vegetable acids. Lemonade, for instance, should not be given to a patient using them.

Hyposulphite of soda is much more purgative than sulphite of soda or magnesia; those salts are commonly rather diuretic. If cathartic, it is without pain. The stools are without fetid odor under the use of these remedies. Dr. Polli recommends especially the sulphite of magnesia, as the most active and having the least taste. The dose is, according to him, fifteen to thirty grains, in powder dissolved in water or an aromatic vehicle, or in troches. He advises *saturating* the system with the medicine; four or five drachms daily for an adult as a minimum. Five to seven drachms of the sulphite of soda are borne well. Polli prefers the sulphites for rapid curative action in acute diseases; the hyposulphites for prophylaxis. Their long continued use may bring on œdema and diseases of debility; otherwise, they show no influence on the system.

Externally, solutions of the sulphites, especially when mixed with a portion of glycerine, are recommended as applications to suppurating surfaces, to sloughing and ulcerated parts. Sulphites of lime and magnesia are somewhat caustic. In septæmia from wounds, &c., Polli administers thirty grains of the sulphite of magnesia, every two hours, internally.

---

[1] Am. Journ. Med. Sciences, Jan., 1865, p. 84.

Chemically, the formula of hyposulphite of soda is given by Pereira as NaO,S_2O_2. That of sulphite of soda is stated as NaO,SO_2+3HO. Dr. G. B. Wood, in the last edition of his Practice, mentions chiefly the bisulphite of soda. H. R. De Ricci, on the ground of experience, insists upon the use of the *sulphite of magnesia*, in scruple doses, *early* in the cases treated.

To show the presence of sulphurous acid in the urine, Dr. Davy half fills a test-tube with the urine, slightly acidulated with muriatic acid, and suspends in the tube, above the urine, a piece of starched paper stained blue with a weak solution of tincture of iodine. If sulphurous acid be present, its evolution will decolorize the paper.

En résumé, we may say that all endemic, epidemic, infectious, and contagious diseases are naturally **self-limited**; and that, so far, we have only reached a certainly *curative* treatment for one class,—viz. intermittent, remittent, and pernicious (classed together as **miasmatic**) fevers; and a **preventive** treatment for another, smallpox.

While, therefore, for yellow fever, scarlatina, pertussis, etc., we are without the possession of any *specific* or *antidotal* treatment, the **palliative** plan is the one for us to pursue. All attempts, by violent measures, to cut short either of these diseases, while they fail to attain that object, will endanger the patient, by lowering his forces, and thus promoting the victory of the depressive toxæmic cause.

Yet, I repeat, we are not to abandon or reject the hope that observation and cautious experiment, guided by the lights of advancing science, will enable us hereafter to discover remedies as potent in the management and control of scarlet fever, yellow fever, and cholera, as quinine is in that of ague, or vaccination in the prevention or salutary modification of smallpox.

Alterative treatment is distinguished, in our classification, it may seem arbitrarily, from the *antidotive*. All antidotes may be said to be alterative, but all alterative medicines are not antidotal; as the latter expression implies at least the *probable*, if not the known existence of a **material cause**, against which the antidote is to act. Yet the distinction is not one upon which we can *insist*, although it appears convenient.

The term *alterative* is by no means a mere apology for ignorance; it involves an important therapeutical principle; viz., the **supplanting** or displacing of a morbid impression, condition, or process in the body, by the **safer** impression and **counteraction** of a medicinal agent. The influence of the latter, physiologically speaking, may be, *per se*, abnormal; yet, having a sanative purpose, it is therapeutic.

This principle may be sufficiently illustrated by allusion to two or three examples. In the peculiar and often violent inflammation of the throat in scarlatina, the free application of a *strong solution of nitrate of silver* to the part will almost invariably arrest (if used *early*) the morbid process: converting it, at all events, from a *specific* and *dangerous* into a *simple* and *mild* phlogosis.

So will the powerful impression of the solid nitrate of silver, or other caustic, upon the surface of the penis affected with *chancre*, *supplant* the *venereal* process, and leave in its place a benignant ulcer.

When erysipelatous inflammation is spreading like a conflagration from part to part, a blister or tincture of iodine, etc.; will form a *cordon sanitaire*, by inducing its own milder irritation in advance of the disease.

The most essential part of the treatment of chronic *diseases of the skin*, is either **alterative** or **antidotal.** *Parasitic* affections, as scabies, favus, mentagra, etc., require the destruction of the epizoon or epiphyte by an antidote. Others, as eczema, lichen, impetigo, lepra, etc., when at all obstinate, are treated in the *same manner* essentially, to whichever class the disease may belong. Why? Because the **principle** is the same in all—the *alterative* principle. The abnormal, perverted nutrition of the cutaneous tissue, whether it be deeply or superficially affected, is (apart from antiphlogistic or sedative treatment, called for in some cases) to be *subverted*, by a decisive change in all its conditions; and, speaking boldly, it is little matter *what* change, so it be considerable. Any means which will hurry the removal of the old, diseased skin, and favor the immediate construction of a new layer, will be *curative*, whether it be only soap, water and frictions, mercurial ointment, vesication, or the actual cautery. And the same principle explains and justifies the *internal* use of arsenic and mercury in the management of so many very *diverse* forms of cutaneous disease— the indication for alterative medication being the same in all.

The administration, for **long periods,** of minute doses of powerful alterative medicines, in the treatment of chronic affections which resist other management, is less common now than formerly, on account of the explosion of some old hypotheses connected with it. It is very possible that in this, as in some other medical reforms, we may have gone too far.

Agents which tend with any degree of constancy to **increase** the rate of metamorphosis of tissue in the body, are few. It is well known that most of the diuretics given for the removal of dropsical accumulations (a treatment often carried to an irrational excess), increase only the **fluid** secretion of the kidneys, affecting little or not at all their *solid* excreta. But there is good reason to believe that potassa and soda, and some of their compounds, as well as iodine and mercury, do hasten the disintegration of tissue. Iodide of potassium has been shown, by Melsens, to be in this way *eliminative* of lead, laid up in some organ (probably the liver), removing it in the shape of iodide of lead.

If any possible measures, beyond attention to hygiene and repose, can benefit cases of *organic* **degeneration,** we might hope for advantage from the combination of tonics or analeptics with alteratives. Dr. Chambers (on *Digestion and its Derangements*) remarks thus:— "In Bright's disease I know of no treatment so advantageous as that which unites **alteratives** (that is, liquefacients of tissue) with those **restorers** of blood *par excellence*, iron and animal food."

Iodide of iron, or iodide of potassium at the same time with **cod-liver oil,** may afford an example of this sort of medication in its simplest form. Its object is, to favor the rapid removal of old tissue, and the formation of good new structure in its place.

Whatever produces a powerful impression, not immediately destructive, upon the system, may act alteratively, and sometimes beneficially, in chronic disease.

Thus, **electricity**, perseveringly used, in certain cases of paralysis and other neuroses, does more than anything else.

At present, while it is clear that electricity must be capable of powerfully influencing the human system, our knowledge of its uses is far from complete. Rash experimentation with it may do harm; but all its appliances may be so graduated as to admit of the mildest and most cautious tentative practice.

Three agencies are included under the term electricity. 1. **Static** or frictional electricity, of the machine, of glass and revolving rubbers. 2. **Galvanic** electricity, or the continuous current of the battery, of metals with acid solutions, &c. 3. **Faradization**, by induced and interrupted currents, electro-magnetic or magneto-electric.

Frictional electricity is least in use. It is best employed by placing the patient on an insulated stool (with glass legs or feet) and charging him from either a plate or a cylinder machine; then withdrawing the electricity by a *pointed* metallic conductor if a mild and general effect is wanted, or by a *round* one of some size to produce sparks and a locally stimulant effect.

Galvanic electricity is developed by chemical action. A battery consists of a series of plates of two materials (copper, zinc, silver, carbon, &c.) in alternation, a liquid in contact with which acts more upon one than on the other. Daniell's and Smee's batteries are especially recommended. Althaus prefers a modification of Daniell's, in which, with copper and zinc plates, a solution of sulphate of copper is used as the liquid. This will run for six months without cleaning; with cleaning, for a much longer time. The *size* of the cups determines the *quantity* of the current of electricity; the *number* of cups, its *intensity*. Quantity is especially powerful for chemical action; intensity, for overcoming the resistance of a slowly conducting medium. The *direction of the current*, through the wire or wires connecting them outside of the liquid, is, in ordinary batteries, from the copper (or platinum, silver, or carbon) to the zinc; the former being the positive and the latter the negative pole.

Faradaic or induced electrical currents are usually obtained in either of two modes. 1. By galvanic electricity (from chemical action) acting upon iron; making it magnetic during the closure of the circuit, which is interrupted momentarily, by a vibrating spring. A current is thus induced, in one direction at the moment of *closing*, and in the opposite at the *opening* of the connection. 2. By causing a magnet or its keeper of iron to revolve so as alternately to approach and recede from contact. A *helix* or coil of wire, wound around the magnet, will then have an induced current at the moment of approach and at that of separation. This is more convenient and manageable than any other sort of apparatus; but some authorities claim greater power for electro-magnetism in therapeutics. It is asserted also (Althaus) that the *constant* galvanic current (as of Daniell's, Grove's, Bunsen's, or Smee's batteries) has a more extended or general stimulant or alterative effect; while the *interrupted* (faradaic) current is more powerful locally. Proof of the difference between the two may be obtained by applying them in succession to the face. The

continuous current produces a flash of light by its influence upon the
retina. The interrupted causes the muscles to contract in proportion
to its force. Remak has greatest confidence in the continuous current.

Duchenne, of Bologne, a high authority, asserts as the conclusion
derived from his experience, the following:—

"In man, whatever may be the direction of the currents, or the
degree of vitality of the nerves they traverse, the same results are
always produced when the conductors are applied to any part over
the course of the nerves—namely, muscular contractions and sensations.

"Various changes in the current-direction produce no appreciable
influence over the sensibility, or capability of voluntary muscular contraction, in man."

On the basis, however, mainly, of Dubois Reymond's experiments,
other electricians insist that a current *towards* the centres of the
body stimulates the sensory nerves, while it lessens the excitability of
the motor nerves; and that a current from the centres out to the peri-
phery stimulates motor and acts as a sedative to sensory nerves.
Pereira says, "in paralysis of sensation only, the current should be
direct or centrifugal. In paralysis of motion, it should be inverse or
centripetal. In paralysis of both sensation and motion, the vibrating
current (faradization) is peculiarly appropriate; for by this the sensitive and motor nerves are alternately excited, while the one current
promotes the restoration of the excitability, which may have been
lessened by the preceding current."

The *precautions* needful in trying electricity as a remedial application are—

1. Always to begin with it very gently, watching carefully its effects;
continuing it therefore at first but for a few minutes at once.

2. To be especially careful in its use at or near the great nerve-
centres, as the brain and spinal marrow.

3. Avoid using it, even locally, during the existence of an inflammation or acute irritation of the spinal marrow or brain.

The idea which Dr. Radcliffe has especially advocated, of using
galvanism, *e. g.* in neuralgia, or convulsions, with such force as to partially or temporarily *paralyze* a disordered nerve-centre, is, I am sure,
unsound in theory and very unsafe in practice. I would not think of
resorting to such a measure in any conceivable case.

The affections in which electricity, in various modes of application,
has been found most positively and frequently serviceable, are—
paralysis, especially *hysterical*, *reflex*,[1] and *lead* palsy; *aphonia*, and
diphtheritic paralysis of the throat; *neuralgia, chronic rheumatism,
amenorrhœa, suppression of lactation*, and *cancerous or other tumors*.
In *surgery, galvano-puncture* is resorted to for aneurisms, &c. In
treatment of *asphyxia narcotic coma*, &c., its powerful stimulation is
sometimes an important means of saving life.

Hydropathy is an example of a most valuable agent misapplied by
exclusivism, which is always quackery. That is, it is quackery to
deny the virtues of all other remedies, no matter how long or well
established, in extolling those of one, made the sole *panacea* of prac-

---

[1] See Mitchell, Morehouse and Keen, upon Wounds of Nerves, &c.

tice. Bathing, local and general, douches, and even packing in the wet sheet, are, used with judgment, potent alterative and often sanative means. Almost hopeless chronic cases, of nervous disorders, dyspepsia, chronic rheumatism, &c., may sometimes have their languid vitality aroused by the revolutionizing action of such processes. The danger is, of unprofessional and ill-judging abuse of an agency of such power.

Hot air-baths have lately attracted attention.

M. Fillet, 1764, and Fordyce, Blagden, and others, 1775, proved, as Chabert the "Fire King" also illustrated, that a *dry air* heat above that of boiling water can be borne by the human body with safety. Within a few years the use of the hot air bath, similar to that of the *calidarium* of the ancient Romans, has been introduced for remedial purposes. It is misnamed the "Turkish Bath," as the latter is a *vapor* bath, at a temperature below 100°.

Erasmus Wilson, the dermatologist, Mr. Urquhart, Sir John Fife, and others in England, have, upon personal experience, lauded highly the virtues of the hot air bath. The requisites are, a heated metallic surface (a common stove will do) to warm the air of the apartment, and an adjoining convenient water bath, with warm or cool water, to plunge into after or alternately with the air bath.

Mr. Urquhart thus speaks of its use in health: "When I come back to it after its absence and the use of other baths, even the best, it is like getting on the back of a thorough-bred after having to ride a cart horse. It is of service at every moment and at all temperatures. You come in from a journey, say before dinner; you go in not heated, when it may stand at 120°; you dress at that charming temperature, with streams of hot or cold water, or the tank to revel in. So, also, you may dress in the morning. My regular practice, when not requiring it for health's sake, is to go in on getting up and on going to bed, dressing and undressing there; five to ten minutes suffice to bring on the flow of perspiration. After that, a plunge in the cold water, and you come out fresh, glowing with a sense of cleanliness, health, and strength, which no other operation can convey to the body. You are then indifferent to the heat of summer and the cold of winter."

Copious perspiration, and the thorough removal of the effete cuticle, as well as of all foreign impurity, from the surface of the body, are the obvious and direct effects of the hot air bath. *Depuration* and functional balance of the excretory processes are thus promoted; while the stimulant influence of heat, so often useful in cases of depressed vitality, is obtained. Renewal of the nutrition of the skin must also be more rapid under such a process. The temperature is made to vary between 120° and 200°; 130° to 140° would be a good standard.

The *diseases* in which the hot air bath has been tried with favorable results, in cure or palliation, are, especially, acute and chronic rheumatism, lumbago, sciatica, neuralgia, gout, dropsy, catarrh, influenza, throat affections, diarrhœa, dysentery, sluggishness of circulation, disorders of the liver, scrofula, incipient phthisis, Bright's disease, ague, obstinate skin diseases, chorea, mania; also, it is said, even cholera and hydrophobia.

10

110   GENERAL THERAPEUTICS.

Mustard Bathing.—Dr. S. Newington, of England, ascertained
by experience in his own person that the extensive and continued
application of mustard to the body is a powerful tranquillizer of nervous excitement and means of restoration of the balance of a disturbed circulation. Use of the same remedy with a number of
patients has confirmed its value. One mode of its application is as
follows : Two handfuls of powdered mustard are tied in a cloth and
placed in hot water, then squeezed in the hand until the strength of
the mustard has been extracted. A thick towel, long enough to
reach around the loins, is then wrung out of this infusion, wrapped
round the body, and covered with a large piece of oiled silk, or gum
elastic cloth. Another plan is that of the mustard bath; that is, an
ordinary warm bath into which have been thrown five or six handfuls
of mustard.

   In *maniacal excitement*, particularly, these applications have been
found usefully sedative. *Sleeplessness*, from any cause, may be so
treated; as well as hysteria, etc. It may be expected that such a
process will be useful also in promoting reaction in cases of internal
congestion; perhaps, in the chill of pernicious fever, in "spotted fever,"
and the incipient collapse of cholera.

   Movement-cure (*kinesi-pathy*) is a name for another kind of practice
(Taylor) founded upon the teachings of Ling, of Sweden; consisting
of *passive exercise* of the muscles, for the purpose of improving the
circulation, innervation, and reparative nutrition of diseased and enfeebled organs.

   Although an elaborate system of particular movements upon a *quasi*-
physiological rationale, has been devised, adapted to each kind of
chronic local or general disorder, careful examination of the treatise
of an authority upon the subject convinces us that this is all *surplus-
age*. What remains to be true is, that, where active exercise is not
practicable, systematic frictions and passive movements of all parts
of the body are very useful in its stead. What is added to this by the
specialist is, faith on the part of the patient, and perseverance in the
attendant; two things which, without extraordinary processes, explain
much, and account for many cures.

   Under any treatment, however, we must not encourage sanguine
hopes in any instance of **organic degeneration**, the origin of which
is, so generally, to be found in a failure of systemic vital power. The
physician will do much for his patient, if he can persuade and instruct
him to adapt his living to the actual condition of his physical resources,
so that, whether his malady be Bright's disease of the kidney, diabetes
mellitus, cirrhosis of the liver, or fatty or other organic disease of
the heart, he may *economize* both the material and the force of his
system, by such a regimen of diet, exercise, and excitement as his
state requires.

   With such management, it often happens that valetudinarians live
longer than those who, with sound constitutions, are less watchful
against causes of disease, and less prompt in taking warning from the
slightest symptoms or approaches of ill health.

   Nor need we look upon the failure of medicine to arrest the process
of organic degeneration as a subject of very humiliating discourage-

ment. As death is the natural result of life, in the human organism as in every other material form, this partial death occurs, also, under **physiological laws**; and, if it be premature in certain instances, we may believe that this, too, may be traced to a near or remote causation, in perfect harmony with the highest interests, moral as well as physical, of man.

INHALATION AND ATOMIZATION.

Although the ancient Egyptians had some knowledge of the effects of drugs whose vapors were inhaled, and Hippocrates, Galen and other Roman physicians, as well as, later, the Arabians, so employed them, they were afterwards long lost sight of. Bennet, of London, in the seventeenth century, seems to have been the first in modern times to use inhalations systematically. The earliest proper instrumental inhaler, was probably that of Dr. John Mudge, an English physician, invented and applied in 1799. St. John Long, the charlatan, used large inhalers, from which a number could breathe at once. Boerhaave and Von Swieten, in the early part of the eighteenth century, employed medicated inhalations. Scudamore, in 1830, issued a work upon the subject which had a wide circulation.

Since that time, an immense amount of experimentation has been made, to ascertain what local effects upon the lungs and air-tubes, and what action upon the system at large, might be obtained by vapors brought in contact with the highly absorbent respiratory membrane. Most important of all, of course, was the discovery of gaseous anæsthesia by Sir Humphrey Davy, and the subsequent practical application of it by Morton. Ether, chloroform and nitrous oxide now take their places not only as alleviators of pain during operations and in parturition, but, also, as occasionally valuable aids to the physician in medical cases. Extreme neuralgic pain is sometimes treated by the inhalation of ether or one of the other anæsthetics. Convulsions are not unfrequently so treated; and, in those at least of a hysterical or merely irritative character, with good success.

For medical as well as surgical use, chloroform is the most prompt, quiet and effectual anæsthetic. In cases of disease, it does not need to be so given as to produce the total relaxation of *profound* anæsthesia. Can its use be justified, in view of the fact that a considerable number of deaths have been traced to it?

Without space to discuss this point, I may say that observation of the use of chloroform, in a method employed by some of the surgeons of the U. S. army during the war, has modified my previous apprehension of it. The great necessity is, as Dr. Sansom[1] has shown very fully, the *dilution* of the chloroform with abundance of air, and its gradual introduction. Most methods exclude air too much. That which I have alluded to above is, letting the chloroform fall, *drop by drop*, upon a handkerchief spread singly over the face of the patient. An instrument is used for the dropping. which will allow only one drop to pass at once. I believe that. with this or any other mode of abundant dilution and slow inhalation of it, watching momentarily its effects, chloroform is safe, in an immense majority of cases.

[1] On Chloroform ; its Action and Administration.

Ether excites some persons too much to be generally relied upon in medical cases. Two or three parts of ether with one of chloroform make a mixture, often used with advantage. Nitrous oxide, although experimented upon by Davy and used by Wells before ether, has only of late come to be highly appreciated for practical use. Its applications by inhalation in disease remain to be tried and studied. Apart from the "anæsthetics," it cannot be said that great success has ever been obtained in the *cure* of diseases by inhalation. *Palliation* of pulmonary and bronchial or laryngeal irritation, or diminution of excessive expectoration, as by simple vapor of water, tar-vapor, or that of infusion of hops, opium, &c., has been often realized. With other aims and agents, disappointment has generally predominated.

For ordinary inhalation, very simple apparatus will suffice. For instance, a wide-mouthed jar or bottle, with a cork in it; the cork pierced by two glass tubes, one straight, and reaching to near the bottom of the bottle; the other short, and bent outside of the cork. The bottle is to be not quite filled with the liquid (more or less heated according to its volatility) ; the bent tube not reaching its surface, the other conveying air into it from beyond the cork. Even this is not necessary, in the case of liquids used with water, at least. We may employ these by pouring boiling water into a convenient vessel of any kind, the medicament being added to it, and then, covering the vessel with a towel, holding the mouth and nostrils under the edge of the latter. Hops, in infusion, or stramonium leaves, or laudanum, &c., may thus be used. Of laudanum, *e. g.*, twenty or thirty drops may be put in a pint of water, for very worrying cough. *Smoking* is a primitive method of inhalation. Tobacco, so used, sometimes relieves in asthma ; but cigars of stramonium leaves, or of paper saturated with nitrate of potassa, are more effectual in the paroxysms of the same disorder.

Recently, first by Sales Girons, extremely minute division or *atomization* of liquids, so introduced into the air-passages, has been substituted for inhalation. Under the natural fascination of novelty, and the imposing appearance presented by instrumental appliances, it is quite probable that a degree of enthusiasm exists now about it, more than will be permanent. Still it is an important addition to our means of treatment of affections of the throat, and, perhaps, of some of those of the lungs. Referring the reader to special works[1] upon it for details, I must give only the briefest account of atomization or nebulization.

The essential idea of it is, the forcing of a fine jet of liquid against either a solid body, or a strong current of air, so as to convert it at once into diffused spray. Bergson, for instance, employed the tubes used for *odorators*, that is to spread perfumed liquids in the air. Two glass tubes with minute orifices are fixed at right angles to each other, so that the end of the upright tube is near and opposite to the centre of the orifice of the horizontal tube. The upright tube being immersed in the liquid to be nebulized, air is forcibly blown through the horizontal one. The current of air, so passing over the outlet of the tube communicating with the liquid, rarefies the air in the latter, causing a rise of the liquid in the tube, and its very minute subdivision

---

[1] See Da Costa, On Inhalation.

(atomization, nebulization, pulverization), as it escapes. Silver tubes may be used instead of glass, but are harder to keep clean. Glass ones may be cleaned with muriatic acid solution, aided by a bristle to remove obstructions. The form of the tubes may be varied, so as to allow of their application to any part of the body.

Richardson's spray-producer (designed for local refrigeration) is constructed upon a similar principle. It consists of a graduated bottle, through whose cork passes a double tube; that is, a tube within a tube. The inner one reaches to near the bottom of the bottle, below, and above to near the extremity of the outer tube. The latter has entering it, above the cork, another tube connected with " hand bellows,"—or, two elastic bags, the one nearest the bottle (protected by silk network) acting as an air-chamber, and the farthest one being compressed by the hand to produce a jet of air into the bottle and tube.

Siegle devised an apparatus for the application of steam-power to atomization. The tubes being arranged upon Bergson's principle. a small boiler is connected with the horizontal one, and in the boiler steam is generated by the heat of a spirit lamp. The jet of steam from the horizontal tube nebulizes the liquid drawn up from the vertical tube immersed in a vessel containing it. Various modifications of this have been made.[1] Though the steadiness of action of the steam-apparatus must be a great advantage, for many purposes the hand-ball atomizer must be more available.

For full effectiveness of any method of inhalation, in chronic or subacute cases, the patient must have the instrument at his own house, learn its management, and use it with regularity for a sufficient time. This of course must limit very much the employment of such medication.

The first inhalations should always be short, and with warm water only, to inure the patient to their use. The distance of the mouth from the tubes may vary from six inches to two feet. When prepared for it, one may inhale "medicated spray" for ten minutes at a time; breathing deeply if we wish the liquid to reach the remoter air passages. It should never be done after a hearty meal; and the patient should remain in doors for a while after the operation.

Proof has been obtained that atomized liquids inhaled do, sometimes at least, pass down into the trachea; constantly, into the larynx. It is probable, indeed almost certain, that a certain portion may even reach the lungs. As to their application, trial has been and is now being made of this process especially in croup, diphtheria. œdema of the glottis, catarrh, chronic laryngitis, hooping-cough, asthma, pulmonary hemorrhage, and phthisis.

False membrane has been asserted by Küchenmeister, Biermer, Geiger, and others to be dissolved, or at least removed from the throat, by inhalation of hot lime-water. Dr. Geiger's method is to make the patient breathe the vapor arising from hot water poured on unslaked lime.

From Dr. Da Costa's monograph upon inhalation I cite the follow-

[1] Gemrig, of Philada., Dr. W. Reed of Boston, and Codman and Shurtleff of Boston furnish improved forms of apparatus for atomization.

ing conclusions, as indicating the present state of experience upon the subject:—[1]

"That in most acute diseases of the larynx, and still more so in acute disorders of the lungs, the value of inhalations of atomized fluids, save in so far as those of water may tend to relieve the sense of distress, etc., and aid expectoration, is very doubtful; though in some acute affections, as in œdema of the glottis and in croup, medicated inhalations have strong claims to consideration.

"That in certain chronic morbid states of the larynx, particularly those of a catarrhal kind, and in chronic bronchitis, they have proved themselves of great value.

"That in the earlier stages of phthisis, too, they may be of decided advantage, and that at any stage they may be a valuable aid in treating the symptoms of this malady.

"That their influence on such affections as hooping-cough and asthma is not satisfactorily proven.

"That they furnish a decided and unexpected augmentation of our resources in the treatment of pulmonary hemorrhage.

"That they require care in their employ; and that in acute affections we should consider whether, as they have to be used frequently to be of service, the patient's strength justifies the disturbance or the annoyance their frequent use may be."

DOSES FOR INHALATION.[2]

Alum	10 to 20 grains.
Tannin	1 to 20 "
Perchloride of iron	½ to 2 "
Nitrate of silver	1 to 10 "
Sulphate of zinc	1 to 6 "
Chloride of sodium	5 to 20 "
Chlorinated soda	½ to 1 drachm.
Chlorate of potassa	10 to 20 grains.
Muriate of ammonia	10 to 20 "
Watery extract of opium	¼ to ½ "
Fluid extract of conium	3 to 8 minims.
" " hyoscyamus	3 to 10 "
Tincture of cannabis indica	5 to 10 "
Lugol's solution of iodine	2 to 15 "
Fowler's solution of arsenic	1 to 20 "
Tar water	1 to 2 drachms.
Oil of turpentine	1 to 2 minims.

HYPODERMIC MEDICATION.

Apparently upon a suggestion contained in Valleix's work on Neuralgia, Dr. Alexander Wood, of Edinburgh, in 1843, experimented successfully with the injection of anodynes under the skin of parts affected with neuralgic pain. Dr. Kursack, of Vienna, disputes priority with him. Mr. Rynd, of Dublin, followed him about a year. Local effects, only, seem to have been clearly recognized by these gentlemen. Mr. C. Hunter, in 1858, proved that general effects upon

---

[1] Op. citat., p. 40.   [2] From Da Costa.

the whole system are produced, in whatever part of the body the injections are made. Since 1855, however, many medical men have studied the subject; especially Béhier, Lorent, Eulenberg, and Nussbaum abroad, and Ruppaner in this country. The practice has now become quite common.

It has been amply proved that hypodermic injection of medicinal substances is ordinarily entirely safe; more rapid, certain, and exact in proportion to the amount, in its effects than medication by the mouth; that it requires one-third or one-half of the quantity necessary when given by the stomach, and produces less complicated and generally less inconvenient results.

The medicines mostly used in this way are narcotics, sedatives, and nervine tonics. It is in diseases or symptoms affecting the nervous system that the greatest number of successful cases has been reported. Pain, most of all, is speedily conquerable by it. Hunter lays down the indications for it thus:—

" When the immediate and decided effect of the medicine is required.

" Where medicines administered by the usual methods fail to do good.

" Where the effect of a medicine is required, and the patient refuses to swallow.

" Where, from irritability of the stomach, or other cause (such as idiosyncrasy, etc.), the patient cannot take the medicine by the stomach."

The following Table is slightly modified from Hunter's:—

The injection of medicines into the cellular tissue beneath the skin may be made

Locally with
   1. Caustics, for nævi, aneurisms, &c.
   2. Anodynes, for local neuralgia.

Generally..

In cerebral....
Insomnia.
Melancholia.
Mania.
Delirium tremens.
Hysteria.
Chorea.
Central neuralgia.

Spinal......
Rheumatism.
Tetanus.
Hydrophobia.
Retention of urine.
Colic.
Convulsions.
Epilepsy.

and
Sympathetic nerve-cases, inflammatory affections, &c...
Peritonitis.
Pericarditis.
Dysentery.
Ophthalmic surgery.
Malarial fever.
Gangrene.
Dysmenorrhœa.
Cholera.
Sea-sickness.
Cancer.
Ulcer of stomach.
Intussusception.

As antidote, diagnostic.....
For opium, belladonna, &c.

Anæsthetic....
With or after chloroform or ether in operations.

The instrument most approved is a small glass syringe, holding about half a fluidrachm, and graduated for drops or minims, with a tube for puncture, of tempered steel, or of silver with a gold point. The end of the tube must be small and sharp. Graduation of the cylinder is not necessary, as it is easy to measure the amount to be taken up by it. Not much pain is usually produced; but sometimes it is quite severe. If the dose of the medicinal agent be not too large, the only danger (unless in an erysipelatous patient) is of a circumscribed inflammation. Repeated injections should not be made at exactly the same spot. In operating, draw the skin tense with the fore-finger and thumb of the left hand, and pass the point of the tube quickly and steadily through it. Then push in, not rapidly, the desired amount of the fluid. Avoid subcutaneous veins; the puncture of one of them may give an excessive action of the medicine.

The agents most used are salts of morphia, atropia, strychnia, and quinia. For anodyne purposes, Dr. Ruppaner prefers *liquor opii compositus*, of which one hundred drops are equal to a grain of sulphate of morphia. Many use the ordinary solution of morphia (gr. j of morph. sulph. in f ℥j) or Magendie's (gr. ij in f ℥j). Doses are as follows :—

Sulphate of morphia	.	.	.	.	gr. $\frac{1}{8}$—$\frac{1}{2}$
Sulphate of atropia .	.	.	.	.	gr. $\frac{1}{80}$—$\frac{1}{30}$
Muriate of strychnia	.	.	.	.	gr. $\frac{1}{24}$—$\frac{1}{8}$
Aconitia .	.	.	.	.	gr. $\frac{1}{30}$
Liq. opii compos.	.	.	.	.	gtt. v—ix
Sulphate of quinia .	.	.	.	.	gr. i—iv

Among the diseases in which palliation or relief of suffering is often important by means of this method of treatment are especially neuralgia, hysteria, cancer and ulcer of the stomach. A case of the latter affection is recorded in which for weeks or months the patient was only able to retain food upon the stomach after disposition to vomit had been allayed by a hypodermic injection of morphia.

Curative effect from anodynes so employed has been asserted in cases of delirium tremens, mania, and tetanus; from quinine (two to four grain doses) in intermittent fever.

Tentative use of the same mode of practice is justifiable in cholera, hydrophobia, poisoning (as the injection of morphia for belladonna poisoning, and the converse), violent hooping-cough (atropia), per-nicious fever, spotted fever, &c.

That the operation is always without inconvenience to the patient is not true. Not only pain, but local inflammation and even suppuration may sometimes be induced. But many patients, suffering painful complaints, have had a hundred or more injections made in different parts of the body, without any disadvantage, and with great relief.

GENERAL CONCLUSIONS.

The following may, in recapitulation, be stated to be the *most general* desiderata in the management of all diseases :—

Rest;
  a, in all *acute* diseases ;
  b, in all cases of *exhaustion.*

Balance;
  a, of the *fluids*, and solids ;
  b, of the activity of *functions.*

Normal blood-change ;
  a, due removal of *excretions ;*
  b, absence of morbid *poisons.*

Support;
  a, in all *asthenic* cases ;
  b, in *later stages* of sthenic cases.

THERAPEUTIC MAXIMS.

1. All **pathology** is but the **physiology** of organic **perturbations.**
2. Never **interfere** actively in disease without a distinct **object.**
3. **Act** only upon *scientific* **reason**, or *well-defined* **experience.**
4. Treat the **cause** of disease *whenever* it is **possible.**
5. **Watch** *always*, and treat, *when* **requisite**, the condition of the *patient.*
6. **Avoid**, especially, **routine** treatment according to the *names* of diseases.
7. Use *no* **violence** with **self-limited** diseases.

I believe that a sound "theory of medicine" may be expressed in a single paragraph, thus :—

Vital **optimism** is the aggregate tendency of all the forces of the living organism, under the controlling influence of *life-force.* But, the *best possible* result in a given case may, from its **conditions** *and circumstances*, fall far **short** of *health.* Medicine, then, is to **favor** or **supply** *those conditions which, under natural laws, allow* or *promote* the **best result.**

In aiming to fulfil this duty, the art of healing must *always* depend, *in part*, upon empirical **observation** (which **every** branch of knowledge requires) and *in part* upon inductive **science.** But in both alike, the physician is, or should be, "*naturæ minister et interpres.*"

SECTION IV.

NOSOLOGY.

DISEASES were by Cullen classified as **locales, pyroses, cachexiæ, and neuroses** (local diseases, fevers, cachectic diseases, and nervous affections). The advances made in pathology since Cullen's time require some *modification* of this arrangement, while retaining its principle. I propose, therefore, that diseases be subdivided as—

Phlegmasiæ: inflammations.
Zymoses: zymotic diseases.
Cachexiæ: cachectic affections.
Neuroses: Nervous disorders.
Ataxiæ: *unclassifiable* diseases.

The following list is intended to present only the *most important* diseases of each class.

Phlegmasiæ:

Laryngitis;	Gastritis;
Tracheitis;	Enteritis;
Bronchitis;	Dysentery;
Pneumonia;	Peritonitis;
Pleurisy;	Hepatitis;
Endocarditis;	Nephritis;
Pericarditis;	Meningitis;
Stomatitis;	Cerebritis;
Pharyngitis;	Myelitis.

Zymoses (diseases produced by a *morbid poison*):—
Not usually included as zymotic diseases—
1. Primary syphilis;   3. Hydrophobia.
2. Gonorrhœa;

Eruptive—contagious—
1. Variola;   4. Varicella;
2. Varioloid;   5. Scarlatina;
3. Vaccinia;   6. Rubeola.

Contagious—not eruptive—
1. Parotitis contagiosa (mumps);
2. Pertussis.

Generally epidemic or endemic—
1. Typhoid fever;   6. Plague;
2. Typhus;   7. Cholera;
3. Spotted fever;   8. Epidemic dysentery;
4. Puerperal fever;   9. Influenza;
5. Erysipelas;   10. Diphtheria.

Endemic—
  1. Yellow fever;   2. Dengue.

"Malarial;" endemic—
  3. Intermittent;   Pernicious fever.
    Remittent;

Cachexiæ :

1. **Diatheses** (general cachexiæ) :
Always chronic—
    a. Spanæmia (anæmia) ;
    b. Chlorosis ;
    c. Leucocythæmia ;
    d. Melanæmia ;
    e. General dropsy ;
    f. Hemorrhagic diathesis ;
    g. Tuberculosis ;
    h. Diabetes ;
    i. Lithiasis ;
    j. Secondary syphilis.

'Acute or subacute—
    a. Scurvy ;
    b. Gout ;
    c. Rheumatism ;
    d. Ichoræmia (pyæmia).

2. **Local** cachexiæ (degenerations) :
    Cancer ;
    Various tumors, cysts, etc. ;
    Goitre ;
    Cirrhosis (of the liver) ;
    Bright's disease (of the kidney) ;
    Addison's disease (of the supra-renal capsules, etc.) ;
    Other organic degenerations.

Skin diseases ; viz.—
    Exanthemata (erythema, urticaria, roseola) ;
    Papulæ (lichen, etc.) ;
    Vesiculæ (eczema, etc.) ;
    Bullæ (pemphigus, etc.) ;
    Pustulæ (impetigo, etc.) ;
    Squamæ (psoriasis, etc.) ;
    Maculæ (fuscedo, etc.) ;
    Hypertrophiæ (elephantiasis, etc.) ;
    Tubercula (molluscum, etc.) ;
    Hæmorrhagiæ (purpura) ;
    Neuroses ;
    Parasiticæ (scabies, etc.) ;
    Syphilida.

Neuroses:

Apoplexy;
Paralysis;
Epilepsy;
Catalepsy;
Hysteria;
Chorea;
Tetanus;
Asthma;
Augina Pectoris;

Laryngismus stridulus;
Convulsions;
Neuralgia;
Delirium tremens;
Insanity; viz.—
 Mania;
 Monomania;
 Melancholia;
 Dementia.

Ataxiæ (unclassifiable diseases):

Hemorrhages;
Local dropsies (ascites, etc.);
Jaundice;
Dyspepsia;

Cholera morbus;
Colic;
Diarrhœa;
Worms.

PART II.

SPECIAL PATHOLOGY AND PRACTICE OF MEDICINE.[1]

HAVING endeavored, in our previous pages, to state, with brevity, the essential *principles* of the science of medicine, we proceed to apply these, in the same condensed manner, to an account of the diseases to be dealt with in *practice*. With the view of simplifying to the utmost our review of the subject for the use of the student (who will find in larger treatises the full discussion of all mooted points), but little reference will be made to authorities, and but slight mention of varieties of opinion. Our purpose will be, to give a brief and clear description of each disease, with its causation, diagnosis, and treatment, according to the light of experience and authority.

The classification of diseases followed in the succeeding pages is chiefly clinical; though based upon the pathological nosology already stated (Part I. Sect. IV.). Such an arrangement finds sufficient justification in its convenience.

AFFECTIONS OF THE RESPIRATORY ORGANS.

PNEUMONIA.

Definition.—Inflammation of the substance of the lung.

Varieties.—According to its *seat;*—single, double, lobular. According to *causation;* idiopathic, from cold and wet; traumatic, from injury; tuberculous, in phthisis; and typhoid pneumonia. Except in phthisis, we scarcely meet with chronic pneumonia; what is commonly *called* so being induration *following* acute pneumonia as an effect, not a continuation of it.

Symptoms and Course.—A chill or stage of depression, followed soon by fever, with oppression in breathing, dull pain (not always present) in the chest, and sometimes short cough. Delirium is common. Temperature of the body high, especially on the 4th or 5th day; sometimes in the evening, reaching 104° or 105° Fuhr. in the axilla. Secretions scant, as in other febrile states. Urine containing an excess of urea, but deficient especially in the chlorides, in the

[1] NOTE to Part II.—The letter F., followed by a *number*, in parenthesis, indicates a reference to a formula, of that number, in the latter part of the book.

11

middle period of the attack. Expectoration commences about the third day usually, the sputa being composed of mucus, lymph, and blood mixed together, making the *rusty* sputum of pneumonia. In this an excess of chloride of sodium will be found by testing with nitrate of silver.

The height of the attack is generally reached between the 5th and the 7th day; after which the temperature declines, and, in favorable cases, all the symptoms subside. In others, oppression in breathing, and prostration increase; cough deepens, and expectoration becomes more abundant, at last purulent. Death seldom occurs before the sixth, and may be as late as the twentieth day.

Stages.—1st, that of congestion or engorgement, and the commencement of exudation; 2d, that of exudation and red hepatization; 3d, that of gray hepatization, softening, or purulent infiltration.

Physical Signs.—These differ in the three stages. In the first, they are, moderate dulness of resonance on percussion over the affected lung, and, on auscultation, after the first day or two, the *fine crepitant* râle.

In the second stage, decided dulness on percussion, no râle, but, instead, *bronchial* respiration and bronchophony; with increased vocal fremitus. In the stage of softening or suppurative infiltration (gray hepatization), dulness on percussion, and coarse crepitant or mucous râle.

When *resolution* follows the second stage, as in most cases of recovery, the bronchial respiration gives way to returning fine crepitation (crepitus redux); and, then, the dulness of resonance on percussion also gradually disappears.

Terminations.—Resolution; death in the second stage from asphyxia; death from exhaustion in the third stage; recovery after the third stage (uncommon); abscess; gangrene of the lung.

Complications.—Pleurisy (pleuro-pneumonia); capillary bronchitis; tubercle.

Sequelæ.—The most frequent is that persistent consolidation of the lung called by some chronic pneumonia. Tubercular deposit, sometimes even acute phthisis, may follow pneumonia, in persons predisposed to it.

Morbid Anatomy.—The *lower* or *middle* lobe is almost always the seat of the disease. Should death take place (as it rarely does) in the first stage, the lung would be found somewhat swollen, dark red, inelastic (splenization), and filled with blood or bloody serum. It will still float in water, though heavier than healthy lung. It is easily torn.

In the second stage, of hepatization, the lung is no longer spongy, but presents considerable resemblance to the liver; although a finger may be easily thrust through it. When entirely hepatized, it will not float in water, the air being displaced from the cells by the exudation of coagulable lymph.

The third stage consists in the degeneration (in the absence of more favorable *resolution* by absorption) of the exudation. This occurs by *granulation, softening,* and *suppuration.* Mostly the latter is infiltrated; occasionally an abscess forms. In gray hepatization, the lung is solid, impermeable to air, with a granite-like appearance of red and white points on section. It sinks in water, but is more easily torn, or crushed into a pulp, than in the second stage.

Diagnosis.—The only affections with which pneumonia is likely to be confounded are pleurisy, bronchitis, and phthisis. In children, collapse of the lung has been mistaken for lobular pneumonia.

From pleurisy, it is known by the absence of the sharp pain belonging to the latter, and by the crepitant râle and rusty sputa. From bronchitis, by the dulness on percussion, râle, bronchial respiration, and bronchophony. From phthisis, by its sudden onset, fine crepitation, and sputa, as well as by the acute violence of the attack. Latent pneumonia sometimes complicates fevers, etc.

Prognosis.—Simple pneumonia, of one lung, in a young and previously healthy person, ought, under favorable circumstances and judicious treatment, always to be recovered from. In the aged, it is dangerous; and double pneumonia is so at all periods of life, though good recoveries do occur. It is double in about one case in eight.

Among the unfavorable signs—most of which are obvious—are expectoration of pure blood in the first stage, and albuminuria in the second.

Pathology and Nature.—Ordinary pneumonia is a phlegmasia, with the usual elements of general pyrexia or fever, local hyperæmia, and local exudation.† As in other phlegmasiæ, the relation of these to each other is not easily determinable. Is the local affection always the first thing, *causing* the fever, or is there a blood disease first, producing both the fever and the local affection? In traumatic and tuberculous pneumonia, it is plainly the former. In other cases, after exposure to cold and wet, we may suppose it to be both; but the primary step of the actual inflammation is probably the local disturbance in circulation, functional action, and nutrition.

Causation.—As already intimated, cold, suddenly or partially applied to the body, especially to the chest, is the most common cause of pneumonia. But the previous state of the health, and especially, also, latent tubercle, may predispose to it. So, in certain regions, does the influence of malaria.

Treatment.—This remains to be a *questio vexata*. Having considered already (General Therapeutics, p. 85), the principles involved, my conclusions may be briefly stated. I am convinced by experience that prompt and moderate antiphlogistic treatment may greatly lessen the danger of pneumonia, if not shorten its duration.

Probably five cases in six would recover without the abstraction of blood; the sixth might die for want of it. I believe that the mortality of pneumonia has increased in Philadelphia since blood-letting has been so generally abandoned.[1] But bleeding from the arm, if resorted to, should be done but *once;* not later than the *third* or *fourth* day; and it may be moderate in amount. Old persons and those of feeble system will neither bear nor require it.

Cupping between the shoulders may, in many cases, take the place

[1] Having no recollection or record in my notes of private or hospital practice, of ever having lost a case of *uncomplicated* pneumonia, the above language expresses my very strong convictions.

of venesection; in some, it may follow this. The early administration of a vigorous purgative, as Epsom salts, or citrate of magnesia, is proper, in the absence of any special contraindication.

Tartar emetic [F. 1][1] in the dose of one-eighth to one-quarter of a grain for an adult, every two or three hours, may be continued during the height of the febrile stage. For this, as for bleeding, the indications are to be found not in the physical signs of pneumonia, but in the general condition of the system; not in the crepitant rále, but in the hot skin, hard or else oppressed pulse, pain and dyspnœa, and more or less darkly flushed face. After the height of the attack, small doses of ipecacuanha [F. 5] may be substituted for the antimonial; or nitrate of potassa [F. 4], gr. x, every two hours.

Asthenic pneumonia requires a different treatment; and the same will apply to the third or suppurative stage of all cases. Support may be required, in a few cases, even from the first; by beef-tea, wine, or spirits (best with nourishment, as in punch), quinine [F. 2] or ammonia [F. 3]. In hospital, I have known more than one case to recover under this plan *alone;* but they are the exceptions. Some cases in which bleeding or cupping will be proper in the first stage, before the fourth day, may require beef-tea in the second stage, and quinine later. A large blister over the affected part is generally useful about the fifth, sixth, or seventh day of the attack.

Varieties of Pneumonia.—When complicated with pleurisy or bronchitis, no important modification of treatment is called for. *Tuberculous* pneumonia requires careful husbanding of the resources of the economy. Loss of blood is there rarely proper; if at all, it must be local only, and in minimum quantity. The necessity for the analeptic treatment of the tuberculous condition is paramount. Dry cups, blisters, and counter-irritant plasters, or croton oil or tartar emetic ointment externally applied, are then suitable. Warm poultices, as of Indian or flaxseed meal, with or without the addition of mustard, kept on day and night for a time, are often very useful, especially in children.

Traumatic pneumonia, following an injury, is not common. It calls for no particular difference of treatment.

Typhoid pneumonia is a term not always uniformly applied. It means, sometimes, or with some authors, inflammation of the lungs complicating typhoid fever; others include under it all cases of asthenic pneumonia. More generally, however, it designates that form of the disease in which epidemic or endemic influence has impressed a peculiar character. Malarial regions especially exhibit this, in the " winter fever" or typhoid pneumonia of our southern states. Early and great debility, out of proportion to the local symptoms, with a tendency to low delirium, and to remittance, mark this disorder. In treatment, it bears little or no depletion, hardly the reduction of excitement by tartar emetic or veratrum viride. Diaphoretics first, as [F. 6] ipecac, ⅓ a grain, with the same amount of calomel and five or ten grains of nitrate of potassa, every three hours; or liquor ammoniæ acetatis [F. 7], or solution of acetate of potassa [F. 8]; then quinine, when the need of a tonic is apparent, which may be very

---

[1] F., with a number, in parenthesis, refers to a formula in the collection at the end of the book.

early; with strong liquid nourishment, and moderate counter-irritation; these are the measures usually proper in typhoid pneumonia. After recovery from an attack of inflammation, the lung will be for some time more than usually susceptible to conditional changes. Exposure should, then, be carefully avoided; and flannel may be worn next the skin. In winter a mild warming plaster, as of hemlock or Burgundy pitch, over the chest, will give good protection.

PLEURISY.

Definition.—Inflammation of the pleura.
Varieties.—Single or unilateral, and bilateral or double; idiopathic, traumatic, and secondary, e. g., tuberculous, cancerous. Chronic pleurisy, so called, is merely the consequence of an acute attack.

Symptoms and course.—Generally, after a chill or cold stage, sharp pain in the side, impeded and accelerated respiration, short, sharp cough, and fever. The pain centres in the infra-mammary or lower axillary region; it is often intense, and is increased by a long breath, by coughing, pressure, or lying on the affected side. The pain and fever lessen after effusion has occurred; but the dyspnœa may then be increased. It is, after that period, most comfortable to lie on the diseased side, so as to allow of free breathing by the other lung. Acute pleurisy is often recovered from without any considerable effusion. When the latter does occur, absorption mostly follows. If not, life is endangered by interference with respiration. At first serous, constituting one form of **hydrothorax**, the fluid may become purulent; this is **empyema**. The term *false* empyema is sometimes given to a collection of pus in the pleural cavity from the rupture of an abscess in the lung. **Pneumothorax** is the accumulation of air in the cavity of the pleura; **hydro-pneumothorax**, of water and air together. Both of these are most common in tuberculous pleurisy, i. e. in the course of a case of pulmonary phthisis.

Stages.—In severe pleuritis there may be, 1, the *adhesive;* 2, the effusive; 3, the suppurative stage. In more favorable cases the 3d stage is that of absorption.

Physical signs.—Of the 1st stage, deficient elevation of the ribs in breathing, feeble respiratory murmur on the affected side, and *friction* sound. 2d stage, dulness of resonance on percussion, bronchial respiration, bronchophony, sometimes *ægophony.* When the effusion becomes very copious, bulging of the side occurs, suppression of respiratory sound and of vocal resonance and vibration, with exaggerated or puerile respiration on the sound side. Displacement of the heart may take place if it is on the left side; of the liver if on the right. There is no physical sign by which empyema can be distinguished from serous effusion; but irritative fever usually accompanies empyema.

Absorption following extensive effusion allows *retraction* and *depression* of the chest on that side, from the slow or imperfect expansion of the lung. Then return, first, bronchial respiration and voice, or ægophony, then gradually the normal respiratory murmur. Sometimes, from adhesions of false membrane over the lung, permanent depression of the thorax on that side is left.

11*

During effusion, its fluid character as well as extent may be shown by percussion in different positions. Sitting up, it falls forward, and rises to a higher line in front; lying on the back, the dulness, from gravitation, may fall much lower in the anterior region. Sometimes adhesions prevent this. *Succussion*, or sudden shaking of the chest of the patient, may produce an audible splashing, if the ear be upon or near the affected side. By ocular inspection and measurement, the changes in the amount of the effusion may be estimated from time to time.

Terminations.—Pleurisy may sometimes be "nipped in the bud" at an early stage by appropriate treatment; that is, prompt resolution of the incipient inflammation may be effected. The other terminations · are, serous effusion, which may vary from an ounce or two to quarts, gradually absorbed altogether; the same, slowly and incompletely absorbed, leaving collapsed lung; death, in double pleurisy, by asphyxia from excess of fluid; and empyema, often, but not always fatal by slow exhaustion.

Complications.—Pneumonia, tubercular deposit, inflammation of the liver (bilious pleurisy).

Sequelæ.—What authors call *chronic pleurisy* is the sequela of acute pleuritis. Its results and terminations have been above named.

Morbid Anatomy.—In the early period, general redness and vascular injection of the pleura, with bands of whitish and more or less translucent or opaque coagulable lymph, causing adhesions of the pulmonary and costal pleura. Later, serous, sanguinolent or purulent effusion, in variable quantity, and sometimes displacement of the heart, lungs, and liver, and bulging of the ribs and intercostal spaces.

Diagnosis.—From pneumonia, pleurisy is known in the height of the acute attack by the sharpness of the pain, the friction sound, and absence of crepitant râle and of dulness on percussion. After effusion, especially by the change of the line of dulness with change of position, sitting and recumbent; by the bulging; and the degree of diminution of vibration of the walls of the chest when speaking.

From intercostal neuralgia, pleurisy is distinguished by the absence of fever and friction sounds in the former, and the non-increase of the neuralgic pain upon inspiration. Congestion, in some rare cases, attends neuralgia; the diagnosis is then more difficult. In intercostal muscular-rheumatism, there is slight increase of pain in breathing deeply, but as much in moving the arms; and the pain is much less acute, and generally without fever.

Prognosis.—Pleurisy is rarely fatal; though death may occur, from very abundant effusion in bilateral pleuritis, or, with empyema in the unilateral, through gradual exhaustion.

Causation.—Exposure to cold and damp is the ordinary exciting cause of "idiopathic" pleurisy. Fracture of the rib, punctured wounds, &c., may cause traumatic pleurisy. In the course of phthisis, it not uncommonly occurs by extension of the disease from the lung. Cancer of the chest may produce it in an analogous manner.

Treatment.—In young and vigorous persons, still more confidence may be placed in early antiphlogistic treatment than in pneumonia. When high fever and constant severe pain occur, bleeding, in such patients, on the first, second, or third day, should be the general rule. Leeches or cups may follow, or be used instead of venesection in doubtful cases. Tartar emetic, after a free purge, may be given, ⅛ to ¼ of a grain every two or three hours, with ⅓ to 1 grain of *opium*. Many practitioners add calomel, ½ grain to 1 or 2 grains every two or three hours [F. 9.] When fever subsides, or vomiting occurs, the antimony should be withdrawn; the opium and calomel may be continued, while the pain lasts,—carefully avoiding over-narcotism by the former, and salivation by the latter.

As soon as the heat of skin has considerably abated, if the pain continues, a large blister should be applied over the affected part. For the effusion, diuretics, as squill [F. 10], juniper berry infusion, or compound spirit [F. 11], acetate or bitartrate of potassa [F. 12], etc., may be used. Iodine, in Lugol's solution, and iodide of potassium alone, are often advised. Repeated blistering sometimes has excellent effect.

When life seems to be threatened by exhaustion from dyspnœa, owing to large effusion not becoming absorbed, *paracentesis*, or puncture of the chest, is proper. Dr. Bowditch's plan is the best for this. He uses Dr. Wyman's apparatus, which is a trocar, with a silver canula having a stopcock, and capable of being connected with a syringe by an intermediate piece, also having a stopcock, both cocks acting the same way. The operation is performed while the patient is sitting up, if able, or lying over the edge of the bed. The puncture is made somewhere between the seventh and tenth ribs, just behind their angles; making sure first of the position of the liver and spleen, so as to avoid them. Insert the instrument near the upper edge of the lower rib, raising its point as it goes in. When the trocar is withdrawn, by the double-cocked syringe the fluid may be *gradually* removed with safety to the slowly expanding lung. The operation may need to be repeated, even several times in the same patient.

The grooved needle or exploring trocar is often used to ascertain the nature of the contents of the chest. When pus is present (empyema) in considerable amount, "drainage" may be resorted to. Following the grooved needle or trocar, a fine long iron probe, somewhat bent, is passed through to the lower and back part of the pleural cavity, against the intercostal space. Being felt there, an incision is made upon it; a strong silk thread is passed through its eye, and then drawn through the first opening. After this, draw in a drainage tube, of India rubber perforated with many holes; both of whose ends hang out and are tied together. Sometimes, by the tube, the cavity may be *washed out*, with water, or dilute astringent or antiseptic solutions.

In chronic cases of pleuritic effusion or empyema, the strength of the patient requires usually to be supported by good diet, and sometimes by tonics. This, in empyema, is often the most important part of the treatment.

ABSCESS OF THE LUNG.

In rare instances, inflammation of the lung, active or *latent*, may terminate in abscess. Before rupture, dulness on percussion, bronchial respiration, and dyspnœa proportioned to the size of the abscess, are present. When an opening occurs allowing the matter to escape into the bronchial tubes, the rather sudden commencement of purulent expectoration should attract attention. Then the physical signs of a *cavity* are discoverable by percussion and auscultation; amphoric or tympanitic resonance on percussion, cavernous respiration, metallic tinkling, etc., varying with circumstances. As is the case with pleuritic empyema, pulmonary abscess may communicate externally by a spontaneous opening.

The principal importance of abscess of the lung consists in the possibility of mistaking it for phthisis. The points of difference will be alluded to in connection with that disease.

PULMONARY GANGRENE.

This may occur in pneumonia from extreme violence of the inflammation, or from a depressed state of the system; also, from cancer within the chest, pyæmia, etc. It is rare, but more common than circumscribed abscess of the lung. Unless very narrowly limited, pulmonary gangrene is always fatal. Its signs are, coarse mucous râle, taking the place of the vesicular murmur in the lower part of the lung, with copious brownish and offensively fetid expectoration, dyspnœa, and great prostration.

In bronchitis, occasionally, temporary fetor of the expectoration ' and breath may simulate gangrene; but transiently, and without the above symptoms.

The treatment of pulmonary gangrene must be, of course, supporting and antiseptic. Alcoholic stimulants, rather freely given, will be proper, with concentrated liquid food, as beef-tea. Sulphite of soda (ten grains in solution every three hours) may be tried; or chlorine water, a teaspoonful or two every two or three hours.

EMPHYSEMA OF THE LUNG.

This is dilatation of the pulmonary air-cells of one or both lungs. It may accompany prolonged asthma, or may follow chronic bronchitis. Sometimes it aids in producing dilatation of the heart. Its symptoms are, dyspnœa, and, when extensive, blueness of the lips, cyanosis, from interference with the circulation through the lungs; in many cases wheezing respiration. The physical signs are, bulging of the chest, increased clearness of resonance on percussion, and feeble inspiratory murmur with prolonged expiratory sound; sometimes displacement of the heart or liver. It is most easily mistaken for pneumothorax. But, in the *latter*, the resonance on percussion is more tympanitic, the inspiratory murmur still feebler, or quite absent, and there is no prolonged expiratory sound; besides which, the *concomitants* of pneumothorax usually serve to distinguish it.

COLLAPSE OF THE LUNG.

In hooping-cough or in severe bronchitis, especially in children, obstruction of a considerable air-tube may lead to an exhaustion of air from the cells supplied by it, and a return of that portion of the lung to the unexpanded condition (atelectasis) of fœtal life. The same state may occur in other conditions, from debility. It was formerly always mistaken for lobular pneumonia. It is usually fatal, unless very much limited.

Signs of it (often difficult of determination, however) are, moderate dulness on percussion, with absence of the murmur of respiration; and, in some cases, an inward motion or recession of the lower ribs during the effort at inspiration.

BRONCHITIS.

Definition.—Inflammation of the mucous membrane of the bronchial tubes.

Varieties.—Acute and chronic; general, capillary; plastic, rheumatic and syphilitic bronchitis.

Symptoms and course.—Systemic depression, followed by fever; tightness and soreness of the upper and anterior part of the chest; cough, at first short, dry and tight, later deeper and looser, with expectoration; the latter being at first mucous, in rare instances pseudo-membranous, in severe cases at a late stage purulent.

Capillary bronchitis is marked by greater dyspnœa and tendency to early depression and prostration.

Chronic bronchitis often is free from febrile symptoms, the cough and expectoration, with slight dyspnœa, characterizing it.

Stages.—Ordinary bronchitis may be divided in its progress into 1st, the stage of *diminution* of secretion; and 2d, that of *increase* and *perversion* of it.

Physical signs.—No dulness on percussion, except in case of collapse of part of a lung from obstruction; sonorous rhonchus and sibilus, generally, though not quite always, on both sides of the chest; varying from time to time, in seat, character, and loudness. In capillary bronchitis, extended mucous, crepitant or subcrepitant râles, closely resembling the fine crepitation of pneumonia.

Terminations.—Acute bronchitis may end in death from apnœa, in the first or second stage; or in chronic bronchitis; but most generally in recovery.

Complications.—Asthma; pneumonia; bronchial dilatation; pulmonary collapse.

Morbid anatomy.—General redness and congestive tumefaction of the bronchial membrane; with more or less obstruction from mucus, epithelium, and, rarely, casts of the tubes, of plastic lymph.

Diagnosis.—No difficulty exists except in distinguishing chronic bronchitis from phthisis. Absence of dulness on percussion and of the signs of excavation, are most important; the expectoration also is whiter and of less weight in bronchitis; and there is no hectic fever.

Prognosis.—Acute bronchitis is dangerous in old persons and young children; seldom fatal in vigorous middle life. The capillary form is always most serious; death taking place sometimes from the 10th to the 12th day. Chronic bronchitis is not often fatal, even by exhaustion; but it may last an indefinite time, even many months.

Causation.—Exposure to cold is the most frequent cause. In some employments, as needle-grinding, cotton-spinning, &c., solid particles inhaled cause bronchitis by mechanical irritation. Transference of rheumatism occasionally induces it in the rheumatic diathesis; and occasionally it is one of the manifestations of tertiary syphilis.

Treatment.—Abortive treatment of a "cold on the chest" may sometimes be effected within the first twenty-four hours, by taking, at bedtime, a glass of hot lemonade or ten grains of Dover's powder after a warm mustard foot-bath. Should this fail or be omitted, a brisk saline purgative should be given, of Epsom or Rochelle salts, or citrate of magnesia. Then, when the fever is high, cough very tight, and breast sore, tartar emetic should be advised, ⅛ to ¼ grain every two or three hours, [F. 1] with frequent draughts of flaxseed tea or some similar demulcent. A large sinapism over the upper sternal region will aid in giving relief; and so will friction with oil of turpentine.

In milder cases, or where the strength of the stomach is doubtful, syrup of ipecacuanha, ¼ to ½ drachm every two or three hours, will answer; and should be continued until the cough softens and the breathing becomes easier. Then syrup of squills [or, F. 13] may follow, in f. drachm doses, every three or four hours. When the cough is troublesome at night, ½ to 1 f. drachm of paregoric [F. 15] may be added at bedtime; or through the day, occasionally, if coughing be very violent or frequent. Opiates do the most good, however, *after* some *loosening* of the cough with free expectoration. When the fever has abated, and especially if dyspnœa continue, a blister may be applied over the sternum.

In capillary bronchitis, or in the ordinary form in the aged and feeble, instead of tartar emetic the more stimulating expectorants may be required, as senega, in decoction or syrup, muriate [F. 15] or carbonate of ammonia, with quinine and beef-tea, wine-whey, or brandy-punch./ Inhalation of steam, alone, or from infusion of hops, sometimes soothes the air-tubes advantageously. (See *Inhalation* and *Atomization*, in Part I., Sec. III.)

Chronic bronchitis requires persevering use of counter-irritation over the chest, by croton oil (3 drops with the same of sweet oil applied nightly till a papular eruption follows), painting with tincture of iodine, or plasters of Burgundy pitch, hemlock, &c., and alternation of stimulating and alterative expectorants, and tonics. Besides squill and senega, ammoniacum, copaiba [F. 17], and muriate of ammonia [F. 16] are most frequently useful. If the system be below par, quinine, iron, and cod-liver oil are important. When secretion is very copious, inhalation of tar-vapor or of creasote should be tried. The former may be used by putting an ounce or two of tar in a cup over boiling water; so as to diffuse the tar-vapor through the chamber. Creasote, 20 or 30 drops, may be put into half a pint of boiling water, to be breathed by means of an ordinary inhaler. When medicine fails, change of air will sometimes entirely cure.

ASTHMA.

Definition.—Paroxysmal and spasmodic dyspnœa.

Varieties.—*Idiopathic* and *symptomatic; dyspeptic* asthma; *hay* asthma.

Symptoms and Course.—Every night, or once a week, month, or year, or at irregular intervals, the attack comes on. Most frequently it is between 1 and 3 o'clock in the morning. Premonitory symptoms often are great drowsiness, or wakefulness, headache, flatulence, itching under the chin. Dyspnœa is then the characteristic symptom. The sufferer sits or stands up, leaning forward, and labors to breathe. The chest is expanded to its utmost, by the accessory as well as principal inspiratory muscles. The countenance is anxious, with pallor, coldness, and in severe cases lividness, of the face and hands. Perspiration is often copious. A wheezing sound accompanies respiration; giving way finally, with relief, upon the expectoration of mucus, usually rather thick, and in pellets.

The attack may pass over in a few minutes, or may last for hours, or, with some remission, days or weeks. Where asthmatic symptoms are persistent, as is not very uncommon, for years, some structural change in the organs of the chest exists; it is then *symptomatic* asthma.

Physical Signs.—Inspection shows unusual elevation of the ribs and shoulders. Placing the ear on the chest, sonorous and sibilant sounds, loud but mostly small in calibre, are found to take the place of the respiratory murmur. These sounds change their locality frequently. As the attack gives way, with expectoration, some mucous râle is heard.

Secretions.—At the beginning of the paroxysm, the urine is abundant and pellucid ("nervous urine"); for some hours after it has ceased, it is more scanty, and is deficient in urea and chloride of sodium.

Complications.—Bronchitis; pulmonary emphysema; dilatation or hypertrophy of the heart.

Diagnosis.—Laryngeal spasm may, without care, be confounded with asthma; but the modification or arrest of the voice ought to distinguish the former. Violent bronchitis is known from it by the febrile condition. Angina pectoris, by the extreme pain, and localization of the distress about the heart. Hydrothorax, by the dulness of resonance on percussion, and absence of rhonchus.

Special exploration is necessary in each case to determine the presence or absence of pulmonary or cardiac complication.

Prognosis.—Death almost never occurs during the fit of asthma. Those subject to it often live to old age. But dilatation of the pulmonary air-cells, and enlargement of the heart, may follow in protracted cases, breaking down the health.

Pathology and Nature.—It has been made certain that asthmatic dyspnœa is owing to a spasmodic constriction of the smaller bronchial tubes, by tonic contraction, mostly reflex, of their involuntary muscular fibres.

Causation.—Asthma is hereditary in a majority of cases. Males have it more often than females. It may occur at any age. Dr. Salter classifies cases according to their causation—1. by agencies acting upon the lungs, as fog, smoke, fumes of various things, as ipecacuanha, mustard, new hay, &c.; 2. by reflex action, as irritation of the stomach from indigestion, loaded rectum, sudden emotion; 3. by pulmonary or cardiac disease. Behind all these there must be a predisposing peculiarity of constitution.

Treatment.—During the attack, our aim must be to give relief, by relaxing spasm. Ipecacuanha wine, with tincture of lobelia, one-quarter to one-half fluidrachm of each [F. 18], every half hour until nausea or expectoration is produced, I have known often to act very well. Hoffmann's anodyne, in one-half drachm or drachm doses, will sometimes do great good. Some practitioners advise hyoscyamus, musk [F. 19], and hydrocyanic acid [F. 20]. Smoking tobacco is relieving in some instances; smoking cigarettes of stramonium leaves in others. More still find comfort in breathing the air in which are burned papers which have been soaked in a saturated solution of nitrate of potassa. Inhalation of ether or nitrous oxide may be carefully used in extreme cases. As an adjuvant, the warm mustard foot-bath may be employed; as well as sinapisms or dry cupping between the shoulders.

Between the attacks, endeavor should be made to rectify digestion and its tributary processes, and to invigorate the nervous system. Some cases will require blue pill, nitro-muriatic acid [F. 21], or taraxacum, bitter tonics and mild laxatives, such as rhubarb, etc. Others need iron and quinine. Iodide of potassium is highly recommended by some; conium, cannabis indica, and arsenic in small doses by others. There is reason for giving trial to the bromide of potassium in obstinate cases; most patients will bear from 10 to 20 grains of this twice or thrice daily for weeks together without inconvenience [F. 22].

Prophylaxis.—No disease is more curiously capricious in its causation than asthma. Some always have a paroxysm if they visit the sea-shore; others are more secure there than elsewhere. One cannot sleep on the first floor; another does better there than higher up. Each must learn his own peculiarities, and be governed thereby.

Most remarkable are the annual attacks of asthma, or asthmatic bronchitis, to which a few individuals are subject. I know one gentleman who for many years was obliged to arrange all his business for such an attack, which was punctual almost to a day, in the summer, and confined him to his room for a week or two. He escaped the paroxysm only by going to the sea-shore before the expected time, and remaining there through the time during which it would have lasted.

In asthmatic persons generally, nothing is more important than prudence and regularity in diet and regimen.

BRONCHIAL DILATATION.

This, of which extreme degrees are not common, is of interest chiefly because it is possible for it to be mistaken for phthisis. There are two forms; the tubular and the saccular enlargement.

In either, slight dulness on percussion may occur, from condensation of the lung around the expanded part. Sonorous rhonchus and coarse mucous râle exist, the latter especially in the saccular form. In this, the signs are almost identical with those of tubercular excavation; but they occur usually at the middle or lower part of the lung, and are stationary, as they are *not* in tuberculization.

Cough, very troublesome, and attended by copious mucous or slightly purulent expectoration, is common in bronchial dilatation. The palliation of this symptom, with care of the general condition of the patient, is all that can be accomplished for it in treatment.

LARYNGITIS.

Slight inflammation or congestion of the mucous membrane of the larynx is very common as the result of cold ; its signs being hoarseness, with a dry, short, harsh cough, and some soreness in drawing a breath. But simple acute laryngitis of severe grade is quite a rare affection.

When it occurs, there is fever, with hoarseness, "brassy" cough, distressing dyspnœa, and difficulty of swallowing. *Œdema glottidis*, or submucous effusion of serum, constitutes the greatest danger in laryngitis ; the swelling obstructing respiration to a degree often fatal. This disorder is almost exclusively met with in adults.

Early purging, the application of leeches, the internal use of ipecac, in doses just short of nausea, with moderate quantities of opium, and the frequent inhalation of the steam of boiling water, constitute the best treatment. If dyspnœa become decidedly serious, threatening asphyxia, tracheotomy is advised. Some account of this operation will be given in connection with Croup.

Œdema of the glottis may be produced immediately by the ingestion of boiling water, or of sulphuric or nitric acid, which has often accidentally happened.

Chronic laryngitis, with ulceration, is a not infrequent attendant of phthisis. Some cases of the latter begin with it; in others it occurs somewhat late in the course of the disease. Syphilitic ulceration of the larynx is tolerably common, as a secondary symptom. This, as well as polypi or other tumors of the larynx, may be discovered, and treated by operation for removal, or with solutions of nitrate of silver, &c., with the aid of the laryngoscope.

My confidence in the utility of very strong solutions of nitrate of silver in chronic inflammations of the mucous membranes, of the throat or elsewhere, has not increased, in fact has not been sustained, by what I have seen in practice. Dr. Horace Green and others make frequent use of it of the strength of sixty grains to the ounce. Except for *ulceration*, which may benefit even by the solid caustic, I believe that from four to ten grains in the ounce of water will do more good, in almost all cases, than the stronger proportions.

The application of *nebulized* liquids, by apparatus for *atomization*, is now much in vogue in both acute and chronic laryngitis. Some remarks upon this have been made already, under *general therapeutics* (Part I.).

APHONIA.

Loss of voice may be transient or permanent; and either functional or structural in its origin. Especially in hysterical females, a nervous shock may produce a *paresis* or enfeeblement of the vocal power, lasting often for days together. I knew of one case in which a young woman could only speak in a whisper for more than three months. A

12

choreic affection of the vocal apparatus is now and then met with ; stammering is, in fact, analogous to this ; depending on a want of command and co-ordination of the vocal muscles.

Faradization, *i. e.* the use of induced electrical currents (as magneto-electricity), carefully applied, has sometimes cured nervous or hysterical aphonia. I have known vesication of the back of the neck to be useful for it.

Congenital dumbness, except in idiots, is due to deafness, making the learning of speech impossible, unless by a recently invented system of instruction by sight.

Organic or structural aphonia is caused by lesions of the larynx, such as ulcerative destruction of the vocal cords, tumors, etc., which are to be diagnosticated by laryngoscopy.

Feigned dumbness is detectable by careful watching, or, in the last resort, by *etherization*. In the stage of early excitement, or when reviving from anæsthesia, the pretender will betray himself by involuntary speech.

The term *dysphonia clericorum* has been applied to an affection of the throat not uncommon among clergymen and other public speakers, called by Dr. Horace Green "follicular disease of the pharyngo-laryngeal membrane." Its symptoms are soreness and irritation in the throat, with disposition to hawk and spit frequently, and hoarseness or partial loss of voice. On inspection, the fauces, pharynx, and glottis are found to be of a reddish granular appearance, with more or less enlargement of the mucous follicles, and, in severe cases, a mucó-purulent secretion about the uvula. Sometimes, however, the membrane is dry.

The conventional treatment for this affection is, the application every day or two of a solution of nitrate of silver, with a brush or probang. Saturated solution of tannin is also used for it. My belief is that, if these local remedies do not relieve in a week or two, the frequent swallowing of small pieces of ice, or gentle gargling several times a day with ice-water, may be substituted with advantage. Counter-irritation over the throat, especially by croton oil, should, if necessary, be persevered in for a considerable time. Three drops of the oil (diluted with as much sweet oil for a delicate skin) may be rubbed over a limited space in front of the throat every night until a papular eruption comes out.[1]

Many cases of this complaint are as much constitutional as local in origin. Where real dysphonia (difficulty or imperfection of vocalization) exists, public speaking or singing must be avoided, to allow the organs repose. Tonics and change of air may often prove the best measures of treatment.

APHASIA.

Loss of speech may occur as one of the symptoms of disease of the brain, either functional and transient, or organic and irremovable. Such a loss of *language* is termed *aphasia*. Importance has been

---

· [1] Patients should be cautioned, of course, against allowing the oil to come near the eyes. I have known a severe ophthalmia to result from neglect of this.

given to it lately by the observations of Trousseau and others, and resulting speculations (Dax, P. Broca) as to the seat of the faculty of speech. Not articulation, as in aphonia, but *expression* is, in this affection, wanting. The power to *write* words from memory, to convey meaning, is lost; but, in some cases, at least, they may be *copied* correctly. *Thinking without words* may go on in such instances; as Lordat recorded, after recovery, in his own case.

Hemiplegia of the right side has in a number of examples coincided with aphasia; and, several times, also, autopsy has shown softening or other lesion of the left anterior portion of the cerebrum. On the suggestion of these facts a hypothesis has been based, that the site of the faculty of language is in the anterior frontal convolutions of the left hemisphere of the cerebrum. This is a very *unphysiological* supposition, in view of the *symmetry* of the cerebro-spinal axis throughout; nor does this objection disappear even upon the conjecture that the "organ" upon the right side may exist always in an undeveloped state. Valvular lesion of the heart sometimes accompanies the disease.

Cases of aphasia are rare. I have never seen one,[1] and am not aware of any special measures of treatment for it pointed out as yet by experience.

LARYNGISMUS STRIDULUS.

This is an infantile affection, consisting in spasmodic closure of the glottis, causing a stridulous or shrill whistling respiration. It is most apt to occur during dentition. but is not very common. Its onset is sudden, and duration brief. Though exceedingly alarming, it is seldom fatal.

The treatment must be prompt;—producing derivation by slapping the back and limbs, and putting the feet into hot water, while cold water is applied to the head. In severe cases mustard plasters (diluted with flour) may be applied to the chest and back. Some advise the momentary inhalation of chloroform. When life is really in great danger from prolongation of the spasm, tracheotomy may be justifiable. Children who have laryngismus are generally anæmic; requiring iron [F. 23] and salt baths.

CROUP.

We understand by croup, an acute cynanche or angina, whose signs are, a hoarse cough,—difficult and audible respiration, and aphonia; the seat of the disorder being the upper portion of the air-passages. Its place in nosology has been empirically or conventionally (rather than systematically) established.

For brevity's sake, the following propositions may be advanced :—

1. The pathological elements of croup are, *a*, spasm; *b*, hyperæmia or congestion; *c*, inflammation, either ordinary or diphtheritic.

The spasm affects especially the muscles, whose action tends to

---

[1] An account of an interesting case has recently been sent to me by Dr Sigmund, of Lehigh County, Pennsylvania. in which aphasia, with right hemiplegia, followed an apoplectiform seizure (preceded for some months by severe pain in the head) ; the patient recovered entirely.

close the rima glottidis; but may involve also the muscular coat of the trachea itself.

The hyperæmia commences in the mucous membrane of the larynx or trachea, but often extends throughout the whole anterior cervical region.

The inflammation may be located in a small portion of the same mucous membrane, or, it may extend downwards indefinitely into the bronchial tubes.

2. We may mentally distinguish between cases in which the croupal dyspnœa results from simple spasm, from simple tumefaction, or from inflammation without any spasmodic constriction of the glottis. But in practice the pathognomonic cough and breathing rarely attend such an isolation of one of these conditions. A certain number of cases, however, occur, of purely spasmodic or nervous croup; now and then substituting more general convulsions; as when worms have been apparently an exciting cause. A purely inflammatory case is at least equally rare. In fatal pseudo-membranous cases, autopsic examination has repeatedly shown that the amount of false membrane was by no means sufficient, alone, to have occluded the larynx or trachea; the result being due to the additional *spasmodic contraction*.

3. The most frequent form of the disease, common night croup, is pathologically characterized by spasm of the glottidean apparatus, with congestion and tumefaction (transient in character), of the laryngo-tracheal mucous membrane.

It is in these respects precisely *analogous* in nature to the asthmatic attack, whose seat is in the smaller bronchiæ. There is no strongly marked line of separation between this form and the *catarrhal croup*, or croupal catarrh, in which more or less active inflammation occurs, prolonging the existence of the symptoms.

4. Looking then, on the hyperæmic state as simply intermediate, we may classify the cases of croup, as they ordinarily occur, clinically, as, 1st, those in which spasm predominates; and 2dly, those in which inflammation is the dominant condition; or, bearing in mind the above expressed qualification, into spasmodic and inflammatory cases.

5. Pseudo-membranous, or "true croup," does not generically differ from inflammatory croup; of which it is only a grade or termination : *i. e.* any case of inflammatory or catarrhal croup *may* end in the exudation of coagulable lymph within the air tubes.

6. Whether this shall occur or not, in any given case, depends, *a*, on the degree of the inflammation; *b*, on the state of the blood of the patient; *c*, on the treatment.

7. It cannot be predicated on the ground of experience, that either vigorous and plethoric, or feeble and anæmic children, are especially prone to the membranous form or termination of inflammatory croup. It may and does occur frequently in both.

8. The ordinarily recognized signs for the diagnosis of inflammatory from non-inflammatory croup, are sufficient, viz. the persistent duration of the croupal cough and voice—the (generally) slow onset—the febrile symptoms—and the *stridulous* inspiration, as the dyspnœa increases.

9. Inflammatory or true croup is, with the above inclusion (as always potentially membranous), not at all necessarily fatal, although

highly dangerous. The presence of the false membrane itself, does not inevitably determine a fatal result.

10. In no disease does more depend on *early treatment,* which is often prevented by the insidious approach of the attack, deluding the parents. The mortality of the disease may thus in part be accounted for.

11. In the treatment of all forms of croup, *relaxation* and *secretion* are the two great desiderata.

12. In the spasmodic cases, emetics and antispasmodics (*e. g.* ipecacuanha, onion, assafœtida, or lobelia) will effect these objects, especially if aided by the warm bath or foot bath.

13. In mild inflammatory cases, saline purging. gentle vomiting, and the use of demulcents, counter-irritation, and pediluvia will relieve.

14. In the more active cases, the loss of blood by the lancet, or by leeching, or by both, will be necessary, and should be early used.

15. The most satisfactory emetic for employment in such cases is a combination of ipecac. and alum [F. 24] ; the latter being used in half teaspoonful doses in urgent cases, until emesis is produced. Nor should the practitioner hesitate to compel repeated vomiting at intervals, in desperate cases. Better for a child to risk being sick for a month, than to die of cynanche (dog-choke, as the Greeks termed it). But the alum is unlikely to do harm.

16. Tartar emetic should not be used as an emetic in croup; in sedative or expectorant doses, it may be advantageous.

17. Calomel [F. 25], freely administered, that is, a grain every hour or two, has the highest authority in its favor, in serious croup.

18. Nitrate of potassa has both experience and reason in its favor. Being a solvent of fibrin, it should tend to prevent the excessive coagulability of the exudation. According to late theories, ammonia might do the same thing ; but the clinical or therapeutic antecedents of ammonia point otherwise. Of the lately asserted value of sulphur in croup, I have no experience.

19. The great evil in membranous croup is the solidifying tendency of the exudation ; why should not, therefore, an abundant imbibition of fluids, even of water, do something towards the counteracting of this? Inhalation of steam, from hot water poured upon unslaked lime, is eulogized by several recent writers. Glycerin, in teaspoonful or half teaspoonful doses, is recommended by others.

20. No clear indication exists for the use of opium in the majority of cases of inflammatory or membranous croup ; although it may become useful, in cases which are protracted, or which are attended by a more than usual disposition to spasmodic symptoms.

21. Blisters are decidedly useful ; but they should not be left on long in croup, a superficial vesication only being desired.

22. The application of a strong solution of nitrate of silver [F. 26] to the fauces (and larynx, if possible), does good in many cases ; in the pre-exudative stage, as a medicament ; in the exudative, as a mechanical operation aiding to dislodge the membrane.

23. Iodide of potassium is too slow in its systemic action to be relied on ; and the same may be anticipated of the bromide, although nothing should forbid their fair trial.

24. Tracheotomy or laryngotomy will, when performed early, succeed in a fair number of cases; but in those very cases it is impossible

to know that they (as well as those in which it fails) might not have recovered without it. Few practitioners, therefore, in this country, can demand the operation early; and in the moribund state, the vascular congestion, from asphyxia, about the throat, renders success extremely difficult, sometimes impossible. Upon the whole, therefore, the number of cases in which the operation may be expected to add anything to our hope in croup, are few indeed; about as few as those in which careful surgery would justify ovariotomy. In 1859, Bouchut, of Paris, introduced tubage (or catheter-like dilatation of the larynx and trachea) instead of tracheotomy, at the same time publishing some statistics very unfavorable to tracheotomy. The Académie Impériale de Médecin, however, decided adversely to the use of tubage as a substitute for tracheotomy. The possible extension of false membrane into the bronchial tubes is an objection to tracheotomy as well as to tubage in croup.

My own experience with tracheotomy, having met with four fatal disappointments, led me to abandon it in practice. I cannot, however, justify this as an absolute principle. With Dr. C. West, who has had but one recovery in sixteen cases, I am obliged to admit its success, in some otherwise hopeless instances; especially in France, where Trousseau and others operate earlier than in England or here. It is most generally fatal in children under three years of age. Where there is reason to suppose the membrane to extend into the bronchial tubes, it is of course in vain. The danger of hemorrhage is least if the operation is early.

If performed, it should be deliberate, making a considerable opening in the trachea, and inserting a tube or canula of good size. Then the patient should be surrounded constantly with a warm, moist atmosphere. The canula should be withdrawn in as few days as possible, upon the return of permeability of the larynx. The wound may then be treated with ordinary mild dressings to exclude the air and heal it up.

Lately, the fact that lime will dissolve false membranes has been applied to the treatment of croup; by making the patient breathe the steam from boiling water poured over unslaked lime. Although the lime is not volatile, some of its minute particles will be raised mechanically by agitation. Several successful cases of its use are reported. I should think the practice worthy of further trial.

To sum up, I would begin the treatment of a case of inflammatory croup with a saline purgative. Then an emetic of ipecacuanha; which may have to be repeated. Leeching, and even venesection will be useful in a robust subject if seen early. Between the times of emesis, there may be prescribed 1 grain of calomel with 5 grains of nitrate of potassa, every two hours; in urgent cases every hour. In children over three years of age, $\frac{1}{12}$ to $\frac{1}{8}$ grain of tartar emetic may be added. The warm bath, prolonged, may be used once or twice daily. Warm poultices, or cloths wrung out of *cold* water (which soon becomes warm when applied) may be applied to the throat; but a blister should follow in a severe case. Inhalation of steam from lime should be tried, early as well as late. Alum must be added to ipecac, if relief be delayed. Nitrate of silver sponging, and tracheotomy, are the last resorts.

PLEURODYNIA.

Synonym.—*Intercostal Rheumatism.*
Symptoms.—Pain, generally rather dull, sometimes quite severe, of one or both sides, oftenest on the left. It is increased by deep breathing or coughing, moving the arms or trunk.
Diagnosis.—From pleurisy, it is known by the absence of fever, and of all modifications of the sounds heard upon percussion and auscultation.
Treatment.—A large mustard plaster over the part; friction with soap or volatile liniment; dry or cut cups; a blister if obstinate as well as severe.

INTERCOSTAL NEURALGIA.

Symptoms.—Severe lancinating pains between the sixth and ninth or tenth ribs, along the intercostal spaces; frequently intermitting, or even regularly periodical. This affection is most generally met with in anæmic patients, or in those who have been exposed to malarial influence. Occasionally the paroxysms are attended by a sort of reflex pulmonary congestion, simulating pleuro-pneumonia.
Treatment.—Liniment of aconite [F. 27], or of chloroform [F. 28], or ointment of veratria [F. 29], may be rubbed upon the side during the paroxysm. Should these not relieve, a moderate or small blister may be allowed to vesicate, and then one or two grains of acetate of morphia diluted with powder of gum arabic may be applied to the surface; or solution of morphia may be used by *hypodermic injection*, half a drachm at a time being introduced by means of the syringe adapted to the purpose, over the part.
General treatment, by iron, quinia or cinchonia, &c., will be determined by the condition of the patient.

THORACIC MYALGIA.

This term has recently been applied to an affection characterized by pain in the superficial muscles of the chest, mostly dependent upon ill nourishment and overwork; sometimes produced by constrained positions of the body, or pressure; as by a desk, or a soldier's belt, &c.
Its treatment consists in the removal, if possible, of its cause; with local calorifacient or anodyne applications, and general invigoration of the system.

PHTHISIS PULMONALIS.

Definition.—Tuberculous consumption of the lungs.
Varieties.—Acute, chronic, and latent phthisis.
Symptoms and Course.—Consumption may begin after a severe acute bronchitis; or, more gradually, with an apparently slight hacking cough; or with a hemorrhage; or with dyspepsia and general debility; or with chronic laryngitis. Increasing, in most cases slowly, the pectoral and constitutional disorder becomes developed. We

have, then, pains in the chest, frequent and severe cough, hemorrhage occasionally (in about two thirds of the cases) and pallor, acceleration of the pulse and elevation of the temperature, with the paroxysms of hectic fever, *i. e.*, chills followed by fever with bright flush of cheek but without headache; emaciation, arrest of menstruation in the female, night-sweats, colliquative diarrhœa; finally, often, though not always, delirium; and death, mostly by exhaustion, but sometimes by suffocation. The spirits of the patient are apt to be cheerful, even hopeful of life almost to the last. Appetite is variable, digestion usually not vigorous; but to this there are exceptions.

The following description of advanced phthisis is from the late Prof. N. Chapman :—

" The cheeks are hollow, the bones prominent. the skin arid, the nose sharpened and drawn, the eyes sunken, with the adnata of a pearl color, destitute of vascularity, the lips retracted, so as to produce a bitter smile, and the hair thinned by falling out, the neck wasted, oblique and somewhat rigid or immovable, the shoulder blades projected or winged, the ribs naked or exposed, with diminution of the intercostal spaces, and the thorax apparently narrowed; the abdomen flat, the joints, great and small, seemingly enlarged from the wasting of the integuments, the nails livid, and occasionally incurvated, the extremities œdematous; the angular points on which the body rests, in several points protruded through the skin,—the whole attended by a most afflicting cough, aphthæ, sore throat, difficult deglutition, and feeble, whispering voice, or entire extinction of it."

The expectoration in phthisis is at first mucous or bloody; later, muco-purulent and bloody, or else *nummular ; i. e.* in roundish masses like coins, not floating perfectly in water; or, abundant and purulent.

Stages.—These are, 1. Incipient phthisis ; 2. The stage of consolidation of the lung; 3. That of excavation or *vomicæ ;* 4. Advanced or confirmed consumption.

Physical Signs.—Is there a *pre-tubercular* stage of phthisis ? If so, it cannot be certainly pronounced upon. The earliest indications upon physical exploration are, a sinking in under the clavicle upon the left side, with prolonged expiratory sound. Not long after, the evidence of consolidation is, increased dulness over the apex of the lung upon percussion (not invariably but *generally* upon the left side) with blowing or bronchial respiration, or interrupted jerking respiratory murmur, and increased vocal resonance and vibration. Dry crackling follows with mucous or coarse crepitant râle.

When softening of tubercular deposits occurs, moist crackling and gurgling become very distinctive signs. The presence of a *vomica* is shown by cavernous respiration and bronchophony or pectoriloquy. Percussion resonance over a cavity will be dull if its walls be thick, and amphoric if they are thin and tense ; if thin and relaxed, the *bruit de pot fêlé* or cracked pot sound. On percussion over a cavity when the patient's mouth is shut, the sound produced will be of a lower pitch than when the mouth is open.

Pneumothorax and hydropneumothorax, *i. e.*, dilatation of the pleural cavity and compression of the lung by air, or air and liquid together, with perforation of the lung, are not uncommon results of tuberculization, although possible without it. Of pneumothorax, the percussion resonance is tympanitic; respiratory murmur lost. Hydropneumothorax may give tympanitic resonance above, with metallic tinkling, on auscultation, and dulness below.

Physical and Microscopical Peculiarities.—Temperature has of late been found to be a diagnostic aid in phthisis. It is asserted that there is a *continued elevation of the heat of the body* in all cases in which tubercle is being deposited; that this may occur for weeks before any local physical sign is discoverable; and that the rise in the heat of the body varies, during the progress of the case, with the greater or less activity of the tuberculization.

When expectoration is copious, some micrologists aver that diagnosis may be aided by its minute characters; arched and anastomosing fragments of pulmonary fibrous tissue, and tubercular corpuscles, being discerned. But it is not certain that the former are only thrown off in phthisis; and the latter may be absent or obscure in character in an otherwise clear case of consumption. Dr. Fenwick, of London, detects minute portions of lung-tissue by boiling the expectoration a few minutes with its bulk of solution of caustic soda (gr. xv in f℥j of distilled water), and then adding cold water, in a conical vessel. The sediment is then examined with the microscope.

Terminations.—The cicatrization of vomicæ, and the cessation of tubercular deposition, have, although exceptional, been often found to occur; and so have the cornification and calcification of unsoftened tubercle. Recovery from phthisis may in such cases be expected to take place, as the arrest of the local disease only attends the presence of a favorable constitutional state.

Death from consumption may come by *asthenia* or by *apnœa*. The first is most common. Suffocation or apnœa may follow—1. from hemorrhage; 2. rupture of a large vomica; 3. pulmonary œdema or hydrothorax; 4. excessive secretion or bronchorrhœa, beyond the power of expectoration.

Complications.—Pleurisy is a frequent concomitant of phthisis. Tubercular peritonitis is much more rare. On account of its duration, however, this disease may be accidentally combined with various affections not specially kindred with it. Asthma is particularly *not* apt to be conjoined with phthisis.

Diagnosis.—It is from chronic bronchitis, cancer of the lung, pleuritic effusion, bronchial dilatation, and pulmonary abscess that phthisis requires the most care for discrimination.

Chronic bronchitis is not common except in old persons; its expectoration is thinner, whiter, and not nummular nor bloody; there is no hectic, although there may be emaciation; and there are none of the physical signs of phthisis.

Cancer of the lung exhibits a marked dulness of resonance on percussion on one side, with blowing respiration, unless a bronchial tube be obstructed, when there is no respiratory sound. There is severe and almost constant pain in the chest. The peculiar auscultatory signs of tubercular disease are absent; and the sallow, cachectic aspect of cancer, and the concurrent existence of carcinomatous tumors somewhere in the body, generally make the case clear.

In "chronic pleurisy," as pleuritic effusion is often called, the dulness on percussion is at the *lower* part of the chest; the side is expanded, unless after the fluid is absorbed; respiratory murmur and vocal vibration are suppressed; and the general symptoms, as irritative fever and wasting, are not so extreme.

Bronchial globular dilatation may give auscultatory signs exactly like those of a tubercular cavity; but there is no hæmoptysis, nor emaciation, nor much loss of health. The expectoration may be more copious than in consumption; but the matter is more liquid, and pus is much more diffused in it. The cough is more constant than in phthisis.

Abscess of the lung is to be distinguished from phthisis by its history, generally following recognized pneumonia; its seat mostly at the base of the lung; its physical signs decreasing instead of increasing; and, as with cancer, the affection being confined altogether to *one* *lung.* The extension of the signs to both lungs is important in most cases in the diagnosis of phthisis.

Syphilitic disorder sometimes affects the lungs and bronchial tubes, with a condition almost undistinguishable from ordinary consumption. The previous existence of venereal disease, and periosteal nodes upon the clavicles, with the slower progress of the decline, will help to enlighten us.

Prognosis.—Phthisis is certainly one of the most destructive of diseases. In no case can recovery be anticipated; but it does occur, as every physician must have witnessed. I have seen several such recoveries; generally from the incipient stage, but even where vomicæ, emaciation, and night-sweats had occurred. Dr. A. Flint has recorded the history of sixty-two cases of restoration from consumption.

Under improved hygiene and medical treatment, the mortality from phthisis appears to be declining. Without referring to statistics (the nomenclature connected with which in past times would be a source of doubt, as chronic bronchitis, &c. were once called consumption), I am convinced that fewer people die of phthisis now than twenty-five years ago, in Philadelphia.

The *duration* of phthisis varies greatly, being least, as a general rule, in the youngest subjects. Eighteen months to two years is the most frequent period. But in some instances life is prolonged under it for twenty, thirty, or even forty years.

Acute phthisis, or galloping consumption, may end life in from six weeks to three months. This sometimes follows pneumonia. Its symptoms differ from those of ordinary consumption chiefly in their rate of progress. Softening of the tubercle and the formation of cavities do not always occur to any extent, apnœa being caused by extensive diffusion or infiltration of the tuberculous deposit through the lungs.

In any case of consumption, the *state of the general system* is of primary import in prognosis. When the patient is gaining in weight and strength, and fever and night-sweats diminish or disappear, there is hope, for a time at least. Spitting of blood (when consumption is proved to exist already) does not increase, but rather lessens the unfavorable aspect of the case. Rapid emaciation, chills, hectic, swelling of the feet, and diarrhœa are always discouraging; as, of course, are, also, all signs of increase in the local pulmonic affection.

Causation.—Hereditary taint of constitution is general; independent origination of phthisis the exception. From 18 to 35 years is the time of life most subject to it; but it is now and then met with even in children, and frequently in the aged. Statistics in Europe and this country show some proportion between the mortality from consumption and nearness to the sea level; the lowest lands having the greatest total amount of it. High, dry, and equable climates and situations, even though cold, are most exempt from it. It is not a disease of the Arctic regions, and there is more of it in South Carolina than in Maine.

Individually, and in families, all causes that depress vitality promote it; but most of all *impure atmosphere*. Sedentary employments and exhausting excesses, with foul air, make large cities most of all productive of it. In constitutions having the proclivity towards it, tuberculization may be brought on by any reducing disease, especially such as involves the breathing organs; as measles, bronchitis, or pneumonia.

Treatment.—*Hygienic management* is, decidedly, more important to the consumptive than medicine. The following precepts are well laid down by Dr. B. W. Richardson:—

1. A supply of pure and fresh air for respiration is constantly required by the tuberculous patient.

2. Daily exercise in the open air is imperatively demanded by the tuberculous patient.

3. It is important to secure for the patient a uniform, sheltered, temperate, and mild climate to live in, with a temperature about 60°, and a range of not more than 10 or 15°; where, also, the soil is dry and the drinking water pure and not hard.

4. The dress of the tuberculous patient ought to be of such a kind as to equalize and retain the temperature of the body.

5. The hours of rest should extend from sunset to sunrise.

6. In-door or sedentary occupation must be suspended; but out-door employment in the fresh air, even in the midst of snow, has been and may be advantageous.

7. Cleanliness of body is a special point to be attended to in the hygienic treatment of tuberculosis.

8. Marriage of consumptive females, for the sake of arresting the disease by pregnancy, is morally wrong and physically mischievous.

Altogether, the *analeptic* principle is now universally adopted for the treatment of consumption. The diet must be nourishing; a "generous" regimen; and the same indication is to be followed in the employment of medicines.

There has been discovered, as yet, no specific to arrest tuberculosis. But cod-liver oil and alcohol, and, in lesser potency, iron, quinine, and other tonics, in a certain number of cases do manifest an important conservative and restorative influence; and palliation of symptoms, as pain, cough, loss of rest, may greatly help the comfort of the patient. My confidence in the *frequent* value of cod-liver oil is based chiefly upon observation. Three individuals in one family, for example, under my care, notwithstanding a well-marked family tendency (shown by the previous death by phthisis of three sisters, a mother, and uncle), recovered from incipient consumption under the use of the oil. Other cases, much more commonly, have life *prolonged* by it. Unfortunately, however, in quite a considerable number of persons the stomach turns against cod-liver oil. When that is the case, it is quite in vain to urge it. It may be taken in the froth of porter or ale, or after rinsing the mouth with brandy, which may also follow it. Some dislike it less when salted. Ammonia added to the oil lessens its taste; but I have never tried the combination extensively. The gelatinous capsules make it much less disagreeable to swallow; but less than two or three tablespoonfuls of the oil daily will hardly suffice. It can always be taken best in cold weather [F. 30, 31, 32].

Alcohol, though variously estimated by different physicians, is, in my view, well established as a remedial or at all events a supporting agent of value in consumption. Not to be used in excess, nor ever to produce excitement in any degree; but simply as a *roborant ;* as an addition to the diet and a supporter of the strength of the invalid. The dose must, therefore, be proportioned to his condition.

Whisky is preferred by many; but ale, lager beer, and wine suit different patients best. A little two or three times daily will be better than a full drink at one time. I would always begin with very small quantities—say two or three teaspoonfuls of whisky, or half a glass or even less of wine, or half a tumblerful of ale or beer. To do good, the stimulant *should not quicken the pulse, flush the face, or be felt to affect the head.* Kept under such restrictions, even when increased to meet greater prostration, I have never known any hankering after excess to be caused by it. One patient of mine, with phthisis, would sometimes, when temporarily much reduced, take more than half a pint of whisky daily for a time; and then, as his strength rallied, would diminish the amount to almost none, without any difficulty or longing for more.

Lately, we hear from abroad of great advantage accruing from the "raw beef and brandy" treatment for consumption; but I am not possessed of the particulars of it. . When it can be done, alcoholic stimulus is best given with nourishment, as in milk, or beaten up with a raw egg, &c.

Beef-tea, as a concentrated nutrient, is very useful when digestive power is low, at any stage of phthisis. One lady under my care, who, with tussicula, hæmoptysis, and emaciation, had greatly the appearance of incipient consumption, and who could not retain cod-liver oil upon her stomach without loss of appetite, was put upon the daily use of a pint of strong beef-tea,[1] for several weeks together; with no medicine except a mild expectorant. She recovered, and has since married and become a mother.

The phosphates and hypophosphites of lime, &c. have been sufficiently tried to prove their inferiority to cod-liver oil. My own experience with them, in the wards of the Episcopal Hospital in this city, as well as in private practice, has been discouraging; and I

---

[1] The mode of preparation of beef-tea is not unimportant. I prefer the following : Cut up a pound of good lean beef into small pieces, pour upon it a pint of cold water, and let it stand two hours beside the fire. Then boil it half an hour. Take off all the scum and oil drops, carefully ; but *do not filter or strain it.* It should have a rich brown color ; and, with salt, is agreeable to the taste.

believe the best phosphate for analeptic use to be the phosphate of iron. Chlorate of potassa has also entirely failed under fair trial. Glycerin will not take the place of cod-liver oil; nor has any other oil been shown to be capable of doing so.

Iron, especially the iodide [F. 33] and the tincture of the chloride, are frequently suitable; and so may be quinine, nux vomica [F. 34], or the simple bitter tonics. But the patient must not be worried and disgusted with much medicine; whatever depresses appetite is likely to do more harm than good.

For this reason, *expectorants* require discretion in their use. Those of a nauseant kind must be very sparingly prescribed in phthisis. The syrup or fluid extract of wild cherry [F. 35, 36] is one of the most suitable. Squills will do when loosening effect is particularly required. Ipecac. and tartar emetic are too depressing to the stomach for the consumptive. Sometimes, at a late stage, carbonate of ammonia will not be too stimulant.

Anodynes and calmatives are almost always wanted as the case advances, to soothe the wearisome cough, and to give rest at night. Lactucarium, hyoscyamus, and finally opium, or morphia, in some form, will be important sources of comfort to the patient, and may economize his strength.

The colliquative sweats seldom demand treatment, they being the *result* rather than the cause of debility. Ablution with brandy or whisky and alum may be practised if they are very excessive. Diarrhœa may require to be held in check, by simple astringents with opiates.

If pleurisy or peritonitis supervene as a complication, the local inflammation must be treated in view of the general condition. Depletion is out of the question at an advanced stage. Dry cups, blisters, and opium are all that we can use in the treatment. For the variable pains in the chest in the course of the disease, mild or moderate counter-irritation, by warming plasters, tincture of iodine, or croton oil, may be used.

It is not, however, to be said that the name or character of phthisis should in *all* cases rule out local depletion in the *incipient* stage. In one of three cases in one family (already alluded to) who recovered, notwithstanding a strong inherited tendency to consumption, from a condition threatening it, great relief and improvement followed the application of two dozen leeches to the side; it was (to borrow an expression of Dr. Condie's) at the time, an acute *tuberculous pneumonia*. Yet I know that such cases are exceptional. The pervading indications in phthisis are economy and recuperation.

Inhalation has often been tried, in phthisis. Not enumerating agents which have summarily failed, I believe the best hope attaches, in this way, to inhalation of the vapor of creasote or of carbolic acid. These agents are styptic, and creasote, at least, by its power of coagulating albumen and albuminoid material, may be expected to aid in arresting the softening and destructive process in the lung. At least, we might hope that it would (and in some cases it has proved so) lessen excessive and exhausting expectoration.

Change of climate is often proposed for the benefit of the consumptive. In an early, or middle, or even a stationary advanced stage, it
13

may be of important advantage. When to forbid, or advise it, may be a very delicate question. More will depend upon the *rate* of progress than upon the period of the case. But the patient must have strength enough to travel, and must be not too dependent upon his home comforts, or he may be made worse instead of better. It is a cruelty to banish one who is already on the verge of the grave to die in a strange place among strangers. Yet I have known life to be prolonged from year to year, in one who was a native of this city, by his spending the winter south.

In selecting a climate for the invalid, equability and dryness are, unless at a late stage, more important than warmth. That climate which will allow the patient the greatest number of days out of doors, will be the best. Minnesota, and other places near Lake Superior, agree extremely well with some. Of southern localities, Florida presents an especially equable, almost maritime, climate. Cuba is often resorted to. A sea voyage (if not subject to exhausting sea-sickness) may do good at an early stage. Across the ocean, consumptives resort to the South of France, particularly to Pau or Biarritz; or to Mentone, or Malaga, or Malta; or Italy,—especially Ischia or Capri, Sorrento or Palermo;—Madeira, and Algeria, the year round, and Egypt in the winter only, are favorite climes. For the winter, nothing could excel in salubrity the atmosphere of the upper Nile.

Our own country affords all the requisites for the migrations of an invalid, to escape the inclemencies of every season, if he can vibrate between Newport, R. I., in the summer, and St. Augustine or the interior of Florida for the winter.

Phthisis in Early Life.—Dr. C. West[1] names the following as characteristics of consumption in children, among whom, however, it is certainly *rare :*—

1st. The frequent latency of the thoracic symptoms during the early stages.

2d. The almost invariable absence of hæmoptysis at the commencement of the disease, and its comparatively rare occurrence during its subsequent progress.

3d. The partial or complete absence of expectoration.

4th. The rarity of profuse general sweats ; and the ill-marked character of the hectic symptoms.

5th. The frequency with which death takes place from intercurrent bronchitis or pneumonia.

The same excellent authority designates the following peculiarities in the auscultatory phenomena of consumption in the child.

1st. The smaller value of coarse respiration, prolonged expiration, and interrupted breathing, owing to their general diffusion over the chest, and to their occasional existence independent of phthisis.[2]

2d. The greater difficulty of distinguishing chronic bronchitis, in the child, from phthisis.

3d. The loss of that information which the phenomena of the voice furnish in the case of the adult.

---

[1] Diseases of Children, p. 404.

[2] The occurrence of harsh respiratory sound as an initial sign in *pneumonia* of children is well established.

4th. The smaller value of inequality of breathing in the two lungs.

5th. The difficulty of detecting minute variations in the resonance upon percussion.

6th. The frequent existence of dulness in the interscapular region, with moderate resonance and tolerably good respiration in the upper part of the chest,—characteristic of enlargement of the bronchial glands.

AFFECTIONS OF THE ORGANS OF CIRCULATION.

PERICARDITIS.

Definition.—Inflammation of the covering membrane of the heart

Varieties.—Simple or idiopathic, and rheumatic pericarditis. The latter is very much the more common. Degrees of violence in the attack also cause variations, from the mildest and almost latent cases, through those of open and active severity, to those attended by rapid effusion and prostration.

Symptoms.—Fever; pain (occasionally absent) at and radiating from the heart; tenderness on pressure in the cardiac region; accelerated, irregular, or oppressed, rapid and feeble pulse; anxiety or delirium; nausea and vomiting in some cases; short hacking cough; towards the end, coldness and pallor or lividity, œdema of the face and extremities, loss of pulse.

Stages.—1st, Acute inflammation; 2d, Adhesion; 3d, Effusion.

Physical Signs.—Before adhesion or effusion, usually, exaggeration of the heart's impulse. Then, pericardial *friction-sounds* (to and fro); the vibration of which is sometimes felt by the hand. After effusion, dulness on percussion, with muffling of the heart's sounds to the ear on auscultation. The friction-sounds disappear during this period, sometimes to return as the effusion is absorbed.

Morbid Anatomy.—In the first stage, there is a rose-redness of the pericardium, diffused, punctated or in patches. Then, deposits of coagulable lymph, white and opaque, sometimes causing local or general adhesion of the two layers of serous membrane. In most fatal cases, effused serum is found in the sac, in quantity varying from ounces to pints. Great quantities of it weigh down the diaphragm below it. *Purulent* exudation is sometimes met with. In scorbutic cases, it may be hemorrhagic.

The muscular tissue of the heart is usually less coherent than usual.

Diagnosis.—From *endocarditis* and from *pleurisy* it is sometimes not easy to distinguish pericarditis. The symptoms of the latter and those of endocarditis are the same; and the *friction-sounds* occur in both. The heart's impulse is more apt to be sustained in strength in endocarditis; and, in the latter, no dulness on percussion occurs, nor are the heart-sounds muffled at any stage; while valvular murmurs follow endo- and not pericarditis.

Friction sounds which are outside of the heart (pericardial) have a *nearer* character to the ear than endocardial sounds; they are more narrowly *limited*, not passing along the vessels; they do not keep exact time with the cardiac sounds, and may vary from day to day; and sometimes the vibration may be felt externally.

Pleurisy causes friction sounds, and afterwards dulness on percussion. But, the former sounds are more diffused, are generally *single*, not "to and fro" or double ; and the dulness extends further over and around the side. Latent pericarditis may possibly, from some symptoms, be taken for inflammation of the brain, or of the stomach. Physical exploration should prevent such errors.

Prognosis.—There is great danger to life in pericarditis ; and its course is sometimes terminated by death in a few days. In other cases resolution may take place promptly ; but more often the heart is clogged for a considerable time (weeks or months) with effusion, or a more protracted interference occurs from adhesion of the pericardial surfaces. This latter is sometimes shown by a dimpling, or sinking in, with each beat of the heart, of the intercostal spaces above and below it.

Causation.—The *materies morbi* of rheumatic fever is far the most common cause of pericardial inflammation, as it is of endocarditis also. Gout is accused of the same thing ; but with much less frequency, or indeed, clearness of proof. Bright's disease of the kidney is occasionally associated with it.

Treatment.—In active cases, and good subjects, *one early and moderate bleeding from the arm* will be proper Afterwards, in some, and instead with feebler patients, whose fever is high and pain intense, leeches over the cardiac region may be used. A brisk saline cathartic, as Epsom or Rochelle salts, or citrate of magnesia, should commence the medication. Calomel, trusted still by some and abused by others, may be confined to open sthenic cases, in previously good constitutions. In such, I would give half a grain of calomel, with half a grain to a grain of opium, thrice daily for three or four days.

Where the rheumatic diathesis is marked, *alkalies* [F. 37] will be indicated. Carbonate or bicarbonate of potassa, or bicarbonate of soda may be given, in scruple or half-scruple doses, with as much of Rochelle salts, three or four times a day. A blister over the heart, as the fever lowers, will often have a very good effect. If effusion occur, blistering may be repeated.

Should no opiate be given through the day, Dover's powder or morphia may be prescribed at night.

For the stage of effusion, or " chronic pericarditis," the usual treatment consists of diuretics [F. 38, 39, 40], as squills, juniper, sp. æth. nit. dulc., etc., varied and continued until absorption occurs. Tonics will often much promote the same end.

A *rapidly depressing* case of pericarditis, with cold, blue skin and feeble, irregular pulse, will require, instead of the above, a supporting or stimulating treatment from the first ; with dry cups and blisters instead of local or general bleeding ; and quinine, ammonia, and brandy, instead of sudorifics or laxatives.

Myocarditis is inflammation of the muscular substance of the heart. It can hardly be said to have other than a nominal existence.

ENDOCARDITIS.

Definition.—Inflammation of the lining membrane of the heart.

Symptoms and Physical Signs; Diagnosis.—These have been sufficiently stated in the account just given of pericarditis and its diagnosis, and need not be repeated. Like that disease, it is most often of rheumatic origin; but may occur in Bright's disease or in pyæmia.

Valvular derangement and its signs give great interest to endocarditis and its resulting changes. Mostly it is the left side of the heart that is chiefly affected. The simplest and most common sign of this is a blowing sound, heard on auscultation. But a bellows murmur is heard also in cases of anæmia, and a blowing sound occurs not rarely in fevers; or it may belong to an organic heart-affection of long standing. This last fact should be ascertained by the history of the patient, as well as by the aid of symptoms; but the *old* murmur is generally rougher and more fixed in its seat. It is *possible*, though very *rare*, for endocardial inflammation to be located so far from the valves as to cause no blowing sound.

Clots sometimes form in the heart, in endocarditis (as well as in some other diseases attended by prostration), obstructing the circulation, even to a fatal extent. Although most clots are *post-mortem* in origin, there is no doubt that sometimes firm fibrinous masses do occlude the valves for some time before death. The symptoms produced are, blueness and coldness of the skin, indistinctness of the heart-sounds, feebleness and irregularity of the pulse, nausea and vomiting, anxiety of expression, and fainting.

Much more often, vegetations or fibrinous deposits of exudation on the valves of the heart are carried in fragments therefrom by the blood into the arteries. Being arrested, as in a vessel of the brain, or a limb, etc., the condition of obstruction designated as *embolism* results; which receives attention in another part of this book. Old valvular vegetations, as well as the recent ones of endocarditis, may give rise to emboli; which may, also, arise from coagulation in a vein, or *thrombosis.*

Endocarditis produces valvular derangement in the *mitral* valve most frequently in the young; in the old (from this cause as well as from degeneration), disease is rather more common in the aortic valve. The forms of disorder, indicated by murmurs, occur in the following order of frequency: 1st. Aortic obstructive; 2d. Mitral regurgitant; 3d. Aortic regurgitant; 4th. Aortic obstructive and mitral regurgitant together.

Enlargement of the heart, either with muscular thickening (*hypertrophy*) or with attenuation (*dilatation*) is a common consequence of endocarditis with valvular lesion. Already (see *Semeiology*) the statement of Dr. Stokes has been adopted, that in every case the important question is, less the state of the particular valves, than the amount of interference with the functional action of the heart. In young persons, remarkable recoveries sometimes take place (as I have seen) from very considerable lesion of the valves. In other instances, *adaptation* of the heart itself, and of the general system, by degrees, is effected, so that quite good health, and even capacity for exercise, may be attained,

13*

while the physical signs of the local organic change remain. Sudden death is less common in heart-disease than is popularly supposed. Some persons having it have lived twenty or thirty years.

VALVULAR DISEASE.

The valves of the heart may be impaired either by inflammation, or by degeneration. (*e. g.* calcareous deposit or "ossification"). The latter, degenerative valvular changes, occur gradually; and mostly late in life. Either form of valve-disease, or at least of valvular alteration, is generally permanent; the *degenerative* form almost invariably so.

Changes may occur, by simple thickening, or by deposits of fibroid, fatty, or calcareous material; or by atrophy, contraction, adhesion, or ulceration of the valves; or gouty deposits, of urates and carbonates of soda and lime. The valve (mitral or aortic primarily, or tricuspid or pulmonary secondarily) may be thus rendered incapable either of perfect closure, or of full opening; in most instances, at least, a permanently half-open state results.

A considerable variety of pathological conditions may exist in organic disease of the heart; while the number of cases in which an exact and unequivocal diagnosis can be made, is comparatively small. We must not confine attention at all to the physical signs alone, but compare also with these the pulse, and force of the heart, with other general symptoms, and the entire history of the case.

Certainty can hardly ever be obtained, unless it be in the diagnosis of one of the following three conditions :—

1. Uncomplicated disease of the mitral valve. Signs of this are,— a permanent murmur, with the first sound, loudest towards the apex and left side, and not heard over the aorta ; the second sound natural. The heart's action natural ; the impulse not excited, the pulse natural.

2. Disease of the aortic valves, with permanent openness. With this, there is no murmur with the first sound ; the second sound is replaced by a double murmur, loudest at the base of the heart, and heard along the aorta. In an advanced stage of this condition, the arteries give to the finger, or even to the eye, an impression of *bounding* pulsation ; with a *jerking*, or abruptly ending pulse at the wrist.

3. Disease of the aortic valve, without permanent openness. Here, the action of the heart is slow and feeble, generally regular, or only occasionally intermitting. A murmur is heard with the first sound, the second sound being healthy ; but a murmur may be heard with the second sound, in the aorta and carotids.

It must be noticed that in *anæmia*, without heart-disease, a bellows murmur is often heard, extending into the arteries. Chiefly by the concurrent signs and symptoms, is this to be distinguished from organic disease of the heart. Anæmic murmurs are more variable, and are not much increased by moderate exercise.

When the aortic valvular orifice is greatly *contracted*, the pulse at the wrist may become very feeble, almost absent; while the heart's impulse is strong.

Advanced mitral or aortic disease is accompanied usually by derangement, sympathetic or obstructive, of the lungs, liver, and other organs; with hæmoptysis, anasarca, cyanosis, irregularity of the pulse, syncope, &c. *Pulsation of the jugular veins* indicates secondary disorder upon the *right* side of the heart, with regurgitation into the venæ cavæ. *Pseudo-apoplectic syncope* may occur in permanent patency of the mitral valve; or in fatty degeneration of the heart, with or without valvular disease.

DILATATION OF THE HEART.

Uncomplicated dilatation of the whole heart, or of either pair of corresponding cavities, or of any one cavity, is very uncommon. **Complicated** dilatation is frequent. It may depend—1. on a debilitated state of the cardiac muscle; 2. on valvular disease; 3. on obstruction beginning in organs remote from the heart.

The commonest form of dilatation forms part of a triple affection, in which the *heart, lungs,* and *liver* are together involved. All this may come, in the first place, from a cachexia, such as gout, or scurvy, or from simple anæmia. Exacerbations in the disorder may occur; as, of pulmonary congestion, enlargement of the liver, cardiac asthma, bronchitis, or dropsy. The prognosis cannot be very favorable in such a case; and only palliative, or recuperative, treatment avails, along with hygienic management, to economize the powers of nature.

Dilatation of the heart is indicated, upon physical exploration, when, with extended impulse of the heart, we have dulness on percussion beyond the usual limits. If **true hypertrophy**, or muscular thickening, be present, the impulse is very *forcible* as well as extended. The heart-sounds are apt to be *clear*, though not loud, in **attenuated** dilatation; rather loud, but dull toned, in enlargement with thickening of the walls. But these differences are hardly to be relied upon.

Hypertrophy of the muscular tissue of the heart is most often induced by valvular obstruction or regurgitation, compelling unusual and continued effort to sustain the circulation.

Sometimes, however, it is more truly idiopathic; following causes of overaction of a heart otherwise sound. Thus violent exercise, self-abuse, coffee, alcohol, tobacco, &c., are, with good reason in predisposed cases, accused of producing it.

In the **treatment** of simple hypertrophy, avoidance of such exciting causes, and particularly of violent exercise, alcohol, and venery, is the main principle. Robust or plethoric patients will bear and be benefited by venesection, at long intervals; or by occasional leeching or cupping over the heart. *Acetate of lead* [F. 41], as an astringent cardiac sedative, is recommended by some, and is worthy of trial (one grain thrice daily), with care to avoid saturnine poisoning.

Digitalis was formerly relied upon as a reducer of cardiac action. Lately the question has been opened widely, whether it does at all tend directly to lower the heart's action; or whether it is not, instead, a *tonic* to the heart (probably through ganglionic influence), lessening rapidity of action only when that depends on debility. The time has hardly come to pronounce upon this question. I think, however, that evidence has been given to encourage us to use digitalis [F. 42, 43] unhesitatingly where abnormal rapidity of the heart's action exists in conditions of debility; and to expect more from *veratrum viride* [F.

44] as a sedative and palliative, in *violent* acceleration of the pulse as in muscular hypertrophy, and in some forms of palpitation.

FATTY DEGENERATION OF THE HEART.

Definition.—Substitution of fatty substance for the muscular tissue of the heart, to such an extent as to interfere with its normal action.

Symptoms and Course.—Though no doubt almost always gradual in its progress, this affection in many instances fails to make itself known by symptoms until a late period; sometimes even till the moment of death. Usually, feebleness and irregularity of the pulse and heart's impulse are observed; with exhaustion and dyspnœa upon exertion. The pulse is slow when at rest; sometimes only thirty in the minute, although the heart beats fifty or sixty in the same time. Attacks of apoplectic syncope or syncopal apoplexy may occur; at first most like syncope, after repetition becoming more apoplectic. These are distinguished from true apoplexy by the feebleness of the pulse, coldness of the skin, sighing respiration, and the slightness or absence of paralytic symptoms, notwithstanding several repetitions of the attack. They are made worse by depletion or reduction of the system; and may be relieved or warded off by timely stimulation; the recumbent posture is most favorable in them. The first attack of this kind may, however, prove fatal.

Physical Signs.—Fatty degeneration is often complicated by the presence of other structural changes of the heart. By itself, it is with difficulty diagnosticated by physical exploration. The heart's impulse is feeble and slow, often irregular, and the sounds weak. A bellows murmur is frequently heard with one or both sounds.

Morbid Anatomy.—True fatty degeneration must be distinguished from fatty *accumulation* about the heart; which may impede its action, but is much less dangerous. In true interstitial degeneration, the heart is, in part or throughout, flabby and pale or yellowish, though it may be more bulky than usual. Minutely examined, the muscular fibrils are found to have lost their transverse striæ, and to have resolved themselves, more or less, into streaks of oil-dots or opaque granules.

Death, sometimes, is shown to have resulted from rupture of the heart. In other instances that organ has, under some exertion or excitement, become exhausted and failed to act sufficiently to keep up the circulation.

Prognosis.—Recovery is not to be expected in cases of fatty degeneration; although life may be prolonged to old age. Much will depend upon circumstances of living, and care to avoid disturbing agencies.

Causation.—In early life this affection is uncommon; its most frequent cause is, then, pericardial or endocardial inflammation. Most cases occur after fifty years of age. It then occurs as one of the local manifestations of waning vital energy; but it may be promoted by any or all exhausting or depressing causes. No special or peculiar line of causation can be pointed out.

Treatment.—This can be only *conservative*, not curative. Tonics, particularly iron, with generous diet, sea or mountain air, change of scene, avoidance of anxiety and exertion, may do much to retard the degenerative process. Violent effort or emotional excitement may be suddenly fatal. Tranquil occupation only should be selected, and all rapid exercise, and even straining at stool, ought to be avoided.

MODES OF SUDDEN DEATH IN HEART DISEASE.

We may briefly enumerate these as, 1. Arrest of the heart's action from debility of the muscular walls; 2. Spasm of the ventricles; 3. Extreme obstruction, or regurgitation; 4. Rupture; 5. Heart-clot. Indirectly, cerebral or pulmonary apoplexy.

ANGINA PECTORIS.

Definition.—An irregularly paroxysmal disorder, characterized by sudden attacks of severe pain, extending from the heart along the left arm, with a sense of stricture in the chest, prostration, and alarm.

Pathology and Causation.—This appears to be a *symptomatic* affection; connected in most, if not in all cases, with organic disease of the heart; especially ossification of the coronary arteries. Gout predisposes to or excites it, but probably not in the absence of heart-lesion. It occurs generally in old people; most often in men.

Prognosis and Duration.—The attack may last from a few minutes to an hour, or even a day. Commonly it is short, going off with perspiration or copious urination. A first attack may be fatal. Returns occur at variable intervals—days, weeks, or months; each one generally sooner and more violent, till one of them ends life.

Treatment.—Stimulants and anodynes [F. 47, 48, 49] are indicated during the attack. Best will be Hoffmann's anodyne, laudanum, Warner's cordial, and brandy, in moderate doses, repeated in a short time if necessary; with mustard plasters over the chest and between the shoulders, and the warm foot-bath. Where gout is present, colchicum [F. 45, 46] and alkalies are important.

THYRO-CARDIAC DISORDER.

Synonym.—*Ex-ophthalmic Goitre.*

Definition.—Enlargement of the thyroid gland in the neck, with over-action of the heart and cervical vessels, and prominence of the eyeballs.

Nature.—This uncommon affection, of which I have seen but one case, is considered by Dr. Stokes to consist in a more or less permanent functional excitement of the heart; which may produce finally dilatation and hypertrophy, with dilatation also of the jugular veins, and an aneurismal condition of the thyroid gland. Although considerable disturbance and prostration of the system must attend such a state of things, yet it has been repeatedly recovered from, as in the case which I saw. The *cause* of the affection has not been made out.

Treatment.—Tranquillization of the heart is the main indication. *Veratrum viride*, in doses not at all nauseating (two or three drops of the tincture every three or four hours), may be persevered in, while watching its effects. Other treatment must depend upon the general condition of each patient. Of course violent exercise and mental excitement must be avoided.

PALPITATION.

All excessive or consciously disturbed action of the heart is commonly thus designated. Over-action, in particular, may have either of the following origins:—

1. **Nervous,** or hysterical; 2. **Dyspeptic;** 3. **Rheumatic,** or **gouty;** 4. **Hypertrophic.**

Nervous palpitation occurs in anæmic persons, especially hysterical females, or in those otherwise debilitated. Alcoholic intemperance, strong coffee, tobacco, excessive venery or self-abuse may produce it.

Dyspepsia is very often attended by palpitation, sympathetic with the gastric disturbance. It usually, in such a case, is worse after meals.

Gouty and rheumatic palpitations are also common. Their nature will be made known by the presence of other signs of the controlling diathesis.

All of the above forms of merely functional disturbance of the heart, and especially the purely nervous, may be known from *hypertrophic* overaction, or the conscious impulse of dilatation of the heart, by the fact that they are not increased by moderate exercise; are often, indeed, much diminished thereby. When the heart is enlarged, especially with valvular change, active movement causes distress and dyspnœa, with great acceleration of the cardiac movement. In palpitation of all kinds, during the attack, it is generally not possible to lie with ease upon the left side; and orthopnœa may occasionally occur, without organic disease.

The **treatment** of palpitation must vary altogether according to its cause. If nervous, invigoration of the system and enrichment of the blood are most probably needed; by iron and other tonics, and regimen. Dyspepsia will require appropriate treatment; as a part of which, exercise in the open air will not be counter-indicated at all by sympathetic palpitation.

Functional overaction of the heart, without organic disease, is in itself not dangerous. It is alarming, however, to the patient, as well as a source of discomfort; and may, if long sustained, bring on true enlargement of the heart. All causes, therefore, of such disturbance ought to be sedulously avoided in the interest of health.

CARDIAC EXHAUSTION.

In our general hospitals during the late war, under my own observation as well as that of other practitioners, quite a number of cases of soldiers presented, who were rendered unfit for duty by heart-symptoms, and yet without signs of valvular or other organic disease. Careful investigation of these satisfied me that the condition was one of **muscular exhaustion** *of the* **heart;** owing to hard marching with deficiency both of rest and food; especially during McClellan's peninsular campaign. Symptoms of this were,—constantly rather rapid though not strong pulse, with less than normal vigor of the impulse of the heart;—the acceleration increased greatly, with dyspnœa, upon even slight exertion. The sounds of the heart were not altered except in the diminution of duration and force of the first sound, making it more like the second. After many months of rest, these men improved, so as to be likely to recover. No special treatment seemed to be required.

ANEURISM OF THORACIC AORTA.

A bulging in the front of the chest, in which pulsation is felt, not continuous or identical with that of the heart, and over which resonance upon percussion is dull,—is probably an aneurismal tumor. If a thrill is also perceptible in it, with or without a murmur on auscultation, we may be still more confident in the diagnosis; and when the signs of pressure upon the air-tubes, œsophagus, sympathetic or recurrent laryngeal nerve, or thoracic duct occur, it is nearly certain.

Murmur may, however, be absent; so may thrill; the bulging may be slight, and the percussion resonance little altered. The sign of most consequence is, the existence of *two* points of pulsation in the chest, the cardiac and the aneurismal; the latter coinciding almost with the diastole of the heart.

The signs of pressure are, chiefly, pain, cough, dyspnœa, loss of voice, difficulty of swallowing; and (as I have seen in *one* instance) emaciation from obstruction of the thoracic duct.

Cancerous or other tumors may produce all these latter signs; but such tumors do not pulsate. In *empyema* the beat of the heart sometimes impels the fluid so as to throb somewhat widely: but this is still a *single* cardiac impulse. Occasionally a consolidated lung, in phthisis, may vibrate forcibly with the pulmonary artery; but other signs then make clear the disease.

The course of aortic aneurism is usually very gradual,—often lasting for a number of years. Death occurs—1. from sudden rupture and copious hemorrhage; 2. from slighter rupture and slow leakage; 3. from slow exhaustion by pressure, interfering with respiration, deglutition, &c.

The causation of thoracic aneurism is obscure. It occurs nearly always in rather elderly people, in whom the process of degeneration of the vessels has commenced; but now and then it is met with before middle life.

The following points may be added in regard to its clinical history:—

1. The effects of the aneurismal pressure may vary from time to time; much more than they do in cancer.

2. The aneurismal impulse may be even stronger than that of the heart; but a feeble impulse in some instances attends a large aneurism.

3. Destruction of one or more vertebræ from absorption under pressure (as shown by autopsy) is not uncommon.

4. Phthisis is often associated with aneurism of the aorta.

ABDOMINAL AORTIC ANEURISM.

Of this, the **signs** and **symptoms** are—deep-seated severe pain (occasionally intermitting) in the back and abdomen, increased by certain movements; unaccompanied by fever, but resisting all treatment; later, muscular spasms of the lower limbs, displacement of the liver, and the manifestation of a pulsating abdominal tumor, felt upon palpation, over which there is dulness of resonance upon percussion. The higher up the aneurism, the more severe are the pains and other symptoms of disturbance.

Aneurism of the aorta may, without careful examination, be confounded with aortic *pulsation* without tumor (common in dyspepsia, &c.), or with neuralgia, rheumatism of the bowels, colic, worms, disease of the liver, caries of the spine, psoas abscess, or cancer. Only the discovery of a distinctly *pulsating tumor* (not a tumor moved by subjacent pulsation) can establish the presence of aneurismal disease.

The **treatment** of either thoracic or abdominal aortic aneurism is, in a word, null. *Hygienic measures* may retard decline, and careful self-management may avert a sudden catastrophe; that is all. Exertion and excitement must, of course, be prohibited altogether.

AFFECTIONS OF THE ORGANS OF DIGESTION.

STOMATITIS.

Definition.—Inflammatory disease of the mouth.

Varieties.—1. Simple stomatitis. 2. Aphthæ. 3. Thrush. 4. Inflamed ulcer or cancrum oris. 5. Gangrene of the mouth. 6. Mercurial sore mouth or salivation. 7. Nursing sore mouth. 8. Scorbutic disease of the mouth.

Simple Stomatitis.—From taking very hot or corrosive liquids into the mouth, it may become inflamed; this condition being shown by redness, swelling, soreness and heat of the tongue, gums, lining membrane of the cheeks, palate, and fauces. Corrosives (as sulphuric acid or creasote) may *whiten* the mucous membrane superficially.

The course of such an affection is generally simple and brief—recovering in a few days under mild treatment. *Glossitis*, however, or inflammation of the tongue, may be more obstinate and serious. I have seen the tongue so swollen as to protrude from the mouth for more than a week, too large to return.

Slight ulcerations and fissures often occur in simple stomatitis, increasing the soreness and pain ; and increase in the flow of saliva is common.

Treatment.—In the beginning, holding ice, iced gum-water or flaxseed tea frequently in the mouth, or if a corrosive agent be the cause, almond oil or dilute glycerin [F. 50], will soothe the irritation. In violent glossitis, leeches may be applied to the swollen tongue; even free *incisions* may be called for to relieve its swelling ; later, solution of alum (ζij in fζvj of water) or sulphate of zinc (gr. j in fζj) may be used as a wash. Remember that such articles ought not to remain long in contact with the teeth, the enamel of which they may impair.

Follicular inflammation of the mouth is recognized by small red elevations over the tongue, soft palate, &c. This is common in infants during dentition; as well as in adults of impaired general health. It requires no speciality of treatment.

Aphthæ.—These are small ulcers, with whitish surfaces, following a vesicular eruptive inflammation of the mouth. The vesicles are small, round or oval, of a pearly appearance, and contain serum. They break in a few days, leaving a sore white ulcer, with redness around it. They may be scattered or confluent. Fever may attend the latter, with disorder of the stomach. Though not common in the earliest infancy, children sometimes have this disease, but less often than adults. Decayed teeth may produce it. On the whole, it is to be considered rare. Its duration is generally a week or two, but confluent cases may last a month, and have occasionally been fatal.

Treatment.—The constitutional condition may require cooling laxatives or saline diaphoretics, and gastric irritation may call for antacids, as bicarbonate of soda or magnesia. Chlorate of potassa should be given, 5 to 20 grains four times daily. Locally, at first, flaxseed tea or gum-water, or a solution of glycerin in rose-water, may be frequently applied. When ulceration occurs, a powder, consisting of equal parts of prepared chalk and pulverized gum arabic [F. 51], may be dusted or laid over each of the ulcers, several times a day. A wash of borax, myrrh [F. 52], alum, sulphate of zinc [F. 53], or acetate of lead, may also be applied. If the ulcers prove severe or obstinate, strong solutions of sulphate of zinc (15 grs. in f℥j of water) or nitrate of silver (20 grs. in f℥j), or solid sulphate of copper, may be used to touch the ulcerated surface every day or two.

Thrush; Muguet.—This is much most frequent in infancy. Its peculiarity is, the occurrence, after a day or two of diffused inflammation, of a number of small whitish points within the mouth, which coalesce and form patches of a whitish curd-like exudation (often confounded with *aphthæ*). In bad cases it may become brownish. This may fall off and be renewed, more than once. The mouth is hot, the stomach disordered; vomiting and diarrhœa may occur, with some fever. The attack lasts from one to two or three or more weeks; being seldom dangerous except in children otherwise in poor health. It sometimes attacks adults.

Nature.—The specific nature of the curd-like exudation appears to be connected with a *microphytic* (minute vegetative) growth, to which the name of *oidium albicans* has been given.

Treatment.—Experience favors the internal administration of *chlorate of potassa* [F. 54] in all severe forms of sore mouth. In the absence of a *rationale* by which its special applications might be definable, I would employ it in thrush as well as in aphthæ, &c. A child under five years of age may take from one to five grains of the chlorate, in solution, several times daily. As a laxative, magnesia will be suitable. Feeble cases may require quinine, beef-tea, brandy, and milk, in quantities proportioned to their condition and age.

Locally, at first, we may use flaxseed or gum arabic emulsion,—then glycerin and rosewater (one part to four or five), borax in solution (2 drachms in 4 ounces) or in powder, equal parts with sugar,— and later, tincture of myrrh in water (f℥ss in f℥ij), alum solution, or

14

sulphate of zinc, or muriatic acid with honey and water (acid. muriat. ʒj, mellis vel syrupi fʒj, aquæ fʒij) ; the latter being applied carefully with a camel's hair pencil, occasionally.

Cancrum Oris.—Canker of the mouth is characteristically ulcerative, from the commencement. It begins on the cheeks, gums, or lining of the lips ; but may reach the fauces. The ulcer is grayish or yellowish-white, with an inflamed border and environs ; the cheek may swell from it externally. It is quite painful. Saliva flows freely, and the odor of the breath is offensive. Fever is often present. The complaint may last for several weeks or even months; but is almost never fatal. It is most common in children, from two to six years of age.

Treatment.—Besides general measures, *adapted to the condition of the patient*, the same local applications, mentioned as appropriate in different forms of sore mouth, may be used. Direct touching of the ulcer with a strong solution of sulphate of zinc (gr. xv vel xx in fʒj), or with the solid blue stone (sulphate of copper) twice daily, will do the most for its cure; especially with the intermediate "dressing" of powder of chalk and gum arabic, and occasional washing with glycerine and rose-water.

Gangræna Oris.—Extreme inflammation or ulceration, in the mouth as elsewhere, may end in gangrene ; but this affection is peculiar, and may be unconnected with any severe inflammation. A morbid state of the system seems to predispose to it. It occurs mostly in children, but has been met with in adults.

There is, at first, an ash-colored ulcer, most often on the gums or inside of the cheek. If the latter, it is accompanied by swelling. Spreading, it assumes a sloughing character; the breath grows fetid ; acrid fluid is discharged, with copious salivation; other like ulcerations are formed, the bones of the face are affected with necrosis, and the teeth fall out. Penetrating the cheek, mortification may go on, rapidly, reaching sometimes even the ethmoid bone. Low fever and prostration attend these local changes ; later, diarrhœa, colliquative perspirations, and death. The only well-marked promotive *causes* of this very serious disease are, bad air (especially *crowd-poison*) and insufficiency of food. When treated early, it is often quite manageable ; but after extensive sloughing has occurred, the prognosis is bad.

Treatment.—Early, I should always try the chlorate of potassa. Quinine, and tincture of chloride of iron [F. 56] will be required on account of the tendency to prostration. Beef-tea and wine whey, or brandy or whisky punch, *pro re nata*, are called for, by the same indication.

To the part, at first, the astringent lotions, mentioned already, may be applied. When the gangrenous condition becomes pronounced, a solution of liquor sodæ chlorinat. in glycerin (fʒj in fʒij) may be applied frequently. Solution of creasote in glycerin, or in water (gtt. iij to gtt. xx in fʒj) may meet the same purpose ; or permanganate of potassa (gr. x in fʒj) ; or chloride of zinc (gr. j in fʒj) ; or sulphite of soda (ʒj in fʒj) ; or bromine (ʒss in fʒij).

Mercurial Sore-Mouth.—Salivation is made known in its approach, by a "coppery" taste, soreness of the gums, tenderness of the teeth when pressed together, with redness and swelling of the gums, and a broad white line just beyond their edge. The tongue also may swell. The flow of saliva increases greatly; the cheeks and even throat may grow sore and painful; the breath offensive. Ulceration of the gums takes place in severe cases, with loss of the teeth. Even sloughing may follow, approaching the state of things in *gangræna oris*. Difficulty of swallowing may be so great as to threaten starvation; and irritative fever may result from the local disorder.

Treatment.—Moderate salivation will always pass away in a few days, spontaneously. A good mouth-wash for it is brandy and water, one part of the former to four of the latter; alum may be added to it [F. 59], or a little tincture of myrrh. Ulcers or sloughs should be treated as in other varieties of stomatitis.

Opium may be called for, at least at night (*e. g.* Dover's powder, 10 grains at bedtime), by the distress of the system. Milk diet, or some other liquid nourishment, must be given during the difficulty of deglutition. In good practice, at the present day, no physician ever seriously salivates a patient.

Nurses' Sore-mouth.—Women who suckle children, and sometimes those who are advanced in pregnancy, are liable to ulcerative stomatitis. It begins with small, hard, painful swellings on the tongue and cheeks, which ulcerate, and are attended by a great deal of local, and sometimes constitutional irritation. When the infant is weaned, the affection subsides soon.

Treatment.—Chlorate of potassa has in this complaint a special curative power. 20 grains of it may be given three or four times daily. Iron, quinine, etc., may be required in subjects of obvious debility. Local treatment, such as has been given for *cancrum oris*, etc., will also have its utility.

Scorbutic mouth affection will be dealt with in another part of the book,—under *Scurvy*.

TONSILLITIS.

When severe, this is commonly known as *quinsy*. Soreness of the throat in swallowing, with pain or swelling of one or both tonsils, and fever, are its symptoms. Unless relieved in a few days, the pain becomes very constant and throbbing, dysphagia extreme, and when the patient begins to be seriously alarmed, a tonsillar abscess breaks, or is opened by the physician, and recovery soon follows.

Treatment.—A dose of citrate or sulphate of magnesia, or other cooling aperient, should be given the first day. Then, wine of ipecac., twenty drops every three hours, with frequent draughts of flaxseed tea or flaxseed lemonade. If the swelling, heat, and pain of the throat are great, apply from 20 to 40 *American* leeches to it. Then, or instead in mild cases or feeble subjects, poultice with flaxseed meal to which lard and laudanum have been added; bathing, when the poultice is changed, with liniment of ammonia, or soap liniment to which aqua ammonia has been added. If still severe, and not certainly suppurating, a *small* blister may be applied, or the part may be painted with tincture of iodine. When an abscess is evidently forming, poultices will be better, until it is ready to open from within.

Lancing the suppurated tonsil requires care, not to open the internal carotid artery. The point of the lancet should be directed towards the middle, not the outside, of the throat.

Not unfrequently, especially in children, repeated attacks of non-suppurating inflammation of the tonsils will leave them inconveniently enlarged. Sometimes persevering use of astringent gargles, or touching daily with strong solution of tannin or nitrate of silver, will make them shrink to the normal size. If not, excision of a part of the tonsil may be proper. With Fahnestock's, or any other guillotining instrument, the operation is easy and safe; at all events if it be not attempted to remove the whole gland, which is not necessary.

PHARYNGITIS.

Slight sore throat is among the commonest of affections, requiring for its treatment only mild gargles (as alum in flaxseed or sage tea), demulcents (flaxseed or gum arabic or slippery elm infusion), fomentation with volatile liniment or spirits of turpentine, and a dose of a saline cathartic, with *slop* diet. With children who cannot gargle, finely powdered alum may be blown into the fauces and throat, through a quill, more readily than in any other way.

Chronic pharyngitis is often a much more troublesome, though not dangerous local disorder. The mucous membrane becomes permanently hyperæmic, almost granulated; with either abnormal dryness or a thickened secretion; and constant soreness. In the treatment of this, all the different astringent, demulcent and alterative applications may be tried—sometimes without success. When nitrate of silver, tannin [F. 60], sulphuric and muriatic acids, sulphate of zinc and acetate of lead have been found to fail, it may happen that ice, or gargling often with ice-water, will prove more useful.

Counter-irritation, with repeated small blisters, tincture of iodine, or croton oil, is always a suitable and important part of the treatment of chronic inflammation of the throat.

Ulcerated Sore-throat.—This may be idiopathic, syphilitic, or tuberculous. The former is most uncommon.

The treatment in the first variety consists of the local application of blue stone or, lightly touched, solid nitrate of silver to the ulcers if within reach. The syphilitic will require, also, iodide of potassium [F. 61] internally (gr. v vel x ter die); the tuberculous, tonics, generous diet, and cod-liver oil.

RETROPHARYNGEAL ABSCESS.

This most often follows fever as a sequela; but is altogether rare. It is shown to the careful observer by dysphagia and dyspnœa, much increased by the recumbent posture; yet not, as in croup, increasing from day to day, or disappearing in a short time. There is also stiffness of the neck, and swelling on one or both sides of it. In such circumstances, a finger passed over the tongue into the pharynx may find a firm projecting tumor occupying its posterior and lateral walls. It may prove fatal, by asphyxia, or by preventing the patient from swallowing food. When diagnosticated in time, the matter may be let out by opening the abscess with a lancet, through the pharyngeal wall.

STRICTURE OF THE ŒSOPHAGUS.

This is uncommon. Its principal causes are, if structural, corrosive poisons, swallowed; or ulceration of the throat involving the œsophagus, and contracting upon cicatrization. Functional stricture may be spasmodic, as in hysteria. Dysphagia, not otherwise accounted for, and obviously low down in its seat, or the rejection of food partly swallowed, may lead to a suspicion of stricture; and examination with a bougie will fix the diagnosis. For the structural affection, I know of no appropriate treatment except dilatation with bougies made for the purpose, applied for a·short period, oiled, once or more daily.

GASTRITIS.

Simple gastritis, in an acute form, is very rare. I have met with but one case of it, in a woman who was kicked over the stomach by her husband. Corrosive poisons almost always involve the intestinal tube with the stomach. The most common form of "idiopathic" gastric inflammation is "gastro-hepatic catarrh," or "a bilious attack," in which the stomach, duodenum and liver are all somewhat involved.

Signs of stomachic inflammation are, epigastric pain and tenderness on pressure, rejection of all food and drink, jactitation, and fever; the pulse however being kept down by the impression made upon the circulation by constant nausea.

Post-mortem evidences of gastritis are—redness, browner or deeper and more livid than natural, and dotted, stellated or arborescent, rather than diffused; moreover, not confined to dependent parts; enlargement of bloodvessels; in acute cases, softening of the mucous membrane: in more lengthened ones, either softening or hardening and thickening; abundance of thickened mucus; rarely, coagulable lymph; almost never, pus.

Gastro-hepatic catarrh may follow any of the causes of indigestion, or exposure to cold and wet. There is nausea, or vomiting of greenish yellow fluid, generally not copious, but very acrid; headache and dizziness; constipation of the bowels, and fever. In the treatment of this, *magnesia* is a good quieting stomachic and cathartic; many will be relieved as soon by a bottle of solution of citrate of magnesia. Ice, melted in the mouth and swallowed slowly, will be comfortable. Rest and abstinence from food as nearly as possible will, with the above, generally complete the cure in two or three, or not many more days.

The best *preventive* or *abortive* of "bilious attacks" is *blue pill*, timely administered. Let the first nausea, constipation, and headache be met by giving at bedtime two or three grains of blue mass in pill (the "*lang syne*" portion was from six to twenty), followed in the morning by a teaspoonful or two of Husband's magnesia. If the bowels are free, bicarbonate of soda (which is a mild cholagogue as well as antacid) will be better; the eighth part of a teaspoonful at a dose.

Sick headache is usually a modification of the above, in so far as the sympathetic cephalalgia is especially severe. In some persons it
14*

is periodic. The treatment above mentioned will be adapted to a majority of cases of it.

Acute softening of the stomach is described by a few French and other writers, as a rapidly prostrating and dangerous affection in children, sometimes epidemic. Its symptoms are said to be, at first, those of simple gastritis; then, with or without diarrhœa, great agitation, prostration, want of sleep, and insensibility,—and death in one or two weeks from exhaustion. I have never met with any such case; although an irregular fever with gastric irritation (gastric fever or infantile remittent) once had a regular place in the nosological catalogue among fevers. It appears to me to be scarcely uniform enough for so special a designation or consideration.

CHRONIC GASTRITIS.

While the same doubt as to the pathological correctness of the *name* (indicating inflammation) exists in the case of this disease as in other "chronic inflammations" (see *General Pathology*), an affection of some distinctness of character, commonly called by the above title, is often observed. With the greatest brevity, we may indicate its symptomatology by contrasting it with that of *atonic dyspepsia*.

In Chronic Gastritis.	In Atonic Dyspepsia.
Much epigastric tenderness.	Little or no epigastric tenderness.
Pain increased by active exercise or stimulating food.	Pain not increased by exercise, lessened by stimulating food.
Vomiting usually.	Vomiting rarely.
Eructation of gas rarely.	Eructation of gas commonly.

Chronic gastritis is apt to be obstinate but not dangerous to life.

Treatment.—Counter-irritation over the epigastrium, by repeated vesication, will be useful. Internally, *nitrate of silver* [F. 62], in pill, beginning with gr. ¼, with gr. ⅛ of opium, and increasing in a few days or a week, gradually rising to 1 gr. thrice daily, with proportionate quantity of opium, I believe, upon experience, to be the most valuable medicine. Sub-nitrate of bismuth [F. 63] is for the same condition lauded by some. Most important is a *bland diet;* lime-water and milk, arrowroot, tapioca, sago, jellies, cracker soaked in ice-water, etc., in small quantities at short intervals. Ice will often quench thirst to better advantage, without disturbing the stomach, than water.

ANTI-EMETIC REMEDIES.

Vomiting is so frequent and troublesome a symptom, in many diseases besides inflammation of the stomach, as to demand much practical study from the physician. For this reason, though quite in deviation from systematic routine, I here introduce an enumeration of the most available medicines used for the relief of the *symptom* of vomiting—the selection among them depending upon the judgment of the practitioner as to the real *cause* of that symptom. At the same time it is clear that many of these remedies prove useful for vomiting when produced by very different and almost opposite causes; the symptoms, *as such*, rationally demanding medication when uncertain or in search of its cause.

Ice.	Cinnamon water.
Lime-water.	Infusion of cloves.
Mineral water.	Hydrocyanic acid.
Effervescing draught.	Aconite.
Champagne.	Chloroform.
Brandy.	Nitrate of silver.
Paregoric.	Oxide of silver.
Solution of morphia.	Subnitrate of bismuth.
Aromatic spirit of ammonia.	Oxalate of cerium.
Comp. tinct. of cardamom.	Enema of laudanum.
Comp. sp. of lavender.	Spice poultice.
Bicarbonate of potassa.	Sinapism.
Bicarbonate of soda.	Blister over epigastrium, vesi-
Magnesia.	cated surface being dressed
Camphor.	with acetate of morphia (gr. ij,
Calomel, small doses.	with gum-acaciæ, gr. x).
Blue pill.	Hypodermic injection of morphia.
Creasote.	

[See F. 64, 65, 66, 67, 68, 69, 70.]

ULCER OF THE STOMACH.

This serious affection is rare after the middle of life. It is most often met with in feeble systems, especially in women.

Symptoms.—Dull, sickening pain in the stomach, extending to the back, with *localized* tenderness on pressure. The pain is increased by motion, and by food, especially by *hot* food, or by sugar. Vomiting occurs, not copious, but rather frequent. Vomiting of *blood* is an important sign; it is impossible to be certain of the existence of an ulcer in the stomach without it. The amount of blood thrown up at once may be very small.

It is often difficult to diagnosticate gastric ulcer from *chronic gastritis*, as well as from *cancer, caries of the spine*, and *aortic aneurism*. No hæmatemesis, however, is met with in the first, third, and last; and a tumor, at some period, will make known cancer. So will angular deformity demonstrate spinal caries.

Perforation, causing peritonitis, and copious hemorrhage, are the most dangerous terminations of gastric ulcer. The signs of the former are, abdominal swelling and diffused pain, with collapse.

Treatment.—Bland diet is very important. Arrowroot, tapioca, sago, corn-starch, rice, and lime-water and milk are suitable. Beef or mutton tea (concentrated) will be better for the feeble than solid food.

Nitrate of silver, in pill with opium [F. 62]; oxide of silver, in 1 or 2 grain doses; and subnitrate of bismuth, are given with the hope of promoting cicatrization of the ulcer. Opium alone, in pill, or laudanum, &c., or conium or belladonna, as anodynes when the pain is severe. When hemorrhage is threatening, creasote (½ drop to 2 drops), tannin, acetate of lead, oil of turpentine (small doses), ammonio-ferric alum. Hypodermic injection of morphia has been used with advantage.

CANCER OF THE STOMACH.

Scirrhus of the **pylorus** is the most common form; occasionally the cardiac orifice is the seat of cancer. The usual symptoms are pain, in rare instances absent or nearly so, often excruciating; epigastric tenderness, about in proportion to the pain; vomiting. of food, mucus, and "coffee-grounds," or mixed blood and mucus, almost never pure blood; acidity and other symptoms of indigestion; fetid breath; decided constipation; emaciation, and cachectic, almost jaundiced, sallowness of complexion; sometimes irritative fever. The diagnosis is made nearly certain by the discovery of a tumor; not absolutely so,— as the tumor may be fibroid, and not malignant.

Cancer of the stomach seldom occurs before forty years of age. Its duration averages about a year; it seldom reaches two years. The patient commonly dies by a slow starvation, the stomach becoming incapable of digesting and transmitting food.

No treatment can avail for the *cure* of such an affection. To nourish by concentrated articles of diet, as beef-tea, milk, &c., and to allay suffering by judicious use of anodynes, will be all that we can do. It is a frequent form of cancer; as of 9118 cases of cancer in Paris in four years, 2303 affected the stomach.

Cancer of the Duodenum, Cæcum, Rectum, and *Omentum* are much more rarely met with. Their possibility must always be remembered in considering the diagnosis of abdominal tumors.

DYSPEPSIA.

Although denied a special place in nosology by writers upon diagnosis, clinical experience calls for a separate recognition of this as a disease, complex as its pathology is, and diverse as may be its symptoms. Of the latter, only a very general account can be given.

Symptoms.—The patient *feels* his stomach all the time, though not nearly always with pain. When the latter occurs, it is often in the breast, causing suspicion of pectoral disease. Little or no tenderness on pressure exists, nor is there much nausea, nor vomiting. The mouth is clammy, or has a sour or bitter taste. The complexion is sallow. The bowels are costive, and stools deficient in color. Other frequent symptomatic affections are cardialgia (heartburn), pyrosis (waterbrash), hypochondriasis, palpitation of the heart, headache, and disorders of the senses, as diplopia (seeing double), &c. Dyspepsia is not a dangerous, but is frequently a very obstinate disease.

Pathology.—The functional disturbances above enumerated have their seat, more or less prominently, in different parts of the digestive apparatus; in the alimentary mucous membrane, glandular organs, or muscular, or ganglio-nervous. The distressing gastro-intestinal irritation, cardialgia, pyrosis, &c., are located in the mucous membrane. Defective action of the liver and enteric glands produces constipation, with its consequences; imperfect secretion of the gastric juice, and pancreatic secretion, like hepatic inaction, impairs the whole process of digestion. So does atony of the muscular coat of the stomach; while deficient power of the peristaltic intestinal contraction is perhaps the most *common* cause of constipation. Insufficient or perverted *innervation* may originate or intensify any or all of these morbid states and actions. Sometimes this is so obviously primary and predominant, as to justify the use of the term, in certain cases, of "nervous dyspepsia."

Causation.—Most briefly, we may assert the causes of dyspepsia to be, either one or several of the following : too much food, or too little food ; imperfect mastication, and hurry in eating ; too little exercise ; too much fatigue ; excessive study, or emotion of mind ; inordinate use of ardent spirits, opium, tobacco, coffee ; or of medicine out of place.

Treatment.—This involves *regimen*, as well as medication : the first is most important. The meals should be regular, and with sufficient time allowed ; and all the food should be simple as well as nutritious ; variety being obtained rather by having a change from day to day, than by a number of dishes at each meal. Some dyspeptics are obliged to eschew variety, and confine themselves to a routine of beef, mutton, and stale bread. Caution should be used not to blame, unjustly, particular articles as " disagreeing," when everything disagrees, because of the state of the stomach. But a sensible person will be able mostly to ascertain what things agree *best* with his digestion ; and others should not be taken.

Most persons even of feeble digestion can eat beef, mutton, chicken, turkey, oysters (not fried nor raw, but roast, panned, stewed or steamed); with stale bread, bran occasionally, as more laxative ; crackers, boiled rice, tomatoes, and young and tender beets. When weak enough to require any stimulant, sherry or Madeira wine, or ale, or in the feeblest, brandy, will agree best at dinner. Let Dr. N. Chapman's precept be here remembered, " whatever we grant, *let it be sparingly.*" Advise, for example, *half* a wineglassful of Madeira or sherry, or half a tumblerful of ale, or one or two *teaspoonfuls* of brandy or whisky, at or after dinner. For breakfast and supper, tea will be the best drink for refreshment ; milk for nourishment ; although some persons do not digest milk with ease. Coffee I have so often known to *produce* dyspepsia, that I would forbid it altogether ; notwithstanding its toleration by some highly respectable authorities. Cocoa is too *rich* for most dyspeptic stomachs ; some find it acceptable. Fruit, especially when fresh, as a general rule is useful ; peaches, in season, are so with few exceptions. Stewed fruit is also excellent for laxative effect. Preserves, cakes, and pies, must be avoided. If a full meal cannot be taken without discomfort, it will be better to appease hunger or sustain strength by a small and simple "bite" between meals. Idleness and emptiness, as well as repletion, in the stomach, promote disorder.

Exercise, daily, in the open air, is very important to the dyspeptic. So is bathing, to maintain healthy action of the skin, with which the stomach sympathizes. But active exercise ought not to be taken just before or just after a meal. " After dinner sit awhile."

Mental states, and nervous impressions, are also of great consequence. Anxious occupation, or harassing responsibility, may increase greatly the difficulty of recovery. Thus travelling, or resorting to watering places, with release from care, may assist the cure.

The *medical* treatment of dyspepsia involves a number of indications, not always exactly the same. Especially are *tonics, laxatives, antacids* and other palliatives, and alteratives, apt to be required.

Tonics.—Pure vegetable bitters, as gentian [F. 72, 73], quassia, and columbo, are most suitable as direct stomachics. Chiretta is a favorite with some. Oxide of silver has had one or two enthusiastic advocates. Where *nervous debility* is prominent and particularly in cases of long-standing, extract of nux vomica [F. 71], or strychnia in very small doses (one-fortieth to one-thirtieth of a grain) will often do much more good than any other medicine. Iodide of iron, in anæmic cases, may be given.

Laxatives.—Rhubarb has been, time out of mind, the stand-by, for habitual constipation [F. 75, 76, 77]. If it be insufficient alone, or lose its effect, compound extract of colocynth, aloes, or resina podophylli may be added, in pill. Senna, magnesia, and sulphur may be used occasionally, for special indications. Saratoga, Cheltenham, and Vichy waters are found sometimes to have excellent regulating effects.

Antacids.—After meals, a pinch of bicarbonate of soda (gr. v to gr. x) or half as much bicarbonate of potassa, or a dessert-spoonful of lime-water, will, in case of acidity, contribute much to the comfort of the patient. Carbonate of magnesia and aromatic spirit of ammonia are preferred by some; and charcoal has useful absorbent powers. Sulphite and hyposulphite of lime or soda, for antiseptic effect, may also be given to allay the after symptoms of indigestion.

Alteratives.—In the commencement of the treatment of a case of dyspepsia, in which derangement, and commonly inaction, of the liver is most generally present, experience justifies the moderate use of blue pill. I prefer to give it in fractional doses, in such a case, say gr. ¼ thrice daily for a week [F. 74]. Occasionally it may require to be repeated, at intervals; but should never be pushed to salivation. Nitromuriatic acid, in 3 or 4 drop doses, acts as a mild tonic both to the stomach and the liver; and may well follow blue mass, where hepatic torpor is believed to exist. The same indication may be met, although with less certainty, by taraxacum. Nitric acid (2 or 3 drop doses) is highly lauded as a tonic by some practitioners.[1]

Cardialgia seems to depend mainly upon acidity, aggravated perhaps by the butyric fermentation. Aromatic spirit of ammonia, tincture of ginger, and camphor water, as well as the antacids above named, may be given for it; or chloroform, in 5 or 10 drop doses [F. 78].

Gastrodynia is a technical name for stomach-ache, common in dyspeptics. Carminatives are appropriate for it; one of the best of these is oil of cajuput, 4 drops at a dose, on a lump of sugar. Spirits of camphor, compound spirits of lavender, compound tincture of cardamom, and essence of ginger are among the most popular preparations for its relief. A mouthful of very hot water will sometimes quell the pain.

---

[1] Dr. Chambers, of London, speaks well of "Boudault's *pepsin*" obtained from the sheep's stomach. I do not know of any reliable equivalent for it in this country. Dr. Pavy, of London, states that a large part of Boudault's pepsin is inert.

Pyrosis is best treated by mild astringents; as oil of amber, catechu, krameria, ammonio-ferric alum [F. 80], creasote ($\frac{1}{2}$ drop or $\frac{1}{4}$ drop doses) [F. 81], tincture of chloride of iron.

After all, the dyspeptic may be able to do the most for his own cure. In the words of the late Prof. N. Chapman, "If he be intemperate, he is to become sober; if he use opium or tobacco, he must relinquish it; if indolent, he must be awakened to enterprise; if luxurious, he must consent to change his scheme of life; if studious, to abandon the midnight lamp; if afflicted, we must cheer him with the light of hope; or, if this be difficult, give him the great consolation of occupation, interest, employment."

The following table is, with slight modification, from Leared:—

EASY OF DIGESTION.	MODERATELY DIGESTIBLE.	HARD TO DIGEST.
Mutton	Beef	Pork
Venison	Lamb	Veal
Hare	Rabbit	Goose
Sweet-bread	Turkey	Liver
Young pigeon	Duck	Heart
Partridge	Wild waterfowl	Brain
Pheasant	Woodcock	Salt meat
Grouse	Snipe	Sausage
Beef-tea	Soups	Hashes and stews
Mutton broth	Eggs, not hard boiled	Mackerel
Milk	Butter	Eels
Whiting	Turtle	Salmon
Turbot	Cod	Herring
Haddock	Pike	Halibut
Flounder	Trout	Salt fish
Sole	Raw or stewed oysters	Lobster
Fresh fish generally	Potatoes	Crabs
Roasted oysters	Beets	Shrimps
Stale bread	Turnips	Mussels
Rice	Cabbage	Oil
Tapioca	Spinach	Melted butter
Sago	Artichoke	Hard boiled eggs
Arrowroot	Lettuce	Cheese
Asparagus	Celery	Fresh bread
Seakale	Apples	Muffins
French beans	Apricots	Buttered toast
Cauliflower	Currants	Pastry
Baked apples	Raspberries	Cakes
Oranges	Bread	Custards
Grapes	Farinaceous puddings	Nuts, pears, plums
Strawberries	Jelly	Cherries, dried fruits
Peaches	Marmalade	Cucumbers, onions
Toast water	Rhubarb plant	Carrots, parsnips
Black tea	Cooked fruits	Peas, beans, mushrooms
Sherry	Cocoa	Pickles
Claret.	Coffee	Chocolate
	Malt drinks	Champagne
	Port wine.	Cordials.

CONSTIPATION OF THE BOWELS.

There is no more frequent source of bodily discomfort than this; and it may produce or increase the tendency to disease. The principal *causes* of constipation are,—neglect of timely attendance upon the call of nature; want of exercise; excess of mental strain; and all the causes of dyspepsia, of which it is an almost constant part. Organic obstructions may also, of course, give rise to it;—as stricture, or cancer, or other disease of the large intestine, or a tumor so situated as to press upon the rectum; or pregnancy. The most remarkable instances of constipation I have met with, in the absence of mechanical obstruction, have been in sea-sickness.[1]

Effects of inaction or non-evacuation of the bowels may be, enteric irritation, or even inflammation; when much prolonged, dangerous intestinal obstruction; in other instances, diarrhœa; sympathetic headache, stomach, or liver disorder; urino-genital irritation; offensive perspiration; and contamination of the blood, by the retention of excretory matter which the bowels ought to remove.

Pathologically, costiveness may depend either upon muscular torpor of the intestinal canal, or defective glandular secretion, or both.

The **treatment** of constipation involves management as well as medicine. It is well to have a regular time to endeavor to empty the rectum. Straining is not beneficial, as it endangers piles or prolapsus ani; but the act of defecation may be facilitated by kneading the lower half of the abdomen with the hands, so as to increase and aid peristaltic contraction.

In diet, fresh and stewed fruits are the best natural laxatives. Prunes are especially opening. Bran bread, by the mechanical quality of the particles, is somewhat of a peristaltic persuader. Molasses occasionally will do, but it is too acescent to be taken constantly.

Of medicines, pills of rhubarb and castile soap, one and a half grains each, may come first; afterwards podophyllin [F. 82], colocynth, aloes, &c., if needed for especial torpor. An occasional dose of senna infusion may give a good start to the bowels. Better than to take medicine every day or two, will be the use sometimes of *enemata*. An injection of warm water alone may answer at first. Then white soap, and afterwards brown soap may be added; salt and molasses, sweet oil or castor oil when the former fail. A *suppository* of soap [F. 83] is less disagreeable to some persons, and will generally act.

Where the torpor of the rectum approaches a paralytic state of the muscular coat, nux vomica will be very important [F. 84, 85]. The addition of belladonna to laxative pills is a not uncommon practice. Electricity may assist in relieving the same condition.

ENTERITIS.

Definition.—Inflammation of the bowels.

Symptoms.—Pain in a portion of the abdomen, increased by motion or pressure; constipation; fever. Later, abdominal swelling, vomiting, and mucous, sanguinolent or even purulent diarrhœa, in bad cases.

Causes.—Blows or other injuries; neglected constipation; exposure to cold and wet. Corrosive poisons, as arsenic, &c. also cause enteric

---

[1] I have, when at sea, passed seven days without any inclination for a movement; and a gentleman told me that he had been eleven days without an evacuation, during a voyage.

inflammation; and it is a part of the results of strangulated hernia or other intestinal obstruction.

Treatment.—A decidedly open and active case may require or bear early venesection in the robust. Leeching should be the rule; and it may be free. After that, poultices, of flaxseed meal or Indian mush, covered with oiled silk to retain moisture. Soft food alone can be taken, as arrowroot, oat-meal gruel, &c., or. in the weak, beef-tea. No medicine can do any good, unless it be opium in moderate doses, to relieve severe pain and promote rest. Cathartics are to be avoided, and entire stillness of the body in bed must be maintained. From slight or moderate inflammation of the bowels recovery may be confidently expected, with care; but aggravated cases of it are frequently fatal.

Typhlitis is inflammation of the cæcum or caput coli. It is rather more common than other forms of enteritis, especially after neglected constipation. **Peri-typhlitis** is a more obscure affection, differing, it seems to me, in involving a local or circumscribed peritoneal inflammation with typhlitis. Pain, tenderness, swelling and dull resonance on percussion, in the right iliac fossa, with constipation, are the symptoms. A number of cases which I have seen have all recovered. With more especial propriety than in most other cases of enteritis, opening the bowels by enemata at least, and even by a mild laxative, as castor oil, has seemed to be indicated. Rest, leeching, poulticing and soft diet, are the other main parts of the treatment.

Abscess may occur notwithstanding; with safe issue if it open outwardly—but fatal if it rupture into the peritoneal cavity.

PERITONITIS.

Inflammation of the peritoneum is one of the most dangerous of the phlegmasiæ, because of the extent and important connections of the membrane involved.

Varieties.—Simple or idiopathic; accidental or traumatic; tubercular; puerperal.

Causes.—Exposure to cold and wet; falls, blows, wounds, or other injuries; abscess of the liver; opening of aneurism, or perforation of gastric or intestinal ulcer (as in typhoid fever); tuberculization; the puerperal state.

Symptoms.—Diffused abdominal pain and tenderness, increased greatly even by *slight* movements, as breathing deeply or raising the lower limbs in bed; vomiting; constipation; tympanites; fever, with *very rapid*, though not full pulse. Later, that is in three or four days, in violent cases, delirium, insomnia, collapse. Its course is usually rapid; from the incipient chill to the fatal end, often occupying less than a week, though sometimes two. Simple sporadic peritonitis, however, even in puerperal women, is, with careful treatment, much more often recovered from than not.

Diagnosis.—The most important point is the discrimination of "simple peritonitis or metro-peritonitis in the puerperal state" from puerperal fever. The main difficulty about this is that the latter disease *includes* peritonitis almost as constantly as erysipelas does dif-

15

fusive inflammation of the skin. We can best discuss this differential diagnosis after considering the fever in question (See *Puerperal Fever*.)

Morbid Anatomy.—After death from peritonitis, the swollen abdomen is found nearly always to contain fluid, often considerable in amount, serous, sero-sanguinolent, sero-purulent, or pus. The latter may form in a very few days; some facts have made it probable, even within forty-eight hours. *Adhesions* are present, with bands and false membranes of coagulable lymph, in various parts of the abdominal cavity; and redness, thickening, and opacity exist to a greater extent.

Treatment.—No disease requires or bears better the early use of the lancet than acute peritonitis. One free bleeding may sometimes, as it were, arrest the conflagration. Yet, apart from epidemic puerperal fever, in which bleeding has upon the amplest trial, proved rather destructive than curative, there are cases in which economy of material makes venesection unsafe. Then leeching may be resorted to, in all but the very feeble subjects. Fifty or a hundred American leeches may be borne upon the abdomen by a patient who would faint if the same amount of blood were rapidly taken from a vein. Exposure of the body, during leeching, may be, with care, avoided. Poulticing with flaxseed or Indian meal should follow the leeching; the poultices should be large, but light, and covered with oiled silk, or changed very frequently to maintain warmth. If no leeches have been used, flannel dipped in spirits of turpentine may be put all over the belly. Later, if the case threaten obstinacy, a large blister should be applied.

Of medicines, *opium* has now the almost universal confidence of practitioners. Except emptying the rectum at first, by mild enemata, no agitation of the bowels by medicine is to be encouraged. Calomel, as an *antiplastic*, has been long valued by physicians of sagacity and experience. Though unable to *prove* that it does lessen the tendency to the *effusion* of coagulable lymph, I bow to the rational empiricism which, not hastily, raised it to the position of reliance for that end. I am not satisfied that its utility as an antiphlogistic, especially in severe inflammations, has been disproved.

With opium, then, I would, in peritonitis, give calomel: ¼ grain to 1 grain of the former, with as much or less of the latter, every 2, 3, or 4 hours, according to the severity of the pain and the urgency of the case [F. 86, 87]. When the stage of debility comes on,—or in very feeble cases from the first,—quinine, instead of calomel, may be combined with opium; and support with beef-tea, and wine, brandy, or whisky, may be required.

When peritonitis follows an *injury*, the treatment may necessarily have to be modified by the concomitant states of other organs involved, or by the general shock of the system. So, also, when perforation of an ulcer of the stomach or bowel, or the rupture of a hepatic or other abscess or aneurism, brings it on,—*collapse* is apt to occur speedily, forbidding any but anodyne and supporting treatment, and affording very little hope under that. Such cases are almost invariably fatal.

Chronic peritonitis is sometimes met with. When not tuberculous, although a very serious affection, it may be recovered from; the tuberculous form, not with any more probability than would be pulmonary phthisis.

Chronic peritonitis should be treated by rest in the recumbent posture (in tedious cases the patient may be *carried* out into the sunshine and air), and resolvent and counter-irritant local applications; as repeated blisters, tincture of iodine, mercurial ointment, cerate of carbonate of lead. The latter, as a local sedative, I have found to have remarkable power. It may be prepared by adding ʒij of carbonate of lead to ʒj of fresh simple cerate [F. 88]. .

COLIC.

Varieties.—1. Flatulent. 2. Bilious. 3. Spasmodic, gouty or rheumatic. 4. Lead colic. Some writers also speak of *nephritic* colic; the pain of which is chiefly owing to the passage of small calculi from the kidney to the ureter; while *neuralgia* of the bowels may also cause pain of similar seat to colic. *Uterine* colic, in females, may be either neuralgic, spasmodic, or obstructive (dysmenorrhœa).

Flatulent Colic.—This is caused by indigestion; as, from excess in the amount, or error in the quality of food; or, from cold and wet, arresting perspiration and disturbing the balance of the "aqueous visceral circulation," which is indispensable to normal digestion. Acrid *irritation* and gaseous distension produce irregular tonic or spasmodic contractions in the intestines; principally in the colon. They are not confined to this, however. In a woman with irreducible umbilical hernia, I have, during an attack of colic, felt a portion of small intestine, several inches in length, grow rigid during the access of pain, and relax when it was relieved. Sometimes the stomach itself is the seat of pain.

In flatulent or crapulent colic the abdomen is distended, but not very tender, except after long continuance of the attack. There is constipation of the bowels; often nausea, with belching of wind. sometimes vomiting; no fever. A sign of the yielding of the attack is audible or palpable rumbling of wind in the bowels; showing a return of the almost arrested peristaltic motion.

Bilious Colic.—The onset in this form is slower. Nausea is greater, and vomiting, of greenish or yellowish (biliary) fluid, is nearly constant. The pain may last, with very slight remission, for a number of days. The bowels are constipated. There may be considerable fever, and some tenderness of the abdomen on pressure. Meteorism is generally present; but less in proportion to the pain than in flatulent colic. In protracted cases, slight or moderate jaundice is quite common.

The greatest suffering in cases of bilious colic is attendant upon the passage of gall-stones from the gall-bladder to the duodenum. Then, the pain is chiefly in the right hypochondriac and lower part of the epigastric region; and sudden relief follows the escape of the calculus from the *ductus choledochus* into the intestinal canal. In other cases, we suppose that the irritant which gives rise to spasmodic pain is acrid, unhealthy bile; which escapes into the intestines, and also, through the pylorus, into the stomach.

Certain persons are particularly liable to such attacks; a large majority of people, indeed, are never subject to them. I have known, in a number of instances, the same patient to have severe bilious colic once in every two or three weeks; in others, the interval may be of months or years. In one case, under my care, the attack was fatal. Autopsy then showed rupture of the gall-duct, under distension from obstruction by an impacted calculus. This must be very unusual. But prolonged bilious colic is never quite free from danger of inflammation of the bowels, or, in feeble persons, exhaustion from continued suffering and inanition.

Gouty Spasmodic Colic.—In the "gouty diathesis," this is one mode in which the disease may invade internal organs. The stomach is the most frequent and dangerous seat of it; the attack being commonly called "cramp in the stomach." It is characterized by suddenness, extreme severity of pain, and tendency to coldness and general prostration of the system. Repulsion of gout from the foot, as by cold applications, may bring it on.

Lead Colic; *Painter's Colic; Colica Pictonum.*—This disease has long been known as the result of exposure to the poisonous influence of lead. The name of "dry belly-ache" has also been applied to it. The abdomen is *shrunken* and rather hard; sometimes *knots* of contracted intestine may be felt. There is no tenderness, the pain being lessened or relieved by pressure. The suffering is often extreme, with restlessness; the face and body being thrown into grotesque contortions. Constipation is obstinate; the feces, when passed, small, dry, and hard. No fever exists. There is a blue line along the edge of the gums. Lead palsy may attend or follow the colic.

Treatment.—In all forms of colic, the indications in common are, 1, to open the bowels; 2, to relieve pain and spasm; 3, to prevent inflammation; 4, to prevent future attacks.

In flatulent colic, we should ascertain if the stomach has just been overloaded, or any very unwholesome food has been taken. If so, a prompt emetic will be proper; as, a teaspoonful of mustard, or a tablespoonful of salt, in a teacupful of warm water—repeated in ten minutes if necessary. Then the antacid laxative, magnesia, may be given; a teaspoonful, with ten to twenty drops of essence of ginger, or ten drops of essence of peppermint, five or six drops of oil of cajeput, or some other aromatic in corresponding proportion [F. 89]. If the bowels are not opened, or relief of the pain not obtained, no great length of time must elapse without an enema, of castor oil, salt, and molasses, or soap, in warm water.

Should the stomach be much unsettled, and the pain violent, we may depend upon the immediate use of an injection to open the bowels; and may give by the mouth antacids and carminatives. Thus, aromatic spirit of ammonia, spirits of camphor, compound spirit of lavender, or oil of cajeput may be given, with bicarbonate of soda [F. 90]. Small doses every few minutes will be better retained than large ones at long intervals, and will act better.

Anodynes come next in order. Extreme and sudden cases of colic belonging rather to the **spasmodic** variety require them *at once.* Other cases, the majority, are better managed by commencing with more *corrective* remedies, as above mentioned. When relief is not obtained without, we must give opium, chloroform, ether, or Hoffmann's anodyne [F. 91]. The first is of all the most certain, although chloroform, internally used, in ¼ drachm to ½ drachm doses, has not disappointed me. Paregoric is a very good opiate for the same purpose. Pills of opium (especially *old* pills) may do better sometimes, where as much as a grain at once may be needed for severe pain. The "chloroform paregoric" [F. 92, 93] combines several good antispasmodics conveniently. Laudanum is the oldest stand-by, and well deserves its place.[1]

It is remarkable how much opium a sufferer with great pain will sometimes bear without narcotism. I have known a teaspoonful of laudanum to be taken at once; not even drowsiness following it. But care must be taken not to overdo this, or to give any more than is really necessary; or, the remedy may possibly prove worse than the disease.

An important part of the treatment of colic is the use of warm external applications. Mustard should come first; a large sinapism, half and half with flour (if the mustard be of good strength) and covered with gauze or thin muslin, over the abdomen. When it is removed, after making a decided impression, let a little lard, sweet oil, or cold cream be rubbed on to prevent further irritation of the skin. A concentrated liquid preparation of mustard has lately been in use. Then, apply a hot flannel, dry, or wrung out of hot whisky and water. For the latter, the best mode is to add to very hot water an equal quantity of raw whisky. Such appliances should be often *renewed*, or they grow cold. Some persons have a tin vessel constructed to hold hot water, and shaped so as to fit over the abdomen. This is very good, if it can be used without its weight causing too much pressure. The feet of the patient should be kept warm; if he is able to sit up, or recline with the legs over the side of the bed, a hot mustard foot-bath will be suitable.

Kneading the abdomen gently with the hand will aid to dispel flatus; but it requires tact not to make it too violent an operation. In every case of violent colic, the possibility of *hernia* must be held in mind; and its presence or absence should be ascertained.

Infants are especially liable to crapulent colic; some, during their first year, having almost daily or nightly attacks. Very simple treatment will often suffice in these; in children, too, over-medication should be even more sedulously avoided than in adults. For infantile colic of slight severity, peppermint water, or infusion of fennel seed, will frequently be enough, with the application of a warm flannel over the stomach. Worse cases may be treated with lac assafœtidæ [F. 98]; which children generally take well, if it be sweetened, in teaspoonful or, for very young infants, half teaspoonful doses. Antacids, as bicarbonate of soda, will assist in giving relief [F. 94]. Keeping the bowels regular, never allowing a day to pass without an evacuation, is most important in young children. For this purpose the simple syrup of rhubarb, manna, and fluid extract of senna are the best medicines. The latter may give pain, but is less apt to do so if one drop of oil of cloves be added to each fluidounce. A *very small* quantity of *resina podophylli* added to syrup of rhubarb [F. 95]

---

[1] *Codeia* and *Narceia* are, of late, getting into use, to some extent, instead of opium and morphia. Their after effects are said to be less unpleasant.

will make it more potent when the bowels are torpid. Glycerin may be occasionally substituted, in teaspoonful doses.

When the food of an infant becomes acescent, lime-water may be added to it when it is taken; e. g., a tablespoonful of lime-water in each pint of milk. Over-feeding an infant is a very common cause of colic.

Bilious colic may be attended with so inflammatory a condition as, in a person of full vigor, to demand early and moderate venesection. Opening the bowels is a cardinal indication in this as in the flatulent form. If the stomach will bear it, castor oil will be the most effectual cathartic. The least unpleasant way of taking this is, in thorough admixture with spiced syrup of rhubarb; two tablespoonfuls of the latter with one of oil. Magnesia may be retained better than oil upon the stomach. Notwithstanding its effervescence, I have known the *citrate* of magnesia solution do very well in colic. The same *antacid, carminative,* and *anodyne* remedies, mentioned for crapulent colic, will be suitable in bilious, and may require more persevering administration. So, also, enemata, mustard plasters, pediluvia, and warm applications to the abdomen are of great service. Besides these, however, a special indication exists for promoting the hepatic secretion, so that by greater fluidity and dilution it may be made less irritating and obstructive. The ordinary treatment, then, is, besides such palliatives as have been named, to give calomel, with opium: e. g., ⅓ to 1 grain of calomel with about as much opium, every two, three, or four hours [F. 86]. Leeches, cups, or later, a blister, over the liver, may be right, if hepatic or cystic inflammation threaten.

When there is strong reason to apprehend that the passage of a gallstone is the cause of the severe pain, the warm bath, if practicable, will be useful by promoting relaxation; and full doses of opium may be called for by the patient's agony. Some prefer to inhale ether, or nitrous oxide.

Gouty, or other **cramp of the stomach,** is generally in need of very prompt treatment; essentially stimulant and anti-spasmodic or anodyne. In moderate cases, Warner's cordial (tinct. rhei et seunæ) has the advantage of being laxative as well as stimulating; from a teaspoonful to a tablespoonful may be given at once, in *hot* water. In worse attacks, brandy, ether, laudanum, and Hoffmann's anodyne are more reliable [F. 97]; with a sinapism over the epigastrium, and a hot mustard foot-bath. Subsequent treatment, prophylactic of future attacks, as with colchicum or other medication, must be pointed out by the nature of each case.

Lead colic, when rapidly produced, may be treated antidotively, with sulphate of magnesia. If slowly brought on, we can do much less in that way; although it has been asserted that the iodide of potassium has an eliminative power over lead combined with the tissues of the body. Alum is confided in by some, for the same end, notwithstanding its astringency. Castor oil as a laxative; the warm bath to relax spasm, and opium to relieve spasm and pain, are the most important usual remedies in this affection. The costiveness being mainly spasmodic, it is not unfrequently found that, contrary to its common effect, opium promotes, in lead colic, the movement of the bowels. Iodide of potassium is believed to exert a decided power in removing from the system lead which has been slowly deposited in various organs.

Prevention of Crapulent and Bilious Colic.—This becomes the duty, if not the interest of the physician; when his patient has been relieved, to aid him in escaping returns of the disorder. To prevent the flatulent form, care in diet and regimen will ordinarily suffice. For the more serious attacks of bilious colic, to which certain persons are subject, prevention is attainable by the same means, along with especial attention to the *abdominal movements and secretions; i. e.,* the state of the liver and bowels. I am sure that I have enabled several persons, who for years had been liable to frequent attacks, to escape them altogether, by a very simple prescription, used upon the first threatening of any of the symptoms. Blue mass and rhubarb were here the sanative agents. A gentleman who has suffered terribly with bilious colic told me that twenty grain doses of extract of taraxacum, taken once or twice, have repeatedly averted it with him. Flowers of sulphur, or *lac sulphuris,* given in teaspoonful doses or less, every day or two for several weeks, have answered the same purpose in another instance. But nothing has so signally satisfied me, as a prophylactic against periodical colic, and also, by a similar rationale, against *sick headache,* as the preparation above alluded to ; which is as follows :—

℞.—Mass. ex. hydrarg., gr. v.
Pulv. rad. rhei et
Ext. gentian, āā ℨss.
Ol. caryophyll. gtt. iv.—M.
Div. in pil. No. 20.

S.—One or two occasionally, as directed; to be continued, if required, thrice daily for several days.

OBSTRUCTION OF THE BOWELS.

Few maladies present so striking a contrast as this, between the facility of pathological explanation after death and the obscurity of diagnosis and uncertainty of treatment during life.

Pathological Varieties.—Dr. Haven has well classified these as follows: 1. Intermural: *a,* cancerous stricture; *b,* non-cancerous stricture, viz., 1, contraction of cicatrices from ulceration, 2, contraction of wall of the intestine from inflammation ; *c,* intussusception ; *d,* the latter with polypi. II. Extramural: *a,* bands of adhesions from lymph ; *b,* twists or displacements; *c,* diverticula ; *d,* tumors or abscesses ; *e,* mesocolic or mesenteric hernia ; *f,* diaphragmatic ; *g,* omental, and *h,* obturator hernia. III. Intramural: impacted feces, calculi, coagula, curdled milk, etc.

Symptoms of Intestinal Obstruction.—These are, persistent constipation ; constant vomiting, partly or altogether stercoraceous ; coldness of the skin, prostration, distressed countenance (facies Hippocratica), collapse. Local evidences, rather more distinctive, are, hardness or swelling in one part of the bowels; arrest of enemata at a certain point, and of borborygmi (gaseous movements) in the same way. If the obstruction be high up, suppression of the urine occurs, with early vomiting. If it be low down, great meteoric distension and stercoraceous vomiting. When blood is passed from the bowels, with such symptoms, intussusception may be inferred.

But, at last, a *probable* diagnosis is all that the nature of the case will admit. The differential discernment of special forms of obstruction during life is impossible.

Treatment.—The simple, primary indication in persistent constipation, with unrecognized cause, is *catharsis*. Castor oil, sulphate of magnesia, croton oil, are, justifiably, given, aided or seconded by enemata of the same or similar purgatives. When the diagnosis of *intestinal obstruction* has been well made out, no more cathartic medicines are to be given; the reliance then being upon nature and opium. The latter drug may be prescribed in grain or half-grain doses every few hours, to sustain a tranquillizing effect favorable to relaxation of the intestinal coats. Besides, we may try *large* enemata of warm water; or inserting a bougie, or stomach-tube, to *catheterize* the bowel, as far as the ileo-cæcal valve : or, the Hippocratic remedy of large *air* injection, to distend and dislodge the intestine. This has succeeded in several cases of intussusception. *Scybala*, or impacted feces, or coagula, &c., may be removed by a spoon or scoop from the rectum. Prolonged use of the warm bath may be tried to relax the system ; and, as in strangulated hernia, the tobacco injection may become allowable as an extreme resort.

COMMON REMEDIES IN COLIC.

The following are put together simply as *memoranda :—*

Peppermint, Fennel, Cajeput ;   Assafœtida ;
Lavender, Ginger ;   Hot water ;
Aromatic spirit of ammonia ;   Brandy ;
Bicarbonate of soda ;   Calomel ;   -
Magnesia ; Castor oil ;   Enemata of oil, or
Warner's cordial ; Olive oil ;   Spirits of turpentine, or
Camphor, Ether, Chloroform ;   Laudanum ;
Opium, in pill ; Paregoric or   Sinapisms and
Laudanum ;   Pediluvia ; Kneading ;
Hot flannel or plate ;   Warm bath.

Remember, always, the possibility of *strangulated hernia* as a cause of the symptoms of colic.

CHOLERA MORBUS.

This very unscientific name has become inseparably attached to what in technical phrase may be most briefly called *idiopathic emetocatharsis ; i. e.* vomiting and purging, neither brought on by irritant poisons, nor by an epidemic influence. The account I shall give applies best to such an affection as we commonly meet with it in this country, especially in the summer. English medical writers describe it sometimes as English cholera ; others, sporadic cholera.

Symptoms.—Nausea, and vomiting of greenish or yellowish fluid, with rejection of all food and drink ; often, but not always, pain in the stomach and bowels ; diarrhœa, with brownish or yellowish stools ; debility, and coldness ; little or no fever. Beginning with such symptoms, if the attack, not relieved, becomes aggravated, cramps in the limbs supervene; the vomiting and purging become more watery; prostration and coldness deepen into collapse,—which may be fatal.

Causation.—Warm weather seems to predispose to it, by relaxing the mucous membranes and exciting the liver. Direct causes often are, indigestible articles of food, as unripe fruit, &c.; excess of ordinary food; sudden change of temperature, checking perspiration.

Diagnosis.—From epidemic cholera, it is important to distinguish cholera morbus; as the prognosis is not the same, nor will the same treatment answer for both. The difference is seen in the *bilious* vomiting and purging of cholera morbus, and *rice-water* discharges of cholera; the greater nausea in the former; much more tendency to collapse, with blueness, dyspnœa, and suppression of urine, in cholera. The presence or absence at the time of an epidemic of the latter may complete the diagnosis by confirming or correcting the evidence of the above signs. It is only in an extreme case of cholera morbus that any real difficulty should exist. During, and before and after, the prevalence of epidemic cholera, an especial tendency to cholera morbus, as well as diarrhœa, often exists. This, called *cholerine*, may present more near resemblance to malignant cholera than our ordinary summer attacks.

Treatment.—A large sinapism should be at once placed over the epigastrium. All theory or *rationale* apart, the following mixture is *admirably useful* in ordinary summer cholera morbus:—

R.—Sp. ammon. aromat. f3j.
Magnes. optim. 3j.
Aquæ menthæ piperitæ f3iv.—M.
To be shaken when taken.
S.—A teaspoonful every twenty minutes.

Few cases will fail to be relieved in an hour or two if this be given *early.*

When the diarrhœa is copious, or the case is seen rather late, paregoric may be added to the above,—f3ij or f3ss in the same mixture. When purging is very urgent and exhaustive, instead of magnesia a like amount of bicarbonate of soda may be used. Infusion of cloves, cinnamon, or ginger may assist to quiet the stomach in an obstinate case. After the sinapism, a spice poultice, of ginger, cloves, and cinnamon, each a full teaspoonful, with a tablespoonful of flour, moistened with brandy, should be applied. Ice may be given if thirst be great.

Extreme prostration may require the use of brandy internally. To check the diarrhœa and vomiting when threatening collapse, a laudanum and starch enema (40 to 60 drops of laudanum in ¼ ounce of starch) may be given; and a blister may be applied over the stomach, the part to be dressed, when vesicated, with 2 grains of acetate of morphia mixed with 10 of powdered gum-arabic.

DIARRHŒA.

Though rather a symptom than a disease, excessive discharge from the bowels often requires express treatment for its relief.

Varieties.—These are, principally, 1. Irritative diarrhœa, as from dentition; 2. Inflammatory, as in enteritis; 3. Symptomatic, as in typhoid fever; 4. Critical, as at the close of remittent fever; Eliminative, as in septic or other poisoning; Colliquative, as in phthisis. The character of the *discharges* varies very much. They may be, 1. Fecal, although liquid; 2. Bilious; 3. Mucous; 4. Serous; 5. Adipose (very rare).

Except in the beginning of attacks, discharges are rarely fecal in . character when much beyond the normal amount. The *gutter-water* discharges of typhoid fever often have nearly the fecal appearance except in consistence. *Mucous* discharges occur in enteritis, and in many cases of summer diarrhœa. *Bilious* passages occur in cholera morbus. *Serous,* or "rice-water," in malignant cholera.

Treatment.—An important point is, that in many cases diarrhœa ought not to be abruptly checked; in some it should not be interfered with at all. The latter is true of the looseness of the bowels in typhoid fever, if the passages are not more than three daily, and are but · moderate in amount. When excessive in that disease, they require checking, not arresting.

Ordinary summer diarrhœa, the most nearly "idiopathic" of all forms, demands *correctives,* generally, before or with astringents. Blue mass or hydrargyrum cum cretâ; magnesia, with charcoal or with aromatic syrup of rhubarb [F. 101]; bicarbonate of soda, with ginger or cinqamon, etc., will often relieve the condition of the alimentary canal in which diarrhœa originates, and thus end it without any astringents.

When the latter are indicated, by continuance or increase of the discharges, *chalk mixture* has long held a routine place as an early prescription. Instead of it some prefer *testa præparata,* or *oculi cancrorum.* In infants, lime-water, with cinnamon or camphor water, will do for mild cases. Kino, catechu, krameria, and hæmatoxylum are familiar as pure astringents. The addition of opium, or camphor, or both (as in paregoric) in small doses to such preparations is generally proper, to increase the binding effect, even in the absence of pain [F. 101, 102, 103, 104].

More obstinate cases should be treated with tannin (gr. iij in pill. with ¼ or ½ grain of opium, pro re nata), or pills of acetate of lead and opium (gr. j of the acetate, with gr. ¼ of opium) every three or four hours; or a mixture containing acetate of lead with acetate of morphia; aided when necessary by enemata of laudanum and starch (30 to 60 drops of laudanum to ½ ounce of starch, cool or cold). As an article of diet in feeble cases, arrowroot with brandy will be especially suitable.

In **chronic** diarrhœa, besides the remedies last mentioned, sometimes enemata of acetate of lead solution, or of some other mineral astringent, will do good. Mention of these will be again made in connection with *chronic dysentery.*

The *food* in cases of diarrhœa always requires regulation. Vegetables and fruits, as a rule, ought to be forbidden; the popular prejudice which makes the blackberry an exception I believe to be a mistake. It has had its origin in the known astringency of the *root.* Boiled rice, and other *farinacea,* will nearly always be suitable. In severe cases, all solid articles of food should be withheld.

Scorbutic diarrhœa, however, from the nature of its *cause,* demands a quite different régime. Officers in the army who were affected with diarrhœa on the Chickahominy in McClellan's campaign, have told me that, when astringents had no effect in checking the complaint, tomatoes, peaches, and lemonade cured it at once.

CHOLERA INFANTUM.

Popularly known as "*summer complaint*," this affection is very destructive to young children in the large cities of this country, in hot weather. The peculiar influence of high heat in an atmosphere contaminated by "town" causes, generates it. In New York and Philadelphia, its prevalence and mortality coincide with the rise of the thermometer above 90^C. The deaths for the hottest week in July, 1866, in New York were over 1200, and in Philadelphia over 700 ; more than in either city during the prevalence of cholera later in the same season, and more than twice the usual mortality.

Symptoms.—These are, diarrhœa, vomiting, rejection of food, languor, debility, apathy ; sometimes stupor. At first the head may be hot, the abdomen swollen ; as the case progresses, coldness and emaciation supervene. In some, with predominance of cerebral symptoms, death may be threatened after a very few days of sickness. In others, copious diarrhœa and constant vomiting endanger the same result. In many, however, without violent symptoms, the child is gradually reduced by diarrhœa and inanition. The period of dentition is particularly liable to this disorder; it seldom occurs after four years of age.

Pathology.—Although decided alteration of the *follicles* of the intestines, with some change in the general appearance of the mucous membrane, has been shown to be generally present after death from cholera infantum, the disease is most probably rather a systemic than a local one. Its seat must be in the whole nutritive apparatus, including the ganglionic nerve-centres. Sanguification is evidently impaired, and an imperfect blood deranges the action of the brain and spinal cord ; hence the stupor, or extreme apathy, and, in some bad cases, convulsions.

Treatment.—*Correctives* are, here, especially important in the beginning. I am, from considerable experience, a full believer in the great value of moderate doses of *calomel* in the early stages of summer complaint. I would always give it, with an antacid [F. 107]. When stomach or head symptoms predominate, with but little diarrhœa, calomel with magnesia will do the best. When there is more looseness, bicarbonate of soda should be used, with the calomel, instead. Spiced syrup of rhubarb may be added to either. Hydrargyrum cum cretâ is the preferred mercurial with many practitioners. I have found it to answer very well, after or even perhaps instead of calomel.

A spice poultice or plaster should be kept over the abdomen so long as vomiting continues ; being renewed or wet freshly with brandy often enough to maintain its strength. Ice (pounded in a rag for young infants) may be given more often than water to quench thirst. The food may be lime-water and milk, arrowroot, farina, chicken-water, beef-tea. After the first stage, many children will require small quantities of brandy (preferably with their food) for support.

In the early stage, if the head be hot and stupor be threatened, a few leeches behind the ears, and the application of cold water, upon a light cambric handkerchief, to the head, may be proper. Such a stage, however, does not often last long.

Later, the two difficulties are, to check the diarrhœa, and to overcome the rejection of food by the stomach. For the bowels, astringents are then called for; especially logwood, blackberry root, geranium, krameria; aided in serious cases by paregoric in small quantities by the mouth, or even the injection into the bowels of one, two, or three drops of laudanum with starch. Sometimes acetate of lead injections (from one to three grains, with starch) may be farther needed, for the same intent.

Protracted summer complaint affords scope for perseverance and contrivance in finding food available for the child. Well-made beef-tea agrees with most children. *Raw beef scraped* or rasped fine, has been found to answer the purpose best with some.

But all medical treatment may fail in some cases of cholera infantum, which will speedily recover on being removed from the city to the country. The immediate effect of a salubrious air is often surprising and delightful.

Prophylaxis.—This is very clear and simple. A child under five years of age ought never to be kept in the close-built parts of a large city, in our climate at all events, through June, July and August, if it can be helped. Next to a *residence* for the summer in a high and open country, will be the benefit of frequent *excursions* or visits; riding or sailing; or even, if nothing else be possible, being carried daily into the squares or parks of the city.

DYSENTERY.

Definition.—An inflammation of the large intestine, involving the muscular as well as the mucous coat.

Varieties.—Acute and chronic; sthenic and asthenic; endemic or epidemic; bilious; ulcerative; strumous or tuberculous.

Symptoms.—Pain in the lower half of the abdomen, with soreness or tenderness on pressure or motion; frequent disposition to go to stool, with small and bloody or blood-marked muco-fecal or mucous passages, sometimes containing shreds of lymph or false membrane; tendency to strain (tenesmus) with griping (tormina); fever in most acute cases.

Severe and protracted cases may be considered as going through, 1st, the inflammatory, and 2d, the ulcerative stages.

Simple **acute** dysentery is commonly sthenic, or open, active, and inflammatory, without early or great tendency to prostration. Endemic or epidemic dysentery (the first name is the more correct) is generally asthenic. In this form fever may be absent, or brief, or of a typhoid character. Vomiting is not rare in this, as it is in the ordinary acute form. Coldness and debility come early.

Sometimes, in malarial districts, dysentery, like all other maladies, may be intermittent ; with daily or tertian exacerbations and intervals.

Morbid Anatomy.—Redness, turgescence, thickening, softening, ulceration, suppuration, and occasionally pseudomembranous deposits, are, after death from dysentery, found, in various degrees, in the rectum, colon, and cæcum; chiefly in the lower bowel. The hemorrhage which makes the typical bloody stools, is due to the congested and inflamed mucous membrane being constricted, in the tenesmus, by spasmodic and irregular contractions of the muscular coat.

Chronic dysentery presents nearly always ulceration of the rectum or colon, or both. The discharges in this may become almost entirely muco-purulent.

Causation.—Predisposition to dysentery is common in the latter part of summer; in this city and neighborhood, from the middle of August to the end of September especially. Relaxation from heat, with sudden exposure to cold and wet, may produce an attack. So, often, will indigestible food; as unripe fruit. Bad drinking water is another cause.

At any season and locality such agencies may produce simple acute dysentery. But in certain regions it becomes at times endemic. This is particularly noticed in many localities having considerable elevation, not subject to malarial fevers, but *within a short distance* of ague districts; dysentery upon the hills, while intermittent and remittent occur in the adjoining or subjacent valleys and meadow lands.

Prognosis.—Either form of dysentery *may* be fatal; but the endemic and asthenic type is much the more dangerous. The other, with good *early* treatment, is generally quite manageable. When allowed to become chronic and ulcerative, the doubtfulness of recovery is much greater. *Bilious* dysentery, that is, the form in which disorder of the liver is a prominent feature, the discharges presenting an excess of more or less altered, irritating bile, is more intractable than ordinary simple dysentery.

Treatment.—Simple acute *form.*—Now and then we may find a robust patient who will require to be bled during the first, active, stage of dysentery. Much more often, leeches over the abdomen, where the tenderness is greatest, will be suitable. After these, warm poultices, of flaxseed meal, mush, &c., may be put on. Later, in obstinate cases, a large blister, in the same region.

At the very *start*, the old practice of beginning with a dose of castor oil, with ten or fifteen drops of laudanum, will do very well. If left for a day or two, it had, as a rule, better be omitted.

Then the first prescription, in a mild or moderate case, may be of blue mass with ipecacuanha. After one or two days (sooner in an urgent case), camphor may be added, in pill. Next, we may substitute, for the blue pill, opium; afterwards, omit the ipecac., continuing the opium and camphor, *pro re nata.* If the disorder be still not checked, we must resort to acetate of lead, with opium, or in solution with acetate of morphia [F. 111, 112, 113, 114].

Perfect rest is indispensable to prompt recovery from dysentery; there is no disease in which this can be more important.

The diet must be bland; as rice-water, arrowroot, or other farinacea; chicken-water, or beef-tea in the feeblest cases. When thirst is intense, iced rice-water or benne-leaf tea, or infusion of slippery-elm bark, may

16

be used as a drink; or, during the active stage, ice in substance may be taken slowly.

Enemata are very important in dysentery. First, of flaxseed-tea, as a demulcent (two to four fluidounces at once); the same with laudanum; or laudanum with starch [F. 115]. In chronic cases, or obstinate acute ones, acetate of lead may be given by enema, with laudanum, in mucilage. So may sulphate of zinc, and nitrate of silver. I have seen some remarkable cures of chronic dysentery by the use of an enema containing ten grains of sulphate of zinc, forty drops of laudanum, and four ounces of flaxseed tea. Such an injection may be painful at the time, and would be too irritating except in an ulcerative case of considerable standing; for which it should be reserved. Solution of tannic acid, in water or in glycerin, will be worthy of trial for a similar purpose.

Asthenic, endemic form.—In this there will be need of the earlier use of opium; and, often, of quinine, and stimulants. No leeching, or little, is likely to be well borne; and ipecac. may be prohibited by vomiting. When it can be taken, in small doses (not more than $\frac{1}{4}$ a grain), I believe it to be a valuable remedy. When malarial influence is obvious, and most of all in the intermittent form, quinia or cinchonia will be the remedy, to which others are adjuvants [F. 117]. *Hope's mixture* will be more likely to do good in this, the adynamic, than in the simple acute form. (R.—Acid. nitric. f3j; tinct. opii, gtt. xl; aquæ camphoræ, f3viij; dose a tablespoonful.)

Bilious Dysentery.—As a distinctive variety, this is not uncommon; and, if it last over ten days, it may be very hard to cure. Ordinary anti-dysenteric medication will not be inappropriate to it—but may disappoint much more than it is apt to in simple acute cases. Without having a very satisfactory recollection of the results of treatment of such cases in my own experience, I should trust most in the *withholding* of mercurials in the first stage, the gradual introduction of one of them in the second week, the application of a blister at the same period over the liver, and, besides opium, acetate of lead, &c., as called for for astringent effect, the administration of nitro-muriatic acid. Of course the chemical incompatibility of this with lead must be remembered; but this will not interfere with saturnine injections while using the acid by the mouth.

HÆMORRHOIDS.

Definition.—Piles; tumors, at the verge of the anus, or within the rectum.

Varieties.—External and internal; varicose and fibrous; dry and bleeding.

Symptoms.—At first, weight and fulness in the rectum; soreness about the anus; pain, increased upon having a stool. The pain may extend up the loins, and down the limbs, even to the feet. As inflammation increases, throbbing and aching may become almost constant. Swelling, and then the formation of one or more distinct tumors, occurs. If without the anus, there may be every variety of painfulness, aggravated at certain times. If internal, the prolapsus of the tumor during defecation, and its constriction or strangulation by the *sphincter ani.* cause great suffering; often the tumor requires to be put back by the hands. Occasionally it cannot be returned, but undergoes mortification and sloughing away. **Bleeding** occurs from internal hæmorrhoids. The amount may vary from a teaspoonful to a pint or more in a day. Cases are recorded by good authorities in which several pounds of blood have been lost in a single night. Commonly it is much less; but may be enough to blanch and reduce the patient to the extreme of anæmia and debility.

Anatomy.—Inspection shows **external** piles to be globate, broad-based tumors, at the verge of the anus, covered by thin integument; livid in color when fresh, losing that hue when old; tense and elastic to the touch, and very tender, at least during inflammation.

The old idea that every hæmorrhoid is a dilated vein, has been corrected by observation. Piles consist of distended skin and connective tissue, with contained extravasation of blood, and deposit and organization of lymph, from local congestion.

Internal hæmorrhoids are described as chiefly of three varieties: 1st. Solid, round or pear-shaped, attached by a peduncle, smooth, and dull in color; composed of mucous membrane, connective tissue, and thickened veins. These bleed very little if any. 2d. Broad-based, bright-red, spongy tumors, villous on the surface, and bleeding readily, arterial blood; consisting of loose folds of mucous membrane, with hypertrophied connective tissue, and enlarged capillary and small arterial and venous vessels. 3d. Florid, very vascular excrescences upon the mucous membrane, not of large size, but bleeding sometimes copiously.

Complications and Sequelæ.—These are, especially, ulceration, abscess, fistula, fissure of the anus, prolapsus ani, and sympathetic irritation of the urethra, bladder, prostate, or testicles in the male, or of the uterus and vagina in the female. *Sloughing* of a strangulated hæmorrhoidal tumor is considered by some to endanger life; but my own observation of its occurrence would lead me to depreciate this danger. Certainly very fine natural cures thus occur.

Moderate bleeding from inflamed hæmorrhoids gives temporary relief. When habitual and not excessive, its sudden arrest may possibly promote some internal visceral congestion,—as apoplexy.

Diagnosis.—Hæmorrhoids may be mistaken for venereal excrescences, or polypi of the rectum, or for prolapsus ani. The first are harder, more abrupt in their elevation and margins, and of a quite different history; in addition to which other marks of the syphilitic constitution exist. Polypi are of slower growth, and unaccompanied by inflammation, or, as a rule, by hemorrhage; and their surface is smoother than that of piles. Prolapsus ought to be easily made out, by examination disclosing the structure of the everted mucous membrane.

The source of bleeding from the rectum may sometimes be in doubt, as to whether it be hæmorrhoidal or not. True hemorrhage from the bowels, other than from piles, is the result commonly of serious and obvious disease; as typhoid fever, yellow fever, &c. Such flow of blood is itself painless, and the blood is dark, clotted, and variously mixed with fecal matter when passed; and the symptoms of piles are absent.

Causation.—Hereditary predisposition sometimes exists. Hæmor-rhoids are uncommon in either sex before puberty; in females they are most frequent at the time of the cessation of menstruation. Warm and damp climates promote them; as in the East and West Indies, &c. The plethoric constitution is the most liable to them; especially with sedentary habits. Pregnancy is attended by them not unfrequently. Other causes are, long standing, or sitting upon hard seats, excessive venery or self-abuse, over-stimulating diet, misuse of purgatives, espe-cially of aloes, ascarides, diarrhœa, dysentery, stone in the bladder. Constipation of the bowels always predisposes to hæmorrhoids.

Treatment.—This must be both *general* and *local;* the former depending upon the constitutional condition, and the cause of the affection. The bowels must be *regulated;* neither over-purged nor allowed to be costive; a *soluble* state is the most desirable. The bleeding of piles must be but cautiously interfered with, if it has been habitual, or if there is a tendency to apoplexy, phthisis, gout, or in-sanity.

The *diet* must be made to consist of digestible and unirritating food. Long standing and sitting, or rough riding, must be avoided; although active exercise in the open air may be very advantageous.

External piles may often be averted in the forming stage, by atten-tion to the bowels, along with the frequent application of the simplest unguents to the irritated and swollen part. Lard, tallow, cold cream, or simple cerate, or spermaceti ointment, will answer very well; but the grease should be applied several times daily, and especially after a stool,—so as to keep the part constantly soothed by it [F. 118, 119, 120, 121].

The laxatives most approved for hæmorrhoidal cases are rhubarb, sulphur, and senna. The confection of senna is a very good prepara-tion for such use. Magnesia is irritant to piles; and so are, though in less degree, the saline cathartics. Enemata are objectionable merely because of the mechanical pressure of the instrument. In internal hæmorrhoids they are often decidedly serviceable. When piles are inflamed, washing with cold water, or a cool sitz-bath, may relieve. Some patients prefer warm water under the same circumstances.

When *bleeding* is so considerable as to need to be checked, cold water injections, or solution of alum, or tincture of iron, may be em-ployed. A piece of alum made into a smooth suppository will some-times do. In really threatening hemorrhage, the patient must lie still in bed.

On the other hand, inflamed non-bleeding piles may require local depletion, by leeches, or, as many prefer, cupping over the sacrum.

Prolapsed internal hæmorrhoids often have to be replaced by the hand. Oiling will of course facilitate such reduction.

Astringent ointments, as of galls, tannin, carbonate of lead, or creasote, with regimen and laxatives, may cure piles even of consider-able standing. But old and obstinate cases demand removal by opera-tion.

External hæmorrhoids should be *excised*, with curved scissors or a probe-pointed straight bistoury; taking off no more integument than what covers the tumors. Good authority, however, pronounces touch-ing carefully with *nitric acid* to be safe and successful.

Internal hæmorrhoids ought, when operated upon, to be removed always by *ligature.* Excision is dangerous, and has several times been fatal, by hemorrhage. Some prefer cauterization with nitric acid. In ligating hæmorrhoids, it is best to apply a double ligature around the base of each tumor. Silk or hemp will answer; Bushe's needle-receiver is a good instrument for the application.

FISSURE OF THE ANUS.

This is a very painful and not uncommon affection, especially in middle life; perhaps most frequent in females. Neglected constipation and hæmorrhoids, with relaxation of constitution and sedentary habits, are its principal causes.

Its symptoms are,—at first, soreness or smarting at one point of the anus at stool. This becomes afterwards very severe; with intense pain, burning, aching and throbbing, and violent spasmodic constriction of the sphincter ani, lasting sometimes for hours.

Examination displays a lesion mostly of the mucous membrane only; though occasionally reaching even to the muscular fibres of the sphincter. In the beginning only a crack, it becomes at last an extended ulcer; and may exist on each side of the anus.

The stools are streaked with pus or blood, and often reduced in size by the spasm of the rectum; suggesting stricture of the rectum; for which this complaint has now and then been mistaken. The suffering of the patient in bad cases is extreme; pain being produced, not only by defecation, but also by coughing, sneezing, stimulating food, or even by the sitting posture.

Treatment.—Most cases, even of long standing, may be cured without an operation. The fissure may be managed as an irritable ulcer; by the constant application of soothing unguents,—as spermaceti or oxide of zinc ointment, lead cerate, unguentum belladonnæ, or lime-water with oiled silk dressing. The latter will be convenient only in the recumbent posture.· Experience in analogous cases would lead me to have especial confidence in *collodion,* to which one-fiftieth of glycerin has been added to lessen its constricting effect. This may be painted upon the part with a camel's hair pencil, as in fissure of the nipple; it makes an excellent artificial cuticle.

Obstinate cases may be treated also with nitrate of silver or sulphate of copper, applied every day or two, lightly, to the surface. Where suffering is great, suppositories of opium and cocoa butter, or of belladonna, may be introduced after defecation. Washing with soap and water, twice daily, will be serviceable.

Should all such measures fail, Boyer's operation, as modified by Copeland and Brodie, should be resorted to. It is, incision through the ulcer, with a bistoury, either from within outward, or from without inwards. It is only necessary to cut through the mucous membrane, not through the sphincter. Mild dressings must follow the incision; which will usually produce rapid recovery.

Dr. W. H. Vanburen's operation consists in the forced dilatation of the sphincter ani, by the two thumbs of the operator; so as to overcome the spasmodic contraction.

16*

PROLAPSUS ANI.

Partial descent of the rectum without the anus is not rare in the adult, but is more common in children. Relaxation of the mucous membrane, or weakening of the anal muscles, may induce it; straining at stool is its usual immediate cause. Tropical countries afford its most numerous examples.

Treatment.—The protruded bowel must be replaced. Commonly, gentle pressure, with lard or oil, and *tact*, will succeed at once. If not, leeches and cooling applications must be applied to reduce congestion and swelling. Sometimes anæsthesia will be a needful aid; but not often.

Having effected the replacement, a pad and **T** bandage will maintain it for the time. The bowels must then be carefully regulated. For the rest, *preventive care* is the main thing. Children affected with prolapsus must not be allowed to strain. The chair or other seat used by them ought to be *high*, so as not to flex the thighs much upon the body. The more nearly erect the posture, the less force in the bearing down.

Old prolapsus in the adult may not be curable without operation; although the air-dilated gum-elastic pessary will sometimes give relief. I refer for the *operation* to works on Surgery.[1]

AFFECTIONS OF THE LIVER.

ACUTE CONGESTION.

This, with deficient secretion of bile, is very common, as the result of exposure to cold and wet in warm seasons or climates, of the chill of intermittent, or of excesses in diet. Its symptoms are, a sense of weight and slight or moderate pain in the right hypochondriac region and under the right shoulder-blade, constipation with lead-colored stools, nausea, a furred tongue, bitter taste in the mouth, a yellowish skin and conjunctiva, and headache or dizziness.

Treatment.—Two or three grains of blue mass at bedtime, one, two, or three nights (two grains only if repeated). When decided constipation exists, one purging dose in the morning of sulphate or citrate of magnesia, or of magnesia. Then ten or fifteen grains of bicarbonate of soda twice daily, with light diet.

CHRONIC CONGESTION OF THE LIVER.

A number of attacks of temporary hepatic congestion, or of hepatitis, or prolonged dyspepsia, or intermittent or remittent fever, may induce a chronic hyperæmia of the liver, with variable disturbance of function. Pain in the right side and shoulder, with sallowness of complexion, constipation, and lowness of spirits, are the principal symptoms.

Treatment.—Supposing blue mass to have been temporarily and sufficiently used, as the leading cholagogue (so established by ample

[1] See Ashton on the Rectum, p. 157.

clinical proof, notwithstanding the failure of a *portion* of the physio-logical experimentation upon the subject), nitromuriatic acid may be then given, 3 or 4 drops twice or thrice daily, for two or three weeks successively. Or it may be used in a bath (f℥vj–viij in each gallon of water). Taraxacum, 10 or 20 grains of the extract twice daily, or a wineglassful, as often, of the decoction of the root, or the fresh leaves in spring or summer, eaten as greens, may follow. Leptandrin (dose, gr. j–iij) is said to be mildly cholagogue and safe. Ordinary laxatives, as rhubarb, etc., may be used to regulate the bowels. Care of the skin, by bathing, proper clothing, and, if chilly, friction with hair gloves or a rough towel (salt bathing will be very good) is important. Exercise in the open air, not violent, should be had every day. Change of air, mineral waters, or sea bathing, may be advised.

HEPATITIS.

The most common form of inflammation affecting the liver is what some writers call " gastro-hepatic catarrh ;" considered on a previous page. There is reason to believe the duodenum, stomach, gall-duct, and liver to be all in variable degree involved in such attacks.

Hepatitis may also be *traumatic*. Whether so or idiopathic, either the parenchymatous tissue, Glisson's capsule, the biliary ducts, or the portal vein, or all together, may be the seat of inflammation.

Some of the symptoms are nearly the same in all cases, and are in part the same as in acute congestion of the liver ; but the pain in the side is greater, with some tenderness on pressure ; there is fever, often vomiting, and sometimes diarrhœa.

In inflammation of the capsule (perihepatitis) the tenderness on pressure, movement, or deep inspiration, is considerable ; the fever, slight or absent ; and there is no jaundice. This may sometimes be confounded with *diaphragmatic pleurisy ;* but there is, in the latter complaint, more severe pain, with cough, dyspnœa, and hiccough.

Inflammation of the portal vein may proceed to suppuration. Then the symptoms are scarcely distinguishable from those of hepatitis with abscess, to which attention will be given presently.

When inflammation is chiefly confined to the gall-bladder and ducts, the points of diagnosis are, the comparative absence of fever, and the considerable degree of jaundice.

Abscess of the Liver.—Although much most common in tropical climates, this may be met with anywhere. Besides the usual symp-toms of hepatitis, when pus is forming, we find rigors, recurring almost regularly as in intermittent, a very rapid pulse, prostration, copious perspirations, and loss of flesh. In a considerable number of cases, however (13 per cent. according to Louis), the disorder is latent ; being made known only by the consequences of suppuration.

The greatest danger attends the escape of pus from the abscess. This occurs spontaneously either through the diaphragm by the lungs, into the stomach, or intestinal canal, into the peritoneal cavity, or, in a minority of cases, through the skin. Any of these may be followed by recovery, except the escape into the cavity of the peritoneum. In this instance, death is almost certain.

The **causes** of abscess of the liver, besides the predisposition belonging to hot climates, are 1. Blows or wounds; 2. Inflammation of the portal vein, with transfer and deposit of pus; or *thrombosis* from some other vein, as the hæmorrhoidal; 3. Dysenteric ulceration; 4. Inflammation and suppuration of the gall-bladder or gall-ducts.

Treatment of Acute Hepatitis.—A highly febrile case in a vigorous subject may be treated by early venesection. Otherwise, leeches to the right hypochondrium will be suitable. All mercurials should be avoided. Saline cathartics are proper, with rest in bed, low diet and cooling drinks. A blister may follow leeches or cupping.

The most serious question occurs when suppuration is known or believed to have taken place. Can we prevent or lessen the dangers of the discharge of the **abscess**? Nature in many cases makes this secure, by adhesion of the liver to the stomach or bowel, so as to allow of the direct flow of the pus into the canal. In other instances, deep-seated fluctuation may be felt below the edge of the ribs. Possibly this might be, a *dilated gall-bladder*, or *hydatids* of the liver. But, if sure that it is an abscess, ought we to open it? The most prudent answer is, not unless we are confident that only the skin intervenes between the pus and the exterior. A very judicious medium between this and bolder practice has been proposed by Dr. Graves: to make an incision about four inches long right over the centre of the tumor, but reaching through the muscle to *within a few lines of the peritoneum.* This, even when the matter is deeply seated, is shown by experience to favor and hasten essentially its escape, without the dangers of a peritoneal incision. Even *acupuncture*, or the use of the exploratory needle-trocar, will be both less safe and less beneficial than this plan.

After the discharge of the abscess, convalescence may be expected; it is sometimes rapid, but may require a month or two.

JAUNDICE.

Icterus or jaundice is a morbid yellowness of the skin, eye, and other parts. It has no uniform pathology, causation, or concurrent symptoms; but is itself so marked an occurrence as to deserve special study. Sometimes it is even epidemic; as in the U. S. Army in malarial districts during the late war; to the extent of over 10,000 cases in a single year.

Varieties.—As to degree,—*yellow, green,* and *black* jaundice. As to causation, jaundice from *suppression,* and from *re-absorption* of bile; and *icterus neonatorum,* jaundice of young infants, of still different origin.

Symptoms.—In ordinary acute cases of jaundice, either suddenly or after some days of *malaise,* the whites of the eyes first become tinged with yellow; next, the roots of the nails, the face, neck, trunk, and limbs. The urine is of a porter color, stains linen yellow, and becomes green on the addition of nitric acid. At the same time the stools are slate or lead-colored, or almost white. The mouth has a bitter taste, and the patient suffers with lowness of spirits and indisposition for exertion.

Pathology and Causation.—Many affections of the liver may induce jaundice; although in some of the most serious of them it may be absent. Most distinctly it is traceable in different cases to the non-removal of the biliary coloring matter, as well as of cholesterin, from the blood by the liver; other organs, especially the skin, then receiving it; or, to obstruction preventing its transit, after secretion, through the intestinal canal, in which case it is reabsorbed into the blood and then thrown out elsewhere.

The remote causes of jaundice of greatest frequency are, malaria, exposure to cold and damp in hot weather, pregnancy, and violent mental emotion.

Diagnosis.—In either form of jaundice we have the yellow conjunctiva and skin; or serum, if a blister be applied, or blood drawn; in both the stools are without color, and the urine yellow or yellowish-brown. But, as Harley first pointed out, in jaundice from *suppression* the biliary *acids* have not been formed, and we find only the bile *pigment* in the urine; while in jaundice from *reabsorption*, that fluid contains *both*.

Harley's test is as follows: "To a couple of drachms of the suspected urine add a small fragment of loaf sugar, and afterwards pour slowly into the test-tube about a drachm of strong sulphuric acid. This should be done so as not to mix the two liquids. If biliary acids are present, there will be observed at the line of contact of the acid and urine, after standing for a few minutes, a *deep purple* hue."

After a time, in cases in which the secretory powers of the liver become impaired, the biliary acids disappear; and then, *tyrosin* and *leucin* are found in the urine. To detect these, evaporate slowly an ounce of the urine to the consistence of syrup, and put it away to crystallize. Tyrosin is known by fine *stellate groups* of needles under the microscope. Leucin, by flat circular crystalline disks, soluble in water but not in ether.

Prognosis.—Acute jaundice is not very often fatal. In the U. S. army, of 10,929 cases only 40 died. When it lasts a month or two, however, as well as when acute yellow atrophy of the liver exists, there is always danger connected with its organic cause. The jaundice of young infants is of short duration, and almost never of serious consequence.

Treatment.—When supposed to be temporary and functional, the great object must be to restore the action of the liver. As remarked already, the large accumulation of clinical experience, sustained by some though not by all of the physiological experiments made by vivisectors, compels me to believe that calomel and blue mass and other mercurials are *cholagogues*. If they be not so *always* in trials upon animals in *health*, they have proved so *generally* in human beings in cases of torpor of the liver. If *obstruction* be the trouble, their action is more doubtful, necessarily. But even then they may promote the solution of a recent gall-stone, if they render the bile more copious and liquid.

Moderate doses of calomel or blue pill may be urged, then, generally, during the first week or more of treatment. These may be aided by saline purgatives, as sulphate or citrate of magnesia, Rochelle salts, or cream of tartar. After them, small doses of *resina podophylli* may be tried, if required by persistence of the disease; or, if the bowels will not bear purging, extract of taraxacum. Bicarbonate of soda, taken before meals, is mildly cholagogue. But, in a case of some weeks' duration, slow to recover, nitromuriatic acid, 3 or 4 drops twice or thrice daily, will often hasten recovery very much. This I saw in a number of cases of malarial jaundice from the Army of the Potomac, in 1862.

ACUTE YELLOW ATROPHY.

This is a generally fatal affection, occurring most frequently in those who have been intemperate, or injured by venereal excesses, or who have been exposed to malaria.

Symptoms.—Beginning like ordinary jaundice, with nausea, constipation, and headache, the skin becomes intensely yellow ("black jaundice"); vomiting comes on, the pulse is rapid, though variable, and delirium occurs. Then, with fever, and often pain in the side, the stomach and head are more and more disturbed. Vomiting of altered blood takes place; not unfrequently also hemorrhage from the bowels. Petechiæ appear on the skin. Prostration, tremors or convulsions, and *coma* end the history, usually in less than a week.

Secretions.—Marked *deficiency* of *urea* in the urine, and the presence of *leucin* and *tyrosin* in that excretion, have been remarked.

Morbid Anatomy and Pathology.—The liver after death is *flattened out* and lessened to perhaps less than half its normal size. Its cut surface has a yellow color like rhubarb; the bloodvessels are empty. The lobules are not distinctly marked, many of the secreting cells being destroyed; in their place are masses or spots of dark bile-pigment, fat, and hæmatin. The kidneys are often found in a state of partial degeneration.

Evidently atrophy, with cessation of functional action, of the liver, is here the cardinal fact. Is it preceded by a violently destructive inflammatory process? Some of the symptoms would point to this. Yet, in the absence of autopsic evidence, uncommon as primary rapid atrophy seems to be in any organ, the precedence of inflammation must not be taken for granted. The cause of death seems to be cholæmic poisoning.

Diagnosis.—From acute hepatitis this complaint is distinguished by the greater amount of jaundice, the occurrence of hemorrhage from the stomach or bowels, the severe headache and stupor; but, most of all, by the *diminution of dulness on percussion* over the hepatic region, in connection with symptoms showing violent disorder of the liver. The urine will also be found after evaporation to contain *tyrosin* and *leucin ;* sometimes in crystalline deposits.

Treatment.—Unless in the earliest stage we are warranted in endeavoring to promote the normal "unloading of the portal circle" by purgatives, it is difficult to see any hopeful indication for treatment in this affection, other than palliation of fever, if there be such, by diaphoretics, aiding the depuration of the blood by diuretics and laxatives, and prolonging life by appropriate *support*. It is doubtful whether any cases recover from acute yellow atrophy of the liver.

PIGMENT LIVER.

Frerichs and others have found after death from remittent fever, or in patients dying from other diseases after exposure to malarial influence, a peculiar condition of the liver. It is steel-gray, or blackish, or chocolate-colored; presenting brown insulated figures upon a dark ground. This change of color is due to the accumulation of pigmentary deposit in the bloodvessels.

The spleen is also somewhat similarly altered; and so, to a less extent, are the brain and kidneys. The blood is deficient in corpuscles, and contains many floating particles or masses of pigment.

Diagnosis.—During life, examination of a few drops of blood will display the abundance of free pigment. The skin is sallow or dull yellow. Enlargement of the spleen, anasarca, albuminuria, diarrhœa or intestinal hemorrhage, and delirium or a tendency to stupor, may occur. There is but little jaundice.

Pathology.—The scientific interest of this affection turns chiefly upon the proof it affords of the effect of malarial poison in disorganizing the blood-corpuscles. This is in accordance also with the remarkable and important influence in *chronic* malarial disease (as obstinate intermittent), of *iron,* as a remedy.

Treatment.—The discovery of pigmentary degeneration or deposit in the blood, or the supposition of its occurrence in the liver or other organs, does not offer any new or special indication for treatment, beyond what the other conditions of the case present. The malarial poison is to be antagonized, and the system aided in restoring the disturbed organs and functions to their normal balance; the appropriate means for which ends will be considered under other heads.

CIRRHOSIS.

Synonyms.—*Hob-nailed liver, gin-liver.*

Anatomy and Pathology.—In its commencement or first stage, cirrhosis is attended by some increase in the bulk of the liver; with increase also of its firmness. When the disease is more advanced, the organ lessens in size, especially the left lobe; the induration becomes aggravated. Knobs or granulations (nutmeg liver) project all over its surface. The capsule of the liver is always thickened.

The character of these alterations is believed to be due to the new-formation of connective tissue, in the ramifications, through the gland, of Glisson's capsule. Bands of this material constrict the lobules, obstructing the bloodvessels and bile-ducts, as well as the gland-cells. Thus diverse effects are produced. Commonly the subdivisions of the portal vein are diminished in size, or obliterated; those of the hepatic artery enlarged; and those of the hepatic vein unchanged. The biliary ducts are at first distended by partial obstruction, causing repletion of the cells; afterwards both cells and ducts may be in considerable part destroyed. The color of the granulations is dark or pale yellow. Along with these changes, in many but not in all cases, fatty or waxy degeneration of the liver-structure ensues.

Inflammation of the capsule of Glisson and its interstitial ramifications is considered by most pathologists to be the primary element of cirrhosis. Without feeling altogether certain of the correctness of this view, I am unable to suggest any other to take its place, without entering upon a discussion too complex for our present purpose.

Symptoms.—Nausea and indigestion, with a furred tongue and slight yellowness of the skin and eyes, are the earliest (of course not pathognomonic) manifestations of this disease. Afterwards, mostly with slow progress, come constipation, vomiting, emaciation, debility, ascites, with or without general dropsy, and enlargement of the superficial abdominal veins. This last sign is especially significant of obstruction of the hepatic circulation. Towards the close of life, hemorrhage from the stomach or bowels, delirium, coma, or convulsions are apt to occur.

Diagnosis.—From acute congestive or inflammatory affections of the liver the slow progress of cirrhosis readily separates it. From fatty and waxy liver, and from cancer, it is distinguishable, though not always with ease, by the continued enlargement of the organ in those affections; while they are also less constantly attended by dropsy and enlargement of the abdominal veins. The spleen is also often enlarged in cases of cirrhosis. This, however, occurs also when the portal vein is inflamed or obstructed, either by coagula or by pressure. There is then, however, apt to be compression of the bile-ducts, producing decided jaundice, with clay or slate-colored stools. Chronic peritonitis is sometimes difficult to diagnosticate from cirrhosis; but in the former there is more abdominal tenderness, and less enlargement of the superficial veins.

Prognosis.—Recovery from cirrhosis of the liver is not to be expected; but its duration varies greatly, and may be favorably modified by regimen and treatment.

Causation.—Although malarial influence and syphilis may predispose to it, the special cause of cirrhosis is believed to be alcoholic poisoning. It is one of the most common results of continued intemperance.

Treatment.—Having the hope only of palliation and delay, we must, most of all, prevent the persistent action of the cause, by enforcing abstinence from spirituous liquors. Nourishing diet is, at the same time, very important. Milk, if well digested by the patient, meat, or concentrated liquid animal food, as beef-tea, chicken-broth, &c., will be suitable. The secretions must be attended to. Saline laxatives, especially the bitartrate of potassa, will often be useful. Bitters or other stomachics may be called for to relieve nausea and strengthen digestion. Dropsy may sometimes require tapping.

FATTY LIVER.

This form of degeneration is not uncommon in intemperate persons, or in those suffering from prolonged debility, as in phthisis. Perhaps its association with the latter disease is the most frequent.

In its **diagnosis**, beyond the fact of enlargement of the liver, with smooth margin and surface, in an enfeebled constitution, unaffected by the symptoms of other hepatic disorders, unless it be slight jaundice, there is nothing positive. The change may go on undiscovered even by a careful observer, until after death.

Anatomically, the liver-cells are gorged with oil; their nuclei being destroyed or obscured. With enlargement, the whole organ presents a pale and flabby as well as greasy aspect; and the latter property is obvious to the touch.

There is no **treatment** especially appropriate to this affection, other than what the constitutional state will point out for itself.

WAXY LIVER.

Synonyms.—*Amyloid, lardaceous, colloid* degeneration of the liver.

This is often an accompaniment of fatty degeneration; but also occurs quite frequently without it.

Anatomy and Pathology.—The waxy liver is pale or mottled in hue, and, when cut, smooth, hard, and dry. It is heavier than natural. The degeneration probably begins in the lobular ramifications of the hepatic artery, and extends to the secreting cells. Under the microscope these are found to have a pearly look, and to have lost their cell-walls and nuclei. The acini or lobules remain very distinctly marked out.

Fatty degeneration may coexist with the waxy; and hence they have been confounded together. The weight of the liver is modified (made lighter) by the presence of fatty degeneration.

The term *amyloid* has been given to the waxy or colloid change because of a starch-like chemical reaction of the degenerated material. It is hardly to be said that the chemical discussion on this point has yet ended.

Symptoms and Physical Signs.—Anæmia, emaciation, and dropsy (with, often, vomiting or diarrhœa, but little or no jaundice), unexplained by other local or general causes, and occurring in a scrofulous, syphilitic, or malarial diathesis, may cause a suspicion of this form of degeneration.

Examination confirms this if we also find the liver uniformly enlarged and firm, with at the same time enlargement of the spleen, and albuminuria.

Diagnosis.—Fatty liver does not exhibit so much increase in size, and it is of a softer consistence upon pressure; splenic enlargement and albuminuria less often attend it; and the same is true of dropsy. Syphilitic inflammation of the liver differs from it in presenting prominent nodules upon the surface of the organ.

Causation.—Syphilis is the most common predisposing cause of waxy degeneration. The tubercular constitution probably comes next. It exists most frequently in males.

There is no especially indicated **treatment** for this affection.

SYPHILITIC LIVER.

Among the organic affections now recognized as displaying locally the effects of the syphilitic diathesis, is a form of chronic hepatic inflammatory degeneration; that is, inflammation followed by a specific organic change of structure.

Anatomically, the liver is somewhat enlarged; with an uneven sur-
17

face, from cicatrices alternating with nodules. This unevenness may be felt upon palpation through the wall of the abdomen. The patient is pale, but not jaundiced; and dropsy is not present as a symptom, unless from other organic causes.

In **diagnosis,** syphilitic liver is to be distinguished from *cancer* of the liver by the smaller size and softer consistence of the projecting nodules in the former, the absence of tenderness on pressure, and, usually, by the signs of general syphilis; as, the marks of cicatrized ulcers in the throat, copper-colored blotches upon the skin, or nodes upon the bones.

CANCER OF THE LIVER.

Mostly in middle life, but occasionally even in the young, cancer of the liver occurs, and has a more rapid progress than most cancers. The *symptoms* are, pain in the right side and shoulder, with tenderness in the right hypochondriac region, disorder of the stomach and bowels, rigidity of the abdominal muscles (especially the rectus), debility, emaciation, a cachectic aspect, and ascites or general dropsy. There is usually little or no jaundice.

Physical exploration shows dulness on percussion below and above the usual limits of the liver; and, on palpation, irregular prominences, hard in most cases, but sometimes, in encephaloid cancer, soft and elastic. The enlargement may become very extensive; and then all the effects of pressure, upon the portal vein, etc., are observed.

This disease is always fatal; affording no room for other than merely palliative treatment. Its duration is often less than six months: seldom more than a year.

HYDATIDS.

These are elastic tumors, consisting of *cysts*, developed around *echinococci.* The latter are the larvæ or immature progeny of a *tænia;* they are found not only in the liver,[1] but also in the brain, muscles, bones, ovary, uterus, kidneys, lungs, heart, spleen, etc. The sac or cyst grows slowly, and may exist for years without great disturbance of the health. If any symptoms occur, they are indigestion, debility, and dropsy.

Hydatids are discovered upon inspection and palpation; the liver being considerably enlarged, so as to press up the diaphragm and right lung, or to sink far down into the abdomen. On percussion, besides an irregular line of extended dulness, a peculiar jelly-like vibration is sometimes perceptible by the finger used to percuss upon. If the tumors be so near the surface and so evidently elastic as to warrant the operation of exploration with a grooved needle, the fluid drawn out will be very characteristic. It is colorless, of specific gravity not much above that of water (1007–1010), and is free from albumen; it contains a large amount of chloride of sodium.

Sometimes the entozoa within the cyst die, and the sac collapses and disappears. In other cases it bursts and is discharged into the

---

[1] Of 508 cases of hydatids, Cobbold and Davaine found the liver to be affected in 216.

alimentary canal, the lungs, or externally through the abdominal walls. Slow recovery may then be anticipated. Danger always exists, however, that the hydatids may open into the pleural or peritoneal cavity, producing pleurisy or peritonitis. In a few instances suppurative inflammation occurs in the cyst.

In the **treatment** of hydatids, some physicians have been disposed to confide in the supposed power of iodide of potassium, and of chlorate of potassa, taken internally, to cause the absorption of the fluid of the cyst, and thus destroy the parasite. But the evidence is not such as to justify such confidence.

Very large and superficial hydatids may, when the diagnosis is clear, be *tapped*, with at least temporary relief to the patient. Should this be safely done without cure, it may be repeated, and then a gum-elastic tube may be introduced and retained in the opening, so as by drainage to induce the shrinking of the cyst and thus the destruction of the *echinococcus.* Dr. Pavy reports success in one case with injection of male fern into a hydatid cyst of the liver ; its anthelmintic or parasiticide power seeming to be thus shown.

TUBERCLE OF THE LIVER.

Primary tuberculization of the liver is never met with. In patients dying with phthisis, not unfrequently yellow or gray miliary tubercular deposits are found scattered over the gland ; they rarely soften, but sometimes small *vomicæ* are met with. It is of course necessary to be aware of the possible existence of such formations, in the consideration of the morbid anatomy of the liver.

DILATATION OF THE GALL-BLADDER.

This may be produced by obstruction of the gall-duct or the common bile-duct, or, more rarely, by a morbid formation of serous fluid within it, allied to a local dropsy. The diagnosis of this may be important, as it may readily be confounded with hepatic enlargement. It is to be distinguished from cancer by the great amount of jaundice (in most cases), the previous occurrence of gall-stone colic (also not invariable), and the more uniform and softer character of the swelling. From hydatids the same signs, except the softness of the tumor, are distinctive ; and the latter grow much more slowly.

For the **treatment** of dilatation of the gall-bladder, the remedies suitable for obstruction of the biliary ducts will be appropriate. Surgical interference would, in any case, be very bold practice.

AFFECTIONS OF THE SPLEEN.

These are necessarily treated of at length in systematic treatises. It will be enough for our purpose to say a very few words of them. The spleen is commonly **enlarged** in *intermittent, remittent,* and *typhoid* fevers, and in *leucocythœmia ;* sometimes, in pregnancy (Simpson). **Rupture** of the spleen, causing death, has been several times reported. Such an affection could scarcely (*i. e.,* rupture of the spleen) be diagnosticated during life.

Enlargement of the spleen is readily ascertained by inspection and palpation. It often increases and diminishes, during and between the paroxysms of intermittent (ague-cake). Piorry asserts its *rapid* diminution under cinchonization. Other affections of the spleen (**inflammation, tubercle, hydatids**, &c.) are so generally difficult of diagnosis as to have chiefly a post-mortem interest; and they present no clearly recognized indications for treatment.

AFFECTIONS OF THE KIDNEYS AND BLADDER.

CONGESTION.

Causation.—Under exposure to cold, overdoses of cantharides or turpentine, or the disturbance belonging to different inflammatory and febrile complaints, *active* renal congestion may occur. *Passive* congestion is more common in heart-disease, or pulmonary obstruction, as in pleuritic effusion or emphysema, or when pressure impedes the circulation of the renal veins or vena cava, as in pregnancy or abdominal tumors.

Symptoms.—Pain in the lumbar region, sometimes with tenderness on pressure on each side of the spine. Scanty urination, the fluid being high-colored, sometimes bloody, or containing albumen. Certain cases exhibit under the microscope fibrinous casts.

Diagnosis.—It is only occasionally difficult to distinguish this condition from Bright's disease. Active congestion begins abruptly, under a recognizable cause. Passive congestion shows a dependence upon some organic affection, and, although variable, is not progressive. They are thus distinguishable from advancing and more or less permanent disease of the kidneys.

Treatment.—For active congestion, cupping the lumbar region, abstracting blood according to the state of the patient. Purgation, by castor oil or citrate or sulphate of magnesia. Then, the warm bath or hip-bath, continued for some time.

URÆMIA.

Definition.—The retention in the blood of the material which it is the function of the kidneys to excrete; from the suppression of their action.

Symptoms.—When well marked, headache, dimness of vision, vomiting, diarrhœa, convulsions and stupor; ending in fatal coma.

Pathology.—The question as to what is the *immediate* toxic agent in uræmia is not yet fully determined; *i. e.*, whether it is urea, or an ammoniacal educt from its decomposition in the blood. In the absence of demonstration of the latter, the former is most probable. A further view has recently been urged; that it is unchanged creatin, creatinin, and other extractives, that contaminate the blood.

Treatment.—This must vary with the circumstances of the production of the suppression; but the great indication is to *depurate the blood*—by the kidneys if they can be restored to action, and by the aid or substitution of the bowels and skin. For this end, the warm bath, or the hot air, or warm vapor bath may be of great service. So may cupping or counter-irritation by mustard or tincture of iodine over the small of the back. Saline cathartics, even hydragogues, may be given to such patients as have strength to bear them; as cream of tartar, Epsom salts, claterium, or croton oil; the last two most rarely. Lemonade drunk freely is often one of the best of diuretics. Others will be mentioned hereafter, in connection with dropsy.

NEPHRITIS.

In the present state of urinary pathology, it is common to merge the topic of inflammation of the kidney (except suppurative pyelitis) as distinct from active renal congestion,—in Bright's disease. If this be questionable as a matter of pathological system, it has at least practically no disadvantage; as the symptoms of nephritis are included in one or other of the affections named; and so is its treatment. We may submit therefore to the usage of authority upon this point, without hesitation. The symptoms of *acute pyelitis* (inflammation of the pelvis of the kidney) are essentially those of renal congestion, intensified; with tenderness on pressure over the kidney, and fever, until suppuration is established; then, purulent discharge for a variable time from the kidneys. (See *Pyonephrosis*). Before pus appears, blood, in small quantity, mucus. and renal epithelial cells may be found in the urine. A tumor in one of the lumbar regions may precede for a while the escape of the pus.

BRIGHT'S DISEASE.

Definition.—Albuminuria, dependent upon structural change in the kidneys; or, to speak, perhaps, more correctly, disease of the kidney, characterized by albuminuria and dropsy.

Varieties or Stages.—Authorities differ as to the discrimination of these. Bright believed there were three varieties. Dr. G. Johnson asserts two,—the desquamative and non-desquamative nephritis. Frerichs considers them to be stages of the same affection, and admits three stages, essentially, of hyperæmia, exudation, and degeneration. Anatomically, we have the *large, smooth, white kidney*, the *small, smooth kidney*, the *granular uncontracted kidney*, and the *granular contracted kidney*. We may safely follow Roberts, in dividing Bright's disease, first, into *acute* and *chronic*. The latter is then divided into, 1. Causes which have lapsed from the acute state (smooth, white, generally large kidney); 2. Cases chronic from the beginning (granular, red, contracted kidney); 3. Cases associated with waxy or amyloid degeneration of the kidneys.

Causation.—Bright's disease is one third more common in males than in females. The greatest number of cases occurs between the ages of 45 and 65. *Acute* Bright's disease is most often produced by cold and dampness; next, by scarlet fever, pregnancy, or violent intemperance. The acute form is most common in early life.

Chronic Bright's disease is also greatly promoted by exposure to cold and wet; and is caused moreover by abuse of spirituous liquors, very often. Other predisposing causes are gout, constitutional syphilis, and affections of the bladder and urethra.

Symptoms.—Acute Bright's Disease.—After exposure to cold, or a drunken fit, or scarlet fever, the patient is seized with chilliness, headache, nausea, vomiting, pain in the back and limbs, checking of perspiration, and oppression in breathing. Fever follows; and the face, trunk, and limbs become puffy with anasarca. Effusion may also occur into the pleura or peritoneum.

The *urine* is scanty, heavy, and dark in color, from the presence of blood; and very albuminous. The disposition to void it occurs more frequently than during health. The deposit from it, under the microscope, shows blood-corpuscles, loose renal epithelium, free nuclei, tube-casts, and shapeless masses of fibrin and *debris*.

After one, two, or three weeks, or even a longer period, the attack proceeds to one of three terminations; recovery, death, or lapse into the chronic state. Death results through uræmia, or from secondary pneumonia, pleurisy, peritonitis, pericarditis,—or hydrothorax, œdema of the glottis, hydrocephalus, or ascites. Probably two-thirds or more of the cases of *acute* Bright's disease recover.

Treatment.—Cupping the loins, hot water or hot air or "blanket" bath, active purging, as with cream of tartar and jalap, or citrate of magnesia, and diaphoretics, as citrate of potash or liquor ammon. acetat. Mercury is not recommended. The diet should be *liquid* and simply nutritious.

Chronic Bright's Disease.—This approaches so slowly as seldom to be detected until after the lapse of months or years. Gradual loss of strength, pallor or puffiness of the face, shortness of breath, and frequent disposition to urinate, are early signs of it. But they are not always present; the *dénouement* of the disease may be by a convulsion, œdema of the lungs, amaurosis, or some violent local inflammation.

Symptoms of a well-marked case (not all present in every instance) are : albuminous urine, deposits of tube-casts and renal epithelium, dryness of skin, frequent micturition, especially at night, general dropsy, or local effusions into the cavities, indigestion, anæmia, uræmic effects (headache,. dizziness of sight, convulsions, coma, vomiting, diarrhœa), enlargement of the heart, and secondary inflammations. Bronchitis is especially common.

The progress of the case is usually interrupted by exacerbations and intervals; each fresh attack leaving the patient manifestly worse than before. Such attacks much resemble acute Bright's disease; they are sometimes referred to known causes; the intervals may last weeks, months, or even years,

In **prognosis**, the tendency is always toward a fatal result. About one-third die of uræmic poisoning. A considerable number die of local dropsical effusions. One-fifth from secondary pneumonia, pericarditis, or pleurisy. The rest, by exhaustion from anæmia, indigestion, and anasarca, or the complications of apoplexy, cirrhosis, phthisis, intestinal ulcerations, &c.

Diagnosis.—The presence of albumen in the urine, with dropsy, not of sudden origin or brief duration, is pathognomonic of this affection. The tests for albumen, by heat and nitric acid, are readily applied. The microscope will show also free renal epithelium and tubular casts in the urine; in advanced cases, the casts are sprinkled with oil-dots. The solids of the urine, especially the urea, are reduced below the normal amount.

Pathology.—Degeneration of the structure of the kidney induces albuminuria, by allowing the serum of the blood to pass almost unchanged through the cortical substance into the *tubuli uriniferi*. The deficiency of urea is due to the same impairment of secreting power. The consideration of the different varieties of renal degeneration would be too complex a subject for these pages. The reader is referred for it to the standard treatises on the subject.

Treatment.—The indications in every case of Bright's disease are— 1. To hinder the progress of structural change in the kidney; 2. To prevent uræmia and secondary inflammation; 3. To palliate concomitant symptoms or states, as anæmia, dropsy, dyspepsia, &c.

Regimen or hygienic management is of the utmost importance for the first of these ends. Avoidance of exposure to cold, wet, or great fatigue; the reform of intemperance, if it has existed, or of other excesses,—will be indispensable. Clothing should be sufficiently warm, with flannel next to the skin. Bathing frequently, at such temperature as is borne without chill or relaxation, is to be recommended. The bowels should be kept regularly open. Nourishing diet, of which milk may generally be part, is of consequence.

Iron will do more good than any other medicine, unless it be cod-liver oil in those of strong stomach. They may be very well combined. The tincture of the chloride of iron is as good as any other chalybeate, as a general rule. With some the citrate of iron in solution, or the carbonate or the iodide, will agree more readily.

It is very doubtful whether astringents ever check to advantage the waste of albumen through the kidneys. If any be worth the trial, it is ammonio-ferric alum. Counter-irritants over the kidneys, unless of the mildest character (tinct. iodin., emplastr. picis, &c.), will not do any important good in chronic Bright's disease.

For the dropsy, warm baths and hydragogue cathartics are advised. Of the latter, cream of tartar and jalap, together, are the favorites : 2 or 3 drachms of the bitartrate with 10 to 20 grains of jalap two or three times a week. If serious dropsical accumulation threaten life, elaterium (gr. $\frac{1}{8}$ or $\frac{1}{4}$ every four hours, in pill, till it acts) may be given. But it is a decided mistake to harass the patient constantly with exhausting purgation. It is to be remembered that it can act only as a *palliative*, removing part of the *effects* of the malady, not the disease itself.

If the warm bath do not agree, or fail to produce diaphoresis, those who have access to it should try the *hot air bath*, at 130° to 150° Fahr. This rarely fails to produce free perspiration. For weaker invalids, the vapor bath is available.

Of diuretics, acetate of potassa, spirit of nitrous ether, and infusion or compound spirit of juniper will be the least likely to disappoint. But all will not unfrequently fail.

Then we have as a resource (where tapping for ascites is not demanded) for the relief of great œdema, the use of incisions with a lancet, or needle, in the swollen legs and feet. I prefer a number of small incisions with an abscess lancet, plunged through the skin of the calf and dorsum of the foot. It is possible that erysipelas may follow;

but this danger will be lessened by repeated warm sponging of the limbs, washing them with diluted glycerin, or inunction with lard or cold cream.

The **complications** of Bright's disease must be treated according to their own indications, on general principles,—bearing in mind always the *degenerative* and *asthenic* tendencies belonging to the malady itself.

LITHIASIS.

Definition.—The formation of calculous deposits (gravel or stone) in the kidneys or bladder.

Causation.—Stone is, by statistics, nearly ten times as frequent, or at least as fatal, in the male as in the female. It destroys life most often after fifty years of age; but is far from uncommon in early life, even under five years. *Locality* has something to do with the causation of stone and gravel. They are common in England, Iceland, France, and Egypt, and uncommon in Sweden, Norway, and Austria. In this country they are not rare; the greatest number of cases probably occurs in the State of Kentucky.

Varieties.—Of these a sufficient account (for our purpose) has been given in the first part of this book. (See *Semeiology.*)

Diagnosis.—Examination with the *sound* is indispensable to determine the presence of a calculus in the bladder. The characters of the urine will aid in determining its nature. If the urine be decidedly acid, the stone is probably uric acid or oxalate of lime, or a combination of both. If alkaline from fixed alkali, it is either phosphate or carbonate of lime (both rare). If alkaline from volatile alkali, whatever its nucleus or central part, the surface must be formed of the ammonio-magnesian phosphate and phosphate and carbonate of lime.

Pain in the bladder and in the back, and pain or itching in the glans penis, retraction of the testicle, and interruption in the flow of urine, occurring at times suddenly, are the most prominent symptoms of stone in the bladder.

Gravel consists of small calculous concretions, which may be voided through the urethra. Pain in the back, with chilliness followed by fever, commonly precedes an attack, or "fit of the gravel;" to which some persons are subject whenever they take cold or suffer from indigestion. *Extreme* pain may attend the transit of a small calculus through the ureter from the kidney to the bladder.

This troublesome affection (gravel) in the large majority of instances is owing to undissolved uric acid and the urates.

Treatment of Gravel.—Under the indication suggested by the last mentioned fact, the dilution and alkalization of the urine are called for. The small calculi often irritate the bladder painfully, inducing sometimes spasmodic retention of urine. Free draughts of a demulcent liquid, as flaxseed tea, will do good; and the secretion may be made more copious, and thus dilute, and the solution of uric acid and its compounds promoted, by the administration of spirit of nitrous æther and bicarbonate of soda, in tolerably full doses, three or four times a day during the attack [F. 122]. The agonizing pain from the passage of a calculus through the ureter will require anodyne treatment, by opium, or inhalation of ether or nitrous oxide, and relaxation by the prolonged warm bath.

Prevention.—Any one inclined to gravel (one sign of which tendency is a pink stain in the urinal left after the urine has been thrown out) should avoid highly animalized or otherwise stimulating food. The urine may be kept dilute by taking a tumblerful of water two hours before dinner, and another at bedtime. The skin must be kept open by baths, frictions, and sufficiently warm clothing. Exercise will generally be beneficial in prevention. If acidity in the urine be positive, *small* doses of the bicarbonate of soda, or of the acetate, citrate, or carbonate of potash may be taken daily.

Treatment of Calculus.—Although the result of much experimentation had been, until of late, to turn over the management of stone to the operative surgeon, new reason has been given for hoping for something in its relief without the knife. Dr. W. Roberts has, in this, made some very promising observations and experiments.

Urinary calculi may be, practically, divided into those soluble in alkalies and those soluble in acids. Of the first, there are uric acid and its salts, and cystine; of the second, phosphatic and mulberry calculi. Solvent treatment affords hope only by alkalizing the urine, in cases of the former, by medicines taken by the mouth, and injecting acid solutions into the bladder for direct action upon mulberry calculi and the phosphates.

Very weak solutions of acetate or citrate of potash, taken often, alkalize the urine most efficiently, according to Dr. Roberts' experiments. He does not encourage the hope that *large* or old calculi can ever be so dissolved. Dilute nitric acid is proposed for injection into the bladder for the solution of phosphatic calculi, especially after their being broken down by the lithotrite; and Sir B. Brodie and Mr. Southam have carried this procedure, in two cases at least, with success into practice.

DIABETES INSIPIDUS.

Definition.—Excessive discharge of almost colorless urine, of light weight, containing neither sugar nor albumen; with *polydipsia* or excessive thirst. **Synonym,** *polyuria.*

Causation.—This is various, and generally obscure. More males have the affection than females. It is most common between five and thirty years of age. Blows on the head, intemperance, cerebral disease, and exposure to cold or drinking cold fluids while heated, are among the supposed causes.

Pathology.—This, too, is various or undetermined. In some instances degeneracy or atrophy of the kidneys has been found after death; in others, renal congestion. Very probably the degeneration may be secondary. Probably the *immediate* cause of the excessive urination is dilatation of the capillary vessels of the kidneys; this having its origin in some remote agency which disturbs the ganglio-nervous influence that controls the circulation.

Symptoms and Course.—Often beginning suddenly, the amount of water passed may reach ten or twenty quarts *per diem.* Thirst is intense, and withholding liquids does not arrest the polyuria. The skin becomes dry and harsh. Debility and emaciation attend, if the attack is prolonged.

The **duration** of the complaint varies from a few weeks to many years—or a lifetime. It is sometimes congenital. An intercurrent attack of febrile or inflammatory disease sometimes suspends, or even cures it.

Treatment.—This has been, so far, tentative only; no specific is known for it. Nitrate of potassa, valerian, ergot, iron, alum, lime-water, tannic and gallic acid, creasote, and bromide of potassium are the medicines most worthy of trial. Blistering the nape of the neck has also been suggested.

DIABETES MELLITUS.

Synonym.—*Glycosuria.*

Definition.—Excessive urination, with the presence of sugar in the urine.

Causation.—Twice as many men as women have this disease. It is most frequent among young and middle-aged adults; the mortality from it being greatest from fifteen to fifty-five. It is more common in cities and manufacturing districts than in the open country. Occasionally it is hereditary.

Exciting causes appear to be, exposure to cold and wet; drinking cold water largely when heated; excessive use of saccharine food; intemperance; violent emotion; febrile diseases; and organic affections and injuries of the brain and spinal cord.

Symptoms and Course.—Beginning insidiously, with malaise and slight loss of flesh, urination becomes excessive, with corresponding thirst, and very often *bulimia* or excessive appetite; emaciation is progressive; the skin is harsh and dry; the tongue, glazed, and furrowed, the mouth clammy; the sexual and mental powers fail by degrees. Lastly, hectic fever, œdema of the limbs, diarrhœa, and often all the symptoms of pulmonary consumption terminate the case.

Complications.—Tuberculization of the lungs occurs in nearly half the cases of diabetes mellitus which last over a year or two. Inflammations of an asthenic type are common in all the organs. Boils and carbuncles are very frequent. Gangrene of the lower extremities has been several times observed. Amblyopia (obscure vision) is present in about one-fifth of the cases. *Cataract* generally forms in cases of long standing; but may be absent altogether in those of less than two years' duration. The *endosmotic* theory of diabetic cataract, suggested by the production of opacity of the lens in frogs by immersion in a saccharine solution, or injecting the same into the cellular tissue, is of doubtful application. Objections to it are, the temporary nature of the saccharine cataract in the frog, the occasional occurrence of diabetic cataract in one eye only, and the late period at which the symptom occurs in the disease.

Morbid Anatomy and Pathology.—Much remains to be done before the pathology of diabetes can be said to be ascertained. In about half of the cases, only, some degree of renal alteration is found. Physiological facts and experiments, in regard to the "glycogenic function of the liver," point to that organ as the probable seat of the disorder. Other observations, as to the production of diabetes in animals by injuring the medulla oblongata or the base of the brain, are also suggestive. But, although in some instances autopsic inspection has agreed with such expectations, in many other cases it has failed to confirm them. The true theory of diabetes therefore remains for the future to discover, or, at all events, to complete.

The most plausible hypothesis, certainly, is, that, under disturbed innervation, the liver modifies its ordinary assimilative process so as to confiscate (to use a bold figure) most of the carbohydrogenous material derived from the alimentary canal through the portal vein, and convert it into glucose or diabetic sugar, which is then eliminated by the kidneys.[1]

Diagnosis.—The detection of sugar in the urine, not temporarily, but for a considerable time, is of itself sufficient to make out the case. The principal modes of testing saccharine urine have been given in another part of this book. (See *Semeiology.*)

Prognosis.—Recovery is not impossible in diabetes; but a large majority of cases end in death. *Amelioration*—keeping the disease in abeyance—is often an attainable end. The younger the patient in whom the disorder begins, the less ultimate hope. In old persons glycosuria seems more often compatible with tolerable health for a long time. Cases traced to mental emotion or to injuries are somewhat more hopeful than those of indistinct origin.

Amblyopia, cataract, and albuminuria, as well as phthisical symptoms, mark the case as incurable. Considerable diminution of the sugar, or of the water, passed, is always a favourable prognostic. But the diabetic patient is much more liable than others to those inflammatory complications which, on slight exposure, may hasten the termination of life.

Treatment.—No direct control over the sugar-forming process in the body has yet been obtainable by medicine. But, although it would seem that simply diminishing the formation of sugar by withholding material for it ought not to be *expected* to do much good, it does prove beneficial. The most important measure yet devised in the management of diabetes is, the prohibition of sugar and starch, and of everything which can yield them, as food. Bread, except bran bread, which is almost free from starch, potatoes, and nearly all vegetables and fruits must be excluded. The safe exceptions are, the cabbage, broccoli, onions, spinach, celery, and lettuce. Of animal diet, milk and liver are forbidden articles. All meats, eggs and butter, and jellies, are allowable. *Gluten bread* is made in France, on Bou-

---

[1] It seems to be established that a natural product of the liver is an amyloid material (hepatin, liver dextrin); whether Bernard's view, of the normal destiny of this being its constant conversion into sugar (and subsequent combustion by oxidation in the blood) be correct, or rather that of Pavy, that such conversion is always morbid, or *post-mortem*. *Artificial* glycosuria may be produced in animals by puncturing the floor of the fourth ventricle of the brain, impeding respiration, thrusting needles into the liver, obstructing the abdominal venous circulation, injecting acid into the veins, poisoning with strychnia, and woorara, and chloroform and ether inhalation. Dr. McDonnell has lately proposed a new theory; that "glycogen" normally in the liver combines with nitrogenous matter derived from food, to make plastic material for tissue; and that this process is interrupted or arrested in diabetes.

chardat's plan, without starch, inflated by machinery with carbonic acid or compressed air. Tea or coffee may be sweetened with glycerin (chemically pure, as Bower's or Price's). Spirits, wines and beer should be avoided unless called for by positive weakness; if that exist, the least saccharine should be preferred, as sherry, claret, or whisky, in minimum quantities. There is no advantage in restricting the amount of water taken to quench thirst. *Variety* of diet, of course, within the prescribed limits, is important, to prevent disgust and loss of appetite.

Of medicines, *none* have been yet shown to do much service in checking the disease. The most positive influence in diminishing the diuresis belongs to opium; but this does not appear to interfere with the progress of the disease. Various drugs have been tried, and lauded greatly by different users; but their effects will not bear scrutiny without disappointment. Among them the most prominent are alkalies, yeast, rennet, pepsin, iron, quinine, creasote, alum, iodine, nitric acid, turpentine, and the inhalation of oxygen. Even free ingestion of *sugar* has been fairly experimented with; but in vain. A therapeutic remedy for diabetes remains to be discovered.

HYDRONEPHROSIS.

Definition.—Renal dropsy; dilatation of the kidney from obstruction of the ureter.

Causation.—Quite a number of the cases recorded have been congenital, from anatomical malformations. Calculus in the ureter is the most frequent post-natal cause; but other mechanical obstructions, from pressure, may occur.

Diagnosis.—Intumescence of the abdomen, usually upon one side, in the hypochondriac, umbilical, and iliac regions, with a soft undulating feel, an outline often lobulated, and fluctuation as well as dulness upon percussion, can, in the male at least, only indicate either hydro- or pyonephrosis. The *symptoms* may be almost null, if only one kidney be affected. When both are so, uræmia finally results. The tumor is commonly quite painless, and not tender upon pressure. This affection is, however, quite rare. It may be fatal, by uræmia, or bursting of the sac into the abdomen; but it has in a number of cases existed for many years.

Treatment.—*Manipulation*, kneading gently, day after day, has sometimes succeeded in dissipating the renal distension. Nothing else should be attempted, unless life is endangered by the pressure of the tumor, or by uræmia. If it be so, tapping is justifiable; and it has been repeatedly performed with success.

PYONEPHROSIS.

This differs from the last-named affection in the production, under similar circumstances, with more or less inflammation, of *suppuration* of the kidney. The symptoms are therefore more active, and the prognosis more grave. Rupture of the sacculated kidney, into the colon, duodenum, or peritoneal cavity, is common, and is nearly always fatal. Renal abscess may occur, also, from "purulent infection," and from embolism. Such an abscess *may* find escape for its contents externally; any appearance of such a tendency should be encouraged by poulticing, and, in fit cases, by incision and evacuation.

CANCER OF THE KIDNEY.

Primary cancer of the kidney is, though rare at any age, *most* frequent in early childhood. Of adults, males have been the most numerous subjects of it. **Secondary** renal cancer may attend any case of the cancerous cachexia, without materially modifying its history. The kind of cancer affecting the kidney is nearly always the **encephaloid**; called, when highly vascular, fungus hæmatodes. It always begins in the cortical substance. The tumor is generally large, and sometimes enormous; reaching, in one case (Roberts), 31 pounds. It is exceedingly rare for both kidneys to be affected.

Diagnosis.—An abdominal tumor, with copious hæmaturia repeated at irregular intervals, is almost certain to be cancer of the kidney.

Beginning between the ribs and the crest of the ilium on one side, the tumor grows forwards, upwards, and downwards, so as to fill in some cases the whole belly. The colon, in this as in all renal tumors, lies in front of it; sometimes also a part of the small intestine. Except over the intestine, percussion resonance is dull.

The swelling is smooth or irregularly lobulated; now and then a sort of fluctuation, and in one instance pulsation, have been observed in it. It is *fixed* in its position.

Bloody urine, usually profuse hemorrhage, is present in about half the cases. No other tumor has this symptom attending it.[1] Its occurrence is, therefore, pathognomonic. The discovery of cancer-cells in the urine by the microscope is of course still more positive; but this sign is very often absent; and the cells are not at all easy of identification when they occur.

Pain mostly, but not always, attends cancer of the kidney; it is sometimes of great severity, shooting down the ureter to the thigh. Tenderness on pressure seldom exists. *Variable* symptoms are those of disorder of the stomach and bowels. Emaciation, and anasarca, show the exhaustion which precedes death.

The **duration** of cancer of the kidney, in children, averages seven or eight months; in adults, over two years. This is a longer period than that of any other visceral cancer.

In **treatment,** as in other malignant diseases incapable of safe extirpation or cure, the judicious management of *regimen*, and of *anodynes*, is all that is possible.

TUBERCLE OF THE KIDNEY.

This may be either primary or secondary. Of all tuberculous subjects, the kidney is found to contain such deposits in from five to six per cent. Among tuberculous children, in from fifteen to sixteen per cent. *Most* of these however were *secondary* cases.

The **symptoms** of primary renal tuberculization are, dull lumbar

---

[1] Roberts mentions one case of great enlargement of the spleen with hæmaturia.

18

pain, frequent micturition, the urine being at first turbid or slightly bloody, afterwards purulent; emaciation, and hectic fever. Almost always other organs, especially the lungs, become also tuberculous; merging the case into one of complicated phthisis. The bowels are very frequently implicated. Death occurs mostly from exhaustion. If both kidneys are affected, it may be from uræmia.

The **duration** of the affection varies—from a few months to two or three years.

Diagnosis.—Only after softening of the tubercle can it be positively proved to exist. Then, the abundantly purulent urine is found upon microscopic inspection to contain also " granular debris, sometimes with tuberculous matter (insoluble in acetic acid) shreds of connective tissue, and beautiful meshes of elastic fibres from the cast-off patches of disintegrated mucous membrane." Great debility and emaciation, with hectic fever, confirm these signs. The absence of tumor, and of hæmaturia, distinguish renal tuberculization from cancer.

Treatment.—Here, again, we must confess the deficiency of our present therapeutics. Indications exist, essentially the same as in phthisis pulmonalis; to the consideration of which we may refer the reader.

HYDATIDS OF THE KIDNEY.

These are more rare than hydatids in the liver or lungs; but more frequent than in other parts of the body.[1] The left kidney is most often affected.

In a majority of cases the cyst formed by the *echinococcus* opens into the pelvis of the kidney. The hydatids then, in part or wholly, are discharged by the urethra. They may however, also, or instead, burst into the stomach, intestines, or lungs.

If no such vent occurs, a tumor is formed in the side (with the colon always in front of it) which has a more or less distinct fluctuation, and, sometimes, the "hydatid fremitus" or vibration to the touch.

The discharge of the contents of the cyst allows the discovery, in some cases, of entire vesicles; in others, of a detritus, in which the microscope detects echinococcus-hooks, laminated shreds, and oil particles.

This discharge is apt to be recurrent or paroxysmal; at intervals varying from a few weeks to one or more years. Before it occurs, chills, nausea, hiccough, and colicky pains often exist; relieved by the passage of the vesicles. These, while in the bladder, may cause pain, irritation, and retention of urine.

After every such an escape, the size of the tumor may be lessened for a time. A vesicle detained in the ureter may, by obstruction, induce a *hydronephrosis*, adding to the hydatid tumescence.

Prognosis.—This is more favorable than in any other seat of hydatids, except the uterus; because of the comparative facility of their evacuation. When, however, no escape by the kidney and ureter is

---

[1] According to Davaine, the order of relative frequency is as follows: liver, lungs, kidneys, pelvis, brain, bones, parietes of the body, heart, and orbit of the eye.

effected, the tumor may become so large as to encroach seriously upon other parts ; or the cyst may suppurate (pyonephrosis) and form a large and dangerous abscess.

Treatment.—Oil of turpentine, iodide of potassium, chlorate and nitrate of potassa, taraxacum, and other medicines, have been asserted by different observers to promote the death and discharge of echinococci. Whether the "post hoc" was "propter hoc" in these cases, larger experience (which ought always to be recorded) will show. Electro-puncture has been tried for the same end ; but without proof of success.

Hydatid colic (passage of vesicles through the ureter) may be treated like that from calculus, by the warm bath and opium. Irritation of the bladder, or obstruction causing retention of urine, will require rest, demulcent drinks (flaxseed infusion), and, sometimes, the catheter. Even in the urethra the escape of the vesicles may be obstructed, and sometimes may require to be aided by pressure for their dislodgment.

A closed renal hydatid tumor, when clearly diagnosticated, and itself endangering life, may be (after exploration by the needle trocar) punctured ; especially if it project *behind*. When in front, Recamier's plan is preferred by some surgeons, of applying caustic potash repeatedly to cause adhesion of the peritoneum to the sac, before making the incision. Safer than this, and in at least one case successful, is *repeated* puncturing with the needle trocar, at intervals of a few days.

CYSTITIS.

Definition.—Inflammation of the bladder.

Varieties.—Acute and chronic; idiopathic, traumatic, secondary.

Causation.—Blows or other injuries ; the presence of gravel, or a calculus, or hydatid vesicles from the kidney; irritating diuretics ; or decomposing urine retained by stricture, may induce acute cystitis. The continuation or frequent repetition of the same causes produces " chronic inflammation."

Symptoms: Acute Cystitis.—Pain in the vesical region ; frequent desire to pass water, with burning in the urethra, and *tenesmus*, or disposition to bear down or strain. There is fever, alternating with chills. The bladder may sometimes be felt as a small round swelling, sensitive upon pressure. In bad cases, there are nausea, anxiety, delirium, and cold perspirations; the scantily passed urine becomes purulent and bloody, alkaline and fetid.

Chronic cystitis has usually much less severity of symptoms ; but it may be very distressing, from the tenderness and irritability of the bladder, and the frequent disposition to urinate, with dysuria. The urine is either mucous or muco-purulent.

Treatment.—Acute cystitis, with perfect rest, may need leeching or cupping above the pubes or (leeching) at the perineum. As a laxative, castor oil is apt to be the best. Warm hip baths will be very soothing. Flaxseed tea may be taken freely. Opium, hyoscyamus or belladonna may be called for by great pain or nervous irritability. Opium or belladonna *suppositories* [F. 124, 125], or laudanum enemata, will answer best if anodynes have to be repeated often. In **chronic** cystitis, local depletion is much less likely to do good. The other measures named may be suitable from time to time ; also injections of lime-water and glycerin, or weak solution of nitrate of silver, or of sulphate of copper, or acetate of lead, in water or in glycerin, may be serviceable. *Catheterism* may at times be indispensable, both in acute and chronic cystitis ; but it should be avoided if possible, on account of the mechanical irritation of the instrument.

RETENTION OF URINE.

Synonyms.—*Strangury, Dysuria, Ischuria.* Although the *mechanical* or *surgical* causes and history of difficult or arrested urination do not belong to this work, it will be proper to speak briefly of its occasional importance, as a symptom in the course of diseases which every medical practitioner must meet.

Retention of urine is either from *mechanical* obstruction, from *spasm*, with or without inflammatory congestion at the neck of the bladder, or from vesical atony or *paralysis*. The first occurs in cases of stricture, calculus, etc. ; the second under the influence of cantharides or turpentine, or in cystitis from any cause ; the third, in typhus, typhoid, and other low fevers and states of debility.

It is very easy in all but the last named cases to distinguish retention from suppression of urine. In low fevers, etc., when delirium exists, it is not at all difficult to make the diagnosis upon examination ; without it, this serious condition may be overlooked. A practitioner must never forget to *ascertain* whether his patient passes water or not. In all serious diseases, indeed, its regular *inspection* is important. In the semi-paralytic retention of low states, catheterism is generally required ; and, when *distension and dulness upon percussion above the pubes*, with absence of urinary discharge for twelve or twenty-four hours, or only dribbling, mark the case, the instrument may be used without delay, and repeatedly ; at least once daily if necessary.

Spasmodic retention of urine, or strangury, with or without the concomitant existence of stricture or gravel, may demand other means of relief than the catheter. The warm hip bath, prolonged for half an hour, is one of the best measures. Cloths wrung out of hot water applied to the perineum and over the pubes may assist. Leeches to the perineum, when there is local tenderness, will often promote relaxation of the part. Laudanum enemata, and opium [F. 124, 125] or belladonna suppositories, will sometimes relieve when other measures fail. Anæsthetic inhalations might be resorted to in an extreme case. Hypodermic injection of morphia has been used.

ENURESIS.

Definition.—Incontinence of urine. Except from paralysis or some local lesion, this troublesome affection is not apt to occur in the adult. In children it is common, especially at *night.*

Treatment.—Withholding fluids for some hours before bedtime, unless in very small quantities, and taking the child up to urinate after two or three hours of sleep, will generally prevent enuresis. Of medicines, those most employed (with variable success) are belladonna, benzoic acid [F. 126], and tincture of chloride of iron.

Moral impressions, acting upon the child's sense of shame or wrong, are only proper to be made use of with great care and discretion ; but sometimes they have much power.

AFFECTIONS OF THE BRAIN AND NERVOUS SYSTEM.

INFLAMMATION OF THE BRAIN.

Synonyms.—*Encephalitis, Phrenitis, Meningitis, Cerebritis.* The last two are not, of course, technically identical; but they are not clinically separable. Inflammation of the membranes derives its importance from the implication of the brain.

Varieties.—*Simple* and *scrofulous* encephalitis or meningo-cerebritis.

Simple Meningo-cerebritis (meningitis). **Symptoms.**—Intense headache, redness of face and eyes, an excited look, dizziness, roaring in the ears, extreme sensitiveness to light and sound, restlessness, wakefulness, wild delirium. Vomiting is common ; the bowels are usually costive. Late in the attack in adults, at any period in children, convulsions may occur. Rigidity of the muscles is frequent in bad cases ; paralysis often follows convulsions.

Stages.—These are generally described as three. 1st. That of active congestion and inflammation ; with hot, hard, *rapid*, full, regular pulse, morbid sensitiveness to light and sound, headache and delirium. 2d. That of commencing effusion and cerebral oppression ; with more moderate heat of the surface, stupor, and *slow* or irregular pulse. 3d. That of cerebral disability or disorganization ; with unconsciousness, convulsions, muscular rigidity or paralysis, and *rapid, feeble* pulse.

Morbid Anatomy.—Except in *traumatic* cases, the dura mater rarely takes part in the lesions of encephalitis. Rather minute hyperæmic injection is found here and there in the arachnoid membrane ; sometimes opacity and thickening occur, with adhesions. In the pia mater, generally with considerable increase of redness, serum has been effused ; or even pus. The pia mater adheres firmly to the brain. The ventricles contain more serum than usual; sometimes several ounces. In some cases it is turbid, flocculent, or purulent. The brain itself is most frequently affected, with redness in the convolutions, and dots of blood in the medullary portion; also, with softening in the gray or white substance, or in both.

Diagnosis.—The distinctions between simple and tuberculous or scrofulous meningitis or encephalitis will be considered presently. Typhoid fever, delirium tremens, and acute mania may be confounded with or mistaken for inflammation of the brain.

Typhoid fever does not have vomiting, long-continued headache, or morbid sensibility to light among its symptoms ; while tympanites, diarrhœa, bronchitic cough, etc., make it known. In delirium tremens, the origin of the affection in alcoholic excess, the usually horrible illusions, tremor and insomnia, *without headache*, are characteristic.

18*

Acute mania is almost or quite without fever; often without headache; and the muscular strength is little impaired; vomiting, also, is absent. **Subacute** or **chronic** encephalitis, now and then met with, presents greater difficulty in distinguishing it from mania. Indeed, the best authorities in psychopathology (study of mental diseases) state that cerebral hyperæmia and inflammation bear a not unimportant part in the pathology of insanity. (See *Winslow on the Brain and Mind*.)

Children afford not unfrequent instances of another question in diagnosis;—how far *symptoms* affecting the brain may or may not depend upon the *stomach* for their causation. "Gastric fever" and "infantile remittent" are phrases applied often to attacks occurring in childhood or infancy; in which, with indigestion and vomiting, there is delirium, stupor, or apathy, with or without convulsions. In such cases, the heat of head and fulness of the carotid and temporal arteries are less, the gastric disorder, fur of tongue, etc., greater, than in cerebral inflammation. *Cholera infantum* is often attended by brain symptoms; but its other features, the time of year, and locality (in a large city almost always) are distinctive.

Prognosis.—Simple encephalitis, under good treatment, is not always fatal; but a majority of cases end in death. I remember several recoveries; two from extremely severe symptoms;—one in a man, the other a girl of ten years of age. In the latter, a convulsion preceded convalescence.

Causation.—Between fifteen and forty-five is the age most subject to this disease. Males are more liable than females to it. Hot climates predispose to it; and so does intemperate living. Exciting causes are, blows or falls upon the head, exposure to the sun, violent or prolonged mental excitement, erysipelas of the head, scarlet fever, metastasis of rheumatic or gouty inflammation, repulsion of eruptions upon the skin, suppression of accustomed discharges. Extension of inflammation from the ear (otitis) to the brain is a possibility, important not to be overlooked.

Treatment.—No disease is more likely to be benefited by early venesection than acute inflammation of the brain. Bleeding should be the rule; omission of it the exception, necessary in cases of debility, anæmia, etc. But its usefulness depends upon its being early; and it should seldom be repeated. Leeching or cupping may follow it. In children, the difficulty of finding a convenient vein to open may cause dependence upon leeches or cups.

Purging actively is very important; by sulphate or citrate of magnesia, or, if dosing be difficult from delirium, croton oil [F. 127] or elaterium. After one free purging, *moderate* catharsis may be, if necessary, repeated every two or three days; and the bowels should be kept open during the attack.

Cutting the hair very short, or, still better, shaving the whole head, will aid in giving relief, and will allow the effectual application of cold. Pounded ice, in a bladder or bag of India-rubber, will do if watched and changed in place often, to prevent too great an impression upon one part. I prefer a linen cloth (as a cambric handkerchief) folded once, dipped in *ice-water*, and laid over the head; it should be wet freshly *every few minutes*, or the good effect is almost lost. Merely wetting the head now and then with cold water produces a *reaction*, not a *sedation*, which is required. If the feet be cold, they should be made warm by mustard foot-baths or sinapisms. In children, the prolonged warm bath may be useful.

The diet in the first part of the attack should be as light and unstimulating as possible. Oatmeal gruel, panada, rice, toast-water, may come first; then milk, chicken-water, mutton broth; later, beef-tea.

Blisters are undoubtedly serviceable after the intensity of the inflammatory excitement has begun to diminish. The best will be a blister *over the whole scalp.*

In a late stage, with secondary debility, concentrated liquid diet, with alcoholic stimulants, and even opium at night, may be required to support the flagging energies of the system.

Convalescence in the best cases may be slow. The faculties may remain feeble, and the brain morbidly excitable, for weeks or months; needing great care as to all mental impressions and efforts, lest a dangerous relapse occur, or chronic cerebral hyperæmia, perhaps insanity, follow.

Scrofulous Encephalitis (tubercular meningitis, acute hydrocephalus).—From two to fifteen years is the age most apt to yield examples of this fatal disease. *Premonitory* symptoms usually occur; dulness, pettishness, and languor; headache; disposition to put the head in the mother's lap, or to lie down; loss of appetite; vomiting; and costiveness. The child sleeps ill, with grinding of the teeth, or sudden starting with alarm. After four or five days, constant headache and anxiety of countenance, heat of head, sensitiveness to light, fever, and drowsiness, alternated with moaning or occasional *screaming*, and delirium at night, mark the case.

Advanced symptoms are, total stupor, strabismus, convulsions, and paralysis. The *pulse* goes through similar changes to those of simple encephalitis: first febrile acceleration, then irregularity and slowness, lastly the rapidity of moribund prostration. The attack terminates on the average in between two and three weeks.

Prognosis is always unfavorable in this disorder. I thought I had met with recovery in one case, the third of his family to be attacked; he remained well, apparently, for a month; and then died in convulsions.

Morbid Anatomy.—Since Papavoine, Rufz, and Gerhard showed the existence of a relation between tuberculosis and "acute hydrocephalus," autopsic inquiry has proved fully, 1st, that tubercle-like granulations, with opacity and thickening of the arachnoid at the base of the brain, adhesion between the hemispheres, and serous effusion, characterize a number of the cases; 2dly, that all of these lesions may be found without any tubercle whatever; and, 3dly, that the amount of such deposit in *most* cases is not sufficient to modify greatly the course of the local disease, at least in such a manner as tubercle acts elsewhere. I conclude, hence (especially in view of such results palpably shown in autopsies under my own eye), that the semi-transparent gray granulations found in the arachnoid after scrofulous meningitis may be rather tuberculoid inflammatory products than tubercles; leaving the *yellow* deposits only to be regarded as of the latter nature strictly; while it is rather the *diathesis* than the *deposits* that make the disease to differ, as in progress and prognosis it clearly does, from simple meningitis or encephalitis.

Treatment.—What can we venture to do in medication for a hopeless disease?   Not to abandon any case of it; for, first, our diagnosis may not be infallible; and, secondly, there is not, as in phthisis, obvious anatomical reason for anticipating a fatal result, in the nature of the case.   Waiving argument, for which we have no space, my judgment is in favor of *treating* this form of inflammation of the brain *on the same principle as simple meningitis*, with more *caution* in depletion and other reducing remedies.   I would not bleed from the arm; but draw blood very moderately by cups or leeches; purge freely, but not exhaustively; blister the head or back of the neck; apply cold with care, and allow liquid nourishment, such as milk and beef-tea, mutton or chicken broth, &c., from an early stage.   If, in this mode, we do not save a patient whom autopsy afterwards shows to have been doomed to die of tuberculization in spite of any treatment, we shall still, according to the indications of clear analogy, have practised rationally; the next best thing to being successful.

HYDROCEPHALUS.

Definition.—Water in the head; dropsy of the brain.   This is almost always an affection of early life.   Sometimes it is congenital. It is mostly a passive dropsical effusion; certain cases show signs of a chronic or subacute inflammatory condition of the arachnoid membrane.

Symptoms.—Languor, strabismus, convulsions, loss of appetite, increase in the size of the head.   This last may be enormous; the fontanelles expanding, and, in a slow case, the bones growing excessively large.   The mental faculties nearly always grow dull.   Bodily emaciation and debility attend.

Although cases are known and recorded in which hydrocephalic persons lived for more than twenty years, the general rule is that they die in a few months; either from cerebro-spinal disability or atrophy, or from some intercurrent disease not endurable by the impaired vital energies of the system.

Treatment.—Small as is the encouragement given by experience in this affection, it is certainly justifiable to *try* measures not out of place in themselves.   Such are, moderate purging, every few days, or once a week, sustaining the strength by nourishing food, and, if it be borne, cod-liver oil; diuretics; shaving the head and rubbing it nightly with mercurial ointment; occasionally blistering the back of the neck; in a child, preferably, by painting it with *cantharidal collodion.*

Is *pressure* by bandages or adhesive straps, or *puncture,* tapping the head, to be advised?   Were I to use either of these heroic measures, I would combine them.   In a case clearly otherwise hopeless, a needle trocar and canula may be introduced through the coronal suture, an inch or a little from the anterior fontanelle; then, during and after the withdrawal of a few ounces of fluid, a bandage may be used for pressure, watching its effects.

SOFTENING OF THE BRAIN.

Pathologists generally recognize two forms of this; 1. Acute red inflammatory softening; and, 2. Slow, white, atrophic softening or degeneration of the brain-substance. Both receive the name of *ramollissement*.

The former of these is farther definable as a local cerebritis; whose symptoms are not nearly always separable, clinically, from those of meningitis or encephalitis, already described. Cadaveric inspection shows not only hyperæmic redness and softening, but, sometimes, abscess, or even gangrene of the brain. This last (gangrene) is probably always the result of injuries. *Induration* of the brain may also follow (especially traumatic) inflammation of the brain. The cerebrum is more often affected with red softening than the cerebellum.

Abscess of the brain is in a certain number of cases latent for a considerable time. Sudden headache is apt to be the earliest symptom. This is attended by feverishness, vomiting, difficulty of speech, numbness, convulsions, paralysis, and coma. *Otitis* and *pyæmia* are said to be, after injuries, the most frequent direct causes of it.

Diagnosis of acute red softening.—The occurrence of imperfect coma, with rigidity of the muscles of the extremities, or of paralysis without loss of consciousness, will make probable this lesion. Most cases die within two weeks; some within two or three days.

White, atrophic softening or degeneration of the brain may take place as a result of old age, or from intense mental labor or excitement, from intemperance, or from *embolism;* that is, obstruction of an artery within the brain by a fibrinous clot carried from some other part. Its approach and progress are more slow and insidious than those of acute inflammatory ramollissement. Neuralgic pains in the limbs, followed by numbness and paralysis; general debility, and dulness of the senses, gradually increasing to blindness, loss of hearing, etc., and a corresponding decline of the mental powers; these are the usual symptoms, which may be extended over a period of many months. Death is sure to be the final result.

Treatment.—If inflammatory red softening can be diagnosticated at an early period, a similar treatment to that named for acute meningo-encephalitis may be advised. Local depletion, at least, followed by counter-irritation by blisters, may be resorted to in a case which appears to be such; the more freely, because apoplexy, which most nearly simulates it, presents very similar practical indications.

Chronic atrophic white softening is not amenable to any such measures; nor, indeed, to any active remedial treatment. Prevention, by avoidance of its causes, and palliation or economy of the waning powers of the system, are alone possible. The management necessary upon such indications must vary with every case.

INFLAMMATION OF THE SPINAL MARROW.

Clinical Synonyms.—*Myelitis, Spinal Meningitis.* The symptoms of this uncommon affection are, constant and severe pain in the back, increased by motion; spasmodic contractions or rigidity of the muscles followed by paralysis, fever, constipation of the bowels, and retention of urine. Authors state that in **myelitis proper**, as distinguished from **spinal arachnitis**, there is no pain nor muscular rigidity, but only paralysis of motion and sensation.

Morbid Anatomy.—Diffuse redness and opacity of the arachnoid, swelling and infiltration of the pia mater, and effusion of serum, communicating freely with the cavity of the cranium, are generally found. Adhesions of the membranes from plastic lymph are less common; and still less so, though repeatedly recorded, is suppuration within the arachnoid. The dura mater is occasionally affected with inflammation, and even ulceration and gangrene, commencing from without. The cord may be reddened from injection of its substance, and softened; more rarely, indurated in parts.

Treatment.—Local bleeding, by rather free cupping or leeching along the spine, followed by a blister, and active purgation with saline cathartics, constitute the essential parts of the treatment of simple inflammation of the spinal cord or of its membranes. If the diagnosis be doubtful, the practice must be disproportionately less bold; this is, of course, a principle of very general application in therapeutics.

Epidemic *cerebro-spinal meningitis* will be considered hereafter, as cerebro-spinal or spotted fever.

SOFTENING OF THE SPINAL CORD.

Rejecting the not uncommon view which refers **ramollissement** of the cord in all cases to inflammation, I have considered softening as one of the lesions produced by myelitis or spinal arachnitis; but separate from this, as in the case of cerebral softening, the **chronic atrophic** degeneration which results in a similar change.

The **symptoms** of spinal softening are, first, numbness in the extremities, with a sense of coldness; pain in a portion of the back, with local tenderness on pressure; then impaired mobility, and gradual loss of sensation in the limbs; or in one limb if only one side of the cord be affected. ·When the anterior columns only are softened, *motor* paralysis prevails; if the posterior columns, *sensibility* is impaired or destroyed. Difficulty in walking, especially on first rising in the morning, is an early symptom. Contractions and rigidity of muscles occur later. At a still more advanced period, loss of control over the bladder and rectum adds to the distress of the patient; who is apt to suffer also from bed-sores, ulceration, and sloughing of the parts upon which the body rests; the system becoming gradually exhausted.

Prognosis and Treatment.—Recovery is not to be expected from atrophic spinal softening. The most unfavorable symptoms are decided paralysis, involuntary urination, and defecation, with *alkalinity of the urine.* **Treatment** must be palliative and supporting only. Passive exercise (as by riding in an easy carriage, sailing, or being carried) in the open air, will be beneficial; and so may salt bathing, and frictions of the surface of the body. Good diet, appetizing tonics, and sometimes alcoholic stimulants very carefully regulated, may retard the decline of the patient.

SPINAL IRRITATION.

Under this term (now discarded by most writers upon nosology and diagnosis) have been included several affections of different pathology, and not always identical in symptoms. Pain and weakness in the back, without proof of any decided or progressive lesion of the cord, or either motor or sensory paralysis, appear to be the common features in such cases. I think a name should be reserved for this combination, for practical or clinical use; however, as in the case of dyspepsia, and some other complex disorders, the term may not define the disease. Some cases included in this account are really **rheumatoid** (chronic non-febrile rheumatism) affections of the sheaths of the spinal nerves; others are instances of **myalgia**; that is, *muscular* pains, from weakness and exhaustion in the muscles. Others again display, with anæmia and general nervous debility, a real irritability of the cord, shown by (hysterical) spasms of some muscles, or general convulsion, under disturbing impressions of various kinds.

Treatment.—The discovery of the *nature* of the case (as above indicated) is important. If, in an otherwise vigorous person, the attack come on after some exposure, the rheumatoid condition is most probable; and then cupping along the spine will do the most good. Afterwards, counter-irritation, as by croton oil, may be used; and wearing flannel next the skin will be important.

Myalgic, or purely muscular pains, follow generally upon fatigue, and are best cured by repose; aided by warm frictions, as with spirit of turpentine, whisky and hot water, whisky and salt, &c.

True spinal irritability (*i. e.* of the cord, not always attended by sensitiveness to pressure along the back) is nearly always an affection of the anæmic and weak. Iron and other tonics, with nourishing food, salt bathing, and pure air, are demanded; and, with these, mild counter-irritation over the spine. Hemlock or Burgundy pitch plasters; repeated dry cupping; painting with tincture of iodine; and the use of croton oil externally, are the best measures of this kind for such a purpose.

INFLAMMATION OF THE EYE.

Although ophthalmology is appropriated as a department of surgery, every medical practitioner meets with cases of affections of the eye, so often as to make it proper to notice here, briefly, its principal acute disorders.

Varieties.—Conjunctivitis (*ophthalmia*, by usage), simple, catarrhal, pustular, and purulent (Egyptian, military, and gonorrhœal ophthalmia, and ophthalmia *neonatorum*, *i. e.* of new-born infants); keratitis (corneitis); sclerotitis (rheumatic ophthalmia); iritis (simple, traumatic, syphilitic); retinitis.

Simple and Catarrhal Ophthalmia: Symptoms.—Bloodshot appearance of the eye, with soreness, pain, and dislike of light, characterize *simple* conjunctivitis. Blotched or irregular injection of the conjunctiva, becoming in severe cases general and velvet-like, with, sometimes, *chemosis* (raising of the mucous membrane in spots, like little water-blisters), and mucous discharge, agglutinating the lids together, especially at night; these are the symptoms of the *catarrhal* variety or grade.

Treatment.—When the inflammation is severe and recent, leeches to the temple near the eye will do good. Iced sassafras-pith water may be applied by laying a light piece of linen, soaked anew every few minutes, over the closed lids. Nitrate of silver solution, two grains to the ounce of distilled water, is recommended to be dropped from a quill or camel's hair pencil into the eye twice daily. A saline cathartic at the beginning of the attack will generally be useful; and so will be, a little later, a fly blister behind the ear. When convalescence has fairly commenced, the use of the injection may be made at longer intervals, once in a day or two; the eye being then kept closed, if the mucous discharge be slight, by a strip of isinglass plaster over the middle of both lids. After recovery, the eyes will be weak for a time, and must be used with caution and moderation. Adhesion of the lids is best prevented, at any stage, by the application of spermaceti ointment, castor oil, or glycerin cream, to their margin. Persistent redness and swelling of the lids will often give way under the use, nightly, as an unguent, of the *cerate of carbonate of lead* [F. 88]. Painting the exterior of the lids, many times daily, by means of a camel's hair pencil, with diluted extract of lead (one drop of Goulard's extract in an ounce of water), followed by cold cream or glycerin cream at night, has, to my knowledge, relieved greatly cases of long standing "weakness" or irritability of the eyes. *Granular conjunctiva*, or "chronic ophthalmia," not yielding to the above measures, may be referred to the resources of the special ophthalmologist. *Pustular* ophthalmia is characterized by the formation upon the conjunctiva of small vascular elevations resembling pustules, although rarely discharging pus. In other respects, the attack resembles the catarrhal inflammation.

Purulent Ophthalmia.—1. Ophthalmia of infancy. Leucorrhœa or gonorrhœa of the mother may produce this; or it may follow exposure to cold or damp air, acting upon a system predisposed by imperfect nutrition. The danger of it is, the possibility of rapidly destructive ulceration of the cornea, producing blindness. It is possible, however, for a considerable ulcer of the cornea to heal, in a child, even without leaving an opaque cicatrix.

Treatment.—Introduce, by inserting the smooth point of a small syringe just within the inner commissure of the lids, several times a day, a solution of three grains of alum to the ounce of water; alternated occasionally with a solution, of one or two grains to the ounce, of nitrate of silver. As very much depends upon the vital energy of the child's system, especial care must be taken as to its nourishment, bathing, and the state of its bowels.

2. Gonorrhœal Ophthalmia.—Produced by contact of the virulent poison of gonorrhœa, this is perhaps the worst form of inflammation of the eye. At the beginning, it (as well as ordinary purulent ophthalmia) may resemble catarrhal inflammation; but its course is so rapid and violent as to become suppurative in one or two days. Haziness of the cornea, and chemosis, followed by ulceration, perforation, or sloughing, may occur. Such ulcers are apt to leave white and opaque cicatrices, even if not very deep, in the adult.

Treatment.—Begin with a brisk purgation. Then apply, at once, and frequently, in alternation, solution of alum, ten grains in an ounce, and solution of nitrate of silver, four grains in an ounce. If the specific character of the inflammation can thus be annulled, the destruction of the cornea may be averted. Sometimes good surgeons have applied the solid nitrate of silver to the ulcerated surface. The prognosis, however, in this form of disease, is generally unfavorable to the safety of vision.

Keratitis.—(Corneitis). A zone of vessels in the sclerotic, immediately surrounding the cornea, with *haziness* in the latter itself, amounting in time to opacity, marks this affection. When developed, we will find a plexus of fine vessels, arranged in a crescent, or semicircle, along the upper or lower edge of the cornea, or both. Intolerance of light is great; any exposure of the eye causes a flow of tears. Children and adolescents are most generally the subjects of inflammation of the cornea. It may be of short duration, the opacity disappearing, in a young person; at a later period of life, it is more obstinate; and if the attack lingers for several months, cloudiness remains.

Treatment.—Most subjects of corneal inflammation are of delicate frame and depressed health. Tonics and good diet are more likely, in them, to be indicated than depletion. Repeated blistering behind the ears will be proper. The bowels should be well opened, and the eyes sheltered from strong light, while photophobia (intolerance of light) exists. In no form of disease of the eye, however, unless for a short time in a very acute attack, should a patient be imprisoned in a dark room. The want of air, and even of sunshine, will do more harm than good. When otherwise in a state fit for it, he may go out, with the eyes protected by a shade or suitable glasses, or a veil. Good authority prohibits the use of nitrate of silver solutions, as *mischievous* in *corneal* inflammation.

Sclerotitis; rheumatic ophthalmia. This is shown by diffused redness of the eyeball, with enlargement of the arteries converging to the margin of the cornea; and *severe pain* in the ball, with intolerance of light.

Treatment.—Not satisfied that there is always proof of the "rheumatic diathesis" in every case of sclerotic inflammation, I should still incline to prescribe for it, as a general rule, a combination of colchicum with an alkali (as carbonate of potassa), after a saline cathartic. A blister may then be applied behind the ear or back of the neck. *Anodynes* are apt to be called for; as belladonna or opiates; so severe is the pain in many instances.

Iritis.—Some writers describe the forms of inflammation of the iris as **simple, traumatic, gouty** and **rheumatic, scrofulous,** and **syphilitic.** The first and last of these are the most important and distinctive.

In either form of inflammation, a *vascular zone in the sclerotic*, near the cornea, *fixedness* and *irregularity* of the *pupil*, with a greenish hue of the iris if it be naturally blue, are the usual signs.

Treatment.—In a robust patient, leeches around the eye; in a weaker one, a blister behind the ear; saline purgation, repose to the eye, and steaming it frequently over hot water, are measures that

19

nearly all will agree upon. More difference of opinion exists as to the use of *mercury* in iritis. Some give it, to retard the effusion of lymph, in all cases. Others, only in the syphilitic. A number, rather increasing of late, in none. I would give calomel in all cases of active iritis, but in none so largely as to endanger salivation. One grain twice daily for a few days will be enough ; stopping it if the gums are at all affected.

Maintaining moderate dilatation of the pupil is considered important in all cases of iritis. For this purpose, *atropia* is locally used. Once or twice daily there may be dropped into the eye two or three drops of a solution of two grains of sulphate of atropia in an ounce of water.

Retinitis.—So obscure is the diagnosis of this, and so greatly has its pathology been modified since the introduction of the ophthalmoscope,[1] that it will be best for us to refer for its consideration to works especially upon the Eye. (See *Mackenzie*, *Laurence* and *Moon*, etc.)

OTITIS.

Definition.—Inflammation of the ear. This is most common in children. *Scrofulous* inflammation and suppuration of the external meatus, with chronic discharge, is frequently met with. *Scarlet fever* not rarely is attended by otitis, extending from the throat ; sometimes ulceration destroys the *membrana tympani*, and even the *ossicula ;* causing deafness.

Otalgia, earache, occurs often without much inflammation, as an almost entirely neuralgic affection. Pain, however, is the first symptom of otitis ; with soreness on pressure upon the meatus or the mastoid process, and heat of the ear. An infant may suffer from this without being able to designate the seat of its distress. It cries or screams at intervals, and buries its head in the pillow, or leans the affected side against its mother's breast or arm. Often in the course of a day or two a purulent discharge gives relief to the intensity of the attack. In other cases pain returns again and again, and the soreness continuing, without discharge, for weeks together.

Extension of inflammation from the internal ear to the membranes of the brain is possible in severe acute otitis ; but it is almost as likely to happen in cases of long-continued *otorrhœa*, or discharge from the ear ; particularly if this be suddenly arrested.

Treatment.—*Earache* may be relieved usually by dropping into the ear three or four drops of olive or almond oil, with one or two drops of laudanum. If continued pain, with heat and tenderness on pressure, show decided inflammation, a few American leeches may be applied behind the ear ; and afterwards a small blister (when the leech-bites have healed) upon the same place. Painting with the cantharidal collodion will here prove very convenient.

Chronic discharge from the ear should be treated with mild astring-

---

[1] The *simplest* ophthalmoscope is a *perforated* hand mirror, which reflects a strong light upon the eye, while the examiner looks through it. In *acute* inflammations of the eye its use is unsuitable. Bouchut, however, has written a volume upon it as a means of diagnosis, even in diseases of the brain, as meningitis, etc.

ents, very gently applied. Syringing is not nearly always necessary; if done with force it irritates, and may cause headache and nausea. Pouring the lotion from a teaspoon, the patient lying upon the other side, and turning over to allow it to run out, will generally do better. Lotions so used should be *warm*.

Castile soap and water; lime-water; glycerin and rose-water (one part to five); and solution of acetate of lead, one or two grains in an ounce of water, will prove the best washes, and sufficiently strong to mitigate without too suddenly checking the discharge.

Deafness (cophosis) may result from, 1. Accumulation of wax in the ear; 2. Inflammatory thickening of the membrana tympani; 3. Obstruction of the Eustachian tube; 4. Perforation of the tympanic membrane; 5. Destruction of the *ossicula* of the ear; 6. Paralysis of the auditory nerve. Only in case of the *last* is the patient unable to hear the ticking of a watch placed *between the teeth*.

HEAT-STROKE.

Synonyms.—*Sunstroke; Coup de Soleil; Insolatio.* Two forms of heat-stroke undoubtedly occur. In one, the direct rays of the sun upon the head induce *cerebral congestion;* in the other, excessive heat, often not under the immediate influence of the sun, affects the whole system with prostration, apparently from a *blood-change;* the chemical operations of the economy being modified by heat in a manner incompatible with the vitality of the blood.

Symptoms.—Genuine *sunstroke* is commonly sudden. Falling unconscious, the head is very hot, the temporal arteries distended; the breathing is apt to be stertorous (snoring), the pulse full and slow. In severe cases, convulsions may precede death.

In heat-stroke of the second variety (more common than the first), almost equal suddenness marks the attack. There is, however, no excessive heat in the head; the pulse is weak; unconsciousness is less complete, and without stertor of the breathing; the whole condition resembles syncope rather than apoplexy.

Causation.—It is remarkable that few cases of heat-stroke occur in the country, among farm laborers; and very few at sea, even in the tropics. *Large cities* afford nearly all the cases. This looks as if the *atmosphere* had much to do with predisposing to it, at least by deteriorating the blood, and lowering the resistance of the vital energy.

It is nearly always, in the case of *heat-exhaustion*, those who have been *fatigued* by exertion, in the sun or shade, who are overcome. Drinking largely of cold water when thus exhausted, much increases the danger. Intemperate persons are particularly liable to heat-stroke.

Treatment.—For sunstroke, or *heat-apoplexy*, cupping or leeching the back of the neck or behind the ears, should generally be the first remedy, after the application of ice or iced water freely to the head. The head and shoulders should be kept raised. A purgative enema should also be administered, and sinapisms applied to the lower limbs.

Heat-exhaustion requires quite different treatment, in part at least. Cold should be applied to the head, and sinapisms to the spine, epigastrium, and limbs, in turn. Local depletion should be avoided. If syncopal symptoms be *decided*, ammonia may be for a few moments applied to the nostrils; and, if the patient can swallow, aromatic spirits of ammonia may be given by the mouth, 10 drops every fifteen minutes at first, gradually increasing the interval. *Mixed* cases of course occur, demanding an intermediate or composite treatment.

INSOMNIA.

Definition.—Morbid wakefulness; impossibility of sleep.   ·

Causation.—Apart from pain, or severe acute disease affecting the brain (as delirium tremens), insomnia may be brought on by intense or prolonged mental labor or emotional excitement. Excessive use of strong coffee or tea, or of belladonna, stramonium, or cannabis indica, may produce it.

Pathology.—Only within a recent period has the correct view been adopted that, during sleep, the arterial circulation of the brain is at its minimum. In sleeplessness, the most certain *error loci* is, an *erythism* (morbid erectility) of the cerebral arteries, which keeps their circulation full and *prevents* sleep. It is not possible to be sure that this is all, as the nature of brain-action and nerve-force is unknown. But this furnishes a basis for rational management.

Treatment.—This must vary with the cause. The overworked brain of the professional, literary, or business man must be withdrawn from his employment. Irregularity of the circulation dependent upon general debility must be met by tonics and generous diet. Accumulation in the head must be diminished by such physical exercise as the strength will bear. Decided *cerebral exhaustion* is apt to be attended by such loss of nerve force as will forbid much effort of any kind; but milder cases of insomnia will be benefited by exercise. The brain should be especially allowed to rest from excitement near the usual hour for sleep. Hence a walk or the use of dumb-bells just before bedtime will be suitable. If the stomach be empty, a little easily digested food, even late at night, will promote sleep; notwithstanding the familiar fact that heavy suppers induce wakefulness or nightmare. A glass of lager beer at bedtime is, as my own experience has proved, one of the best of hypnotics.

The warm bath or pediluvium, with cold to the head, will be serviceable in abstracting blood from the brain. *Position of the body* is important. The sufferer from insomnia may often be very sleepy before lying down—yet, once in bed, he becomes wide awake. Several persons, in such case, to my knowledge, have found it best to recline with the head and shoulders raised. Thus, by gravitation, the flow of blood to the head is retarded, and sleep is promoted.

Of medicines, for simple insomnia, in the absence of pain, opium and other powerful narcotics are not appropriate. Hops, lactucarium, and hyoscyamus are strong enough. Bromide of potassium, in ten or twenty grain doses, is just now most relied upon as a cerebro-vascular sedative. [F. 134.]

NIGHT-TERRORS.

Dr. C. West' gives the following description of an attack which I have seen a number of times, occurring in infants or children under ten years of age. "A child who has gone to bed apparently well, and who has slept soundly for a short time, awakes suddenly in great terror, and with a loud and piercing cry. The child will be found sitting up in its bed, crying out as if in an agony of fear, 'Oh, dear! oh, dear! take it away! father! mother!' while terror is depicted on its countenance, and it does not recognize its parents, who, alarmed by its shrieks, have come into its room, but seems wholly occupied with the fearful impression that has aroused it from sleep. In from ten minutes to half an hour, as the terror abates, it may become quiet at once and fall asleep; but frequently it bursts into a fit of passionate weeping, and sobs itself to rest in its mother's arms. In some instances a quantity of limpid urine is voided as the fit passes off, but this occurrence is by no means constant. Usually the remainder of the night is passed in tolerably sound sleep; two attacks do not often occur in the same night." "Seizures of this kind may come on in a great variety of circumstances, and, according to the cause whence they have arisen, may continue to return for many weeks together, or may occur but a few times. As far as I have had the opportunity of judging, they are never the indications of primary mischief in the brain, but are always associated with some disturbance of the intestinal canal, and more or less obvious gastric disorder. In the majority of cases constipation of the bowels exists."

My experience with such cases confirms that of Dr. West, as indicating that these attacks do not prove disease of the brain. But the nervous system of a child so affected must be morbidly susceptible; and signs of indigestion, constipation, or irritation of the bowels are not always present.

During the attack, the child should be at once gently lifted up from the bed, and either carried for a few moments or laid down in a different position. Washing the face softly with a rag dipped in cool or cold water may arouse thoroughly. If any medicine be suitable, it will be a teaspoonful of camphor water. Care is needed to *prevent* the attacks. *Violent exercise* and *mental excitement* are almost as apt to bring them on as indigestion or constipation. The bowels should, however, be kept open, as by fluid extract of rhubarb, or senna, &c. To promote tranquil sleep, some one should remain with the child, if timid, for a while after it goes to bed; or a light should be left burning low. A child liable to night-terrors ought to be allowed to finish its morning sleep undisturbed. Abundance of sleep is sedative to an over-excitable brain. Neglect of such precautions may convert a mere transitory functional disturbance into a serious attack of brain disease.

' Diseases of Children, p. 210.

APOPLEXY.

Definition.—Sudden coma, produced neither by injury nor poison.
Varieties.—Some terms once used have been shown to be without pathological justification; as *serous* apoplexy, *nervous* apoplexy. Good authority still sustains, however, the mention of two forms at least of genuine apoplectic seizure : *congestive* and *hemorrhagic*.

Symptoms. —Congestive Apoplexy.—Premonitory symptoms often seen are, flushed appearance of the face and eyes, heat of head, throbbing of the carotids, distension of the temporal arteries and jugular veins ; constipation, languor, dulness, drowsiness; dimness of sight, vertigo, headache. The attack is marked by sudden stupor; with slow and sometimes snoring respiration, full and slow pulse, dusky or turgid appearance of the face. The total loss of perception may be brief, its partial absence or deficiency continuing for some time. Slight convulsive movements are not uncommon. Paralysis of the muscles occurs only for a short time after the attack, if recovered from.

Symptoms of Hemorrhagic Apoplexy.—Generally no clear premonition is given, the attack being very sudden ; a *stroke*, literally. Unconsciousness is complete, for some seconds, minutes, or hours. After this, general or local paralysis, most often hemiplegia, is left; the mental powers also, in many cases, being impaired, at least temporarily. During the coma, the breathing is commonly stertorous, and the pulse slow, and somewhat full, the head hot, the face more or less dark or flushed. But the fulness of the bloodvessels and heat of the head are much less, as a rule, than in congestive apoplexy.

Anatomy and Pathology.—In the congestive form, excessive cerebral hyperæmia produces coma by pressure upon the brain; the extremest degree of which (vascular pressure) is met with in strangulation.

In hemorrhagic apoplexy, from the rupture of a degenerated artery, either in the substance of the cerebrum or cerebellum, in the ventricles, or under the arachnoid membrane, effusion of blood occurs, and a clot is formed. If this be small, it may be gradually absorbed; autopsic inspection sometimes shows the remains of such, where another hemorrhage has caused death.

Fatty degeneration of the arteries of the brain has been repeatedly, but not always, observed. The parts of the brain most liable to extravasation of blood are (Aitken) stated in this order: 1. corpus striatum, thalamus, and hemispheres above them; 2. corpus striatum alone ; 3. hemisphere above the centrum ovale ; 4. thalamus alone ; 5. lateral lobes of cerebellum; 6. mesocephalon; 7. posterior lobe of cerebrum ; 8. before the corpus striatum; 9. pons Varolii ; 10. middle lobe of cerebellum; 11. meninges ; 12. peduncles and olivary body. The age of the clot may be ascertained in part by the discovery, with the microscope, of *blood-crystals;* which are not found until after seventeen or eighteen days from effusion.

Diagnosis.—Apoplexy is to be distinguished from *uræmia, alcoholic intoxication* (dead drunkenness), *narcotic poisoning* (as from opium), *compression of the brain,* or *concussion,* from blows or falls, *asphyxia* (suffocation), *sunstroke, catalepsy, cerebral hysteria, acute* *softening of the brain,* and *spotted fever* or "cerebro-spinal meningitis;" as well as from all forms of *syncope.* From uræmic coma it is only to be known by the history of the case, showing a renal origin for the symptoms, in partial or total suppression of urine. Alcoholic intoxication is revealed by the odor of the breath, and the attendant circumstances. Similar aid exists sometimes in cases of narcotic poisoning; in opiate narcotism, moreover, the pupil is *contracted ;* in that from most other narcotics, it is as firmly *dilated.* Concussion and compression of the brain are generally suggested by the position of the body (if found without a history), and the external marks of injury. Asphyxia is also usually pointed out by the condition of things surrounding the patient.

In asphyxia, blueness of the lips, and embarrassment of respiration, with coldness of the surface, show the origin to be in the function of breathing. Sunstroke is attended by feebleness of the pulse, at least in the majority of cases ; in some, it *is,* identically, a congestive apoplexy. In catalepsy, there is rigidity of the muscles, with rapidity of the pulse, susceptibility of the pupil to light, brief duration and repeated recurrence of the attack, without any paralysis. Cerebral hysteria is rare, and occurs only in females, whose previous disorders of the nervous system will aid in interpreting even coma as belonging to the same category. Acute red softening of the brain may be very difficult to distinguish from apoplexy. It is, however, seldom if ever so sudden in its invasion ; there is more slobbering or flow of saliva, and watering of the eyes; and there is not the partial or entire restoration of the faculties which an attack of apoplexy, not fatal, allows so often. Spotted fever, or "cerebro-spinal meningitis," will be especially described, and its diagnosis considered, in another place.

Syncope, of any form or origin, is always marked by *pallor, coldness,* and *loss of pulse.*

Prognosis.—This is always alarming; most so as there is the most reason to believe that cerebral hemorrhage has occurred ; and, therefore, especially in those advanced in life. In younger subjects, where stertor of breathing is absent, under proper treatment, congestive apoplexy may be entirely recovered from. So may a single attack of the hemorrhagic form, with a small clot only, and limited, transient paralysis. Each succeeding attack becomes more dangerous ; a third is seldom survived. The *immediate* danger connected with an attack of apoplexy should not be considered over for ten days at least after the stroke itself. Very seldom, indeed, after a hemorrhagic attack, are the mental or bodily powers so good, for the rest of life, as before.

Causation.—*Age* is the most constant promotive cause of apoplexy. Cases are on record, though of extreme rarity, in children; between thirty and fifty it is much more frequent; but after fifty it is one of the most common modes of death. Arterial degeneration is here the general occasion of the catastrophe; some mental excitement, or bodily shock or effort, as danger, or joy, or a few glasses of wine, or the stooping posture, or straining at stool, causing a rupture of the weak vessel, and fatal cerebral hemorrhage. Neither sex seems to be more liable to this disease than the other.

Full living, especially with alcoholic intemperance (even moderate) and indolent habits, predispose to it in a marked degree. So does

excessive brain-work. Florid, short-necked, big-bellied people are
most exposed·to it. Hypertrophy of the left ventricle of the heart is
believed also to promote it. After dinner and during sleep are the
two most likely times for the attack to occur.

Treatment.—The younger the patient, and the more vigorous his
antecedent health, the more probable is the existence of the congestive
form : and, also, the better the prospect of recovery from hemorrhage
within the cranium, if, only, the effects of pressure be averted at the
time. If, then, in a person under fifty, not before of broken constitution, we find the head hot, face turgid and flushed, the arteries and
veins of the neck and temples full, the pulse also strong, and the
heart's impulse so (or the heart's action vigorous though the pulse
at the wrist be oppressed), *bleed.* Watch the effect, with the hand
on the pulse. If the breathing improve, and the pulse rather gain
than lose in naturalness and force, take out ten or twelve ounces.
Should the improvement progress, but still a relapse into deeper
stupor afterwards threaten, either the lancet again, or cupping to the
nucha, may be used.

Older or more doubtful cases may be treated tentatively, with cups
alone, aided by mustard plasters to the legs, back, and epigastrium
in turn ; with laxative injections into the rectum during the attack,
and saline purgatives afterwards. The head should be kept raised,
and cooled with wet cloths until its temperature become normal. If
the hair be thick, it should be cut very short or shaved off entirely.

When, however, there is reason, as usually is the case in really *old*
or *broken down* patients, to believe that *structural degeneration*,
arterial or that of ramollissement, is the source of the attack, loss of
blood will be *out of place*. It may even, by exhausting the enfeebled
system, hasten death. Such cases, if they survive the first apoplectic
fit, require rather nourishing diet, and sometimes even tonics, to
support strength, favor repair, and prolong life. Great delicacy of
judgment. of course, is necessary in deciding, in different cases,
between these apparently so opposite modes of treatment. It is right
to add, that the tendency of medical opinion, for the last ten or
twenty years, has been towards the curtailment, to a great extent, of
the use of the lancet in apoplexy.

Where a moderately plethoric condition is present, and the taking
of blood, generally or locally, is not decided upon, purgation is safe
and likely to be useful. Jalap, resina podophylli, or croton oil, in small
doses, will have the advantage of convenient administration [F.
135, 136].

PARALYSIS.

Varieties.—*According to the proximate cause :* 1. Cerebral palsy ;
2. Spinal ; 3. Reflex paralysis ; 4. Toxæmic (*e. g.*, lead palsy); 5. Hysterical palsy. *According to the extent of the affection :* Facial or
other local palsy ; Hemiplegia ; Paraplegia; General paralysis. *According to its nature :* Motor (acinesia), and Sensory paralysis (anæsthesia).

Facial Palsy.—This is an affection of the *portio dura* of the
seventh pair of cephalic nerves, the motor nerve of the face. It occurs

at any age, usually from rheumatoid inflammation of the sheath of the nerve at its escape from the cranium through the stylo-mastoid foramen. One side of the face is without change of expression; and the eye on that side is not closed (in severe cases) from the paralysis affecting the *orbicularis palpebræ* muscle. The tongue is not affected in its movements.

The facial motor nerve is not often involved in the much more serious cases of *cerebral* palsy. Absence of disturbance or of incompleteness of control over the tongue, while the power over the eyelid is partly or wholly lost, with the absence also of severe cerebral symptoms, will, especially in a young person, make the diagnosis easy as well as important. The **prognosis** is, generally, of recovery in a few days or weeks. The **treatment** of this form of local palsy may be by repeated small blisters behind the ear; followed, when convalescence has begun, by some warm covering (cotton wadding, flannel, or silk) to protect the part from cold.

Other local Palsies.—Pressure upon a nerve may cause its paralysis, generally temporary. I remember the case of a man whose hand was rendered powerless for about three weeks by sleeping all night with his arm bent under his head. Frictions, the endermic application of strychnia, and galvanism were used in that case. *Writer's cramp,* or scrivener's palsy, is the result of exhaustion of certain muscles from over use. Its cure is rest.

Palsy of the optic nerve is designated as *amaurosis;* of the sense of hearing, *cophosis;* of taste, *ageustia;* of smell, *anosmia.* Except the first, however, these terms are not much used.

Hemiplegia.—Brain-lesion is most often the cause of this affection; either an apoplectic clot, a tumor, or softening. Spinal disease may, however, produce it; and some cases are, by writers upon the subject, referred to a peripheral or reflex origin. There may occur, also, transiently, *epileptic, choreic,* and *hysterical* hemiplegia. Owing to the decussation of the anterior pyramids of the medulla oblongata, lesion of one side of the brain produces paralysis of the other side. In spinal lesion the palsy is usually on the same side. Brown-Séquard, however, has shown decussation in the upper part of the cord also.

Symptoms.—Suddenly, almost always, but not always with loss of consciousness, the patient loses the power of motion, and more or less of sensation, on one side. In *complete* cases, the parts involved are the arm and leg, the muscles of mastication (with the buccinator), and the half of the tongue. In trying to protrude the tongue it is pushed out towards the affected side; in retracting it, the reverse happens; that is, it is drawn towards the sound side. The palsied cheek hangs; but the eye can be shut or opened at will. The third, fifth, and ninth nerves are especially apt to show implication by disturbance of the actions under their control; of the fifth, those of the muscles already mentioned, as well as of facial and lingual sensation; of the third, loss of power to lift the eyelid, strabismus, and dilatation of the pupil; the ninth, one-sided movement of the tongue, affecting also the speech.

Hemiplegia may be attended either by rigidity or relaxation of the muscles; and the former may be early or late. Where there is decided relaxation in cerebral paralysis, it is probable that white softening, or atrophy from embolism of the brain is the lesion, with or without a clot; where early rigidity is marked, an apoplectic clot may be inferred. Late rigidity is probably due to an atrophic state of the muscles; a "*rigor mortis in vitæ.*"[1] Contradictory accounts are given by authorities as to the susceptibility to galvanic excitation of the muscles on the sound and on the *paralyzed* side. It is probable that the loss of excitability of the muscles is in proportion to their atrophy.

The **Prognosis** in hemiplegia depends greatly on the ascertainment of its causation. If it follows an epileptic fit, or attack of chorea, or occurs in a hysterical subject, it may be of comparatively brief duration, ending in recovery. If an apoplectic attack precede it, or if any lesion of the brain be inferred from the history of the case, the prospect is bad. Partial improvement may occur, not often entire restoration; and renewed attacks or "strokes" are likely to follow.

Treatment.—Essentially the same principles are applicable to this as have been mentioned in connection with apoplexy. The younger the patient, the more vigorous his or her previous health, and the fuller the circulation, the more appropriate will be the general or local abstraction of blood, to diminish pressure upon the brain. Where softening is apprehended, bleeding should be exceptional and cautious. Epileptic, choreic, and hysterical hemiplegia indicate little or no depletion as a rule. Rest, regulation of the bowels, and counter-irritation by dry cups to the upper part of the spine, and afterwards a blister; with frictions, as with brandy and red pepper, or whisky and hot water, or salt and spirits, to the affected limbs; these are measures of general utility. A seton in the back of the neck is sometimes recommended. As to strychnia, it is not safe where cerebral or spinal irritation is likely to exist, as near the commencement of most attacks. Even at a late stage, it should be used with extreme caution, watching its effects [F. 137]. Precisely the same statement may, upon the best authority, be made as to electricity, in cerebral paralysis. In the *hysterical* form, if it last long, electricity may be applied locally, with safety and advantage. In any curable case, *passive exercise* of the weak limbs will be very useful.

Paraplegia.—This is paralysis of both the lower extremities. *Spinal* disease or injury is its source; with or without cerebral implication or complication. It may come suddenly or gradually; generally its beginning, at least, is sudden. *Reflex* paralysis as described by several authors, is sometimes paraplegic.

Symptoms.—In organic or spinal paraplegia, as well as in the reflex form, numbness in the feet and pain in the back are apt to be early signs. The power of motion is lessened or lost in the lower limbs. The muscles may be either relaxed or contracted. The lesion of the spinal marrow, if progressive, is productive finally, in many cases, of loss of power over the bladder and sphincter ani. Bed-sores, with deep ulceration and sloughing, may occur in protracted cases.

Treatment.—When *myelitis* is believed to exist, at an early stage, local depletion to a moderate extent, in otherwise good subjects, may

---

[1] I have above, purposely, avoided alluding to the complications introduced of late into the special pathology of paralysis, by the vivisections of Brown-Séquard and others; because, brilliant as they are, while they have unsettled much, they do not appear to me to have positively *settled* anything.

be advised. In any case, counter-irritation (not vesication, in a bedridden patent, unless he can lie well on either side), by repeated sinapisms, or stimulating liniments [F. 138, 139], will be proper.

While inflammation or active irritation of the spinal cord is made apparent by the symptoms (pain, cramps, muscular twitchings, or rigidity) strychnia is not suitable. After these have subsided, it may be given,—not more at first than the thirtieth of a grain twice daily. If it produce jerking movements of the hands or feet, or nervous restlessness, or any marked uneasiness, it should be suspended. Electricity may be used, with similar caution, in a secondary or relatively late stage of paraplegia. Moderate (at first very gentle) shocks of the interrupted circuit are preferred.

Hysterical Paralysis.—In females, this is among the many forms of functional disorder which that strange and not yet clearly defined disorder, hysteria, may produce. It is diagnosticated by the aid of the history of the patient. Dr. Todd stated that, in it, the affected limb (it is most often hemiplegic) in walking is dragged after the other, as if a dead weight; while in cerebral hemiplegia the palsied leg and foot are brought round in a curve, the body being bent toward the sound side at the time. I am doubtful of the universality of this sign.

Treatment.—Tonics, good nourishment, and change of air (in a word, analeptic management), are most needed in nearly all hysterical cases. For the paralysis itself, electricity has been found useful. Mild shocks for a few minutes twice a day may be given with the magneto-electric apparatus.

Reflex Paralysis.—From the times of Whytt and Morgagni, occasional instances of palsy, of motion or sensation, caused by an injury at a distance from the affected parts, have been recorded. Since Stanley's paper (1833) asserting the production of paralysis, sometimes, by disease of the kidney, a number of medical writers have added to the list of supposed cases of "paralysis without apparent lesion." Worms, dysentery, diarrhœa, uterine irritation, teething and external injuries are all thought to induce reflex paralysis in certain instances. Diphtheritic and scarlatinal palsies have by some been placed in the same category. The simplest and clearest cases are those of wounds; e. g., Morgagni's case, in which amaurosis was suddenly produced by a blow upon the eyebrow, affecting the supraorbital nerve. I would exclude most of the asserted instances of *visceral* reflex paralysis.

The *pathology* of this form of palsy is a subject of much controversy. To my judgment (after reading considerably upon it) the best explanation is that of Handfield Jones and S. W. Mitchell; expressed in the term proposed by the former.—"inhibitory action." Denying, against no matter what present authority, the existence of *inhibition* or repression as ever proved to be a *normal* function of any nerve (such as some assert on the ground of experiment in regard to the pneumogastric), I consider it most reasonable to admit it here *pathologically.* In other words, a *morbid* impression, from injury or disease, in one part of the body, being transmitted along a nerve to a nerve-centre, overwhelms or paralyzes it; this effect being shown, of course, in the parts to which it distributes nervous branches.

Treatment.—In true reflex paralysis, of short or moderate duration, the removal of the irritant cause produces instant relief; as in M. Jones' case, where strabismus from palsy of the external rectus oculi muscle disappeared after a piece of dead bone was extracted from a whitlow on the thumb; or Lawrence's in which blindness of one eye (of thirteen months' standing) was cured by the extraction of a carious tooth, with a splinter of wood projecting from one of its fangs. When the nature of the case does not admit of such prompt relief, if the diagnosis be clear, the same indication remains; to address our remedial measures to the seat or source of peripheral irritation. Palliate, if we cannot cure, the trouble there, and we will obtain palliation, if not relief, of the reflex disability. Electricity has proved signally useful in the subsequent treatment. This form of disorder is, however, very rare.

Diphtheritic Paralysis.—After the termination of an attack of diphtheria, commonly within three weeks, the muscles used in swallowing and speaking, less often those of the upper and lower limbs, and the sense of sight, may be partially paralyzed. Loss of sensibility usually accompanies the loss of motor power. This condition of things may last for weeks or even months, but is generally recovered from. Whether the immediate cause of the paralysis is the peripheral lesion of the nervous terminations (in the pharyngeal and laryngeal affection) or the toxæmic influence, upon the nerve-centres, of the morbid poison of diphtheria, cannot yet be decided. In extended palsy as a sequela, I believe the latter to be the more probable explanation.

Treatment.—Passive exercise, stimulating frictions, and electricity, sometimes with change of air, and sea-bathing, are suitable measures for this affection.

Syphilitic Paralysis.—The most unequivocal instances of this nature are accounted for by periostitis within the cranium, involving the dura mater, or, by nodular exostosis, pressing upon the brain. The most remarkable fact connected with such cases is the recorded experience showing the prompt curative effect upon it of *iodide of potassium* [F. 140]. Obscure paralysis, without apoplectic symptoms, and in a syphilitic constitution, may be tentatively so treated, on the basis of such experience.

Lead Palsy.—Considerable time of exposure to the influence of lead is generally necessary to cause this. So commonly does it first affect the extensor muscles of the forearm, that the cognomen of "wrist-drop" is often applied to it. When it lasts for some weeks, the muscles waste away. A blue line is observed to form along the edge of the gums. Pain precedes the palsy, and attends recovery of power. During the attack, the muscles have their excitability by electricity considerably diminished or lost.

Mostly, though after a long time, lead palsy is recovered from. Iodide of potassium appears to act as an eliminant of the lead accumulated in the system. Ergot is asserted by some to be curative also. Faradaic electricity has been found decidedly beneficial; used in moderate strength for a few minutes two or three times a day. (See *Medical Electricity*, in Part I. Sect. III.)

Mercurial palsy is occasionally met with in those who work in the metal. Mostly *tremor* is a predominant symptom. Early withdrawal from the influence of the cause, and the continued use of the iodide of potassium, are the principal measures of treatment.

Paralysis agitans or shaking palsy is described as a more or less constant involuntary and uncontrollable shaking, of the hands, arms, head, or, progressively, of the whole body. Slight or moderate degrees of such tremor are common enough, from general nervous debility. Extreme cases evince the wreck of the cerebro-spinal system, and are therefore incurable. No especial treatment can be pointed out for this affection.

Wasting Palsy.—(Cruveilhier's). A few of the muscles of one limb, or the voluntary muscles of the whole body, may lose their power, and then waste away almost to nothing. The shoulder and the ball of the thumb are frequent points of commencement for the palsy and atrophy. Insidious in its approach, the affection may last from six months to several years. It may end in recovery, in permanent arrest at a certain stage of the disease, or in death. Twelve months is the earliest recorded period for the occurrence of a fatal end. This end is the result always when the *trunk* is invaded. After death, the spinal marrow has been examined in but a few cases. No lesion has been found in most of them; in a certain number it has. But our methods of inspection of nervous tissue are yet too imperfect for it to be pronounced that such an atrophic disease is independent of the nervous centres. It may be the *ganglia* which regulate *nutrition* that are in fault.

Duchenne's Disease—"Ataxie locomotrice progressive." Rheumatoid pains, in this affection, precede loss of power. Occasional strabismus and incontinence of urine may occur. Then there is an awkward, unsteady gait; the sensibility of the feet becomes blunted, and walking is insecure. If the patient shuts his eyes he falls down; and even with them open he reels as if drunk. Co-ordinative power over movements is lost. The duration of this progressive disease varies from six months to ten or twenty years. It is most common in males of middle age. That this is a spinal affection is obvious, even without many autopsies to prove it so. Lesion of the posterior columns of the cord has been several times found. In its **treatment**—hygienic management, general tonics, electricity, and very careful use of strychnia, may be tried, without much hope.

General Paralysis of the Insane.—Only a minority of insane persons have this affection; which comes on at an advanced stage of chronic mania, melancholia, or dementia. Difficulty of speech, and general tremor, characterize it, followed by the gradual loss of all muscular and sensory power. It is incurable.

INFANTILE PARALYSIS.

Under this name (perhaps not very well chosen), Dr. C. F. Taylor designates what Handfield Jones would call *paresis* of the nervous centres, so far as, in infancy, to arrest nutrition as well as abridge power in the limbs. Dissipation or bad health in the parents will predispose to this in children. No violent symptoms attend the attack. Mostly, with care, it tends to recovery. But want of knowledge or attention may allow *deformity* to result from it; especially

20

club-foot. "Talipes equinus" (in which the heel will not touch the ground), says Dr. Taylor,[1] "is the first, and simplest, and most natural sequence of the paralysis—the weight of the foot being all that is necessary to produce it—and no other form of talipes is likely to occur while the patient lies in bed. The bending of the ankle outward (talipes varus) is the result of weight on a foot with a shortened tendo-Achillis; bending inward (talipes valgus) of the ankle is the result of weight *partially* overcoming the gastrocnemius, soleus, &c.; and talipes calcaneus (where the toes are raised so as to be unable to touch the ground at the same time with the heel) of weight entirely overcoming those muscles." The author just quoted concludes from special experience that all such deformities are *preventable* by proper care as to the position and use of the limbs and muscles of all parts of the body during the paralytic or paretic state. The *treatment* of club-foot is a subject for surgical treatises.[2]

In some cases, fatty degeneration of the muscles takes place to such an extent as to render the case almost or quite incurable. Brodie observed that a case is likely to recover, in which, when the child is lying on the back, there is power to draw the limbs up by flexing the thigh towards the body.

Treatment of infantile paralysis should consist of general recuperative management, including tonics (strychnia in some cases, with care) and cod-liver oil, salt bathing, passive exercise of the affected limbs, and galvanism. The latter must be cautiously conducted in children. Local application of *heat* is advised by Drs. Taylor and Hammond. The former prefers dry heat; seating the child before a fire and thrusting its legs through a screen, so as to be thoroughly warmed for hours together. Dr. Hammond immerses the paralyzed limb in hot water, at 140°—160°.

EPILEPSY.

Definition.—Periodical convulsions, with unconsciousness during the attacks.

Varieties. *Grand mal* and *petit mal* of the French; the latter is the *eclampsia minor* of some writers; in which unconsciousness occurs with scarcely any convulsion.

Symptoms.—*Premonition* occurs in a minority of cases before an attack; headache, dizziness, terror, spectral illusions, or the epileptic *aura.* This is a creeping or blowing sensation, like that of a current of air or stream of water, beginning in a hand or foot, and extending toward the trunk. It (if it occur) immediately precedes the paroxysm. Then, often with a scream, the patient falls down, and is violently convulsed. Foaming at the mouth, grinding of the teeth, and biting of the tongue, are common; the face is flushed, the eyeballs roll, the pupils are unaffected by light, sometimes vomiting, or involuntary urination or defecation takes place; and respiration may be very laborious.

---

[1] Infantile Paralysis, p. 83.
[2] *Adhesive plaster* has lately been successfully used for gradually rectifying congenital club-foot, in early infancy.

The fit lasts on an average from five to ten minutes. The interval between the attacks may be from several months down to a few hours. Old cases (as in lunatic asylums) may have two or three paroxysms daily. They vary much even in the same individual.

The condition after the attack is also various. Generally, drowsiness or deep sleep follows it; or headache, debility, or delirum; sometimes maniacal frenzy. Homicide has been committed in this state; for which, of course, the person is not criminally responsible.

Anatomy and Pathology.—Epilepsy is not often the immediate cause of death. Autopsies of epileptics (Schrœder van der Kolk) have shown changes especially in the medulla oblongata; dilatation of the bloodvessels being prominent. Exaggeration of reflex motor excitability, with loss of the controlling power of the brain over the spinal axis, would seem to be parts, at least, of the morbid condition. Marshall Hall's idea of "trachelismus," or temporary partial asphyxia from spasm of the muscles of the neck, has been exploded. Brown-Séquard's notion of the importance of the *aura*, as indicating a peripheral irritation at its seat, has, after causing the tentative amputation of a few limbs, suffered the same fate.

Diagnosis.—From *hysterical* convulsions, which also may be periodical and violent, those of epilepsy are distinguished by the total loss of consciousness,—which is partially retained during the hysterical paroxysm. Curability belongs also much more to the latter than to the epileptic disease.

Prognosis.—Few cases of genuine epilepsy recover. The younger the patient, and the longer the interval, the more hope. Life may last indefinitely with the disease. Gradually, in most cases, the mental faculties are impaired. Yet several great men have been epileptic; Cæsar, Mahomet, Petrarch, Napoleon, Byron.

Causes.—Hereditary transmission of this disease is common. Intemperance, venereal excess and self-abuse, blows on the head, and fright, are among the most frequent exciting causes.

Treatment.—During the paroxysm, when habitual, little or nothing is to be done. Place the patient so that he cannot strike his head or limbs against anything hard; loosen the clothing about the neck to favor free respiration and circulation; and insure fresh air about the patient; that is all. An *occasional* convulsion requires treatment; of that more will be said hereafter. (See *Convulsions.*)

To break up the recurrence of the fits is the problem, for which a vast number of remedies has been tried, in vain. To name them would be to go over almost half the materia medica. Prominent, since nitrate of silver was abandoned as useless, in this disease, have been belladonna, arsenic, valerianate of zinc, digitalis, and bromide of potassium. I have known valerianate of zinc to postpone the paroxysm for considerable periods. Beginning with one grain twice daily, it may be gradually increased to three or four times that amount. A case of recovery occurred under my knowledge in which rather large doses of digitalis were persevered in for several months. Bromide of potassium[1] is now the favorite medicine with many; upon the belief

---

[1] In this country, at least, the introduction of bromide of potassium as an anti-epileptic may be credited to Dr. C. E. Brown-Séquard.

that it is a direct sedative to the excito-motor susceptibility of the medulla oblongata and other nerve-centres. From ten to twenty grains twice or thrice daily may be given, and continued for an indefinite length of time. *Bromide of ammonium* (dose 10 grains) is spoken favorably of by some who have used it.

Self-management is very important to the epileptic. Temperance, with *nutritious* diet, as the disease is one of *asthenia*, is necessary. Regularity of the evacuation of the bowels is a *sine quâ non*. Abundant exercise in the open air, short of exhaustion, does good ; systematic gymnastics have even *cured* some cases. They are worth trying always. Avoidance of, or the extremest moderation in sexual intercourse must be insisted upon. Self-abuse will make recovery impossible. Tobacco ought not to be used, unless by smoking only a single pipe or a segar, or two, in the day. Coffee should not be recommended.

It has already been implied, in referring to pathological views, that *tracheotomy*, suggested by Marshall Hall, and amputation of the limb in which the aura is felt, are useless although severe measures in epilepsy.[1]

A seton kept in the back of the neck is well worth trying in every case. I have known it to promote recovery.

CATALEPSY.

This is a periodical disease, in which the attack is marked by unconsciousness, and fixed rigidity of all or many of the voluntary muscles. It is rare. The attack generally lasts but a few minutes. Sometimes, in lunatics, a semi-cataleptic state of the muscles is permanent.

I am not aware of any special treatment appropriate for this affection. Management like that suitable for the epileptic will be in place also in catalepsy. Both are now so well understood to be asthenic disorders, with impaired *hæmatosis* (blood-making) as an important element, that all reducing measures are properly omitted from their treatment. This must be essentially tonic and *analeptic* or restorative.

CONVULSIONS.

These may be classified as, principally, **infantile, epileptic, parturient** and **puerperal, hysterical** and **occasional** convulsions. During infancy, causes which in an adult would cause delirium, produce convulsions; excito-motor action having in early life the predominance. They are, usually, of less serious prognosis in the infant than in the adult.

The *exciting* causes of infantile convulsions are numerous. Constipation of the bowels; indigestion; worms; irritation of the gums in teething; and excitement of the brain, as by fright, are about the most frequent. Many acute and chronic diseases of infancy (*e. g.* scarlet fever, meningitis, hooping-cough, etc.) have convulsions among their occasional symptoms or complications. Sudden drying up of eruptions on the scalp may bring them on, also.

---

[1] Another operation, *clitoridectomy*, practised by Baker Brown, of London, in certain cases in females, does not meet with general favor in the profession.

Premonition of a fit is often observed, in the child's fretfulness, or restlessness, or gritting of the teeth in sleep. When a fit comes on, the muscles of the face twitch, the body becomes rigid at first, then in a state of twitching motion; the head and neck are drawn backward, the limbs violently flexed and extended. Sometimes these movements are confined to certain muscles, or are limited to one side. Nurses call by the name of "inward fits" cases in which the limbs move but little, but the countenance is affected, the eyes are unnatural in expression, or roll spasmodically, and the body is more or less rigid. During a fit, consciousness is absent. The eye shows no sign of sight, though open; a finger passed over it does not make it wink. The pupil is immovably contracted or dilated; the ear is insensible even to loud sounds. The pulse is small and very frequent; breathing hurried or labored; skin wet with perspiration, often cold and clammy. After this condition has lasted a few minutes it mostly gives way. The child falls into a quiet sleep; or, it becomes conscious and bewildered; or gradually resumes its ordinary healthy state; or dies in the fit. Sometimes one attack is followed by another, with intervals of conscious or unconscious quiet between, for many hours. These are the most serious cases, although recovery often happens even from them. *Salaam* convulsions, or nodding convulsions of infants (eclampsia nutans), are a rare form of disease, usually the precursor of epilepsy.

Treatment.—Ascertain, if possible, the *cause* of the convulsion. If the gums are swollen, or have been tender and irritated, at the time of teething, lance them freely; dividing the tense gum with a sharp gum-lancet down to the coming tooth. If the bowels have not been moved, or if the abdomen be swollen and tense, give at once an enema, of castor oil, soap, and molasses [F. 141], or some other laxative material, with warm water. When the head is hot, apply cold water all over it, by wet cloths, renewed every two or three minutes. If the fit lasts long enough for it, place the child in a warm bath; supporting, of course, the head while the body is immersed. Then mustard plasters may be applied, to the back, epigastrium, and legs, at once or successively.

Bleeding from the arm is to be recommended only in a child of known vigor and fulness of system, the attack being severe, and not habitual. But a moderate amount of blood should be taken. Cupping the back of the neck, in doubtful cases, where time is allowed by a protracted fit, may be resorted to.

Etherization, so much used by some practitioners in puerperal convulsions, requires certainly more caution in its use in infants. I have never tried it in the convulsions of childhood; but would regard it as justifiable in an obstinate case at any age; watching its effect.

Convulsions of Pregnancy.—Probably about one pregnant woman in fifty has more or less albuminuria; principally from the pressure of the womb upon the renal veins producing congestion of the kidneys. About one in ten of these will have epileptiform convulsions, either during gestation, while in labor, or after delivery.

Pathology.—All convulsions of pregnant women are not uræmic; this has been proved. There are (putting aside instances of Bright's disease already existing) several conditions possible: 1. Uræmia, as

20*

above stated; 2. Cerebro-spinal reflex irritation, of uterine origin; 3. Cerebro-spinal (apoplectic) congestion; connected especially with the bearing down efforts of labor itself.

Treatment.—It is important, particularly during gestation : 1. That plethora should be avoided; 2. That free action of the kidneys, as well as regularity of the bowels, should be maintained. For the first, care of the diet is proper, that, in women of full habit, it be not too highly animalized or stimulant. If headache, with a full, hard pulse, occur, a mild cooling laxative may be given; if not relieved, cups to the nucha or bleeding from the arm will be a safeguard. When urination is not free and copious, even if no albumen appear in the urine, cream of tartar, a teaspoonful every day or two, or acetate of potassa, may be a useful prophylactic, by favoring free excretion from the kidney.

When convulsions actually occur, in the pregnant or puerperal state, the question is to be considered—are they reflex, uræmic, or simply congestive or apoplectic ?

When they come without previous signs of cerebral disturbance, but in a woman of delicate and impressible nervous organization, and without much heat of head, or snoring respiration, the pulse being rapid and feeble, it is probable that *reflex irritation* is the nature of the case. Counter-irritation, by dry cups to the spine and sinapisms to the epigastrium and limbs, and etherization, may be here used.

When plethora has existed before, and the head is hot, its vessels distended, the coma profound, with snoring respiration and full, rather slow pulse, either uræmia or simple congestive apoplexy is to be concluded upon. In either case, but especially in the latter, bleeding from the arm, or by cups from the back of the neck or temples, will be advisable. Laxative enemata may also be used. After bleeding, if the convulsions are protracted, while the coma is less intense, careful inhalation of ether may be tried ; but it is less hopeful here. The prognosis of the apoplectiform convulsion is always one of great danger. The *uræmic* condition, if labor be survived, generally passes off spontaneously, soon after delivery.

Occasional convulsions in adults, from whatever cause, should be studied and treated upon the same principles essentially as those just laid down for the convulsions of pregnancy. *Hysterical* convulsions will be considered under *Hysteria*.

CHOREA.

Synonym.—*St. Vitus' Dance.*

Symptoms.—Incessant and irregular movements of the voluntary muscles, over which the will has but partial control. Walking, in severe cases, is difficult or unsafe ; the hands cannot be regulated enough to write or work ; speech may be affected ; the muscles of the face often twitch grotesquely. During sleep all these movements cease. The pupil is, in some cases, unnaturally dilated ; palpitation of the heart may occur ; and also constipation and indigestion. The urine is of great density.

Prognosis.—The mean duration of chorea is about four weeks ; but it may last for several months. Recovery, if the attack be uncomplicated, may always be anticipated.

Complications.—Endocarditis and pericarditis have been observed in connection with chorea in a number of cases. Generally, however, the affection of the heart precedes the chorea; both probably depending upon the same cause, *rheumatism.*

Paralysis complicating chorea increases greatly, of course, the seriousness of the case. Although it may be of the transient, hysterical form, yet the danger exists that it may be the result of organic lesion (as softening) of the brain or spinal cord.

Causation.—From six to sixteen, in both sexes, especially often, however, in girls, chorea occurs. Nervous debility is nearly always present before the attack. Fright is a frequent cause. Over fatigue, or mental excitement, blows or falls may produce it. Rheumatic fever is sometimes followed by it.

Treatment.—Good diet, salt bathing, and systematic gymnastic exercises (light gymnastics or calisthenics) will suffice for mild cases. Where marked anæmia exists, iron (citrate, phosphate, [F. 142], or pyrophosphate, tincture of chloride, liquor of iodide) is important. Obstinate cases may be treated with Fowler's solution of arsenic, in small doses, gradually increased. Cimicifuga has been a good deal used, perhaps with benefit. Cod-liver oil should be given if great debility exist. *Calabar bean* has recently been introduced as a remedy in chorea; f ʒss of the tincture, or from gr. j to gr. vj of the powder thrice daily.

It is well to separate a child having severe chorea from other children; both because of the annoyance of their curiosity, and because *sympathetic irritation* sometimes extends the affection from one to another. This has been repeatedly observed.

TETANUS.

Definition.—A disease characterized by continued tonic contraction of the voluntary muscles generally.

Symptoms.—Stiffness of the muscles of the jaws commonly begins the attack. This extends to the throat and neck, face, trunk, and lastly to the limbs. Though never ceasing entirely, the spasm of the muscles is paroxysmally increased. Sometimes *opisthotonos* occurs, *i. e.,* arching of the body upon the back and heels, the abdomen projecting; or *emprosthotonos,* arching forward, the face approaching towards the toes. *Pleurosthotonos,* or lateral curvature, is much more uncommon.

Chewing of food is impossible; swallowing nearly or quite so; respiration becomes very difficult. The patient suffers dreadfully, and cannot sleep; but delirium scarcely ever occurs. Death in most cases takes place within a week.

Varieties.—These are, tetanus from cold (idiopathic), traumatic tetanus (from an injury), and *trismus nascentium,* or tetanus of infancy. The first is the least certain to be fatal.

Causation.—This is principally included in the above. Much the greater number of cases results from lacerated and punctured wounds; but amputations and other operations may be followed by tetanus.

Irritation (not inflammation) of the ends of sensitive nerves, transmitted to the spinal cord, produces the reflex spasm, whose general extension and continuance proves fatal. Strychnia, in poisonous doses, causes a very similar state. While there can be no doubt that the spinal marrow is the seat of the disease, no characteristic organic change has been found in it; sometimes not even congestion.

Treatment.—In two cases which I have seen to recover, opium and brandy were the remedies used. A tablespoonful of brandy (to an adult) every two or three hours, with milk or beef-tea, and a grain of opium every three or four hours, may be given. The opium may be, if needful, increased to a grain every hour at night, and every two hours through the day. Beyond that I would not go.

Chloroform and other anæsthetics, by inhalation, have been tried, with variable effect; nearly always without success. Belladonna, aconite, hydrocyanic acid, cannabis indica, tobacco, woorara, and quinine, are among the many medicines favored by different practitioners. In so desperate a disease it is excusable to give them all further trial. My father, Dr. Joseph Hartshorne, used vigorous counter-irritation all along the spine, by the decoction of cantharides in turpentine (linimentum cantharidis).

HYDROPHOBIA.

I have known a physician of distinction and of many years' practice to deny the existence of hydrophobia because he had not seen it; asserting that the cases so called were tetanus. I have seen two cases of it; and no one who has observed it can fail to perceive the wide distinction between it and tetanus.

Symptoms.—A month or more after the bite of a mad dog or other rabid animal, the wound having healed, irritation is felt in it. Nervous restlessness also exists; which increases (in most cases) to violent, angry delirium. Then difficulty of swallowing occurs, from a spasm of the muscles of inspiration (gasping) taking place at the moment of deglutition, making the patient choke. The same spasmodic gasping is brought on by any sudden impression; as of sound, a flash of light, or even a current of air passing over the face. Insomnia exists; the patient grows prostrate, and must die for want of food and drink, even if the affection of the cerebro-spinal axis were not itself fatal. There is intense thirst, and no dread of water, except that the attempt to swallow it causes distress. Death occurs in from four to eight or ten days.

I have not met with satisfactory evidence that a case of genuine *rabies canina* or hydrophobia has ever been cured. By statistics, however, only one in eleven (some say one in five) of those bitten by mad dogs have the disease, even when no precaution is taken.

Treatment.—If we cannot cure, what can or should we do? We may certainly promote at least *euthanasia*, by allaying the wretched sufferings of the patient. In the case of a boy of eight years of age under my own care, I administered chloroform freely by inhalation; continuing it nearly all the time (with short intervals and equally short applications) for two days and nights. It mitigated the spasms and quieted the delirium. That it did not itself cause death (as might have been suspected from the quantity used) was proved to my satisfaction by the fact that after the chloroform was finally withdrawn, the boy was made to gasp spasmodically by waving the hand to and fro over his face. Reflex excitability of the medulla oblongata was thus shown still to exist.

Hypodermic injection of atropia or morphia might, perhaps, more effectually quiet the suffering, and even afford a possibility of cure, than inhalation of anæsthetics.

Prophylaxis.—The only perfect safety to one who is bitten by a rabid animal (and the bite of a much *enraged* dog, not rabid, is said to have also caused *hydrophobia*) is in immediate and total excision of the part. While awaiting this, forcible suction will aid in removing the poison; and ligation with any kind of bandage above the part will retard the absorption of it. When excision cannot be safely performed or is refused, cauterization is the next best thing. Free application of lunar caustic is recommended. Even if the person bitten is not seen until a day or two afterwards, excision or the use of the caustic is to be recommended, as lessening the danger of this horrible disease.

It is well to know that canine madness is not restricted to, nor even especially frequent in, hot weather.

HYSTERIA.

From its occurring nearly always in females, and from a supposition of its originating in some affection of the womb, this name has been given to a variable disorder, of which the main characteristic is, *morbid excitability of the whole nervous system.*

A "fit of hysterics" is a paroxysm whose nature may vary, from mere uncontrollable laughter or crying, to a severe epileptiform convulsion. This last, however, differs from epilepsy, in there being less complete loss of consciousness, and in its curability. It is often preceded by a sensation (globus hystericus) like that of a ball rising towards the throat.

Simulation of other diseases, indeed the assumption of severe functional disorders of different organs, is a common trait of hysteria. Thus I have seen hysterical amaurosis; hysterical insanity is not uncommon; nor is hysterical paraplegia or coma rare. Retention of urine, cough, aphonia, &c., are often thus produced. "Phantom tumor" is among the most curious of such things. I had in my care a woman who had been laid out by a surgeon in another city for exploratory gastrotomy, upon the supposition that she had ovarian tumor. When she was etherized, however, the tumor altogether disappeared! "*Bed case*" is the name given to the complaint of a hysterical valetudinarian, who believes herself to be ill or powerless, while there is really nothing the matter, except the morbid *neurosis* itself.

Treatment.—Much skill and care will often be required in the management of hysteria; as each case has peculiarities of its own. Generally, a tonic regimen is demanded. Iron and cod-liver oil are most often the appropriate medicines. Bromide of potassium is sometimes quite useful. For a paroxysm of "hysterics," assafœtida [F. 143] is universally safe and suitable; in pills of three grains each, *pro* *re nata.* Sinapisms and pediluvia are also proper. Menstruation is often irregular in hysterical women; it should be regulated as far as possible. Exercise in the open air is, as a rule, very important for such persons. Mental and emotional excitement should be avoided; but tranquil, even engrossing *occupation* will be beneficial. For hysterical paralysis, electricity is said to be promptly useful. Cold bathing, especially the shower bath, or sea bathing, when followed by reaction, will do good. Feeble and delicate persons should, however, be careful not to remain in the bath too long. In the surf, for example, a bath of five or ten or fifteen minutes may be of great service, when a longer time would do real harm.

NEURALGIA.

Definition.—Pain, without inflammation or other disorder, except that of the nerve or nerve-centre involved; literally, *nerve-pain.*

This may affect any of the sensitive nerves. It is, also, sometimes referred to parts which have, in health, no sensibility; as the heart, stomach, &c. Different names are given according to its site. Thus, *tic douloureux* is facial neuralgia; *hemicrania,* that affecting one side of the head; *sciatica,* that of the hip; *gastrodynia,* neuralgic pain in the stomach; *pleurodynia,* in the side. Angina pectoris is, chiefly, a neuralgic affection of the heart.

The pain is generally acute, shooting, or darting; with tenderness of the part upon pressure. There is, however, no heat nor swelling, nor throbbing of the bloodvessels in pure neuralgia. Complicated cases occur, in which inflammation and neuralgia exist together; and inflammation of the fibrous neurilemma may be the immediate cause of the neuralgic pain.

Pathology.—At least three sources of this sort of pain are possible. 1. Local disease affecting a *nerve;* 2. A morbid state of a sensorial *nerve-centre;* 3. A morbid condition of the *blood.* Neuralgia always fixed or returning in the same spot, is likely, although not certain, to depend .upon a fault in the nerve itself; as *e. g. neuroma* (tumor of a nerve). Radiating pain (although possibly of reflex origin) must involve at least part of a nerve-centre. Flying pains, never long seated in one part of the body, mostly are due to a defect, or morbid poison (as that of gout, or malaria) in the blood.

Treatment.—This must, of course, depend upon the cause or nature of the case. *Tic douloureux* often depends upon decay of the teeth; if so, they must be attended to. Other purely local neuralgias require local treatment. Even division of the affected nerve is sometimes, but should rarely be, resorted to. Laudanum or paregoric, applied by saturating a rag and laying it upon the part, covered by oiled silk to prevent evaporation, is an efficient local anodyne. So is chloroform, similarly applied; it is very pungent, burning like mustard. Sinapisms will sometimes relieve promptly. Mere warmth, as of flannel steeped in hot water, will do so in some instances. Rubbing for a few minutes with saturated tincture of aconite root, until the skin tingles; or the application of ointment of veratria (gr. xx in ℥j of lard), may be used in severe cases. In the most obstinate ones, a blister may be applied, dressed, after removal of the cuticle, with two grains of acetate of morphia, diluted with ten grains of gum Arabic. Or, most prompt usually of all, solution of morphia may be hypodermically injected, to the amount of one-fourth drachm to one drachm at once. Sometimes the inhalation of ether, nitrous oxide, or chloroform is resorted to, for the relief of intense neuralgic pain. Debility predisposing to it, in some cases moderate doses of some alcoholic stimulant will give relief.

Of anodynes internally used, belladonna has, for neuralgia, the greatest reputation. It will not quell suffering so directly as opium or morphia, but it has been thought more entirely to do away with the neuralgic state. For this, however, *iron*, especially in combination with quinia or strychnia, is the most effective medicine. Cases of neuralgia which will not be benefited by iron are decidedly exceptional. Larger doses of it are generally recommended for this than for other cases requiring chalybeates. Quinine is particularly wanted in neuralgias of malarial origin (very common); and strychnia or nux vomica in those whose obstinate persistence depends upon great loss of nervous energy. Everything that recuperates, as generous diet, change of air, sea-bathing, &c., will assist in curing neuralgia, when it is connected, as it so often is, with anæmia and broken health.

Odontalgia, toothache, is sometimes purely neuralgic. More often, it results from exposure of the nerve by the decay of the tooth. Again, it may attend *inflammation* of the jaw, or abscess at the root of the tooth affected. For toothache from *exposed nerve, creasote* is a certain remedy. Insert carefully into the hollow a plug of cotton, wrapped over the end of a knitting needle and dipped in pure creasote. If the latter run out into the mouth (which should be avoided if possible) rinse it at once with cold water.

DELIRIUM TREMENS.

Synonym.—*Mania a potû.*

Symptoms.—Sleeplessness, debility, tremors, horror, hallucinations; often with loss of digestive power. The *insomnia* is a cardinal symptom; if the patient sleeps a whole night he recovers. Debility varies in degree in different cases; in a first attack it is not always great. Tremor is nearly always present. The illusions of the patient are wonderfully real, and generally dreadful. He is pursued by demons or beset by mortal enemies; he cannot bear to be alone, especially in the dark. Sometimes, however, the visions are indifferent, or even amusing. The patient may suppose himself to be well, and engaged about his usual avocation; going through all its movements in pantomime, though with empty hands.

After several days and nights of sleeplessness, prostration usually increases; the skin grows cold and clammy, the voice feeble, and the patient no longer inclines to move about. Death must result, if sleep be not obtained, within a week, or, at the most, two weeks. In favorable cases, a sound sleep of many hours comes on within three or four days; the patient then wakes up rational and well.

Causes.—There is no room for doubt that this affection may come on under two different conditions or circumstances: 1. where stimulants are suddenly withdrawn from one accustomed to them; and 2.

while their use in excess is continued. The second class, according to my observation, furnishes the most dangerous cases.

Treatment.—Old as this disease is, it is yet the subject of great difference of opinion. The practice which early training led me to adopt, consisted in the *moderate* use of stimulants ("tapering off") and of opium, with concentrated liquid nourishment. If the patient was not much prostrated, I would give only ale or porter, a bottle or two in the day; with hop tea *ad libitum*, and a grain of opium every three or four hours. The latter would be increased, if sleep were delayed, to a grain every two hours; or, as a maximum, a grain every hour. Very weak cases, accustomed to spirits, might have a table-spoonful of whisky or brandy every four, three, or two hours, accord-ing to their condition. Beef-tea and mutton-broth, &c., seasoned with red pepper, are preferred as diet. In an obstinate case, I have seen sleep follow the raising of a blister upon the back of the neck. Sub-stituting valerian for opium, or combining the fluid extract or tincture of valerian with morphia solution, has answered well in some cases. [F. 144, 145]. Injection of laudanum into the rectum is occasionally resorted to.

Other modes of treatment have recently been urged. 1st. The *expectant* treatment, of Drs. Dunglison and Laycock; giving only strong food, without stimulants or opium. 2d. The treatment of tablespoonful doses of tincture of digitalis. 3d. That by the internal use of chloroform, in one or two drachm doses.

The expectant treatment will no doubt do very well in mild or moderate cases. From what I have seen, I should fear to trust to it in severe or threatening ones.

The digitalis treatment, bold as it seems, has a good deal of posi-tive testimony in its favor. Why not try, as some do, less immense, and yet large, doses; as half a drachm or a drachm, instead of half an ounce, of the tincture, every three or four hours?

Dr. E. McClellan and others have recently reported excellent suc-cess with one or two drachm doses of undiluted chloroform. The corrugated stomach of a spirit drinker will probably bear the pungency of chloroform better than another's. Generally only one or two such doses of it are said to be required. My experience with the internal use of chloroform leads me to believe such practice perfectly safe, at least. It is well worthy of further trial.

The large majority of first attacks of mania a potû are curable. Third and fourth attacks are often fatal, or are followed by permanent insanity.

METHOMANIA.

Definition.—The disease of uncontrollable or irresistible intem-perance.

Synonyms.—*Dipsomania; Oinomania.*

Varieties.—*Periodical* or paroxysmal, and *chronic* or persistent methomania. The subject of the first may be temperate for weeks or months, and then will abandon himself to violent excess for some days or for a week or two. The persistent methomaniac is constantly intemperate, so long as the opportunity exists.

Causes.—Hereditary proclivity exists in many cases. Wilful or unwise excess is the cause, of course, of intemperance in every case. To designate it as a disease is not at all to deny the accountability of those who voluntarily incur it; only thus its true character of *uncontrollableness* (in many instances) by the will is indicated. That any intrinsic power exists in alcohol, employed for its *proper needs* as a medicine, and in proper quantities, to bring on intemperance, I do not believe. I have known too much of its use in practice in low fevers, in phthisis, and many other conditions of debility, not to be sure that it is only when used in *excess*, or out of place, that any hankering or slavish demand for it is begotten.

Treatment.—No safety exists but in *seclusion, for a year or two*, where the individual cannot obtain stimulus, and is not made, by company or opportunity, to desire it. Laws should be made by which every person, proved upon inquiry before a commission, to be habitually intemperate, should (like a lunatic) be deprived of the control of his liberty and property. Then in every community there ought to be institutions where a safe and home-like retreat could be had, for a sufficient time to restore self-control; which, I repeat, ought to be never less than a year; better, two years. Such institutions exist now in New York and Massachusetts; and, with the encouragement of recent improvements in legislation, one is about to be established in this State, near Philadelphia.

INSANITY.

Definition.—Loss of control of the will over the mental faculties or impulses; intellectual, emotional, or sensorial derangement.

Varieties.—1. **Mania**; acute[1] and chronic; also divisible into intellectual insanity or *delusion*, emotional or *moral* insanity, and illusional derangement or *hallucination*. 2. **Monomania**, or partial insanity; *e. g.*, homicidal and suicidal; *kleptomania*, or insane propensity to steal; *erotomania* (satyriasis, nymphomania), or uncontrollable amatory desire; *pyromania*, morbid propensity to commit arson, etc. 3. **Melancholia**. 4. **Dementia**; *i. e.*, total wreck of the faculties, or imbecility. **Idiocy** is congenital imbecility.

Premonitions.—By noticing these, often *prevention* may be suggested and effected. Hardly any of them alone may be sufficient, while all together they become so. 1. Headache, not accounted for by ordinary causes, and continuing for days or weeks together. 2. Irritability of temper, not previously habitual. 3. Unnatural hilarity, without occasion. 4. Depression or gloom, not justified by any event. 5. Alternations of excitement and despondency, both extreme. 6. Any great modification of the natural temper or habit of mind, so that the individual becomes the opposite of his usual self. 7. Dislike or distrust of near friends and family, without any reason for it.

Diagnosis.—Alienation from his own accustomed character, and disruption from rational and harmonious relations with persons and things around him; these are the cardinal elements of the insane state. This, all authorities admit to be more easily detected or discriminated

[1] Puerperal insanity is one form of acute mania.

than defined. The old legal test, that the lunatic must be incapable of knowing right from wrong, must be given up; as very many cases of emotional or "moral" insanity are proved to exist, in which, with full knowledge of right and wrong, the morbid impulse is irresistible by the will. There is no *physical* test of insanity, by the pulse or otherwise; as in chronic mania, &c., all the organic functions may go on normally. The expression of the face is, it is true, nearly always unnatural. Perhaps the greatest difficulty sometimes exists in monomania, unless one knows the peculiar delusion or morbid proclivity of the patient; as, upon all other matters, he may be sound. *Feigned* insanity is generally over-acted; sometimes it may require the skill of experts to expose it.

Prognosis.—More than half of first attacks of insanity, under good management, are recovered from. With each repetition, the hope grows less; and so it does, also, in proportion to the *duration* of chronic mania. Sometimes, however, cures occur of those who have been insane for years. Dementia is a common, and generally hopeless, termination of prolonged chronic mania or melancholia. Puerperal mania is curable in a large majority of cases. Ordinary acute mania varies in duration from a week or two to several months. It may end either in recovery, in lapsing into chronic mania, in dementia, or even in death during the attacks. *Periodical* insanity is occasionally met with, especially in females.

Causes.—These are numerous. The principal ones are, hereditary predisposition, injuries of the head, intemperance, reverses of fortune, loss of friends, and domestic troubles.

Pathology.—Much yet remains to be learned of this. Subtle alterations of the brain structure are still to a considerable extent unrecognizable, even with the aid of the microscope. Two elements in the pathology of insanity have, at least, been distinctly made out; cerebral *hyperæmia*, which predominates in the more acute cases, and *atrophy*, which is (either quantitative or qualitative) present in nearly all those which are chronic.

For the **treatment** of insanity it is proper to refer to special treatises upon the subject. (See Bucknill and Tuke on Insanity.) The advice of a physician, in nearly every case, ought to be, early removal to a well conducted asylum or hospital for the insane. There, security, and the prospect of recovery, will be much better than at home, though amongst the kindest of friends. In the treatment of insanity, in recent times, while medicine (especially tonics and anodynes) is not neglected, the tendency is to confide a great deal in moral or mental treatment; *i. e.*, the aggregate of personal, local, and circumstantial influences, which, in an asylum, can be arranged especially with a view to the most favorable effect upon its inmates.

HEMORRHAGES.

Varieties.—1. Active; 2. Passive; 3. Traumatic; 4. Symptomatic; 5. Critical; 6. Vicarious. Local hemorrhages are also classified according to the organ from which the blood escapes.

Active hemorrhages are those in which determination of blood in excess to the part precedes the bleeding. **Passive** hemorrhages, those in which, from inaction of the circulation, or passive dilatation of bloodvessels, congestion occurs; or in which the coats of the vessels give way too readily, partly from the blood itself being incapable of maintaining properly their nutrition. The idea of bleeding by " exhalation " without rupture at least of capillaries, is now abandoned.

Traumatic hemorrhages are, of course, all produced by wounds; coming thus under the department of surgery.

Symptomatic hemorrhages are met with in many diseases; e. g., epistaxis in typhoid fever; hæmoptysis in consumption; vomiting of blood in cancer of the stomach; bleeding from the bowels in piles, &c.

Critical hemorrhages are occasional terminations of febrile disorders; as yellow fever, remittent fever. **Vicarious** hemorrhage is that which substitutes one which is normal or habitual; e. g., spitting of blood when the menses have been suppressed; or bleeding at the nose following arrest of the bleeding of habitual hæmorrhoids.

Epistaxis.—By usage, this term is applied only to bleeding from the nose. In young persons, especially from ten to fifteen years of age, it is common, and, if moderate, harmless; seeming often to relieve a temporary congestion and prevent a headache. It is more often seriously troublesome in older persons. Generally it is from one nostril only, but not always.

Treatment.—When slight, it may be allowed to stop of itself; only not blowing away the clot that forms as a natural plug. If it continue so as to threaten an injurious loss of blood, applying cold water to the forehead and nose, or *ice*, there or to the back of the neck, or to the roof of the mouth, will generally stop it. If not, a plug of dry cotton may be introduced and left in the bleeding nostril. Wetting the cotton first in strong alum-water, or dilute tincture of chloride of iron, or dipping it in powder of tannin or matico may make it more effective. When these measures fail, the posterior nares must be plugged. Either the watch spring canula may be used, or an elastic catheter, having a piece of waxed ligature or twine passed through its eyelet hole, may be carried back from the nostril to the pharynx. Then the string should be drawn out of the mouth with forceps, a plug of cotton fastened to it, and the other end of the string drawn out through the end of the catheter till it forces the cotton plug against the posterior orifice of the nares. Raising the arms high above the head is a popular mode of endeavoring to stop nose-bleeding.

Bleeding from the Mouth.—This, unless when ulcerative, is generally from the gums; as in scurvy. It is, in itself, scarcely ever serious in amount. Considerable bleeding, sometimes hard to stop, may occasionally follow the extraction of a tooth.

Treatment.—Borax in solution, or tannic acid, or myrrh and rosewater, will be suitable washes for the bleeding and spongy gums of scurvy. For hemorrhage after the removal of a tooth, it may be necessary to plug the cavity with lint or cotton dipped in tincture of chloride of iron, or creasote.

Hæmoptysis.—This term (spitting of blood) is generally applied to hemorrhage from the lungs, bronchial tubes, trachea, or larynx. Ulceration of the larynx, trachea, or bronchi may produce it, not often dangerously. I remember one case, in which ulceration of the larynx extended so as to open the carotid artery, with fatal result. More often the source of the blood is the lungs. The diagnosis of this is of great consequence. I have known much alarm to be produced by the spitting of blood whose source examination proved to be the posterior nares. This was not supposed by the patient, because there was no bleeding anteriorly from the nose. Between pulmonary hemorrhage and that from the stomach, the following contrast of signs exists:—

From the Lungs.	*From the Stomach.*
Dyspnœa.	Nausea.
Blood coughed up.	Blood vomited.
" florid, sometimes frothy.	" dark, not frothy.
" mixed with sputa.	" mixed with food.

In a majority of instances, spitting of blood from the lungs is a symptom of phthisis. Cases occur, however, sometimes, especially during adolescence and early maturity (from 18 to 30 years of age) of more or less active pulmonary hemorrhage, whose subsequent history disproves a tuberculous origin for it. In these cases, there may be immediate danger, more probably than in the frequent bleedings of consumption. Aneurism of the aorta may also cause hæmoptysis, by rupture of the tumor, which must cause death. This of course is rare, and is made known by signs already considered.

Treatment.—For active, congestive pulmonary hemorrhage, in a young and robust person, it was formerly the common practice to take blood from the arm, as a *derivant* measure. I have known this to succeed perfectly, with no subsequent disadvantage. But, dry cupping over the chest and back, with sinapisms to the legs, and ice, salt, or alum, swallowed slowly, the patient being at perfect rest in bed, with the head and shoulders raised, will be sufficient treatment at the start for most cases. Then we should prescribe, if the bleeding continue after the first gush, acetate of lead with opium in pill; say a grain or two of the former with half a grain of the latter every four, three, or two hours as the case needs, for a day or two.

In passive, or tuberculous hæmoptysis, rest, with the head and shoulders propped, is also necessary. Ice, salt, and alum, alone or together, may be held in the mouth and swallowed very slowly, till the bleeding has stopped for the time. For medicines, in the anæmic, gallic acid (gr. x to gr. xxx, in solution with aromatic sulphuric acid) [F. 146], oil of turpentine (gtt. x to gtt xx, in mucilage) [F. 147], and ammonio-ferric alum (gr. v to gr. x), or tincture of chloride of iron, are most recommended. But dosing with these styptics in consumption is not proper for every trifling discharge of blood. They are suitable only when the hemorrhage itself is, or threatens to be, a source of additional debility.

Pulmonary Apoplexy.—This is the extremest degree or result of congestion of the lungs; hemorrhage occurring into the air-cells, and obstructing respiration, sometimes to a fatal degree. Disease of the heart predisposes to this. Its attack is apt to be somewhat sudden; there is great dyspnœa, with a purple countenance, and skin rather cold. Percussion resonance is dull. On auscultation, at first, a bubbling or mucous râle is heard; after the blood coagulates, no respiratory sound at all.

Treatment.—If diagnosticated early, in a person of tolerable strength, venesection should be performed at once. Then (or instead, in a feebler subject) dry or cut cups should be applied extensively between the shoulders; followed by a large sinapism over the anterior part of the chest, and a hot pediluvium. At the same time the reaction which should aid in unloading the oppressed lungs (the object of venesection, cupping, &c.) may need to be favored by hot drinks, as hot lemonade, carbonate of ammonia, or, if coldness be decided, whisky punch.

Hæmatemesis.—Vomiting of blood may result from cancer, or ulcer of the stomach, congestion of the liver, aneurism of the abdominal aorta, &c. We have given, above, the distinguishing signs between it and hæmoptysis.

Treatment.—Of course this must be varied according to the cause. Slight ejections of blood from the stomach may not of themselves require treatment—having only a diagnostic importance. In ulcer of the stomach the greatest danger may occur, except from rupture of an aneurism. In copious hæmatemesis, with absolute rest in the horizontal position, ice, creasote (one or two drops, *pro re nata*), in solution, or pills [F. 81], gallic acid [F. 146], oil of turpentine [F. 147], ammonio-ferric alum, or tincture of chloride of iron, may be prescribed. Food must be given in small quantities, and concentrated.

Hæmaturia.—This may be either from the kidneys or from the bladder. If the blood is thoroughly mixed with the urine, it is probably renal. If the water flows off nearly pure, and the blood follows or accompanies the last portion, it is vesical. When it follows the use of a catheter or bougie, independently of urination, and flows in a stream or in fresh drops, it is urethral and traumatic.

Renal hemorrhage may attend congestion or inflammation of the kidney; or cancer; or scarlet fever (generally a late stage); or the irritation of a calculus; or that of cantharides or turpentine; or, in old persons, it may be passive. In Egypt, a parasite sometimes produces it; the *distoma hæmatobium*.

Treatment.—For hemorrhage from the kidney sufficient to deplete at all seriously, astringents, as gallic acid, tincture of chloride of iron, alum, or acetate of lead, may be used. Rest is important, in this as in all hemorrhages, during the attack. Bleeding from the bladder may be treated by the injection, through a catheter, of solution of alum or dilute solution of creasote (gtt. j in f ℨj of water) or tannic acid (gr. x in fℨj).

Intestinal Hemorrhage.—The causes of this are, especially, typhoid fever, of which it is sometimes symptomatic, and occasionally critical; *i. e.*, the commencement of convalescence. The same may occur in yellow fever, or in remittent fever (less often). Aneurism of the aorta, congestion of the liver, abdominal cancer, may cause it. Blood is passed, commonly in small quantity, with the discharges of dysentery. Aged persons not unfrequently have passive hemorrhage from the intestines. Internal piles are very often productive of it. The blood from the latter is bright red; other bleeding from the bowels is darker and more mixed.

Treatment.—Acetate of lead, by the mouth, with opium, or by enema; tannic or gallic acid, in pill or by injection in solution; oil of

turpentine; creasote, and tincture of chloride of iron, or ammonio-
ferric alum, are here, as in the other hemorrhages mentioned, the most
reliable astringents. For bleeding piles, special treatment has already
been alluded to.

Vicarious Hemorrhage.—The most frequent instances of this are
in connection with suppressed menstruation. Epistaxis, hæmoptysis,
hæmatemesis, renal or intestinal hemorrhage may occur, but it is most
apt to be from the stomach or lungs. The **prognosis** in this form of
hemorrhage is much less serious than in the same of other origin. Its
treatment should be addressed mainly to the regulation of the disturbed or interrupted uterine function. Warmth to the lower extremities and back, with such *emmenagogues* as each case may indicate, will
generally be required. Astringents are to be avoided in vicarious
hemorrhage, unless it be in excess of the ordinary menstrual or other
suppressed discharge.

Uterine Hemorrhage.—Besides simply excessive menstruation,
uterine hemorrhage may be from placenta prævia ("unavoidable hemorrhage"); abortion; subsequent to delivery; uterine cancer; ulceration
of the os or cervix uteri; tumors within, or in the walls of, the womb.

Treatment.—In considerable uterine hemorrhage, of either variety,
ergot, in substance or the wine, is likely to be of use by promoting
contraction of the womb. Ammonio-ferric alum is also a good medicine to give by the mouth in the same case. Locally, ice or iced
water may be (with care not to chill too much) applied for a short
time over the hypogastric region, or thrown into the vagina. Tincture
of chloride of iron, in strong solution, will have a powerful effect.
Tannic acid or matico may be likewise applied; or the "styptic rod"
of tannic acid and cocoa butter, shaped to fill the vagina. But
threatening cases (except *post partum*) may require the actual *tampon*,
or plug of lint for the whole vagina, or the sponge-tent inserted into
the os uteri itself. Stimulants may at times be called for to prevent
fatal exhaustion under large hemorrhage, either from the uterus or
from any other organ. Pressure upon the aorta has been sometimes
resorted to, through the abdominal walls, in uterine hemorrhage.
Other measures, suitable after delivery, belong to the department of
obstetrics.   ·   ·

Habitually excessive menstruation requires that the patient so
affected should maintain absolute rest, from the beginning of the flow
till its cessation. Iron is nearly always indicated in such cases,
through the interval; particularly the tincture of the chloride of iron.

DROPSICAL AFFECTIONS.

Varieties.—1. *Œdema,* local infiltration of connective tissue with
serum. 2. *Anasarca,* general cellular dropsy. 3. *Hydrocephalus.*
4. *Hydrothorax.* 5. *Hydropericardium.* 6. *Ascites.* 7. Other local
dropsies; as *Ovarian* dropsy, *Hydronephrosis, Hydrocele* of the
testis, etc.

Causation and Pathology.—Obstruction to the venous circulation,
arrest of excretion and absorption, and excess of water in the blood,
are the three cardinal elements of the pathological causation of dropsy.

Either one may induce it. Disease of the heart or of the liver brings on dropsy by venous obstruction. Disease of the kidney, or the action of cold and wet upon the skin, may produce it by checking excretion. Wasting diseases are liable in their advanced stages to œdema and anasarca, on account of the watery state of the blood.

Acute general dropsy results from the powerful impression of cold and wet, or of the scarlet fever poison, upon the system; suppressing both the action of the kidneys and that of the skin at once. Its most common form is anasarca; but it may take that of ascites, hydrothorax, or even hydrocephalus. When from cold and wet, it is much more curable (especially anasarca or ascites) than similar dropsy of *visceral* origin, e. g. from disease of the heart. Albuminous urine is quite common in acute general dropsy.

Hydrocephalus, Hydropericardium, and *Hydrothorax* have been already sufficiently considered.

Ascites: peritoneal dropsy; accumulation of water in the abdomen. The **causes** of this of greatest frequency are, cirrhosis of the liver, and disease of the kidney. It may also follow obstruction of the portal vein by cancer, or general obstruction of the circulation from disease of the heart, aorta, or spleen; and it is sometimes ascribed to chronic peritonitis.

Symptoms and Diagnosis.—Often with emaciation of the face, neck, and arms, there is great enlargement of the abdomen. When this is far advanced, *orthopnœa* exists, from pressure upon the diaphragm. The patient is generally weak, with poor appetite and deficient rest at night.

On *inspection,* in the upright posture, the fulness is greatest in the lower part of the abdomen; when recumbent, it spreads evenly; on one side, it falls over that way. *Palpation* will make evident *fluctuation,* especially when one hand is placed on one side of the abdomen and the other strikes gently, at a distance of a few inches. *Percussion* discovers resonance above and about the umbilicus, the intestines rising there upon the fluid to the surface under the abdominal walls. Elsewhere, the sound is dull, even flat.

The amount of fluid in ascites is sometimes immense; as much as twenty-five pints have been withdrawn at once by tapping. It is generally clear, pale yellow or colorless, albuminous and alkaline.

Ovarian Dropsy.—Leaving the history of this, as belonging to the special department of diseases of women, it is right to state that its diagnosis is important, but not always easy. Like ascites, it produces abdominal enlargement, with dulness on percussion and fluctuation. The most-nearly constant points of distinction are, that the ovarian tumor begins somewhat on one side, and only by degrees becomes symmetrical; its shape is, throughout, more globular and coherent, and altered less by changes of position; and the intestines do not float up above the umbilicus so as to make a clearness of percussion-resonance there. The progress of ovarian dropsy is usually slower, and attended by less proportionate depression of the general health.

Treatment of Dropsy.—Acute general dropsy, from suppression of the action of the skin and kidneys, should be treated by active purgation and the use of diuretics. Jalap and cream of tartar (gr. x. of the former with ʒij to ʒiv of the latter) every day or two, will answer well for catharsis. The diuretics most satisfactory are the infusion of juniper berries (a pint daily), acetate of potassa, citrate of potassa, squills, and sweet spirits of nitre [F. 37, 38, 39, 40]. When the patient is hard to purge, elaterium may be given, in gr. ¼ doses, every four hours till it operates.

Ascites, or other dropsy, from disease of any of the great organs, kidneys, liver, or heart, being less curable, and attended by greater general debility, needs more economy of strength. No doubt exists that real harm may be done by the routine of severe purging and plying with diuretics. The one may render the blood thinner and aggravate the constitutional disease, while the others, failing to remove the fluid by secretion, may even irritate the kidneys to the point of suppression of their action. Nourishing concentrated food, tonics, anodynes, &c., may, in visceral dropsy, be of more importance than diuretics. Of course it is desirable to lessen the accumulation of fluid; but the effects of the remedies used must be observed, and one symptom must not be allowed to overshadow the rest.

When enormous distension makes rest impossible, and almost prevents breathing, it is necessary to relieve it by any possible means. Then, purging, as, by elaterium, should diuretics fail, must be resorted to. Or, if the patient's stomach or general strength will not bear that, paracentesis, tapping, is called for. Some patients require this many times.

The operation is best performed while the patient is lying down, upon the side, near the edge of the bed. A trocar and canula are introduced half way between the pubes and the umbilicus, and the fluid is drawn out through the canula. Then a bandage (with a compress) is applied firmly around the abdomen. Some practitioners favor keeping open the orifice, with a slip of lint, to maintain drainage. If no local irritation occur, threatening peritonitis, in consequence, this may be a serviceable measure. If the bolder practice of injecting iodine after tapping (as in hydrocele) should be thought of in any instance, it must be that of simple peritoneal dropsy, uncomplicated by serious visceral disease.

Sometimes œdema of the lower limbs and scrotum becomes so great as to cause great inconvenience. Then the fluid may be let out by making a number of small punctures with an abscess lancet or small pointed bistoury. The only drawback to this is the possibility of erysipelatous inflammation about the punctures. Such danger will not be at all great if, immediately after the operation, the parts be soothed by bathing or anointing the skin with diluted glycerin (fℨj in fℨj of rosewater), or cold cream (ung. aq. ros.), or glyceramyl (glycerin and starch) [F. 148].

For the treatment of ovarian dropsy, the reader is referred to works upon surgery. I only venture the opinion, that the place of ovariotomy has hardly yet been defined clearly and with certainty by experience. If compelled to decide upon it in a doubtful case, I should incline towards the views of those who make it a very rare operation.

ZYMOTIC DISEASES.

VARIOLA.

Synonym.—*Smallpox.* **Varieties**:—Discrete and confluent; also, *varioloid* or modified smallpox, after vaccination.

Symptoms and Course.—*Stages:* These are, incubation, primary fever, eruption, secondary fever, and desquamation. The incubation (period between exposure to the contagion and beginning of the attack) lasts about twelve days. The first symptoms are languor, headache, vomiting, and severe pain in the back; soon developing into fever. On the third day of this, pimples, at first small and red, appear, first on the face, then on the neck, arms, trunk and lower limbs. These papules become vesicles and then pustules; suppurating perfectly by the ninth day of the fever. Then they flatten and scab. Four or five days later, about the fourteenth day of the fever, these scabs begin to fall off. Desquamation is commonly completed by the end of the third week of the attack. To recapitulate; there are, after about twelve days of incubation, three of primary fever, six or seven for the coming out and maturing of the eruption, four or five for its scabbing, and six or seven for desquamation.

These periods vary somewhat; and the severity of the disease depends mostly upon the amount of the eruption. This makes the difference been the discrete (scattered, separate) and confluent smallpox. Even the primary symptoms are generally worse in the latter. The secondary fever, connected with the full development of the eruption (about the eleventh day of disease), is much the most severe in the confluent. The suffering of the patient is great, even extreme, in this form, the whole surface of the body being covered with inflamed pustules. Even the eyes, mouth, and throat may be invaded. Blindness sometimes follows; and I knew of one case in which the eruption in the throat proved fatal by obstruction of the breathing and swallowing. A peculiar and disagreeable odor emanates from the body in confluent cases.

Malignant smallpox is simply a violent form of it characterized by rapidity, and extreme prostration, with or without extensive pustulation. The eruption, in it, is sometimes attended by lividity of the skin. Delirium is common, and a typhoid stupor may exist.

After smallpox, abscesses in various parts of the body, hard glandular enlargements, ulceration of the cornea, suppuration of the ear, pneumonia, or pyæmia may occur.

The danger to life in this disease is always serious. Before vaccination, thousands died annually from smallpox.

Causation.—There is no disease more certainly contagious than variola. Generally either contact or approach within a few feet seems necessary for its conveyance; but I have met with one instance in which it must have traversed the high walls of an inclosed public institution, attacking an inmate who had not left the house for ten years, and without the admission of any one who could have brought it. In the large majority of cases, smallpox occurs but once in a lifetime. Exceptions are well known, however ; some in which the same person has had it three,—it is *said*, even five times.

Treatment.—The preliminary symptoms of smallpox do not differ from those of most other acute disorders, except that the headache, pain in the back, and vomiting are apt to be more severe. In that stage, rest in bed, after a warm mustard foot-bath, and drinking hot lemonade to promote full reaction, will be enough to do. The fever calls, first, for a cooling laxative dose, as Rochelle salt or citrate of magnesia. Then, refrigerant diaphoretics will be in place ; as *neutral mixture* [F. 149, 150], *effervescing draughts*, or *liquor ammoniæ acetatis ;* the first if the stomach be good and the bowels slow to act ; the second if nausea or vomiting continues ; the third if the bowels are free and the fever low in type. No *cutting short* of smallpox is possible ; it is a self-limited disease. There is no specific remedy for it ; we can palliate it only, and conduct the patient through it.

So decided is the tendency to exhaustion of the system in severe smallpox, that early support by concentrated liquid nourishment must be the general rule. As in other acute illness, appetite and digestive power are almost lost. Milk, however, in small quantities often (one or two tablespoonfuls every two or three hours) and chicken or mutton broth, or beef-tea may be given. Other sick diet, as gruel, arrowroot, toast-water, &c., may do during the primary fever. But a good many cases will require even wine-whey or brandy punch in the second and third weeks ; malignant cases, perhaps, in the first. Quinine should go with these, in tonic doses ; *e. g.*, one or two grains every three or four hours. An opiate at night is often serviceable, especially in the confluent form. *Sarracenia purpurea* is of no use whatever in small-pox.

An important object often is, to prevent the *pitting* of the face. Three plans are resorted to: 1. To abort the vesicles. 2. To soothe and mitigate the inflammation connected with them. 3. To exclude air and light during the scabbing and desquamation. The first of these ends is sought by touching each pimple, on the face, on the fourth or fifth day of the attack, with a point of lunar caustic. Soothing inflammation is aimed at by covering the whole face during the first week with a soft poultice of bread and milk, flaxseed meal, or slippery elm bark. Exclusion of the light may be attained by gold leaf; of air, by mercurial ointment, or collodion, softened by adding $\frac{1}{50}$th part of glycerin before it is painted upon the face.

How are we to choose ? I would begin by touching the *worst* papules, on their second day, with nitrate of silver. Then poultice the whole face for four or five days, till the pustules flatten and umbilicate. Lastly, apply the collodion, softened by the addition of glycerin, with a camel's hair pencil, over each pustule, thickly enough to make an artificial cuticle ; which may be renewed, every day or two, until desquamation has been completed.

The *sequelæ* of smallpox must be treated as they arise, by the opening of abscesses, improving the tone of the system by iron, &c. Great care is needed in convalescence from this, as from other acute (especially eruptive) diseases, not to be exposed to sudden changes or extremes of temperature. The danger of pneumonia, pleurisy, or bronchitis is, at such times, much greater than usual.

Varioloid : Modified Smallpox.—In those who have been vaccinated, while the liability to be affected by the virus of smallpox is in most cases removed, in a few the disease is taken, on exposure, in a milder former. The primary fever is rather less severe, the eruption is more scattered, the pustules are not so deep nor so much inflamed, they scab sooner, and very rarely *pit ;* and there is no secondary fever. Varioloid is seldom fatal. Its treatment should be essentially the same as that of smallpox ; only there is less often need of special measures to prevent marking of the face.

VACCINATION.

The ancient practice of inoculation[1] with smallpox, while it was, by the mildness of the attack, nearly always protective of the individual, at the same time propagated the disease, multiplying the amount of its virus. Jenner's introduction into professional practice of inoculation with the virus of cow-pox, known before his time among dairymen, has greatly abridged not only the destructiveness, but the prevalence of variola.

Whether " vaccinia" or cow-pox is smallpox affecting the cow, or is a different disease whose virus is protective against smallpox, is not yet determined to the satisfaction of all investigators.[2] Experiments have been tried repeatedly, with conflicting results. Either way, the fact is plain, that most persons are, by one good vaccination, protected for life ; that modified smallpox, occurring in the vaccinated, is very seldom indeed fatal, and hardly ever pits; and that repeated vaccination, after an interval of years, will make protection almost always complete.

Vaccination may be performed either with the fresh lymph, the same dried by keeping, or the scab; and, either directly from the udder of the cow, or from a human being inoculated with cow-pox. In Europe the lymph of the vesicle, before maturation, is generally preferred. In this country the scab is used, and is found reliable, when fresh enough. No matter how it is kept, after a month it is uncertain ; although it has sometimes been found efficient after being sealed up for a year.[3]

Direct inoculation from the cow makes a very sore arm, with considerable fever. For infants, unless rugged in health, this is an undesirably severe process. It is, at the same time, probable that many transits through human bodies may somewhat modify the virus.

---

[1] Inoculation was introduced into England from the East by Lady Mary Wortley Montagu, in 1718. Dr. Boylston, of Boston, brought it into practice in this country in 1721. Dr. Jenner's first vaccination was performed in 1796. Vaccination was first performed in America in 1799 ; in France, in 1800.

[2] Dr. Cutter, of Boston, inoculated 50 cattle with virus of smallpox, without producing any definite pustule. He then vaccinated the same cattle with cowpock virus ; they all took it regularly. This looks as if vaccinia is a distinct disease peculiar to the cow ; but other facts are opposed to these.

[3] Recent observation, authentically reported, seems to show that *glycerin* will preserve vaccine virus for several months. The method used in England is, to take matter from the sore on the eighth day of the vaccination, on quill points, and mix it with ten times its bulk of glycerin diluted with an equal quantity of water.

Renewal, by inoculating healthy children, not too young, every now and then, from the udder of the cow, is to be recommended. Cattle with the cow-pox may be found in almost any agricultural neighborhood.

In the absence of smallpox, the second month of infancy will be time enough for vaccination. But under danger of exposure, a babe should be vaccinated at any time after birth. Matter only from healthy children ought ever to be used. While it is unlikely that any constitutional disease (as syphilis[1] or scrofula) can be so introduced, there should, in practice, be no room left for any doubt of the kind; and some cutaneous diseases might certainly be transmitted. Unless on account of risk from exposure, the existence of an eruption on the skin, or any other indisposition of the child itself, may be a reason for postponing the operation. The excitement produced by it may aggravate an existing inflammatory affection. Vaccination has often been blamed for the breaking out of eruptions, supposed to be transmitted, when their cause was really the state of system of the patient.

For the operation, the outside of the arm, near the shoulder, is commonly selected. The exact method used is not important. A small, wedge-shaped lancet, or even a sharp-pointed penknife, will do. Various slides have been contrived for the purpose. I prefer to cut or push out a very small flap of cuticle, under which a thick paste made by pressing and mixing a portion of the scab with a drop of tepid water may be inserted. The art of the operation is, to pierce the skin without drawing enough blood to flow; it is most successful when there is no blood at all. Besides the flap, it is as well to scratch the skin, and puncture it, at a little distance, giving three chances of taking instead of one. No disturbance of the arm must be allowed for twenty minutes or half an hour afterwards.

If it be successful, no sign of it is distinctly visible for two or three days. On the fourth day a decided, small red pimple is to be seen and felt. This is a vesicle of some size on the fifth day; it grows large and cylindrical, or hat-shaped, and by the tenth or eleventh day is fully umbilicated, or depressed like a navel in the centre. Before that, about the eighth day, the bright red ring or *areola* forms around it. This fades after the eleventh day, and the vesicle dries up into a round and flat, but rather thick, mahogany-colored scab, which falls off about the nineteenth day. All of these particulars are important, as showing the genuineness of the vaccination. So is the appearance of the cicatrix left; which should be large in proportion to the vesicle, and *dotted* or marked with subdivisions. This is owing to the vesicle being composed of several small cells or compartments.

Slight fever, with restlessness, is not unfrequently observed during the first few days after the vesicle appears; but there is rarely anything requiring treatment.

Re-vaccination.—Experience shows that a small number of persons, after several years, reacquire the susceptibility to smallpox. As the only test of this is exposure either to the latter or to *vaccinia*,

---

[1] At Rivalta, Italy, and Morbihan, in western France, a number of cases of syphilitic disease (primary and secondary), following impure vaccination, have been reported; the last instances by H. Roger and Depaul.

the renewal of the latter, at least once after puberty, is always advisable. On the occasion of epidemics of smallpox, it may be repeated again and again. There is no pain of any consequence in this operation, nor danger, and, if a genuine vesicle form, making a sore arm, that discomfort for a few days cheaply purchases immunity from the terrible disease. I have sometimes thought it possible that the system may be protectively affected by re-vaccination, even when no local effect, or only a "spurious" sore follows. Certainly smallpox is extremely rare in re-vaccinated persons.

The virus from a second vaccination should not be relied upon for use.

VARICELLA.

Synonym.—*Chicken-pox.* This is a mild exanthematous disease resembling smallpox or varioloid considerably. After an incubation of four or five days from exposure to the contagion of one having it, pimples form, generally scattered widely. In the second day they become vesicles filled with lymph. Two or three days more find them scabbing; they dry and fall off soon, without pitting, except in rare instances. There is little or no fever or other indisposition. The disease is attended with no danger to life, and requires only precautionary treatment, *i. e.* to avoid exposure to cold and wet, to keep the bowels regular, and, if needful, promote action of the skin by a diaphoretic, as neutral mixture.

The eruption of varicella differs from that of variola in coming out in successive crops; in not suppurating or becoming umbilicated; and in not deeply involving the true skin.

SCARLATINA.

Synonym.—*Scarlet fever.*

Varieties.—Scarlatina *simplex, anginosa,* and *maligna.*

Symptoms and Course.—After an incubation supposed to be about five days after exposure to its cause, lassitude, anorexia, headache, and pains in the back and limbs mark the beginning of the attack. Soon these are followed by fever; on the first day, very often, the throat is sore. On the second day, usually, a punctated red eruption appears on the face and neck, and in ten or twelve hours has covered the whole body. It is of a scarlet, or sometimes a brick-red hue, uniformly diffused, with a swollen appearance, and great heat; reaching by the thermometer even 106° Fahr. Occasionally miliary vesicles are seen. There is also a sense of burning and some soreness or irritation of the skin. The tongue has a strawberry-like look, from the projection of enlarged red papillæ through a whitish fur. The throat is very red and swollen, generally, with a hue not unlike that of the skin. Fever runs very high, with an extremely rapid pulse, great thirst, headache, perhaps delirium, costiveness, in some cases vomiting. Bad cases may have stupor. By the fifth day, mild examples of the disease show already an abatement. Most have passed the height of the pyrexia by the ninth; although *sequelæ* may protract the attack much longer. Malignant cases may be fatal in a day or two, or even in less than

22

twenty-four hours. Desquamation of the skin follows the fading of the eruption; often large masses of cuticle coming away at once. At this stage more or less decided albuminuria is common.

Scarlatina Simplex.—In this the eruption comes out early and well, with moderate fever, little inflammation of the throat, and an even course throughout. Sometimes there is hardly any febrile disturbance; and the child may play about without having to go to bed.

Scarlatina Anginosa.—Here the violence of the disease falls upon the throat chiefly. The tonsils swell greatly, suppurating either early or late, or they are covered by pseudo-membranous deposit, white, gray, or dark-brown, whose coming away leaves an ulcerous surface, with in some instances an acrid, offensive discharge. The extension of the ulcerative inflammation may pass the Eustachian tube to the tympanum, and even may destroy the auditory apparatus so as to cause permanent deafness. After the rash has disappeared, abscesses in the neck may form and discharge, exhausting the patient.

Scarlatina Maligna.—This term designates an overwhelming toxæmic impression of the morbid cause of the disease. Depression in the first stage becomes intense, without reaction; or, after the eruption has partly come out, it recedes, or grows livid in appearance; or the brain is oppressed with coma. Coldness is sometimes present, or unequal temperature of different parts of the body, instead of the usually diffused febrile heat. The throat may be much or little affected. In some instances the patient seems almost as if struck by lightning, —so sudden and deep is the general prostration. In this condition death may take place in a few hours. Otherwise, with continued prostration, hemorrhage from the stomach or bowels, or vomiting, or diarrhœa, threatens an untoward result.

Sequelæ.—Abscesses about the throat have been mentioned; similar local affections may take place elsewhere after the attack. Ozæna is not uncommon; neither is suppurative inflammation of one or more of the joints, or of the testicle; nor vaginitis. Endocarditis or pericarditis may occur. So may paralysis; either hemiplegia or paraplegia; generally it is partial, and it is often slowly recovered from. Dropsy, from arrested action of the kidneys, with imperfect action of the skin, is the most common and in many cases the most serious of the sequelæ of scarlatina. It comes most frequently within a week or two after desquamation has commenced. Mild cases are almost as likely to be followed by it as severe ones. Exposure to cold is the generally observable direct cause; but cases happen in which no such exposure could have existed. Anasarca is the least dangerous though most frequent form of this dropsy. There may, instead or in addition, be ascites, hydrothorax, or hydrocephalus. Albuminuria, and often hæmaturia, accompanies either form.

Diagnosis.—From measles scarlet fever is known by the eruption coming out on the second day, without catarrhal symptoms but with sore throat,—and by its being of a brighter red color, and uniformly diffused instead of being in patches.

From roseola, it is distinguished by the fever and sore throat, and by the rash, in the latter, being in irregular blotches, damask rose color instead of the brick or scarlet-red hue.

Prognosis.—This is proverbially *uncertain* in scarlet fever. The simple form is, however, the least dangerous, and a very large majority of cases get well. The anginose is more threatening and serious. But the malignant variety, as its name indicates, is far the most so ; recovery from it is the exception, although it does occur. I have known two cases of such recovery ; one, in which coma was complete for thirty-six hours. Adults are, when affected with scarlet fever, in somewhat greater danger than children ; and so, especially, are puerperal women.

Causation.—Although most (not all) authorities agree that this disease is contagious, it is certainly very capricious or variable in its manifestation of this quality. That is, many persons who are exposed escape it. It is true, that several children in a family often have it in immediate succession. But the escape of all but one is, so far as my observation has gone, as common. It rarely occurs twice in the same person. I remember but one or two instances of this in my own practice.

Treatment.—Mild cases require no medication at all, other than to make sure that the bowels are well opened. If fever is high, after a saline cathartic, as citrate or sulphate of magnesia, or Rochelle salts, neutral mixture or effervescing draught, or liquor of acetate of ammonia may be given. Sweet spirits of nitre may be added, in small dose (¼ to ½ fluidrachm for an adult, and proportionately less for a child) if the kidneys act slowly. Drinking cold water freely is to be encouraged ; as it is demanded by thirst. If the throat be much inflamed, the frequent melting in the mouth of ice, in small pieces, will do good.

Venesection is prohibited now in scarlatina by nearly all writers. In the early part of my practice, I bled from the arm in six cases, all of which did remarkably well. They were examples of violent fever with severe sore throat and headache, in children of good constitution. I do not *advocate* the practice, simply in deference to the common opinion of the profession. Certainly it should be always ruled out in asthenic cases, and in all unless at the early stage of the fever.

For the sore throat, which is *specific* in character, besides the use of leeches externally if the inflammation be great and the case sthenic, local alteratives may be used. An old and popular gargle is one of red pepper, vinegar, and water. More powerful in changing the character of the inflammation, I think, is a strong solution of nitrate of silver (gr. xxx in f℥j) applied with a large hair pencil. When pseudo-membranous patches appear, with a tendency to fetor or ulceration, muriatic acid, with honey, equal parts, may be applied in a similar way ; or, diluted, used as a gargle. Sulphate of zinc (gr. xv to xx in f℥j) is also a good application ; and so are solutions of creasote in glycerin ; chlorinated soda ; and permanganate of potassa.

For the irritation of the skin connected with the rash, according to my experience the best relief is to be obtained from sponging with cool or tepid water, two or three times a day. Inunction with lard, or glycerin, is preferred by some. Cold *affusion* is unnecessarily violent and troublesome.

The diet in scarlet fever should be, as a rule, liquid, but need not be *low*, generally, in the sense of dilution or exclusion of animal material, unless in the first few days. Sooner than in most diseases, the tendency to debility is manifest. Then, milk, chicken broth, mutton tea or beef-tea, &c., will be suitable. At the same stage, some patients will require a tonic treatment, by quinine, or, as some prefer, nitric acid in small doses.

Malignant scarlatina is a disease of terrible depression from the outset. Deficient reaction is one of its characteristics. To promote this, external stimulation is primarily important. The hot salt or mustard bath is a powerful agent for the purpose. *Urtication, i. e.,* the direct application of fresh nettles, has been sometimes employed. Mustard plasters may be applied energetically; and so may hot bottles, or bags of hot salt, &c. Internally, ammonia, quinine, and capsicum are the most prompt and reliable stimulants, although we may add to the same list, Hoffmann's anodyne, and brandy, whisky, or wine. Where a tendency to stupor exists, *free purging* will be the main hope. Jalap is a convenient article for the purpose [F. 151].

The sulphite of soda is now, very reasonably, under trial in various zymotic diseases, as an antagonistic of morbid blood-changes. The dose for an adult (perhaps not yet well settled) may be about ten grains every two or three hours [F. 152]. Chlorine water, in fluidrachm doses for an adult (ten drops for a child of two years) is sometimes given in scarlet fever with a similar view; and so is chlorate of potassa.

Other modes of treatment for severe cases are, the use of tincture of chloride of iron freely; of infusion of digitalis; of diluted acetic acid (ʒj to ʒiv of the officinal acid in fℨiv of water, the dose of the solution being a tablespoonful, sweetened, every few hours); and of diluted nitric acid.

Of the *sequelæ* of scarlet fever, each has its own indications for treatment. That of dropsy is the most frequently important. If, during desquamation, the kidneys show any threatening of inaction or suppression, the greatest care of the state of the skin must be maintained. It is, indeed, a good rule of precaution, for fear of some carelessness or exposure, not to allow a patient recovering from scarlet fever to leave his chamber for three or four weeks at least from the beginning of the attack, nor the house for four or five. Lemonade as a drink, if the urine be scanty, may be freely used. Cream of tartar and acetate of potassa are approved in the same case as diuretics. Digitalis has the confidence of many. *Quinine,* in doses large enough to cinchonize, is reported very favorably of, in scarlatinal dropsy, by some practitioners. Dry cupping to the lumbar region, and the application there of a large mustard plaster, are measures suggested by the known congested state of the kidneys. Purgatives must not be omitted when diuretics fail; the principles governing their use being the same as in other varieties of dropsy.

Prophylaxis.—Belladonna has been asserted to have a protective power against the contagion or infection of scarlet fever. But the evidence in its favor does not appear to me to warrant our giving any confidence to it, or to any prophylactic.

MEASLES.

Synonym.—*Morbilli.* Formerly, with all writers, and still with many, *rubeola* is a synonym for measles. Some English writers, however, designate by the name of rubeola only a hybrid or blending of measles with scarlatina.

Symptoms and Course.—After an incubation of from ten to fifteen days from exposure to its contagion, measles begins with a slight or obscure stage of depression, passing into fever. With this there are all the symptoms of a cold; running at the nose, redness and watering of the eyes, and a cough. On the fourth day of the attack the rash begins on the face, and extends over the body and limbs. It is not so bright in color as the eruption of scarlet fever; and is irregularly distributed in patches, more or less crescentic in shape. By about the seventh day the rash begins to fade, and about the same time or before, the fever has begun to decline. Desquamation is much less extensive than after scarlatina.

No such intensity of febrile movement, nor severity of any kind, as is common in the last named disease, exists, except very rarely, in measles. *Camp* measles, during the late war in this country, often assumed a typhous character, with a considerable mortality; due to the conditions under which it occurred among the soldiers. Otherwise measles seldom threatens life.

The **sequelæ** which are of the most consequence are, ophthalmia, diphtheria, chronic bronchitis, and phthisis. Very severe inflammation of the eyes sometimes follows measles; but blindness from this cause is rare. Diphtheritic sore throat is not infrequent, and may be fatal in children. Chronic bronchitis is common, especially when care is not taken during convalescence to avoid exposure. Phthisis, under the same circumstances, is to be apprehended only where the constitution suffers under a predisposition to tubercular disease.

Causation.—Measles is one of the most contagious of diseases, beyond all doubt. Singularly enough, however, I once attended with it one of a pair of twins at the same mother's breast, the other escaping it altogether. A second attack is exceptional, but not very rare.[1]

Treatment.—Beginning with a moderately active saline cathartic, diaphoretics, expectorants, and demulcents are next in place. Syrup of ipecacuanha with neutral mixture (¼ drachm of the former, for an adult with each tablespoonful of the latter) every two, three, or four hours, would be an average treatment for the first week; flaxseed lemonade being freely used as a drink. After that, the continuance or relief of the bronchial symptoms must determine whether some other expectorant (as squills or wild cherry) shall follow. Or, debility may require tonics during convalescence.

HYBRID BETWEEN SCARLATINA AND MEASLES.

This, called *Rubeola* by some, is not common, but undoubtedly does occur. I have seen a case in which the symptoms of the two disorders were so nicely balanced that two physicians of similar experience pronounced it, the one scarlet fever, and the other measles. In severity it is more like measles; although dropsy and albuminuria may

---

[1] Not long since, Dr. Salisbury of Ohio produced measles-like symptoms in several persons by exposing them to the influence of fungi growing upon damp straw. The identity of the affection with measles is not, however, certain. Drs. Hammond and Woodward, at Washington, repeated the same experiments without result.

22*

follow it, as well as bronchitis, &c. Its treatment requires no special
consideration, being involved in what has been said of the two diseases of which it really seems to be a combination.

MUMPS.

Synonyms.—*Parotitis contagiosa ; Cynanche parotidea.*
Symptoms and Course.—This is generally a mild affection, of a few
days' duration. The parotid gland swells and becomes hot, painful,
and tender to the touch. Some inconvenience in swallowing may
result. There is little or no fever, but some general malaise ; and the
attack is generally at an end within a week. One or both parotids
may be affected. There seems to be reason to believe that attacks
may occur at considerable intervals, even of years, involving first one
gland and afterwards the other. Suppuration is rare ; I have seen it
but in one case. Mumps is undoubtedly contagious.
Diagnosis.—As the parotid gland, as well as other glands about
the neck, may inflame from cold, salivation, or scrofula, it becomes
sometimes a question whether a swelling in that region be mumps or
not. When the parotid alone is affected, it is impossible to decide,
unless immediate exposure to another case of mumps be known. The
parotid is, however, not apt to inflame under other causation, even
from salivation by mercury ; the submaxillary glands are much more
liable to swell from that cause. The suddenness of the attack, and
its brief duration are generally also quite diagnostic of mumps, as
compared with scrofulous or other inflammations of glands about the
neck.
Complications.—*Metastasis* of mumps, to the mamma or testicle,
or, it is asserted by good authority, even to the brain, occasionally
occurs. In either of the first two, a somewhat similar inflammation
of the gland attacked takes place : usually more protracted than that
of the parotid. If the brain be the seat of the transfer of the morbid
element or action, meningitis, or coma, may follow ; and even death
is said thus to have resulted. I have never met with any case of the
kind. Otherwise, mumps are free from danger to life.
Treatment.—Care to avoid being chilled, lest metastasis or greater
severity of the attack be produced, is important. No general treatment is necessary, nor does the patient usually need to remain in
bed. Perhaps a mild laxative may be given on the first or second
day. A poultice of flaxseed meal is a good local application for the
gland. It may also be bathed night and morning with soap or volatile
liniment.

HOOPING-COUGH.

Synonym.—*Pertussis.*
Symptoms and Course.—After an incubation of about six days,
with symptoms much like those of acute bronchitis, including fever
of variable degree, the attack commences ; soon showing its peculiar
character. This is, a spasmodic and paroxysmal cough. For hours,
the patient may be apparently well ; and then, often with a premonitory sensation which leads the child to run to its mother or nurse, or,

if at night, to sit up in bed, a fit of coughing begins, and lasts for several seconds or minutes. It consists of a rapid succession of short but violent expiratory efforts, with scarcely any intervals of inspiration; at the close of which air is taken in by force through the contracted glottis, making a whooping sound, whence the name of the disease. All who have it do not whoop; but the paroxysmal character of the cough is pathognomonic.

Expectoration is often copious, of thick mucus, sometimes even of lymph and pus. Vomiting occurs very often, during the spells of coughing. The child may become very much exhausted, even to a fatal end; but unless from complication or previously feeble constitution, death does not very often occur. I never knew asphyxia to be fatal during the paroxysm; though it is sometimes very threatening. There may be many variations of severity in all the symptoms in the course of an attack.

The *duration* of hooping-cough is seldom less than six weeks, although cases have ended within three weeks. Often it lingers for three or four months; in one case I knew it to last a year.

Complications.—Pneumonia, collapse of the lungs, and (as a sequela) phthisis, are the most liable to occur. Deafness from rupture of the membrana tympani during the violent coughing, has been known. Sometimes the eyes become blood-shot from the same cause. Convulsions occasionally increase greatly the seriousness of the disorder.

Causation and Pathology.—There is no question of the contagiousness of hooping-cough. Generally it occurs but once in the same person; but second attacks are not very rare. Like scarlet fever, measles, &c., it is much most often met with in children; but this is merely from their susceptibility under exposure; as adults also have it.

Belonging with the zymotic diseases, caused by a specific morbid poison, the spasmodic nature of the cough points to the nervous system as in main part the seat of its action. Yet the expectoration as well as early (and occasional, afterwards) febrile symptoms, show that bronchial inflammation exists, secondarily at least.

Treatment.—Mild cases need only care to avoid exposure to damp and cold. After the first few days, if there be no fever nor soreness of the chest, the patient need not be kept in the house in good weather. Indeed, he will cough least when most out of doors. When the cough, at first, is tight and painful, with little expectoration, syrup of ipecac, or squills, may be given. As soon as the spasmodic character of the cough declares itself, with some violence, the "milk," or the tincture, of assafœtida may be given; with or without other expectorants according to the case [F. 153]. Severe cases may be quieted by belladonna or musk; but I have been especially satisfied with the effect of the fluid extract of hyoscyamus, in the dose of from four drops in a child of ten or twelve years of age, down to a fraction of a drop at a time, in a young infant [F. 154, 155, 156]. Hydrocyanic acid, bromide of ammonium (from two to twelve grains at once for a child), nitric acid, alum, clover-tea, and benzoic acid are among the other remedies often employed to allay the violence of the paroxysms. Application of strong solution of nitrate of silver to the larynx has some advocates.

In protracted cases, counter-irritation to the chest and back of the neck may be required. I once met with great relief upon the application of a small blister to the nucha. *Tonics* are also not unfrequently called for toward the end of the attack in a feeble child; especially quinine, or tincture of bark (Huxham's), iron, or cod-liver oil. There is very seldom need to restrict the diet in this disease, unless during the first week.

DIPHTHERIA.

Synonyms.—*Pseudo-membranous Angina ; Putrid Sore Throat ; Diphtheritis.*

History.—Though the name diphtheria (from διφθερα, a skin or membrane) was only given to this disease by Bretonneau of Tours about forty years ago, it appears to have been described by Aretæus of Cappadocia as a disease of Egypt; and was mentioned also by Macrobius and Cœlius Aurelianus among early writers. Hecker gives account of its prevalence in Holland in 1337; Carnevale, at Naples, 1620; Tamayo, at Madrid, called it *garotillo*, in 1622. Ghisi first clearly described the pseudo-membranous formation, at Cremona, 1740. In France, Chomel saw it in 1743–9; in England, Fothergill, in 1754; Douglas of Boston, in this country, in 1736, and Samuel Bard of New York in 1771. Huxham, Cheyne, Rosen, Albers, and Guersent also described it under different titles. Bretonneau most fully made it out as a distinctive disease, in 1826. Since that time it has been recognized and treated of by nearly all medical authorities.

Late epidemics of it have been, principally, those of Paris and Boulogne of 1855–7, passing to England in the latter year; and of our own country beginning in California in 1856, and in the Eastern States a little later, gradually increasing in prevalence until 1860. Since that time it has declined in frequency, although still existing, and sometimes attended by great local fatality. Bretonneau, not unreasonably, supposes Washington and the Empress Josephine to have died of diphtheria. Stephanie, the beautiful queen of Portugal, and Valleix, the eminent French physician, were victims of it.

Varieties.—1. Simple; 2. Croupous; 3. Ulcerative; 4. Malignant diphtheria.

Symptoms.—Premonitory, but not distinctive, are general *malaise*, slight sore throat, and swelling of the lymphatic glands behind the jaw. Then, in the **simple** form, fever occurs; with headache, furred tongue, constipation, and difficulty of swallowing. On examination, a swollen and very red or purple appearance of the fauces will be observed, as well as of the palate and tonsils. Over one or both of the latter, there may be seen, often as early as the second or third day, a whitish or yellowish white membranous deposit. All the symptoms continue, in this form, from five to nine days; when, in favorable cases, convalescence follows.

The **croupous** form has caused the greatest number of deaths, especially in children. This seems especially prone to follow measles or scarlatina. In it, after the same early symptoms as those above described, but sometimes violent from the beginning, increase of discomfort in the throat is complained of. Then an abundant yellow or brownish leathery exudation is found to cover the tonsils and fauces; which, under the exudation, are much swollen. Often quite early in the attack, the pseudo-membranous inflammation extends to the larynx. This is shown by the usual symptoms of croup; the barking cough and voice, and difficult inspiration, becoming whistling or sibilant when the obstruction to breathing is the greatest. A fatal termination may occur, by asphyxia, in a very few days. This can only be averted by the detachment and expulsion of the membrane, without its re-formation.

The **ulcerative** variety is not common. When destruction of the palate and tonsils has attended it, with copious dark-colored and pulpy exudation, and some extravasation of blood, it has been mistaken for, and described as, gangrene; whence the old name, "putrid sore throat." The occasional existence of true gangrene cannot be altogether denied.

Malignant Diphtheria.—At the commencement of this, there is, with intense headache, not unfrequently *vomiting*, which is uncommon in the milder varieties, and hemorrhage from the nose, mouth, stomach, or rectum. Great dysphagia soon exists, and enormous engorgement of the submaxillary, parotid, and cervical glands. The tonsils, pharynx, and palate are covered thickly with a leathery deposit, at first yellowish, but soon becoming ash-colored, brown, or almost black, and of an offensive odor. The tonsils may suppurate or even slough. The nostrils are also sometimes involved; being swollen, lined with false membrane, and emitting an acrid and fetid discharge. Extreme prostration comes on, at a more or less early period; it may be from the first day. The pulse becomes very rapid, the face lividly pale, morbid heat of the skin being followed by clammy coldness. Coma often precedes death. The latter may take place in three, four, or five, occasionally in one or two days; sometimes from the constitutional impression of the disorder, before the local affection has been fully developed.

Special Symptoms and Complications.—*Albuminuria* is present in most severe cases of diphtheria, from an early time in the attack. A diphtheritic affection of the *skin* has been now and then observed. A blistered or otherwise abraded surface will usually, in the course of the disease, be covered by false membrane. *Pneumonia* is an occasional and dangerous complication.

Sequelæ.—These are, especially, long-continued debility, paralysis of the soft palate, and general paralysis in various degrees. In the last of these, deglutition, articulation, vision, and locomotion may be involved. A fatal result may occur after a few weeks, or recovery after a longer period; sometimes from two to eight months.

Morbid Anatomy.—The pellicle or deposit, formed upon the highly injected and tumefied mucous membrane of the fauces and throat, constitutes the anatomical peculiarity of the disease. Minutely examined, the false membrane is found to vary from $\frac{1}{10}$ to $\frac{1}{8}$ of an inch in thickness, and to be fibro-laminated; *i. e.*, of layers of fibrinous network, including epithelial cells, and having on its free surface exudation corpuscles or "pyoid globules," and granules; these forms appearing to be only stages of degeneration. No process of organization or development occurs in the mass; it is aplastic. In some cases only a granular superficial infiltration of the mucous membrane is observed, without even distinct fibrillation.

The common deposit of diphtheria differs from the false membrane of simple inflammatory croup, and still more from the "coagulable lymph" of inflamed serous membranes, in being thicker, more tough, yellower, and less capable of anything like organization.[1] (Dr. B. Sanderson asserts that he discovered evidence of development of the exudation in one or two specimens of the simple form of diphtheria.)

Pathology.—Excluding from the title of diphtheria all instances of accidental or merely inflammatory "diphtheritic" or pseudo-membranous formations, as they occur, for example, in croup and scarlet fever, we must admit that there is a special zymotic or "enthetic" disease, for which that name is appropriate, and should be reserved. It is a toxæmic or "dyscrasial" affection, in which the morbid change in the blood has its main and characteristic local manifestations in the throat.

Causation.—Not doubting the existence of a special material cause, yet unknown except by its effects, we can only say further that the disease is generally epidemic or endemic, with a special tendency to limited localization. It acts with intensity in confined centres; as, a small village, a crowded school, a numerous family; inflicting therein often a terrible loss in proportion to the members attacked; a sort of domestic pestilence.

Is diphtheria transmitted by contagion? I incline to believe that it sometimes is so, although clearly not dependent upon that mode of propagation in its epidemic migrations. The certain examples of its extending from one person to another are few; but I think I have known of one at least.

Children are much more liable to diphtheria than adults. Climate and season do not seem to affect its prevalence. Nor does it show any very decided preference for otherwise unhealthy places, where filth or crowd-poison abounds. Even its promotive causation, then, seems to be at present less known than that of most other diseases.

Diagnosis.—From *scarlatina*, diphtheria is distinguished by the absence of the eruption, and of the peculiar punctated or brick-dust like flush of the throat, and "strawberry" tongue. That scarlet fever *predisposes* to diphtheria, as a subsequent attack, is a well established and not unimportant fact.

With *membranous croup*, it is contrasted in the following manner. That disease is a sporadic and sthenic local phlegmasia, whose general symptoms are, as much as in any inflammation, dependent upon the local affection: while diphtheria is a constitutional disorder, usually epidemic, in which the local symptoms are secondary. More directly, in practice, we may mark the commencement of the pseudo-membranous deposit, in diphtheria, about the tonsils and pharynx; in croup, in the trachea or larynx. That of diphtheria rarely extends, in any

---

[1] Bretonneau long since, and Dr. Sanderson a few years ago, imitated the diphtheritic exudation, by injecting oil of cantharides into the throats of animals. The principal difference was in the manifest tendency to organization in the cantharidal pseudo-membrane

case, below the larynx ; that of croup, not unfrequently even into the bronchial tubes. After the laryngeal complication or extension has occurred in diphtheria, the croupal symptoms are really the same as those of any other laryngeal obstruction, and thus are not different from those of croup.

From *thrush*, and *aphthæ*, diphtheria is known by the deposit being much larger and thicker, never vesicular, and mostly duller in color; and attended generally by more severe constitutional symptoms. Thrush begins in the mouth ; it is, moreover, much more uncommon in adults than diphtheria ; and is never epidemic.

Prognosis.—*Simple* diphtheria is not very dangerous to life. The croupal form is decidedly so; and the malignant is fatal in a large majority of cases. *Insidiousness* is a trait often belonging to the disease in children ; a name which has been applied by some, for that reason, is "creeping croup."

Treatment.—No specific remedy having been discovered for this disease, we must be governed in our tentative treatment of it by our idea of its nature ; while concluding upon its therapeutics, finally, through experience. Nothing, it may be confessed, is very satisfactory, as yet, in the management of bad cases of it. All agree that it is not a mere local inflammation, but a systemic affection primarily ; and that its type is most generally asthenic. Much depletion is therefore not to be thought of. I would never bleed from the arm in diphtheria. In simple, open cases, I have used leeches to the throat, with seeming decided advantage, within the first three days. Even their use, however, must be exceptional. Moderate purgation, as with citrate of magnesia, or Rochelle salt, at the very beginning, is well in the simple and croupal, though not in the malignant form.

Chlorate of potassa is a favorite medicine with many in this disease. My best results in bad cases have attended its early and free use. An adult may take twenty grains in solution every three hours ; I have given five grains every two hours to a child five or six years old [F. 156].

Tincture of chloride of iron is relied upon by some ; from ten to twenty drops every three hours for an adult ; with or without the chlorate of potassa [F. 157]. Sulphate of quinine is also given, alone, or at the same time with the above remedies, by a number of practitioners ; say, of quinine, for an adult, a grain every two or three hours.

Besides these, or instead of them, for internal use, permanganate of potassa has, after some trial, the recommendation of one or two observers. A drachm of it may be dissolved in a pint and a half of water, a fluidrachm of this being taken every hour. Sulphite of soda, ten grains every two or three hours, is worthy of trial in this, as in other zymotic diseases.

Concentrated liquid food must, as a rule, be given throughout an attack of diphtheria ; milk, beef-tea, and very often wine whey or brandy or whisky punch ; in small quantities at short intervals, according to the degree of prostration present.

Local treatment is, by most physicians, regarded as very important. Experience has shown, I think, that it ought not to be violent. Ice in small pieces melted in the mouth slowly, is probably as useful as any application. Muriatic acid and honey, equal parts, applied freely with a large camel's hair pencil; or diluted with water and used as a gargle, I believe to be serviceable. Creasote dissolved in glycerin [F. 158]; lime-water; chlorinated soda dissolved in twenty parts of water; and permanganate of potassa, a drachm in a pint, make also appropriate gargles. In a young child ice is often the only local application possible without a struggle so disturbing as to make the benefit of it doubtful. Cold water compresses may be applied outside of the throat in the early stage, while there is excess of heat. Later, flannel wrung out of hot water to which an equal amount of spirits or vinegar has been added, will give more comfort.

Inhalation of the steam of lime-water is worthy of trial in diphtheria, especially in the croupous variety; or, the *atomization* of lime-water by the *nephogene* or some other apparatus constructed for the purpose.

But, I believe the local treatment to be, after all, secondary. And especially is the effort (which I have seen practised) to remove the patches of exudation by force, as by excision or actual cauterization, to be deprecated, as likely to do harm rather than good.

GLANDERS.

Synonym.—*Equinia.* Though not common in the human subject, it is important to know that this affection can be taken from the horse. It is said to occur either in the *acute* or the chronic form; generally the former.

Symptoms and Course.—After an incubation of from two to seven days, with febrile symptoms, the nostrils become inflamed, and at the same time pains in the joints occur, like those of rheumatism. Over parts of the body the skin becomes red in patches, which may grow dark and even gangrenous. Crops of pustules also appear, one after another, on the face and limbs. In the course of a week or so, a muco-purulent discharge comes from the nostrils, which are swollen, ulcerated, or gangrenous. The fauces, pharynx, larynx, even the lungs, may become seriously involved. The face and eyes also inflame and become œdematous. Throughout, fever of a low form continues, with great thirst, delirium or coma, a fetid odor from the skin, and diarrhœa. Death almost always occurs within three weeks; sometimes one or two weeks later.

Chronic glanders is rare; it is described as milder than the above, and much less fatal.

Treatment.—This must be purely tentative. Most worthy of trial are the sulphites, as sulphite of soda. Locally, I would use creasote dissolved in glycerin,—dilute chlorinated soda, and lime-water.

INFLUENZA.

Synonym.—*Epidemic Catarrh.*

History.—Although, among persons exposed to the same weather, catarrhal affections are of course common at certain times, there is evidence that, apart from the conditions of humidity and temperature of the air, *epidemic catarrh* sometimes occurs as a zymotic disease.

It is recorded as having been quite fatal in France in 1311 and 1403; in 1570 also it prevailed, and in 1557 spread over Europe and extended to America. It occurred again in 1729, '43, '75, '82, 1833, '37, with notable violence. In the United States, one of the most remarkable epidemics, for extent, was that of 1843. The local prevalence of influenza may occur at very irregular periods, and sometimes so mildly as not to be distinguished from common sporadic catarrh.

Symptoms and Course.—The ordinary symptoms of "a bad cold" are those of influenza; but the illness is somewhat more severe, and prostration is generally greater. Of this there are all grades, however. Bronchitis, sometimes capillary, and pneumonia, are not rare complications. Old people are especially apt to be carried off by influenza. Its mortality is very small among persons in early or middle life. The *duration* of an attack is commonly from three to ten days.

Causation.—The hypothesis has been entertained, in consequence of the irritating effect of ozone upon the air-passages, that an excess of it in the atmosphere may be the cause of influenza. But no facts raise the supposition beyond conjecture.

Treatment.—Mild cases require housing, and little more. A warm mustard foot-bath at night, followed by a large draught of hot lemonade if there be chilliness, or the same taken cold if fever exist— and a dose of solution of citrate of magnesia or Rochelle salt or senna tea in the morning, will generally suffice. Sweet spirit of nitre may be added to the night-draught if the skin be dry and the urine scanty.

Great prostration, especially in old people, may call for support, by quinine and stimulants. Hot whisky punch is, for such a case, not out of place. The *abortion* of an attack of influenza is sometimes practicable within the first two days, by giving quinine, in four grain doses, thrice daily. Bronchitis or pneumonia, as complications, will require treatment as in other cases.

DENGUE.

Synonym.—*Break-bone Fever.*

History.—Frequently in the Southern United States, occasionally in the Northern (at least Dr. Rush seems to have described it at Philadelphia in 1780), and in the East and West Indies, this disorder has occurred. English writers regard it as a variety of scarlet fever; naming it *Scarlatina rheumatica.*

Symptoms and Course.—Usually after a chill, fever comes on, moderate in degree, but attended by considerable debility, and severe pains in the head, back and joints; the latter being somewhat swollen. In about two days, or less, the fever subsides, and the pains lessen, though they do not disappear. Toward the end of a week from the commencement of the attack, a rash breaks out, resembling that of scarlatina, or duller and more in patches. The fever returns, often, about the fourth or fifth day, and lessens or ceases after the eruption has come out. All the symptoms gradually subside, leaving the patient well but very weak, by the beginning or middle of the second

23

week of the attack. This disease, without complication, is never fatal; nor does it leave any sequela except debility.

Its **causation** is not known, beyond what is comprised under the term "epidemic influence." It is noticeable that it affects more persons at one place and time than almost any other epidemic : nearly all the population may have it in one season; all ages and both sexes being alike attacked.

In **treatment**, dengue requires merely good nursing—regulating the bowels, and relieving or mitigating the pains with Dover's powder or other opiates, especially at night; or by the local application of landã-num, &c.

MALARIAL FEVER.

Varieties.—*Intermittent, Remittent*, and *Pernicious* Fever. These may all be properly regarded as grades or modifications of the same type of disease; agreeing in the nature of their cause, the periodicity of their symptoms, and their mutual convertibility. Each will, how-ever, require a separate description.

INTERMITTENT FEVER.

Synonyms.—*Ague ; Chills and Fever.*

Varieties.—*Quotidian*, when the paroxysm occurs every day; *tertian*, when it is every other day; *quartan*, on the first and fourth days; also, *quintan, sextan, septan*, and *octan*. The quotidian and tertian are common; the octan, or weekly return of the attack, is not unfrequently met with; the others are very rare. The time between two paroxysms is called the *intermission* (apyrexia); the period from the beginning of one chill to the beginning of the next is the *interval*. Paroxysms are sometimes *double :* as, double quotidian, with two paroxysms on one day; double tertian, with a paroxysm every day, but those of every other day corresponding in time or character; &c. These also are rare. I have, in a large number of cases of malarial fever, in the suburbs of Philadelphia, never met with a double parox-ysm of either type.

Symptoms and Stages.—No disease has ordinarily so regular a succession of definite stages as intermittent fever; viz., the *cold*, the *hot*, and the *sweating* stage.

Cold Stage, or Chill.—Beginning with languor and yawning, a sensation of coldness comes on, often creeping and shivering, with chattering of the teeth and *rigors* or tremulous movements. The skin has a sunken appearance, and the lips and finger-ends may be blue. The *sense* of coldness does not prove a low temperature of the body; which the thermometer sometimes shows to be even hotter than natu-ral. Thirst exists, with loss of appetite: occasionally, vomiting. Headache, depression of spirits, and drowsiness are common. Per-spiration is absent, but the urine is abundant and nearly colorless, with a low specific gravity. The duration of a chill varies from ten minutes to two or three hours; averaging not more than three quarters of an hour.

Hot Stage; *pyrexia.*—Gradually warmth is felt to return; the shivering ceases; a flush succeeds the pallor or lividness of the face.

A real increase of the heat of the surface is found by the thermometer; reaching 105 to 110°; seldom more than 108°. The mouth becomes dry, the tongue furred; vomiting is common, with total anorexia. Headache is apt to be violent; but delirium is rather exceptional. The pulse is accelerated, and generally strong and full. The bowels are constipated; the skin dry, the urine scanty and high-colored. The hot stage may last from an hour or two to sixteen or eighteen hours.

Sweating Stage.—This also comes on gradually; the face first becoming moist; then the trunk and limbs. This is attended by increased comfort; the headache lessens, the stomach if disturbed becomes quiet, the patient often goes to sleep and sweats profusely all over. After this, the fever has gone; the pulse is slow and soft, the skin cool. The urine now again is passed freely, and deposits a brick-dust like (lateritious) sediment. There is no definite length of time to be assigned to the sweating stage.

Of the three stages, now and then one or two may be wanting. There is then only a chill, or a fever, or a sweat, occurring daily, or every other day, at the same hour. Or, a paroxysm of pain may occur, in one part of the body, with the same regularity. One form of this is called "brow ague." *Dumb* ague is a popular name for an attack in which the chill is absent or obscure, the other symptoms recurring periodically. There seems to be no doubt that a single limb, or even a single finger, may go through all the three stages— cold, hot, and sweating; the rest of the body being unaffected. Intermittent *neuralgia* is very common in malarial districts, especially after chills and fever. In the same regions, *all* complaints are apt to take on periodicity; so we may have intermittent dysentery, pneumonia, etc.

The *intermission* is often a time of apparent health, except some debility, and perhaps headache and want of appetite and of good digestion.

The greater number of paroxysms of intermittent occur in the day time. An attack which began as a tertian, may become a quotidian; or the converse may happen. Intermittent sometimes passes into remittent fever; though much less often than remittent becomes intermittent.

Sequelæ.—Protracted intermittents are often accompanied or followed by anæmia, of a marked character, and by enlargement of the spleen and liver; especially of the former. Dropsy is a quite frequent result of these visceral affections and of the anæmia.

Morbid Anatomy.—*Melanæmia*, or pigmentary degeneration of the blood-corpuscles, with deposit of pigment granules in the liver, spleen, kidneys, brain, etc., is almost a characteristic of malarial disease. Enlargement and softening of the spleen, and engorgement of the liver, with a bronzed appearance of it, are the only other peculiar changes of structure.

Diagnosis.—*One* chill can hardly ever be certainly pronounced to be malarial, because very many acute disorders begin with a cold stage. Two, with a distinct apyrexia, cannot often be confounded with anything else, except hectic fever. In the latter, there is usually a known *cause* for the symptomatic febrile symptoms; the patient is weak and emaciated, the paroxysm is irregular in time and duration, there is a bright roseate flush upon the cheek, and headache is usually absent.

Prognosis.—Left to itself, intermittent will sometimes get well as early as the seventh, eighth, or ninth paroxysm; more often it will last ten weeks; sometimes for as many or more months.

When under treatment, it is almost always possible to *break* the chills by cinchonization; but they frequently return; especially at the end of one, two, or three weeks. It is a good sign for the paroxysm to occur later and later in the day, and to become shorter and shorter. Tertian ague is generally the most readily cured; quartan the most intractable, though comparatively uncommon. Death, in modern times, since the discovery of the properties of Peruvian bark, almost never happens from the ordinary type of intermittent; the *pernicious* form is very dangerous.

Pathological Nature.—As to this, it is possible only to speculate at present. It is most probable that ague is a *toxæmic neurosis*.

Causation.—Upon the origin of malarial fevers, the following facts seem to be established :—

1. They are reasonably designated as *autumnal* fevers, because very much the largest number of cases occur in the fall of the year. Spring has the next greatest number of cases.

2. They are always strictly localized in prevalence.

3. They never prevail in the thickly built portions of cities.

4. An average summer heat of at least 60° for two months is necessary to their development. Their violence and mortality are greatest, however, in tropical and sub-tropical climates.

5. They prevail least where the surface of the earth is rocky; and most near marshes, shallow lakes and slow streams. The vicinity of the sea is free from them, unless marshes lie near it.

6. The draining of dams or ponds, and the first culture of new soil, often originates them.

7. Their local prevalence in the autumn is always checked by a decided frost.

Upon these facts, it was a legitimate hypothesis (urged especially by the late Prof. J. K. Mitchell of this city) that the material cause of malarial fevers is a minute vegetative organism, whose substance or emanations enter the body. Professor Hannon of Brussels relates that he learned in 1843 from Prof. C. Morsen, and verified the statement in his own person, that the exhalations of certain fresh water algæ would produce ague.

Dr. Salisbury of Ohio has recorded in the January number of the *American Journal of Medical Sciences* for 1866, some observations and experiments, tending to show that minute cryptogamic plants of the family of Palmellæ, abounding over the surface of marshes, can generate intermittent fever, when transported to localities otherwise free from it. Such results require of course repeated investigation to make them actually matters of demonstration. When confirmed by such means, they will make a very important contribution to etiology.

Treatment.—One remedy, in this disease, overshadows all others; *cinchonism*. By this we mean, the production of the constitutional impression of the cinchona bark, or of one of its essential constituents.

At any stage it appears to be safe, unless it be the very height of the pyrexia. Nor, as a rule, is any special preparation necessary.

It is well, always, during the chill, to promote speedy reaction by external warmth, and perhaps by hot drinks, of a not too stimulating character. The bowels ought to be opened well; and the stage of fever may be palliated by the free drinking of cold water, made more diaphoretic by the addition, if necessary, of neutral mixture or effervescing draught. Then, as soon as sweating fairly begins, the quinia, or cinchonia, or bark in substance, may be prescribed.

The sulphate of quinine has the most universal reliance. Some give it in doses of several grains each, twice daily. I think experience warrants me in preferring to give one grain every hour [F. 2, 159]. The amount required in the intermission of ordinary intermittent is about fifteen grains. Less may often cure, but can hardly be depended on. The quinine may be given in pill or in solution. I direct that, in tertian ague, the patient begin early on the day of the intermission, and take one grain every hour till he has taken twelve grains. The next day let him begin at the same rate and, if no chill occur, take ten grains. The third day, nine; and so diminishing daily until six grains are reached. Let this be continued till a week from the last chill, when a greater tendency to a return will exist; on that day let ten grains again be given. After that time, if no paroxysm has occurred, he is, for the time at least, well.

Sulphate of cinchonia, in doses one-half greater (gr. jss instead of one grain) [F. 160, 161] has always succeeded with me, in a considerable number of cases; as it has with others. It generally produces much less ringing in the ears than quinine, and can be taken by some whose heads do not well bear that medicine. Quinidia, quinoidine, and other extractives of bark I have not tried, and would prefer not to trust; though some favorable experience with them is recorded. Bark in substance, especially Calisaya bark (an ounce in the intermission), is of course perfectly reliable; but it is disagreeable and oppressive to the stomach, and should only be used when its derivatives cannot be obtained.

Other remedies in considerable number, have obtained more or less reputation in the treatment of ague. Opium, given in full dose (say 60 drops of laudanum) shortly before the time of an expected chill, has been found generally to abort it. Arsenic (10 drops of Fowler's solution thrice daily) is considered to approach very nearly in certainty to the preparations from cinchona. Sulphate of copper is asserted by some (in ¼ grain doses) to be antiperiodic; and so is nitric acid (10 drops thrice daily, diluted); and common salt (a drachm at a dose, half an ounce during an intermission). Dogwood bark; pepper, and its extractive, piperin; willow bark, and salicin obtained from it, have also some reputation of the same kind. Chloroform, taken by the mouth, has recently been used with success by Dr. Merrill. He gives f℥j at once, at the beginning of the chill. It may be diluted with mucilage.[1] A strong impression of almost any kind, upon the

<hr>

[1] Pollacion and others in Spain have found the internal use of chloroform successful in intermittent. Bonafont reports the cure of fifteen cases by *inhalation* of chloroform.—*Dublin Quar. Jour. of Med. Sci.*, Feb. 1867, p. 187.

system, during the apyrexia, may arrest or prevent the paroxysm.
So may act the drawing of a blister upon the spine; or a cold shower
bath. I have known one case to be cured by the patient being
solemnly assured (without medicine) by a quack that "he would never
have another chill."

But the *breaking* or interruption of chills, though generally curative of a first attack, is not nearly always so in a second or third.

Chronic intermittent may maintain a constant tendency to relapse,
in spite of cinchonism. In such a case, *anæmia*, and the malarial
cachexia, are usually present. Here the great remedy is *iron*. This
has never disappointed me; that is, I have never failed to cure a case
of uncomplicated chronic intermittent, even of several months' duration, by breaking the chills first with quinine, and then causing the
patient to persevere for a month or two with iron. I prefer the pill
of the carbonate, Valleix's mass; with a grain of quinine in each pill
[F. 162].

REMITTENT FEVER.

Synonym.—*Bilious Fever.*

Varieties.—Simple and malignant. The latter, however, will be
described under *Pernicious Fever.*

Symptoms and Course.—Although the premonitory stage is usually
short, and not unfrequently wanting, its general occurrence is well
established. Its symptoms are those of general *malaise*, with some
headache, slight nausea, and furred tongue. These increase until a
chill, not violent, but lasting sometimes half an hour or an hour, fairly
begins the attack. Or, an ill-defined cold stage, with a feeling of
chilliness, languor, and debility, and perhaps cerebral oppression and
gastric disorder, may occur.

After this, the febrile condition is developed. The skin becomes
hot, dry, and harsh; the *pulse* rises in force and frequency, although
less hard and tense than in some diseases, and not exceeding generally, during the first exacerbation, 110 or 115 beats in the minute.
The face is flushed; *headache* is throbbing and severe; the faculties
being unfitted for any mental exercise. *Violent pain* is almost always
felt in the back, and very often also in the limbs. *Epigastric uneasiness* is nearly universal; nausea and vomiting extremely common.
Bilious matter is in many instances ejected from the stomach. The
bowels are costive; when opened, however, the stools are colored with
bile. The *urinary* secretion is scanty. *Thirst* is always great; cold
drinks being much preferred. *Respiration* is hurried, although free.

After a continuance of from eight to twenty hours, these symptoms
abate more or less, even without treatment. The feelings of the patient are more comfortable; he sleeps; and wakes with a skin less
hot, and moist, perhaps even with considerable perspiration. Headache, however, and some pain in the back remain; and the pulse does
not subside to the natural standard. In some instances it is little
altered. The stomach, however, is less disturbed, and thirst is somewhat less intense.

There is reason to believe that a few cases of genuine miasmatic remittent may, by prompt treatment during the hot stage, be quelled, so as not to advance beyond the first exacerbation and remission. We ascribe their facility in yielding, chiefly, to a less degree of intensity in the morbific cause.

Mostly, in from six to twenty-four hours, the patient's discomfort again increases; the skin becoming even hotter than before, and quite dry; the pulse rises to 120 in the minute; thirst is great, although sometimes less than in the first paroxysm; the headache returns, and with it usually severe pain in the back. The tongue is now thickly furred, often with a yellowish hue. Nausea and disgust for food are again felt, and in a large number of cases vomiting returns; the stomach rejecting everything, even cold water. The stools, when obtained, are sometimes slate-colored; but more often decidedly colored with bile. Diarrhœa is uncommon, and is most apt to accompany a later stage. *Delirium* is common only in violent cases; restlessness is almost universal. *Yellowness* of the skin appears in a majority, in various degrees.

The advance of the disease, after the second paroxysm, is exceedingly various. The periodical character, however, is maintained throughout. The remissions may occur at any hour,—in moderate cases being as often in the afternoon as in the morning; in the protracted, more commonly in the morning, the fever lasting through the night. Quite frequently a *double tertian* type is observed; the exacerbation occurring one day in the morning, and the next in the afternoon; and sometimes with different degrees of violence.

Duration.—Favorable cases often terminate in six or seven days in an intermission, which in some becomes a cure even without any antiperiodic treatment. The more violent, especially if ill-managed or in an abnormal constitution, may be protracted for three, four, or occasionally five or six weeks. We should distinguish, however, between the true periodical disease and its *sequelæ*. The average duration of a case of remittent fever may be stated as about fourteen days.

Complications.—These are usually dependent on *local inflammations*. The *brain* is perhaps the organ most frequently affected, with cerebritis or meningitis. In late autumn, or other cool weather, *pneumonia* is not uncommon. *Gastritis* and *enteritis,*—diarrhœa and dysentery of an obstinate character sometimes occur. When any of these affections exist, they partake to some extent of the periodical character of the fever; and are often lessened or removed by the treatment adapted to it. In other cases, however, they remain in a subacute or chronic form; and, when death occurs, in a majority of instances the immediate cause is a violent phlegmasia of some organ. *Hepatitis* and *splenitis* are more common in the chronic form than in the acute,—and as sequelæ rather than complications of the attack.

The Typhoid State.—At any time after the fourth or fifth day, but particularly near the end of the second week, a patient suffering with remittent fever may pass into the condition designated by the above term. Its features vary somewhat; but it is usually marked as follows: Pulse 120 to 140, and rather deficient in strength; skin harsh, varying, however, with the slight remissions in dryness and temperature; face dark or flushed; head hot; delirium, active more frequently than comatose; bowels occasionally affected with diarrhœa, but as

often costive ; tongue heavily coated with sordes, brown or black, and
with cracks or fissures across it. Muscular debility is usually great.
Hemorrhages from the bowels, lungs, or stomach occasionally increase
the danger.

The chief causes of this condition are, 1. Neglect of treatment in
the early stage. 2. The premature and improper use of stimulants,
or even, in some cases, of tonics, without previous depletion. 3. The
existence of cerebral inflammation. 4. In the opinion of some, a particular epidemic tendency to the typhous condition, in all diseases, at
certain times. 5. In the view of others, the concurrent existence of
a true continued fever, making a sort of hybrid.

Modes of Termination.—These are, either, 1. Recovery in a week
or two directly from the febrile state ; 2. Conversion into a distinctly
intermitting fever ; 3. Cessation of the fever with remaining organic
inflammation or other disease ; or, 4. Death during the progress of
the fever.

The first of these occurs sometimes even when the onset has been
violent, and the circulation much disturbed. How frequent its spontaneous occurrence *might* be, is difficult to determine under ordinary
practice in miasmatic regions ; where the first intermission is made
use of to introduce anti-periodic remedies. But it appears that
remittent fever may much oftener be cured by antiphlogistic treatment alone, than intermittent.

The rule, however, with many historians of the disease is, to consider
that favorable cases terminate in curable intermittent. This change
is generally accompanied by a discharge from some excretory organ
or surface, with propriety termed critical. Copious perspiration ; free
urination, with a *lateritious* or other thick deposit ; the discharge of
abundant, dark, offensive stools ; one or all of these, may precede or
accompany the commencing improvement of symptoms.

A local inflammation, as pneumonia, enteritis, cerebritis, or hepatitis,
may survive the attack which kindled it, and proceed as if it had been
an original malady.

Where death occurs within the first three weeks, it is almost always
the result of some inflammatory complication. Remittent fever rarely
proceeds to a fatal termination, *in this vicinity*, by mere exhaustion
of the powers of nature. In feeble or aged persons, however, this
may occur.

Sequelæ.—A slow and imperfect convalescence not unfrequently
follows a violent attack ; attended with sallowness of the skin, feeble
digestion, muscular and nervous debility. The only organic alterations at all constant are enlargements of the liver and spleen.

Morbid Anatomy.—The most striking observation upon this was
that made at the Pennsylvania Hospital by Dr. T. Stewardson, in
1841, of the unusual color of the liver ; bronzed without and olive-
green within. Subsequent confirmation of this has been afforded ;
although Dr. Drake of Cincinnati failed to find it in his autopsies.
The spleen is almost always enlarged, congested, and softened. Inflammation of different organs (making fatal complications), especially the brain, lungs, or bowels, may exhibit their usual results. Such
lesions, however, are sometimes absent in the most malignant cases.

Causation.—This has been considered already, under the head of intermittent fever.

Diagnosis.—Yellow fever has by some physicians been regarded as identical with remittent, differing mainly in the grade of its violence. The correct view is, that they are specifically distinct diseases   To prove this, we might be satisfied with the simple facts of the different localization of the two fevers. Remittent is always a country fever; yellow fever almost invariably, a disease of towns and the vicinity of the sea. The latter is restricted much more narrowly, also, in its actual geographical limits.

But there are symptomatic differences also ; which may be best pointed out after giving a description of yellow fever. Among the important points is, that one attack of the latter disease commonly gives immunity from it for life; but this is not at all the case with remittent fever.

When the typhoid state supervenes, there may exist very considerable similarity to the true typhoid fever. It is asserted that a coexistence of the two diseases occurs. Some, upon the same facts, ground the opinion that they are not specifically different;—but that *typhoid fever is merely* a protracted remittent of low form. This is however contradicted clearly by at least two facts; 1, the comparative rarity of typhoid fever in regions where remittent most abounds ; and 2, the frequent prevalence of the typhoid where remittent fever is almost unknown ; as in some of the Eastern States.

The mode of onset in the two, moreover, is usually quite different ; in typhoid, insidious and almost imperceptible at first ; in bilious fever, after a day or two of malaise, a chill abruptly ushers in the attack. Vomiting is extremely common in the one,—quite rare in the other ; the converse is true of diarrhœa,—and still more particularly of tympanites and abdominal tenderness. The *deafness,* and *sleeping* stupor,—and *livid* countenance of typhoid fever, are almost entirely peculiar. Epistaxis, bronchitis, and the rose-colored eruption, so nearly constant in the latter, are rare in the typhoid remittent ; the last mentioned is perhaps never observed. The yellowness of the skin, also, and the *distinct* remissions, mark well the remittent attack. In dissection, we find more *gastric* and *hepatic* change after bilious fever, and more *enteric* and *splenic* alteration in the typhoid.

Prognosis.—Recovery may be anticipated in a majority of instances. The writer is of the opinion that the typhoid prolongation of the attack ought scarcely ever to occur, under proper treatment from the beginning. Before the use of cinchona, remittent was often quite fatal. Alexander the Great, Emperor Charles V., James I., and Cromwell are said to have died of it.

Favorable signs are, the earlier occurrence and prolongation of the remission, and its becoming more and more complete ; moistening and clearing of the tongue ; copious perspiration ; turbidness of the urine, from increase in the amount of its solids ; tar-like and offensive stools; and the appearance of vesicles about the lips.

Unfavorable, of course, are, the shortening and postponement of the remission, and its indistinctness, dryness, and blackness of the tongue ; retention, or still worse, suppression of urine ; extreme frequency, with weakness of the pulse ; hiccough ; and other important evidences of the victory of disease over the vital functions,—not, however, peculiar to the fatal termination of this disease. The supervention of the usual symptoms of *inflammation of the brain* are always very alarming ; *gastritis* may occasionally threaten to wear out the patient's strength ; and *pneumonia* is attended with more danger when occurring as a complication of fever, than when an original disease.

Treatment.—In sections where it is very prevalent, this disease has been subjected to a variety of experimental practice—pushed, in some States, with a boldness and energy characteristic of border populations. At one time, the early use of large doses of tartar emetic to produce vomiting ; at another, of calomel, administered by the half ounce or ounce ; and, recently, of quinine with corresponding extravagance—have been the methods used, until fairly proved to be needless or improper in violence.

The other extreme, however, of trusting all to nature, would meet with more signal disappointment in this than in many other affections. A decided treatment is imperatively required ; what then are its best and most promising weapons ? It would be interesting and instructive to collate and compare many authorities upon this point; but we will discuss the subject in reference, chiefly, to the experience of our own physicians. It will be proper to state the valuable testimony of Dr. Drake, to the mode of practice which the separate judgment and observation of medical men throughout the great Western Valley now converge upon. "Its fundamental principles," he states, "are—that autumnal fever is the product of a specific cause, and, therefore, consists in a morbid action of a peculiar kind, requiring a specific remedy ; that we possess such an antidote for the intermittent variety of the fever ; and that we have only to abate all the causes and points of difference between the two varieties, to render the sulphate of quinine as efficacious in one as in the other."

No clearer or more correct expression need be demanded, we believe, for the safest and best plan of management of our own fall fevers. Yet many voices may demur at this assertion.

Some will quarrel with the *terms* of the above paragraph ; objecting, and with some force, that we go beyond what is known in proclaiming the specific nature of the cause of these fevers—and still more in awarding the name of *antidote* to the sulphate of quinine or Peruvian bark. But this is a verbal question. We *do* know that miasmatic fevers are *quite peculiar*—in locality—in periodicity, and in other characters ; and we do know that the salts of the alkaloids obtained from cinchona control and arrest them, as few if any other remedies can—and with a power which those salts do not exert over other fevers.

This power, however designated or explained, is now fully acknowledged amongst us ; the danger is, in fact, of its causing us to lose sight of other important points.

In some tropical latitudes, especially, in which venesection is not, comparatively, well borne—and in districts *poisoned* with malignant miasm—it has been proved that quinine is often required in liberal doses —is borne in very large ones—and acts favorably even without any of the depletory and sedative preparation of the system, once thought indispensable. These facts have been fully proved. But the remaining questions to be settled are—is not success greater in remitting fever, even in those regions, if some evacuant (and reducing) treatment at least *accompanies* the use of the great remedy?—and—is not a modified treatment at least in this respect necessary in less malignant cases, and different climate.

The weight of evidence favors the affirmative of both of these questions.

It remains, then, to state in a few words, what is the plan of treatment proper to be adopted.

The physician is rarely called until the febrile condition has fairly set in.

In a person of robust constitution, if the headache be very severe, skin hot and pulse full as well as rapid, venesection will be *safe* at least. But it is much more common now to administer *first* a saline cathartic, and decide on the grade and resistance of the fever by its effect. Epsom salts will be the best when the *stomach* is little disturbed—the Seidlitz powders in repeated doses under contrary circumstances may answer. But many commence the treatment with a dose of calomel or blue pill with rhubarb, to be *followed* by a saline purge. If obstinate vomiting prevail, as will frequently happen, no purgative will suit so admirably as the effervescing solution of the citrate of magnesia. If a second exacerbation occurs with increased violence after a slight remission, in a robust person, *venesection* may be performed; but the amount taken should be *moderate*. The utility in many cases of *leeches or cups* to the nucha, and sometimes to the epigastrium, is undoubted.

As a refrigerant diaphoretic, the citrate of potash solution, with or without effervescence, may be constantly given.

Special treatment may often be called for by the great intractability and distress of stomach. Lime-water or *magnesia* in small doses with ammonia and an aromatic will frequently relieve.

Sinapisms and pediluvia are of course useful adjuvants. *Ice* will answer better to quench thirst than water, where gastric irritability is great; otherwise free dilution by drink is an advantage.

As soon as the violence of systemic excitement has been moderated,—without waiting for its entire subjugation,—if the pulse has *begun to subside,*—lowering for instance from 110 or 120 to 90 or 100, and the headache is less intense,—*the bowels freely moved,*—we may begin with quinine; but it is unnecessary *here* to give large doses generally. Unless where some malignancy is suspected, or the remission is very complete, *a single grain every two hours* will be sufficient at first. Under this, after the reducing measures, we may find the pulse continue to subside,—the skin to moisten, and all the symptoms to improve. At all events in the next remission the dose should be increased to *a grain every* hour,—not, as a general rule, however, awaking the patient from sleep. Two grains every hour for eighteen hours, is the freest administration I have ever seen to be necessary in a case even threatening malignancy. This term, it need hardly be said, is used to express the existence of a state of prostration attended with signs of visceral congestion, increasing dangerously with each paroxysm ; reaction being deficient, as we believe, from an unusual intensity of the morbid cause,—or defect of constitution. Such cases do require a large amount of the special remedy; and such cases are no doubt much more frequent in warmer southern States than here. We have no difficulty in believing in the toleration,—or even the propriety of *considerably* larger doses than are here given ; but there is a limit even there, to go beyond which, is excess. Perhaps we should allow somewhat in the estimate in some remote places, for the immense *adulteration* of valuable drugs which prevails.

After two or three days of constant *" quininization,"*—the amount, usually, may be diminished to six or eight grains distributed through the day. In rather smaller quantities it should be continued even through the period of convalescence.

The treatment of inflammatory or other *complications* must of course superadd modifications appropriate to each. We have named in the above sketch all the *main elements* of the plan which is found successful in such cases as ordinarily occur.

The existence of local inflammations, in a genuine miasmatic case, does not contraindicate the use of quinine. Being lit up by the fever-poison,—and aggravated by its febrile state, the treatment which annuls or removes these will often lower or check the phlegmasia. But this maxim should be applied with caution and some exceptions, in cases particularly of *cerebral inflammation*, or great pulmonic oppression.

In slow convalescence, with sallowness and deranged digestion, the daily administration for a few days of minute doses of blue mass generally proves useful. And, to improve sanguification, as well as lessen the danger of relapse in some form, the *protocarbonate of iron*, in pill with a portion of sulphate of quinine, will make a very valuable termination of the treatment. Arsenic may also sometimes be required.

PERNICIOUS FEVER.

Synonyms.—*Congestive Fever; Malignant Intermittent; Malignant Remittent.*

Symptoms and Course.—Unlike ordinary intermittent, a paroxysm of the pernicious form may commence either in the day or at night. At first, however, in many cases, it begins like the common type of chills and fever, or remittent fever ; after one, two, or three days becoming more alarming.

Then, the skin grows lividly pale, shrunken, and sometimes clammy with cold sweat ; the countenance anxious ; the tongue either pale, furred, or natural ; in the worst cases it is cold. Thirst is intense, with a sense of internal heat. The stomach is excessively irritable, and vomiting very common, of mucus, or a muco-serous or even bloody fluid. The bowels are in most cases loose, the dejections resembling bloody water. The pulse is usually small, weak, and rapid or irregular ; in a few instances corded. The respiration is interrupted and sighing, with a sense of oppression.

Restlessness is common ; but the mental faculties in many cases are clear. There are, however, many others in which the weight of the attack falls on the brain. Then, the early symptoms are drowsiness and hesitation of speech. Stupor marks the depth of the paroxysm. The breathing may be stertorous; or tetanic spasms may occur. The pulse, in the fórmer case, may be slower than in the other form described; but it is still weak, and, even if the head be somewhat warm, the vessels of the neck and temples are not apt to be swollen, and the skin of the body is cold.

Partial, or, it may be, complete reaction in most instances follows after three or four hours of the above symptoms; though death may, instead, take place in the collapse. Again the fever may intermit, or remit;—and, at the same or an earlier hour the next day, another paroxysm occurs. This is more dangerous than the first. If a *third* be allowed to take place, it is generally fatal.

Morbid Anatomy.—*Congestion*, of the brain, liver, spleen, and alimentary mucous membranes, is so prominent an autopsic phenomenon as, with the symptomatic appearances of the same, to have seemed to justify the older and more common name of the disease. We have good reason to believe, however, that the toxæmic impression of malaria, and its effects upon the nerve-centres (either of organic or of animal life), are primary, and the congestion secondary.

Diagnosis.—The *intensity* of the symptoms, and the general prostration, or coma, will distinguish this from ordinary intermittent or remittent. The condition of a severe case is not unlike an attack of epidemic cholera; but the discharges are different; and the locality and season, unless in the presence of that epidemic, will point directly to malarial causation.

As pernicious fever is rare in the latitude of Philadelphia (much more common farther south, especially near the rice plantations of the Southern States), I have seen but two or three cases of it. One of them gave me difficulty in diagnosticating it from apoplectic coma; as it occurred in a lady over sixty years of age. The distinctness of the cold stage at the beginning of the attack, and my knowledge of the patient's history, and the possibility of exposure to malaria, led me to prescribe quinine with some freedom; and the result established the nature of the case, as entire recovery followed.

Prognosis.—Without appropriate treatment, a large majority of cases would be fatal. There are few diseases displaying a greater tendency to death. Under cinchonism, and other proper management, not more than one in eight probably die.

Treatment.—As above implied, quinine is our great reliance in this disease. Larger doses are required, also, than in ordinary intermittent. While opinions differ, the best evidence I can obtain convinces me, that from thirty to sixty grains of quinine in twenty-four hours will do all that the remedy can do; more will be wasteful and dangerous.

But, in *most* cases, other means must be employed, sometimes before quinine can be kept upon the stomach, to promote reaction. External stimulation is foremost among these means. Direct heat may be applied, by hot-water bottles or tins, hot bricks, or bags of hot salt laid along the spine. Thirst should at the same time be quenched by cold water, or, if the sense of heat is great, and vomiting occur, with

24

ice. Mustard plasters may be placed upon the spine, epigastrium, or limbs; or the extremities may be rubbed with brandy and red pepper. The opposite of this plan is preferred by some, upon asserted favorable experience, viz., the pouring or dashing of cold water quickly upon the naked body. Extensive dry cupping along the spine is recommended by others. From what I have seen of the good effects ·of dry cupping along the spine in the collapse of cholera, in one case, I should have some confidence in it for this analogous condition.

Internal stimulation also is demanded under the same circumstances. Most used have been camphor, opium, ether, oil of turpentine, ammonia, and capsicum [F. 163]; besides wine and brandy or whisky. The best testimony is in favor of camphor and opium, with quinine [F. 164], in moderate doses every half hour during the chill, when no comatose symptoms are present. If these exist, oil of turpentine, by the mouth or rectum, has its decided advocates.

Calomel has been largely used in the same cases. My experience in pernicious fever has not afforded me data for an opinion about it; but I respect the evidence of those who think they have seen it to be beneficial. I should give it, in the dose of a grain, every three hours, at least.

Alcoholic stimulants seem to be indicated in the collapse. A tablespoonful of brandy or whisky every half hour or hour until *reaction* occurs would be a suitable average.

After reaction has been established, even imperfectly, and an intermission or remission exists, the "sheet anchor" is quinine. Then, if the stomach bear it, five to ten grains may be given every two or three hours, until cinchonism is fully established. When the quinine is rejected by the stomach, hypodermic injection may be resorted to. Ten grains or more may be introduced at once, in solution in water with sulphuric acid enough to dissolve it perfectly.

In the *cerebral* cases, calomel is particularly appropriate. A blister to the nucha may be recommended in the same case. Purgatives are also apt to be required; and, if the heat of the head be great, iced water may be kept-applied over it, while hot bottles or sinapisms are put in contact with the legs or feet.

When the critical period in pernicious fever has been passed, it will need treatment like an ordinary case of intermittent or remittent, according to the type which it assumes. A modification of this affection, sometimes called "winter fever" in the South, has been already considered under the head of *typhoid pneumonia*.

PROPHYLAXIS OF MALARIAL FEVER.

When avoidance of malarial localities is impossible, during the season of fevers (from July until frost in some parts of the United States, from the first of August at least, in this vicinity), exposure should be especially guarded against at night, and just before and after sunrise and sunset. Residents in such places should have a fire burning to dry the house whenever the weather is damp, whatever the season. Going into a marshy place with an empty stomach is very exposing.

Quinine may be used as a prophylactic. Livingstone and Du Chaillu have tried it in Africa; the former too sparingly to succeed perfectly;

the latter with better results. In the United States army during the
late war it was found useful. Six grains daily is the least amount to
be relied upon.

TYPHO-MALARIAL FEVER.

Trusting that this, having had its origin in the circumstances
of the late war, may be now altogether a matter of history, it yet requires a place among recognized diseases. It was the result of a threefold causation; the elements of which were *malarial influence, crowd-
poison*, and *scorbutic taint*. According to the predominance of one
or the other of these, its character in different cases was determined.
During service in a United States General Hospital, in the summer
and fall of 1862, as well as in the Episcopal Hospital, I saw many
cases of this, called from its local origination the Chickahominy
fever.[1] Of the form in which the malarial element predominated, the
somewhat abrupt commencement, gastric disturbance, and icteroid
skin and tongue, with remissions, for a while at least, tolerably distinct, were prominent features. The lenticular spots of typhoid fever,
and the sudamina and tympanites were often wanting altogether.

A slower onset, less distinct remissions, more cerebral disturbance
and diarrhœa, with epistaxis and bronchitis sometimes, but with both
less constantly than in civil life, marked the predominance of the
typhoid pathogenetic element. Deafness, under my observation, was
less frequent than in civil life, but was sometimes very well marked.
The aspect of the countenance, and the character of the somnolence
and delirium, were precisely the same as in ordinary typhoid fever.

The scorbutic complication was recognizable, in the third group of
cases, by the peculiar mental and bodily prostration which preceded
and followed the disease—the remarkable irritability of the heart, the
state of the gums, tendency to hemorrhage, discolorations and petechiæ, pallid, large and smooth tongue, and extremely protracted convalescence.

Morbid Anatomy.—Most of importance was the intestinal lesion,
similar to that of typhoid or "enteric" fever, though not identical.
The following account of this is from Dr. Woodward.[2]

"In the earlier stages there is little to distinguish the intestinal
lesion from the corresponding process of ordinary enteric fever, except
perhaps the great tendency to the deposit of black pigment in the
enlarged follicles. In the later stages, certain peculiarities are often
distinctive enough to enable the anatomist to recognize typho-malarial
fever by the *post-mortem* appearances alone. The tumefaction in
typho-malarial fever rises very gradually from the surrounding mucous
membrane, and attains a moderate degree of thickness (three to six
lines) on the edges of the ulcer. In this it differs materially from the
ordinary typhoid ulcer, in which the enlarged patch rises abruptly
from the mucous membrane in such a way that the summit is often
larger than the constricted base, giving rise to the comparison made
by Rokitansky, who likens the shape of the tumefaction to that of

---

[1] A full account of all varieties of typho-malarial fever is given in Dr. J. J.
Woodward's "Camp Diseases of the United States Army."
[2] Op. citat., pp. 102-3.

flat, sessile fungi. The umbilicated depression, so frequent in the ordinary typhoid patches prior to ulceration, has never been observed in typho-malarial fever. The ulcer itself presents ragged, irregular edges, which are often extensively undermined in consequence of the erosion extending more widely in the submucous connective tissue than in the glandular tissue of the mucous membrane. This characteristic undermining of the edges is much more extensive in these than in ordinary typhous ulcers."

Pathology.—Doubting not at all the presence of the malarial element, the question occurs, was the modifying "febrile" cause of the *typhous* or of the *typhoid* character? Granting, that is, that these are pathogenetically distinct, we should expect that the typhous or "crowd-poisoning" element must result from the circumstances, as from those which have made typhus or "camp fever" the scourge of armies in Europe. Only, against this, we have the local lesion, of the glands of Peyer and mucous membrane of the bowels, recalling enteric or typhoid fever.

But, as, where typho-malarial fever occurred, causes of intestinal irritation (bad water, deficient food, &c.) were present,—I am not satisfied that such an appearance (not, as we have seen, *identical* with that of typhoid fever) should exclude the idea of the action being that of the *typhous* cause. In that opinion, as a probability, not, of course, now demonstrable, I rest.

Treatment.—From the above view of the hybrid and threefold nature of the disease, came its rational treatment. More *quinine* than in typhus, more *alcohol* than in remittent, more *fresh vegetable food* and fruit than in either. Experience justified this plan. In our hospitals, in Philadelphia, few died from fever who were not moribund on their arrival from the seat of war.

YELLOW FEVER.

Only certain localities have ever been subject to this disease ; and of those, most have had it but occasionally. In Philadelphia. it first appeared in 1699 ; then in 1741, 1747, 1762, 1793, 1819, 1820, 1853, 1854, and 1855 ; the last visitations being to a very moderate extent. The worst epidemic at New Orleans, where it has been frequent (almost annual), was in 1853. Sanitary measures, under General Butler's military rule, in 1862, appeared to avert it, in that city, under circumstances which might have been expected to promote it. A very severe visitation of yellow fever occurred at Norfolk and Portsmouth, Virginia, in 1855.

All the places which it has ever visited are upon the borders of the Atlantic Ocean, or its tributary waters, the Gulf of Mexico and the Mediterranean Sea. Although with like climatic conditions, it is common in the West Indies and west Africa, but unknown in the East Indies, the eastern shore of Africa, and the Pacific coast of America.

Symptoms and Course.—With an abrupt beginning, or an indistinct cold stage, with pains in the back or limbs, commencing often in the night, a febrile stage occurs, of long average duration ; sometimes three days without remission. Violent cases have it shorter ; sometimes lasting only a few hours.

The skin, at this period, is hot and dry. Thirst is extreme; the tongue is generally furred. Nausea and vomiting are common on the second day, with great epigastric tenderness. The bowels are costive; if discharges occur they are very offensive.

A flush of the forehead, with a fiery look of the eyes, is character-istic. Delirium is frequently present. Violent headache is nearly universal.

The stage which follows this pyrexia is a sort of remission or in-termission. All the symptoms abate except the epigastric ten-derness. The flush of the face and other portions of the skin is succeeded by yellowness, which grows deeper as the disease advances. The pulse grows slower, heat abates, respiration becomes natural in frequency, the patient sits up and feels better. This state of things lasts for a variable time, averaging about twelve hours.

Sometimes convalescence now takes place. Much more often a third stage succeeds, of prostration or collapse. Muscular debility becomes great; the pulse is rapid, irregular, and compressible; the capillary circulation sluggish; the skin deep yellow or bronzed; the tongue brown; the stomach excessively irritable. It is at this time that the *black vomit* occurs, which is pathognomonic of this fever. Hemorrhages may also occur from the mouth, throat, or bowels. The mind grows apathetic, or low muttering delirium exists. In bad cases, which are many, hiccough, clammy sweats, convulsions, and involuntary discharges precede dissolution. Death most frequently occurs on the fourth, fifth, or sixth day.

When reaction from the collapse takes place, there follows a second-ary fever, of very variable duration, and which may terminate in a tedious convalescence, an almost equally prolonged typhoid condition, or death by exhaustion.

Black Vomit.—This has been found, upon chemical and micro-scopical examination, to consist essentially of blood, altered by action of the fluids of the stomach. It is usually acid to test-paper.

The *urine*, in yellow fever, is scanty and high colored at the begin-ning, and especially deficient in amount from the third to the fifth day. About the fourth day, it becomes cloudy and deposits a sedi-ment. Granular tube-casts from the kidneys may be discovered in it on the fifth day. Blood and bile may also appear in it; as well as large amounts of creatin or creatinin. Urea is apt to be below the normal quantity.

Morbid Anatomy.—Congestion of the brain is not uncommon; inflammation of the stomach is usual. The liver is most frequently dry, pale yellow, and anemic; but occasionally it is engorged. Fatty accumulation in the liver has been repeatedly observed; and exudation into it is asserted. The spleen is little altered; the kidneys are always congested.

Diagnosis.—The only doubt likely to be entertained is as to its identity (or that of an example of it) with bilious remittent fever. As already remarked, the latter is a disease of the country, in any warm quarter of the globe. Yellow fever is restricted geographically, and is but seldom met with except in towns and near the sea. The order of stages in the two diseases is different; remittent never has a pyrexia lasting over twenty-four hours without mitigation. There is more epigastric tenderness in yellow fever. The jaundiced hue of the skin is more commonly met with, and more decided, in that disease. The black vomit, when it occurs, is decisive. Possibly, even probably, in a few localities, the combined causation of the two fevers may produce hybridity between them. Immunity for a lifetime after one attack is common with yellow fever; not at all with remittent.

Prognosis.—This is a very dangerous disease; the deaths from it averaging about one for three cases. A long and moderate febrile paroxysm, without excessive irritation of the stomach, is favorable. So is the occurrence of secondary fever instead of collapse, after the remission. Black vomit is almost always a fatal sign. Some instances of the disease are called *walking* cases, because their early symptoms are slight, only the countenance and pulse betraying the danger, until near the end.

Pathology and Causation.—There seems no room to doubt that yellow fever is a zymotic disease, whose cause is generated by certain local conditions. That cause must be itself material; and, *probably*, being slow and limited in transportation, it is a microscopic vegetation.

The local conditions observed are, 1. Continued high heat; about 80° for one or two months. 2. Excess of moisture in the air; a high dew-point. 3. Vicinity to the sea, or to a large river emptying into the sea. 4. Organic, especially vegetable, matter in a state of decomposition. This is furnished not only by the offal, etc., of cities, but by decaying wharves and causeways (as at Norfolk, Va.), and by newly upturned earth. Mobile was almost relieved of danger from yellow fever by paving the city with oyster shells.

But it is very remarkable, as already stated, that certain quarters, only, of the globe ever have this disease, though presenting all the above conditions. It *never* visits the Pacific coasts either of America[1] or Asia. Canton, Calcutta, Bombay, Alexandria, Constantinople, and Athens, have never seen it. Nor have any of the interior cities of either continent. It visits *often* the western coast of Africa, the tropical islands of the Atlantic, the north coast of South America, Vera Cruz, the West India islands, New Orleans, Savannah, Charleston; *occasionally*, Rio Janeiro, Natchez, Mobile, and other United States ports and cities, as far north as Boston and Providence; also, Gibraltar, Marseilles, and other places on the Mediterranean, as far as Sicily.

The *contagiousness* of yellow fever, from person to person, is disproved by the immense preponderance of facts incompatible with it. A very few apparent examples of transmission by individuals, if admitted to have occurred, are otherwise explained. *Transportation by ships* is admitted, because a ship may carry a section, as it were, of a locality, with all its conditions and atmosphere. But, then, the port to which the ship goes must have all the conditions rife for the propagation of the disease, or its "germs" will not be maintained so as to cause an epidemic.[2] More will be said of this, under the head of *Prophylaxis.*

---

[1] A single instance of its occurrence on ship-board off the west coast of South America has been asserted. If so, no doubt it was a case of limited *transportation.*

[2] On the whole subject of yellow fever, see La Roche's exhaustive treatise.

Treatment.—No specific has been found for yellow fever, and no abortive treatment. All kinds of remedies have been tried for this in vain; especially bleeding, calomel, and quinine. I say in vain as to cutting it short; but in palliating and conducting it through its stages with safety, those and other remedies may be of use. Bleeding is suggested by the relief often attending spontaneous hemorrhages in its course; but, as in other malignant affections, the cases for it must be well selected, the time early, and the amount moderate. Much the greatest number will gain only by the application of leeches or cups to the epigastrium or back of the neck.

Many authorities approve of the use of calomel as a cholagogue cathartic, at least in a single dose (say of three or five grains), followed by a saline laxative, as citrate of magnesia, near the beginning of the attack. All the result of the use of *quinine* of which I feel sure is, that it is not likely to do good at any early stage, but only when prostration begins to appear; and then in tonic or supporting, not *cinchonizing* doses. It is undoubtedly of service during convalescence.

Attention to the *stomach* is demanded by urgent symptoms. Ice, by the mouth, is refreshing and useful. So is mineral water, or iced champagne, a little and often; lime-water, charcoal water, and hot coffee have sometimes done service in arresting vomiting. A mustard or spice plaster over the epigastrium, or a blister dressed with acetate of morphia, may have an important effect upon the same symptom.

During the hot stage, cold sponging to the face, body, and limbs, will sometimes promote perspiration better than any other measure. Enemata of cold water (with care not to chill too powerfully) has been used for the same end.

In the collapse, stimulation will be needed, by wine, brandy, or whisky, etc.; along with concentrated liquid food, in small amounts at short intervals.

The experiments with anti-septic and anti-zymotic substances, as chlorine and the sulphites, made with other affections analogous to yellow fever, might be properly tried with it also. I am not acquainted with any such trials as yet.

Prophylaxis.—Besides what has been said, the following statements will indicate the principles of prevention of this disorder:—

1. The infection of yellow fever is rarely diffused over regions of great extent; mostly its limits may be measured by fractions of a square mile.

2. The removal of the inhabitants of an infected spot will inevitably put an end to an endemic or epidemic of it.

3. Sanitary police may effectually prevent it (as at New Orleans under General Butler), and will mitigate it even after its outbreak.

4. The material cause of yellow fever is never generated or multiplied in the bodies of those having the disease; they may be taken anywhere without fear of communicating it, any more than well persons.

5. The germs of the disease are extremely seldom, if ever, transported by *fomites; i. e.*, clothing, bedding, merchandise, etc. If it exist in any such material, it is certainly destructible by cleansing and disinfecting measures.

6. A ship may carry yellow fever on board of it for a length of time

(during warm weather) and to a great distance; but the disease will not spread far from the ship; at least unless favorable local conditions add their propagating influence.

7. Thorough airing, cleansing, and disinfection of ships (especially by dry heat or superheated steam) will always deprive them of the power to generate or transport yellow fever.

8. Against yellow fever, the true prophylactic method must be that of sanitary police; a part of which should be, the inspection, near ports liable to it, of all vessels arriving during warm weather.

9. At the place of such inspection, all foul vessels should be detained until cleansed, being first emptied of their passengers and cargo. The cargo also should be inspected, and, if unwholesome, destroyed or disinfected.

10. No *personal* detention whatever, other than of those ill, for necessary hospital treatment, should be imposed upon the passengers or crew of vessels which have yellow fever on board. There are no facts which give reason for any such detention.

RELAPSING FEVER.

Never yet met with in the United States, unless a few times in emigrants, but little need here be said of this disease. By excellent British authority, it is ascribed directly to famine; sometimes being called " famine fever."

After an indistinct premonitory stage of a few days, or else abruptly, the attack begins with a cold stage, often having decided rigors; a chill. There is then fever, with pain in the stomach, vomiting, and prostration; sometimes petechial spots over the body. On the fifth or seventh day great amendment appears; this seems to progress favorably, until, about the fourteenth day from the beginning, a relapse occurs, with fever and prostration. This again gives way generally in three or four days, and is followed in a majority of cases by convalescence. The mortality in Great Britain is estimated at one in forty.

Treatment.—No specific being known, mild palliative management is called for. This may be described, in the words of a late author, as consisting of " gentle aperients, refrigerating drinks, liquid diet, perfect repose," and quinine, in tonic doses, from the middle of the attack, though with no hope of its acting to prevent the relapse. Other palliatives will be suggested by particular symptoms in the course of the disease.

CEREBRO-SPINAL FEVER.

Synonyms.—*Cerebro-spinal Meningitis; Spotted Fever.* The name adopted above is preferred by me, in the absence of sufficient preponderance of authority or reason in favor of either of the other names. The disease is a *fever*, or systemic disorder; not a mere local phlegmasia. It has no more claim to be called cerebro-spinal meningitis than typhoid fever has to be called *enteritis*, or scarlet fever *faucio-pharyngitis*. Yet the term spotted (or petechial, Wood) fever is not fully justified as distinctive—because, only in a minority of cases it exhibits any eruption, and something like the same is also at times seen in typhus.

History.—Often obscurely described, this disease appears to have been known in France in 1310 and 1482; over Europe, or parts of it, in 1503 '10, '16, '17, '28, '45, '59, (Sicily) '64, '68, (Paris) '69—'74. In 1580, it was at Rome, Venice, and Madrid, with great mortality; again over Europe in 1582; at Trent, 1591, Florence, 1592, at various places, 1616 and 1624. Sydenham described it in 1661. In 1691 and '93 it was in Italy; and in England, 1698, 1710, 1741; in Prussia, 1704. Other years named for it are 1720, '60, '61; 1757 and 1778. A well-known outbreak of it occurred at Geneva, 1805; one in the Prussian army, 1806, '7; in Sicily, 1808; at Dantzic, 1811; Brest and Mayence, 1813, '14; Grenoble, 1814, and the same year at Paris; 1815 at Metz; elsewhere in 1816 and 1823. Afterwards in Europe its historians (under the names *méningite cérébro-spinale épidemique, cerebral typhus*, and *tifo apoplettico tetanico*) speak of it in 1832, '37, '39, '40, and almost every year till 1850, extending over many places in succession as far as Gibraltar and Algiers at the south, and Scotland and Ireland at the north. From 1854 to 1861, in Sweden, Norway, and Holland. In north Germany and Russia, it is said to have prevailed in 1863, '4, '5.

In the United States, its first recorded visitation was in 1806, in Massachusetts. Then it gradually spread through the New England States, New York, and Canada, from 1807 to 1812, when it had reached Philadelphia. After that it was met with at various places until 1820; but not with great frequency. Between 1840 and 1850 it was epidemic in several of the Middle, Western, and Southern States (Kentucky, Indiana, Illinois, Michigan, Missouri, Tennessee), also in 1852 and 1858. Next we hear of it in 1862-3; most clearly in the descriptions of Dr. Gerhard in the latter year, as it occurred in the neighborhood of Philadelphia. Since that time (at which cases were seen especially at Frankford, Falls of Schuylkill, Manayunk, and Norristown, but only a few in the city) it has been observed in a number of places in Pennsylvania, New York, Ohio, Indiana, Michigan, Missouri, Rhode Island, Vermont, Massachusetts, Maryland, and the District of Columbia.

Symptoms and Course.—The attack is nearly always sudden. Chilliness, terrible pain in the head, extending to the back of the neck, nausea and vomiting, are the earliest symptoms. Delirium follows; ending not unfrequently in coma. Tetanic spasm or rigidity of the muscles of the back of the neck (and sometimes of the back and limbs), is common. Convulsions are much less so, but do occur, particularly in the young. Painful sensitiveness (hyperæsthesia) of the whole surface of the body is present in most cases, where there is no coma. Loss of sight and hearing may take place during the middle period of the attack. The pulse is at first slow, then accelerated, but diminished in volume and strength. Respiration is slower than natural in most, but not in all cases. The tongue is usually at first white and moist; sometimes natural; in prolonged cases it may become yellow or brown. The bowels are costive or natural.

The skin has almost always at the beginning an abnormally low temperature. When reaction occurs it does not become very hot as a rule. Dryness of the surface is most common, although late in the attack profuse perspiration may occur.

In a minority of the cases, though varying in proportion in different epidemics, *spots* (petechiæ) appear, on the second or third day, or later; on the neck, breast, or limbs; seldom on the face. They are of different dimensions, from the size of a pin's head to three-quarters of an inch in diameter, and distinct; but not elevated nor disappearing on pressure. Their color is red, purple, or black. Sometimes they remain after death. They are either congested portions of the skin, or subcutaneous extravasations of blood.

The *duration* of fatal cases of this disease is generally short. Some die in three or four hours; many within twelve or twenty-four. That much time overpassed, the danger becomes less, but a fatal result may still occur, even after a number of days. The first four days are the most perilous to life.

Morbid Anatomy.—The blood, during life, is found to have an excessive proportionate amount of fibrin and corpuscles. After death, where it has taken place on the first or second day, no anatomical changes, even in the brain, have, in several instances, been found. Most generally, however, the brain and spinal cord show some alteration. It is the *pia mater* especially in which congestion, at least, is nearly always present. At the base of the brain, most of all, is this, often with serous and plastic exudation, observed. The surface of the hemispheres may also be diseased; and, next in frequency, the pia mater of the cervical portion of the cord. The ventricles of the brain have usually an excess of fluid in them; serum, either clear or mingled with blood or pus. The substance of the brain is more or less injected or congested; the spinal cord occasionally so. Softening of the brain is reported in protracted cases.

No other lesion or appearance is shown to be usual in this disease. A few observers record the presence of rather firm fibrinous clots in the heart.

Diagnosis.—From typhus fever, this is known by the suddenness of its onset, the early period of danger, and, in favorable cases, the rapid recovery; as well as by the peculiarity of the eruption. From ordinary inflammation of the brain, while the diagnosis may be very difficult, it differs in the unexplained abrupt attack, severe from the start; in the lowness of temperature during the first day or two; in the early tetanic tendency; and the eruption in many cases. Malignant scarlet fever resembles it considerably at the onset; and so does the chill of pernicious intermittent. Locality and season will designate the latter; age and exposure, especially, the former. Fortunately, the principle of treatment is not essentially different in these affections, at the stage which may present a doubt.

Prognosis.—More than half the cases die. Those who survive three days, have a fair, though not certain, prospect of recovery.

Causation.—Of either sex, more children, and of adults, more males, die of this disease. Coincident with the circumstances of war, or military *régime*, most of its epidemics have been, though not all of them. The analogy which it presents to typhus suggests a probable relation of the disease in causation to local or atmospheric contamination. I can think of only one plausible hypothesis; that it depends upon a peculiar zymotic material, or "morbid poison," generated by a slow change in human or animal emanations, such as, in camp or garrison life, the long un-washed clothing of soldiers may particularly engender.

There is no proof whatever of personal contagiousness in cerebro-spinal fever.

Treatment.—We must lament the unsatisfactory condition of the evidence upon this subject. Almost all agree that *asthenia* characterizes the disease, most of all at the beginning. The resemblance to pernicious fever has suggested the use of quinine. And several very positive statements of success are made, with it in large doses; as, two to four grains every hour or half-hour until cinchonism is produced, or until from thirty to sixty grains have been taken; afterwards, a grain or two every two or three hours. Some other practitioners, upon trial, have abjured quinine altogether in this disease. Were my diagnosis sure in every case from the commencement, I should feel compelled to renew the trial of it, from what has been reported, in this city particularly, of its success.

Opium has equally enthusiastic advocates and opposers. *Early*, if it be given, must be the time. The idea of those who urge it is, to give of it a grain every two or three hours, until an *opium sleep* is produced; then withdraw it or give it in much less doses.

Stimulation with brandy or whisky is generally employed in the first stage, with freedom. External stimulation is also, of course, indicated by mustard, direct heat, friction with red pepper and brandy, or hot whisky and salt, &c. Dry cupping, or in some cases cut cups (when reaction occurs) to the back of the neck, will be proper; followed by a blister at the same place.

Cantharides (20 to 40 drops of the tincture, every hour till reaction), camphor, chloroform, sulphite of soda, and bromide of potassium, have each had laudation from some who have used them. But more positive experience is needed to give the profession much confidence in the treatment of this affection.[1]

TYPHUS FEVER.

Synonyms.—*Ship Fever; Camp Fever; Jail Fever.*

Symptoms and Course.—For a day or two, premonitory weakness, headache, and loss of appetite occur. Then a cold stage, of variable distinctness, begins the attack. In rare instances, it is said that death takes place in this, without reaction. Much more commonly, fever follows; with severe headache, great heat of skin, pulse 120 (110 to 130), but compressible, tongue whitish or yellowish, bowels costive. Delirium is common, especially at night. The temperature in the axilla is from 102° to 108°; generally, after the third day, 105°-6° in the morning, 106°-7° in the evening. Muscular debility is very decided.

---

[1] See J. S. Jewell, M.D., **Report on Cerebro-Spinal Meningitis**, Chicago, 1866; and J. J. Levick, M.D., "Report on Spotted Fever, so called," Trans. Am. Med. Assoc. 1866.

For a number of days this condition lasts; the patient lying in a stupid half-sleep much of the time, muttering to himself, easily roused, but soon lapsing again; the face having a dusky flush of redness. Hardness of hearing is present in most cases. Positive coma is a very bad prognostic, but is not infrequent. Suppression of urine may take place in the worst cases; retention occurs in many severe ones. The tongue grows darker as the attack progresses; brown, even black; often cracked or fissured; and it as well as the teeth may be covered with sordes.

Towards the end of the first week, in most cases, a rash appears, of little and numerous red papulæ (miliary eruption), all over the chest, abdomen, and upper parts of the limbs. They are accompanied by *sudamina* (minute vesicles) in many instances, by *petcchiæ* in a few. Sometimes a strong odor comes from the body; but I have never noticed this, even in the cases of ship-fever from Ireland in the Pennsylvania Hospital, in 1847-8, at which time I took the disease myself from them.

The urine is scanty. Generally it contains an excess of urea and uric acid, with a deficiency of the chlorides. Sometimes there is actually less than the normal amount of urea eliminated; when excreta may be supposed to accumulate in the blood, promoting coma. *Costiveness* is the general rule in typhus.

The *dicrotous* or double pulse, and *subsultus* or twitching of the tendons at the wrist, are common. Weakness of the impulse of the heart is often noticeable; sometimes so much so as to justify Dr. Stokes' diagnosis of "typhous softening." *Hypostatic* pneumonia (*i. e.*, beginning with passive congestion of the lungs posteriorly), is the most frequent complication of the fever.

The *duration* of an attack of typhus is generally three weeks. The critical period is usually about the eleventh day; after which *defervescence* (the decline of the fever) may be looked for. Occasionally death may take place within five days, or recovery within fifteen, from the commencement.

Morbid Anatomy.—Absence of lesion of the solids has been repeatedly noticed. The blood is always altered during life; after the early stage, it is less coagulable and darker in color than in health. Passive congestion in various organs is observed, as in the lungs, brain, liver, &c., but without anything characteristic.

Pathology and Causation.—No disease affords more reason for pronouncing it a disease of the blood than typhus. Its cause, demonstrably in many cases, is *ochlesis* or crowd poison; the effluvia from human bodies, accumulated, especially in cold weather, in small and ill-built dwellings of the poor, and most of all in filthy towns, ships, jails, or camps. Having once been thus generated, it becomes contagious; one patient having, in his morbid emanations, the poisoning power of a whole crowd. Yet the contagion is not very strong; many who are exposed often escaping the disease.

Diagnosis.—After the first two or three days (during which there may well be doubt as to its character) the only probable question will be between typhus and typhoid fever. All medical authorities are not yet agreed as to the non-identity of the two forms of slow continued fever.[1] A large majority, however, regard them as quite distinguishable during life, and separated pathologically by the absence in typhus of the morbid alterations of Peyer's glands, and those of the mesentery, characteristic of typhoid fever. I have many times seen typhus and typhoid cases in the same ward, lying side by side, and should feel confident of being generally able to point out which was which by the countenance alone. Under the head of *Typhoid Fever*, the clinical differences will be enumerated.

Prognosis.—Murchison states the mortality in the hospitals of Great Britain, from typhus, to be one death in five cases. Cheyne and others in private practice have found it but one in twenty or more. I have not seen many deaths from it, in private or hospital practice. Probably one in ten or fifteen would be a fair general estimate. Bad signs are, great feebleness or extreme rapidity of the pulse; profound coma; hiccough; suppression of the urine; involuntary defecation. Pneumonia complicating the attack increases its danger, though I have known several recoveries notwithstanding this.

Treatment.—More than half the cases of typhus, according to my observation, require alcoholic stimulation, as well as concentrated nourishment, after the fourth day. But not all the cases; as my own, among others, proved. I was bled on the second day, the diagnosis not being made out; and leeched on the third day, freely, on the back of the neck; yet no stimulus was required, after the typhous nature of the attack was clearly shown; recovery following at the usual time.

We may begin the treatment of an ordinary case of typhus with a mild laxative,—*e. g.* a *moderate* dose of solution of citrate of magnesia, on the second day. The diet at first may be of gruel, toast-water, etc.; but very soon must milk and beef-tea or chicken or mutton broth (or an alternation of these) be given to support the strength. Before the first week is out, half the cases will need wine in moderation; some, brandy or whisky. In the second and third week, more than half the cases will require steady support of a positive kind. In such cases, the proper routine is, a tablespoonful of brandy or whisky punch (one part of spirit to three, two, or one of milk) every two hours, and, in the alternate hours, a tablespoonful or two of beef essence or beef-tea.

Of medicines, quinine has had the most extended trial in typhus. It acts well as a tonic, in one or two grain doses, four or five times daily, after defervescence has begun; *i. e.*, after the tenth or twelfth

---

[1] Dr. J. Hughes Bennett, for example, still maintains their identity; and some German writers call typhoid "abdominal typhus." The definite history of typhoid or "enteric" fever began with Prost, of Paris, 1804. Louis, 1829, studied it elaborately, showing the constancy of the intestinal lesions. In 1833, Dr. Enoch Hale, of Massachusetts, described two forms of continued fever. Dr. Gerhard, upon careful autopsies in the Philadelphia Hospital, announced evidence of the distinctness of typhus from the typhoid or "dothinenterite" of Louis, in 1835. Dr. A. P. Stewart, of Glasgow, published similar conclusions in the same year. In 1846, Dr. W. Jenner commenced an investigation into the subject, whose results most physicians have accepted as decisive. He concluded that typhus and typhoid fevers are clinically and anatomically distinct, as well as different in causation. Dr. Gairdner has lately recorded cases in which patients convalescent from typhoid fever have taken typhus upon exposure to its contagion.

25

day usually.   Dr. Dundas' plan of treating typhus early with *large*
doses of quinine is, I am satisfied after seeing some trial of it, futile
and even dangerous.
  Mineral acids have acquired much reputation in typhus.   Nitro-
muriatic acid I have known to produce an excellent effect in the
depression of the middle stage.   Large doses are not required; but
the acid should be given several times in the day.   Some prefer dilute
nitric acid [F. 165, 166].   *Chlorine water* is lauded highly by some.
The sulphite of soda may be worthy of trial.
  But the great point of skill will be to determine when and how far
to stimulate.   Delirium favors the probability of its being needed;
especially a low, muttering delirium.   Of course a very feeble pulse
indicates it.   On trial, if the pulse grows slower, the skin more moist,
and the restlessness or delirium is quieted, the stimulus has done
good, and should be continued.   If, on the contrary, a more hurried
or a *harder* pulse follow, with heat of head and dryness of skin, and
wilder delirium or deeper stupor, it should be stopped, for awhile at
least, or if given be diminished in amount.
  Catheterism may be needed for retention of urine.   Inquiry and
inspection should determine every day the state of the bladder.
Constipation, through the attack, may be overcome by enemata, or
by small doses of oil, Rochelle salt, or other mild laxative.
  When the coma is very deep, a blister to the back of the neck may
do good; as well as sinapisms to the extremities.   Great heat of the
head may render proper, especially in the first week, the application
of cold water to the head.   Sponging the whole body daily (best at
night) with whisky and water, warmed, is extremely comforting and
beneficial.
  Hypostatic pneumonia, in typhus, cannot be treated actively.
Even abstraction of blood by cups is hardly ever to be ventured upon.
Dry cups, between the shoulders, and a blister upon the breast, are
about all the special treatment allowable.   It is, however, possible
generally to *prevent* hypostatic pneumonia, by not allowing the patient
ever to lie for many hours together upon his back.   Let him be turned,
once in awhile, upon one or the other side.
  Prophylaxis.—Thorough *ventilation* is the one security against the
generation of typhus fever; and this is capable also of almost disarming
its contagion.

TYPHOID FEVER.

  Synonyms.—*Slow Nervous Fever ; Common Continued Fever ;
Enteric Fever ; Abdominal Typhus* (Pythogenic fever of Murchison).
  Symptoms and Course.—After a more gradual approach than
that of any other fever, with languor and debility, anorexia and headache, for several days,—bleeding at the nose, and a bronchial cough
are almost pathognomonic early symptoms.   The patient takes to
bed, with fever of considerable violence.   The face acquires a dark
purple flush.   He lies dozing. perhaps muttering, unless disturbed, all
day ; but is more or less wakeful and delirious at night.   Hardness of
hearing is common from the middle of the second week.   Swelling of
the belly (tympanites) comes on towards the end of the first week ;

diarrhœa about the same time. Rose-colored lenticular spots (*taches rouges*), disappearing on pressure, are discoverable, few in number, and on the abdomen only, toward the end of the second week; they continue a week or two. Tenderness on pressure in the right iliac region, with gurgling under the hand, generally exists. Sudamina over the chest are not unusual. The duration of the typhoid pyrexia is seldom, from the start, much less than two weeks, and it is often more; the whole attack of typhoid fever may be protracted, as I have seen it, to two or three months. One month may be considered the average time, from taking to bed to leaving it convalescent.

Late symptoms in severe cases are, the dicrotous pulse, subsultus tendinum, retention (perhaps suppression) of urine, hemorrhage from the bowels; and, if death be imminent, hiccough, cold sweats, involuntary discharges.

In protracted cases, great emaciation and bed-sores may supervene. Even during convalescence, abscesses in various parts of the body may give trouble. These usually affect the glands or connective tissue, but may occasionally involve the long bones.

Danger of perforation of the intestine, from deep ulceration of the glands of Peyer, exists always after the first week, until late in convalescence. Patients out of bed for a week or two have sometimes died, after imprudence, from this cause. The occurrence of perforation is recognized by symptoms of severe peritonitis, with collapse. The result of this is almost inevitably fatal; the only recorded exception being reported by Prof. G. B. Wood. I saw a case of suppurative peritonitis, opening externally, which recovered, in the Philadelphia Hospital, several years ago; but I was not able to learn the antecedents of the case.

Temperature.—This has, of late, been made a special study in typhoid fever. The rise from 98.5° (the normal degree) is gradual, during the first four or five days; reaching 104° on the evening of the latter; sometimes 104.5°. An attack of disease in which on the second day the heat in the axilla is as high as 104°, is not typhoid fever; and the same exclusion applies if from the fourth to the eleventh day the temperature falls below 103°. A difference of 1° or 1.5° between morning and evening (greatest heat, the latter) is usual; the reverse is not a good sign. Toward the end of the second week, lowering of the heat below 103° is always favorable; persistence at 104°, 105° or 106°, shows a severe case; the higher the worse. Sudden increase of temperature indicates a complicating inflammation; as pneumonia.

Discharges.—Liquidity of the stools is a characteristic of this disease, even if there be but one daily. Generally, after the middle of the first week, there are two or three passages, brownish with a slight yellowish tinge every day. From the very beginning of the attack, the bowels are unusally susceptible to the action of purgatives; a teaspoonful of castor oil operating readily. Excessive diarrhœa, at a middle or late stage, not unfrequently adds to the prostration of the patient. Hemorrhage from the bowels, when it occurs, is most apt to be met with in the second or third week.

The *urine*, through the attack, is commonly scanty, high colored, excessive in the amount of urea, deficient in the chlorides, and sometimes albuminous in severe cases.

Complications.—*Pneumonia,* especially the hypostatic form (as in typhus) is the most frequent. It has been, by some writers, denied that true pneumonitis, anything more than passive congestion, occurs in these cases. But, in the analogous instance of typho-malarial fever, especially when the scorbutic diathesis was also present, I have seen, after death, more than once, suppuration, as well as hepatization, confined altogether to the posterior portions of both lungs. I do not doubt the same happening in typhoid as well as in typhus fever.

Inflammation of the brain may complicate typhoid, more often than typhus; but still it is not common.

Peritonitis follows always when perforation of the ileum takes place. Examples of its occurrence without that accident are said to have been, though very rarely, observed.

Sequelæ.—Prolonged debility, or a very slow convalescence, is common. The mental faculties are sometimes enfeebled for weeks or months. Paralysis is an occasional sequela. Abscesses have been mentioned. Periostitis, followed by necrosis, of the tibia, femur, or humerus, may happen, though I have known of but two such cases. Perforation of the bowel may, as already stated, occur after convalescence has seemed to be established.

Morbid Anatomy.—Omitting variable and unessential or occasional appearances, the parts characteristically affected in typhoid fever are, the agminated glands or patches of Peyer in the small intestine, the mesenteric glands, and the spleen. Careful study of Peyer's glands, by many observers, has shown that, at first, the glands thicken and become elevated from one to three lines above the membrane around them. They are generally at this time reddened; but with variable depth of hue. Sometimes, after this, a sort of induration occurs; in other instances, softening. Later, ulceration affects many, though not all, of the altered glands; and this process may go on until, as above said, it may perforate all the coats of the intestine. This, however, is exceptional. The healing of the ulcers by granulation is the general rule.

The *solitary closed glands* of the small intestine are also commonly enlarged, and often softened or ulcerated. The *mesenteric* glands are almost uniformly enlarged, congested, and softened; occasionally they suppurate.

The *muscles,* especially the *recti abdominis,* in protracted cases, have been shown to undergo a granular, or sometimes a waxy or amyloid degeneration; resulting, in the rectus, occasionally, in rupture of its fibres.

Pathology.—Typhoid fever is believed by most authorities to be a general or systemic disorder, with a characteristic secondary local lesion in the intestines. How far the matter deposited in the patches of Peyer before ulceration is *specific,* is a question. Rokitansky and Carl Wedl believe it to be peculiar,—the former comparing it to that of encephaloid cancer, the latter to tubercle. I do not believe that there is anything properly to be called *specific* in its nature.

Dr. G. B. Wood holds the opinion that an inherent predisposition to the disease exists in many persons, analogous to the tuberculous, gouty, and rheumatic diatheses. This seems to me very probable.

Another view is, that the affection of the intestine is primary; and that the "typhoid" symptoms result from the absorption into the blood of morbid, putrescent material from the glands of Peyer, producing a *septæmia* or *ichoræmia*. This does not appear to me to be a reasonable hypothesis, in view of the order of events in the disease.

Causation.—More doubt exists as to this in typhoid fever than in any other common disorder. Depressing causes of all kinds seem to promote it; foul air, removal from home, fatigue, anxiety, &c. Yet it will occur in the entire absence of all such causes. No locality limits it; all climates allow it; from the Arctic regions to those bordering upon the tropical; from the cities of the East to the Rocky Mountains. The "mountain fever" of hunters in the far West was found in the autopsies of Dr. Hammond to present the lesions of Peyer's and the mesenteric glands.

Such universality is very much in the way of the "pythogenic" theory of Murchison (*i. e.* its reference always to foul air, as that of sewers), or that of Budd, that its only cause is a specific matter, passed from the bowels of those having it, and, by water or air, conveyed into the systems of others.

Contagion of this kind is, nevertheless, widely believed in now, especially in England. Some facts asserted in proof of it are hard to explain without admitting such a mode of propagation (*e. g.*, by the discharges of a patient getting into a well, &c., so as to contaminate drinking water). But the large majority of cases allow of no such explanation; most of all those occurring in the open country.

There is no doubt that typhoid and typhus fevers may *coexist* as epidemics; sometimes affecting the same patient, the one fever shortly after the other (Gardiner); and occasionally together, as a hybrid disease. This may help us to account for some instances in which foul air has appeared to generate typhoid, and where the latter has seemed to be contagious. My own experience leads me to adopt the view expressed by Dr. Anstie, that "typhoid fever is certainly not contagious in the same sense as typhus is."

Typhoid fever is rarest in old age; not frequent in childhood; most common between fifteen and thirty years. Few have it under ten or over forty; almost none beyond fifty. It scarcely ever (relapses apart) occurs a second time in the same person.

Diagnosis.—From *remittent* fever, typhoid is known by the absence of vomiting and sallowness of the skin, the slower onset, more protracted course, the *hebetude* or mental dulness and drowsiness, and the abdominal symptoms.

From *typhus* fever, the distinctive points are as follows:

IN TYPHUS:	IN TYPHOID;
No epistaxis or bronchitis;	Epistaxis and bronchitis;
Bowels constipated;	Diarrhœa;
Belly seldom tympanitic;	Tympanites, gurgling, &c.;
Miliary eruption, 5th to 7th day;	Lenticular rose spots;
Progress moderately slow;	Progress very slow;
Death often within ten days;	Death rarely within fourteen days;
Countenance dusky red;	Countenance purplish red;
Causation mostly obvious;	Origin obscure;
Anatomy not peculiar;	Lesions characteristic.

Cases called "febricula," or "irritative fever," (formerly "synochus") are described by some writers, and met with once in a while in practice, which give a good deal of trouble in diagnosis. Some of these, probably most of them, are mild examples of typhoid fever.

Prognosis.—The mortality from this disease varies greatly under different circumstances. The possibility of perforation of the ulcerated bowel gives an element of uncertainty to every case. Probably one death in twenty cases will represent its average mortality. The favorable and unfavorable symptoms, other than those common to typhus or other febrile affections, have been indicated sufficiently already, in our account of the disease. The state of the tongue especially at the period of defervescence (end of second week, about) should be always noticed, as it aids our observation of the abdominal symptoms in concluding upon the progress of the intestinal lesion.

Treatment.—Self-limited as typhoid fever is, no *cutting short* of it is possible. We must *conduct* the patient through it as safely as possible. For this, little medication, perhaps none, will suffice, with good nursing, in many cases. I have treated the disease with so little medicine that it might be said to have been left to nature, supported by regulated liquid nourishment alone. Yet this is not always proper or safe.

The course of treatment which I learned in the Pennsylvania Hospital twenty years since, has been followed by me throughout my practice, with successful results. My only deviations from it have been in the direction of diminishing the amount of medicine given. It was, upon the average, as follows :—

In the course of the first few days, if the bowels were costive, a teaspoonful of castor oil was given ; after that, no laxative. During the first week, while the fever was highest, the tongue furred and often dry, skin hot and without perspiration, small doses of blue mass with ipecacuanha were prescribed, with the view of favoring freedom of the secretions. Afterwards, or at the same time, spiritus mindereri (liquor ammoniæ acetatis) was given, a tablespoonful (diluted) every two or three hours, from noon till midnight, as a diaphoretic.

Liquid food is necessary from the first. Oatmeal gruel, toast-water, rice-water, the first three or four days ; then milk may be added, one or two tablespoonfuls every two or three hours. *Less than half* the cases of typhoid fever which I have seen have required alcoholic stimulation at any stage ; not more than one-fourth of the cases need it before the middle of the second week, when the fever begins to decline. After that time, many require it, first in wine whey, half a wineglassful about every three hours ; later, when weaker, brandy or whisky punch ;—a tablespoonful of brandy, for instance, every four, three, or two hours, sometimes every hour, with the same or twice as much of milk. Beef-tea is indispensable in nearly all cases, from the second week. It may alternate with punch, hour by hour. As in typhus, a patient prostrated with severe typhoid fever should be waked from sleep to take the required nourishment, night and day ; otherwise he will sink for want of it.

Quinine, I am satisfied, has no place as a *curative* of this fever. It is useful as a tonic, after the critical period of the passing of the height of the fever; not more than eight or ten grains (in one or two grain doses) in twenty-four hours [F. 2].

In the first ten days, headache and heat of the head may call for the application of cold to it; sometimes for leeches to the temples or back of the neck. Dryness and heat of the surface of the body may be best allayed by sponging all over (one part only uncovered at a time) with tepid whisky and water. This operation, done in the evening, will promote sleep. Great tenderness of the abdomen may be treated by application of large poultices of hot mush, with which one-fourth part of mustard has been stirred.

Diarrhœa being a symptom of the disease, it needs not to be checked unless the passages number more than three or four a day, or are uncommonly copious. Then, a pill of tannic acid and opium (3 grs. of the former to gr. ¼ of the latter), *pro re nata*,—or *small* doses of paregoric or laudanum, will generally reduce it. Rarely is it necessary to use laudanum and starch enemata, or to add acetate of lead to opium in pill. Hemorrhage from the bowels is not apt to continue long, or to be dangerous. If it should, astringents, as lead and opium, by enema or by the mouth, must be used.

Shall we attempt to *medicate* the affection of the glands of Peyer? This also being symptomatic, its palliation only appears to be indicated. I am not satisfied that any special treatment for it is demanded in mild ordinary cases. But if, after the tenth or twelfth day, the *defervescence* does not take place, and restlessness is great, with abdominal tenderness, a dry tongue, and considerable diarrhœa, oil of turpentine is recommended by authority and experience. The dose should be not more than ten drops four times daily, in mucilage, with a few drops of laudanum. Nitrate of silver is used instead by some; I have had no experience with it; but I have often seen the good effects of turpentine. It seems to act as a local alterative to the ulcerated surface of the bowel.

Attention to the state of the bladder, day by day, to prevent or relieve retention of the urine, is important. Long protracted cases may demand a great deal of care to avoid severe bed-sores. In anticipation of these, when threatened, frequent changes of position should be made, and the parts should be bathed with whisky, spirits of camphor mixed with olive or lard oil, or soap liniment. The bed-clothes must be kept smooth under the person. Adjustment of pillows, with the addition of small ones made for the purpose, may do much. When a part is unavoidably pressed upon, it may be protected by a piece of kid spread smoothly with soap plaster. Actual excoriations must be treated like ulcers,—with simple cerate, lime-water, poultices, adhesive plaster, etc., according to their condition.

PLAGUE.

Of this oriental disease, now fast being extinguished, little need be said here. It is a zymotic affection, allied to the fevers, of rapid course and great mortality. Its symptoms are debility, restlessness, fever, dyspepsia, vomiting, glandular swellings, especially in the axilla, or carbuncles. Death often takes place in two or three days.

Plague was once thought to be the most contagious of diseases.

Excellent reasons are given, however, for believing it not personally contagious at all; but locally infectious. Not quarantine, but sanitary police and hygienic improvements in the great cities (Cairo in Egypt, for example) have almost put an end to it.

In treatment of plague, diaphoretics, opiates, and mineral acids are best reported of. Polli's treatment with the sulphites might be tried in it with propriety.

ERYSIPELAS.

Synonyms.—*St. Anthony's Fire; Rose.*
Varieties.—*Traumatic* and *idiopathic.*
Symptoms.—These are both local and general. Sometimes the former and sometimes the latter appear first. *Idiopathic* erysipelas generally begins with an ill-defined cold stage, followed by fever. The eruption most often commences on the face, with soreness to the touch, and redness; which spread like a slow conflagration, from part to part. This character of continuous diffusion or *spreading* is pathognomonic. Heat and moderate swelling attend the eruption. It may extend almost all over the body. It may also be superficial and transient, or the inflammation may involve the subcutaneous cellular tissue (especially on the limbs), causing suppuration and sloughing.

The fever of erysipelas has no special features, nor has the disease any definite period of duration. When the scalp is the seat of the eruption, delirium is common, and inflammation of the brain, or fatal coma, may follow. Otherwise, the danger of the disease seems to be from suppression of the function of the skin, and exhaustion. Traumatic or secondary erysipelas combines the danger of the disease itself with that of the injury, abscess, or other local affection from which it starts. *Sthenic* and *asthenic* forms or types of the disease may be discovered, according to constitution and circumstances.

Erysipelas is often destructive in surgical hospitals, as an endemic or infectious malady. Ventilation and cleanliness will do much toward its prevention. Absolute contagion is not proved of it; but the theory of "continuous molecular change" (Snow) applies very well to it. The immediate promotive cause of it would seem usually to be *accumulation of effete material* thrown off from the human body in connection with *inflammation.*

Treatment.—As above remarked, erysipelas may be more or less sthenic or asthenic. Thus we may account for the diverse views and results of treatment. It is very common now to treat erysipelas with free stimulation. And yet I do not remember ever to have lost a case of erysipelas, in which life was not already in serious danger from a previous injury,—either in private or hospital practice. Nor have I, in more than a very few out of a large number of cases, found occasion to give any alcoholic stimulant whatever.

I have commonly begun the treatment of an attack of the disease with a mild saline cathartic,—as a *small* dose of Epsom salts, or one of Rochelle salt or citrate of magnesia. Then blue pill with ipecac, ·if the stomach be good (gr. $\frac{1}{2}$ of the former with gr. $\frac{1}{3}$ or gr. $\frac{1}{4}$ of the latter, every three hours) and neutral mixture or liquor ammonia acetatis. *Asthenic* cases appear to gain by the free use of iron,—twenty drops of the tincture of the chloride every three hours.

Locally, mild emollient applications are the best, except as *cordons sanitaires*, or lines of demarcation. At the very start, lard, tallow, or cold cream may almost " put out the fire" at once. Mucilage of slippery elm bark, or of flaxseed, and diluted lead-water, are all that my experience justifies for application to the eruption itself.[1] I would not try to *suppress* it. I think I have seen one death result in the practice of another physician from the attempt to do this with nitrate of silver over a large surface; cerebral congestion and coma took place.

To *head off* the eruption is perhaps only worth while when, from the face, it is spreading to the head. Tincture of iodine, or strong solution of nitrate of silver may, for this purpose, be painted in a line of half an inch in width; or a narrow strip of fly blister may be put on.

When, in traumatic erysipelas, a limb is greatly swollen and inflamed, threatening destruction of the subcutaneous tissues, long incisions through the integument to relieve pressure and congestion may be justifiable.

A milk diet is usually suitable in this disorder.

PUERPERAL FEVER.

In the time succeeding confinement, liability always exists (besides the transient "milk fever" about the third day) to *metritis*, and, more often, *peritonitis;* also, but with much less frequency, to that asthenic febrile affection, to which the designation *puerperal fever* is best given.

As this belongs rather to *obstetric* practice, I propose only the briefest allusion to it. Beginning with a chill, its symptoms are, fever, with an extremely rapid pulse, pain in the abdomen, and tenderness on pressure, or on motion, as in drawing up the knees; tympanitis, often; and, a day or two later, vomiting, delirium, and a tendency to collapse. Death may occur within a week; and more than half the cases are fatal. Sometimes the pain and tenderness of the abdomen are slight or temporary only; the general debility proceeding still to the fatal end.

Autopsy shows in much the larger number of cases the manifest lesions of peritonitis; serum, lymph, with extensive adhesions or pus. In a few cases, however, these are absent entirely. Inflammation of the uterine veins has been met with.

In **causation,** it is observed that nearly all the cases of this disorder (distinct from simple *peritonitis* of the lying-in room) take place in towns, or in hospitals, especially those which have surgical as well as obstetrical wards. Puerperal fever is many times *endemic* in such localities. Physicians have been beset with it, in some instances, in practice, as a " private pestilence;" every woman attended by one practitioner, for months together, being attacked by it; when others have none of it. Hence we infer two or three things.

[1] Solution of *sulphite of soda* has lately been used, with great asserted advantage.

One, that this fever has a material zymotic cause, which may be localized. Another, that this *materies morbi* seems to be conveyable by hand from person to person. Although disputed by eminent authorities, the evidence preponderates in favor of this opinion. Further, several morbid poisons appear, in the peculiarly susceptible, *quasi-traumatic* state of the womb and abdomen after delivery, to promote the disease. *Erysipelas* does so, or at least the conditions productive of erysipelas ; also, the typhus poison ; perhaps that of smallpox and scarlatina, &c. As to erysipelas, it presents a close and striking analogy with puerperal fever. Thus :—

| *Erysipelas* is an acute febrile disease, occurring most often in surgical hospitals, in which a peculiar diffusive inflammation is a prominent characteristic ; the seat of this inflammation being the *skin* and connective tissue. | *Puerperal fever* is an acute febrile disease, most common in lying-in hospitals, in which a peculiar diffusive inflammation is a prominent characteristic ; the seat of the inflammation being the uterine veins and *peritoneum.* |

Pathologically, some questions are yet not entirely decided. Is puerperal fever a special disease, with one specific morbid material cause or virus ; or, is it a *cachæmia*, which *any* morbid poison has power to produce during the lying-in state ; or, again, is it an *ichorrhæmia*, from absorption of foul matter from the cavity of the uterus by its semi-patulous sinuses ; or, a *pyæmia* from inflammation and suppuration of the uterine veins ?

I am not ready to answer these questions. Perhaps the *ichorrhæmic* theory has the most of evidence at present in its favor ; adding to that, the hypothesis of "continuous molecular change," alluded to already in another place.

Practically, sanitary measures of precaution are clearly indicated to *prevent* puerperal fever. Lying-in hospitals must be great evils, rather than benefits, unless they have the best possible situation. construction, and administration. And no such hospital, or ward, should ever be under the same roof or in the same inclosure with a surgical ward or hospital. Moreover, in private practice, attendance in delivery by a physician who is visiting at the time a case of puerperal fever or of erysipelas, is at the risk of the patient ; if the danger of conveyance of the disease be removable, it is only so by the most careful and thorough cleansing and disinfection. The clothes should be changed, and the hands washed in strong solution of chlorinated soda, before making such a transit from the one patient to the other. Several physicians in this city always decline confinement cases under the circumstances named.

The **treatment** of puerperal fever has often proved unsatisfactory. I had intimate knowledge of the experimentation to which it was subjected in the wards of the Pennsylvania Hospital, by Drs. Meigs and Hodge, between 1845 and 1849. Venesection, purging, and mercurials, &c., were tried amply, and failed most signally.

Quinine, in tonic doses, with beef-tea, and, if collapse be threatened, alcoholic stimulation, has, though far from always successful, had at least better results. Leeching the abdomen freely, at the beginning of the attack, in the least asthenic cases, does important good. After the leeches, for a day, warm poultices may be applied; then, a large blister. Sulphites are worth trying in this disease.

For the personal prophylaxis of puerperal fever, obstetricians of authority confirm from experience the reasonable view, that it is of great importance to empty the womb, and, if possible, the vagina, thoroughly, after child-birth. Good uterine contraction is indispensable as a safeguard. Washing out the vagina, within a few hours of delivery, with lime-water or solution of glycerin, may also be recommended for a similar end.

CHOLERA.

Synonyms.—*Epidemic, Spasmodic, Malignant, Asiatic, Indian Cholera; Cholera Algida; Cholera Asphyxia.*

Symptoms and Course.—Premonitory diarrhœa, mostly painless and watery, occurs in most, but not in all cases. Its duration varies from an hour or two to two or three days. The worst epidemics of cholera, however, have been marked by some cases of fearful rapidity. In India, in a few instances, death has resulted, by collapse, in ten minutes.

Commonly, the diarrhœa increases in frequency and copiousness, and, in a few hours, vomiting commences. The discharges are colorless or "rice-water" like, and are spirted out with spasmodic force. The skin grows cold by degrees, and great debility comes on; with cramps, in all the limbs, usually.

If not checked, *collapse* arrives; with intense thirst, oppression in breathing, loss of voice, disappearance of the pulse, suppression of urine, cold, *blue*, and shrunken skin, sometimes bathed in sweat, and, at last, cold breath; ending in death. This occurs, on the average, in about eighteen hours.

When reaction takes place, recovery may immediately become complete; or, a low fever may supervene. The termination of this may be in death, within a few days, or recovery in a week or two.

Appearances after Death.—*Rigidity* occurs *soon;* sometimes in less than an hour; generally within two hours. Startling *movements* of the corpse have been several times noticed; as of a patient, dead with cholera, slowly lifting both hands over the chest and joining them; opening the eyes and rolling them downward, etc. *Increased heat* of the body, cold during the attack, has been sometimes observed after death. Internally, several of the great organs, the brain, spleen, and kidneys, at least, are commonly gorged with blood. So are the *right* cavities of the heart; but the left side of the heart empty or with but little blood, and firmly contracted. The lungs are almost bloodless. The liver varies in appearance; but the gall-bladder is almost always *full of bile.* The urinary bladder is, constantly, greatly contracted. The stomach and intestinal canal are congested and swollen; the late Prof. Horner observed the frequent throwing off of the "epithelial"

---

[1] See, for the fuller statement of the author's views upon this disease, his Essay, entitled "Cholera: Facts and Conclusions as to its Nature, Prevention, and Treatment."

lining of the canal; Böhm, of Germany, confirmed this; Drs. Parkes, Gull, and Lindsay assert it to be a *post-mortem* occurrence.[1] The intestinal glands are found considerably enlarged. The *blood* has been carefully examined by Drs. Garrod, Schmidt of Dorpat, and others.[2] Its water and salts transude into the alimentary canal, with some of the albumen and fibrin; also *the contents of the blood-cells transude into the serum.* The blood drawn from a vein during life is (as I have seen it) dark, thick, and tarry, scarcely capable of flowing. Schmidt found the amount of oxygen in the blood-corpuscles less than half the normal proportion. The blood is *acid* sometimes in cholera; the reverse of its natural reaction.

The *ganglia* of the " sympathetic" system have been often examined, and are frequently changed in appearance ; congested, softened, altered in color; but no *special* change has been *shown* to belong to them in cholera.

Diagnosis.—Common cholera morbus alone (*absurd* name, hybrid of Latin and Greek; as absurd also is cholera, from the Greek for bile, a cognomen for a disease in which the excreta are remarkable for the *absence* of bile), when severe, resembles epidemic cholera so much as to be easily mistaken for it. The *collapsed* stage of the one, preceding death, is almost identical in appearance with the collapse of the other. But cholera morbus is *caused* by some irritant of the stomach and bowels, and is clearly an affection of *those organs*, not a *toxæmia* or systemic disorder ; it is sporadic, *not epidemic ;* in it the discharges are always *bilious* at first, and mostly so to the last; collapse in any degree is *rare*, and death, under judicious treatment, very uncommon. In all these things, it differs greatly from Asiatic cholera.

History.—Putting aside some possible resemblance to this disease in descriptions of Arctæus and one or two other ancient authors, probably the epidemic in France of 1545, " *trousse-galant*," came more near to it. The earliest distinct account of cholera was that given by Bontius, a Dutch physician of Batavia, 1629. Willis (1684), Morton (1692), and others, described epidemic fluxes and " dysenteries" in England in such terms as strongly to remind us of cholera; and so did Degner of Nymwegen, in the Netherlands (1736), and Morgagni in Italy, in 1733. Some British physicians (Greenhow, Aitken) now believe that cholera may have repeatedly visited England. It appears to me more probable, however, that this opinion is due to an overestimate of the resemblance between the autumnal cholera morbus of Great Britain (like our own) and the pestilential disease.

Certainly cholera must have existed in India for an indefinite time. From 1781-2 dates its extended prevalence, in a most destructive form ; at Calcutta, in Madras, on the Coromandel coast, and in Ceylon.

In August, 1817, Jessore was the birth-place of the first great migratory epidemic. Shortly after, in Calcutta, 36,000 were attacked in three months. At many military stations, it was very severe. Roads were covered with dead and dying, unable to reach their

---

[1] Edinburgh Med. and Surg. Journal, Jan. 1855.
[2] Brit. and For. Medico-Chirurg. Rev., July, 1854.

homes. In November, the grand army of the Marquis of Hastings was devastated by it. Of 90,000 men, in twelve days 9000 had died. On marching the army across a river to dry and elevated ground, the commander was relieved of this otherwise invincible enemy. In 1818, the Birman empire was invaded by cholera; and there and elsewhere in Asia, its ravages were fearful. In 1819, 150,000 died of it in the Presidency of Bombay. It also reached Mauritius, 20° S. latitude, three thousand miles from any place before visited by it. The Island of Bourbon was visited in 1820; as well as the Philippine Islands. In 1821, Borneo and Java were affected; and a large Persian army was repulsed by it from before Bagdad, without a battle. In 1822 its limits were much narrowed, and its destructiveness abated.

India almost escaped in 1823, but China was ravaged by it; and it extended northwestward, in that year, to Orenbourg, on the Ural, near the borders of Europe and Asia. In 1826 it passed the great wall of China in its northward progress; but almost left Western Asia. It reappeared in Persia in 1829.

Orenbourg was revisited in that year, and the epidemic there lasted from August to near the end of February. This city had a population at that time of 11,000, of whom 6000 were soldiers. Those first affected had no communication whatever with any infected place.

1831 saw the cholera in the north of Europe, as far as Archangel, near the Arctic Ocean, more than 64° N. latitude. It reached Warsaw in April, during an insurrection, and was very fatal. Hungary suffered from May to September; losing 100,000 of its population. In June, St. Petersburg, and in September, Moscow, were reached by the pestilence. Berlin had it also for three months and a half, beginning in August. Mecca was attacked during the visitation of throngs of pilgrims, in May; of 50,000, as many as 20,000 are said to have perished. In this year, while Hungary was infected, the Austrians surrounded Vienna by a double *cordon militaire;* but in vain. The disease began there in August and continued for three months. The southern provinces of Austria and the Rhineland were exempt. Constantinople was affected by it, but not with very great severity. The Turkish government, that year, maintained no quarantine. Cairo suffered dreadfully in 1830-31; and so did Smyrna.

Attacking Hamburg on the 11th of October, 1831, it was officially announced at Sunderland, England, October 26th. It had occurred in several cases in England months before. Three or four weeks later it appeared at Newcastle; and in December, at Haddington, a Scottish town on the Tyne.

Edinburgh and Glasgow first had cholera in January, 1832; London in February; Dublin and Paris in March. London then suffered but moderately; Paris terribly—especially in April and May; 20,000 deaths.

On the 8th of June, it first invaded our continent, at Quebec; and within a week, at Montreal. In the same month it was in New York and Albany. Philadelphia had its first cases in July. Between the 1st of July and the 18th of August, New York had reported 5337 cases, with 2068 deaths. That city lost 3513 in all.[1] From the 27th

[1] Dr. A. Clark, Lect. on Cholera. In 1834, New York lost 971; in 1849, 5071; in 1854, 2509.

of July to August 18th, Philadelphia had 1610 cases, with 615 deaths. Boston and Baltimore were moderately affected in August.

Detroit, Buffalo, Elizabeth City in North Carolina, Wilmington and Newcastle, Delaware, Norfolk and Portsmouth, Virginia, and New Orleans were the principal of more than fifty towns in the United States reached by cholera in 1832. It had entered twelve different States before September.

Havana and Mexico were attacked in the spring and summer of 1833. The *City* of Mexico, notwithstanding its great elevation above the sea, did not escape.

Portugal was also first visited in that year; Spain but slightly until 1834. Northern Italy was affected in the autumn of the same year. In 1835, Alexandria and Malta; in 1836, Rome, Naples, Egypt, and Central America especially suffered. North Germany, South France, Rome, Naples, Sicily, Malta, Egypt, and Syria, in 1837. After that cholera disappeared from Europe and America for nearly ten years. It still existed, in variable violence and extent, in India.

In 1847, it ravaged a Russian army west of the Caucasus; and in September returned to Moscow. In 1848, Turkey, Russia, Austria, Prussia, Belgium, Holland, Great Britain, and France (though not Paris) were successively attacked. Then the cholera showed its power to traverse the sea without human aid or agency, by attacking two emigrant ships, a thousand miles apart, one sixteen and the other twenty-seven days out from Havre, *when no cholera was prevailing at the port.*[1] The *cholera-cloud* itself also reached New Orleans about the same time, and progressed up the valley of the Mississippi. New York was not affected by the visit of the infected ship; the disease not occurring again there until May, 1849.

Paris was reached by it in February of that year, but suffered the worst in June. Lyons now had it for the first time. Tunis and Algiers were visited toward the end of the year.

In January, 1849, after Memphis, St. Louis, Missouri, was attacked. Chicago, Buffalo, and other towns on the lakes, in May. New York and Philadelphia in the same month. Baltimore had this year only a local epidemic, in July, in the Almshouse; the restriction of which to one side of the building was very remarkable. As in 1832, the mortality in Philadelphia was much less for the population than in New York: 1022 deaths occurred in our city; New York had a mortality 450 per cent. greater.[2] Canada was reached this time from the westward.

In 1848–9, the number of deaths from cholera in England and Wales was over fifty-four thousand (54,398) ; in 1832–3, nearly thirty-one thousand (30,924). In London,[3] probably owing to greater attention to sanitary means, the mortality was two-fifths less the second time than the first. Some parts of southern Rhineland were visited in 1849; especially the filthy city of Cologne.

Cholera lingered in various places almost sporadically, in Europe and America, from 1850 to 1854. Canada and the far West (Indiana

---

[1] Report on Cholera in the United States, by Dr. James Wynne ; and Dr. Gavin Milroy, Brit. and For. Medico-Chirurg. Review, Oct. 1865, p. 444.
[2] Dr. J. H. Griscom, Medical Record, March 15, 1866, p. 35.
[3] London had 13,098 deaths from cholera in 1849 ; in 1854, about 10,000.

also had cases every year) suffered the most in this way, on our continent. In the West, emigrants' camps and military stations seemed especially to furnish its required local conditions.

In 1853, Persia had it severely; also some parts of Northern, Central, and Southern Europe (Copenhagen, Hamburg, Berlin, Piedmont, Lyons, Paris, and Southern Portugal). Before the end of the year it was again in New York, New Orleans, and the West Indies. Mexico had been visited in the spring, and through the summer.

1854 was still more a cholera year in Europe and in this country. Scarcely any European state or kingdom was exempt. The French, English, and Russian troops suffered from it much in the Crimea. Greece, Italy, Germany, France, Spain, Portugal, in short, all Europe was traversed by it: 150,000 died of it in France alone ; in England and Wales about 20,000. Newfoundland, on our side of the ocean, was reached for the first time in 1854. This was the year of the epidemic at Columbia, Lancaster County, in this State ; so remarkable for the absence of some of the usual promotive conditions of cholera. Our great cities, however, did not suffer nearly so much as in 1849.

In 1855, the disease was widely spread in Europe, though not very malignant except near the seat of war, before Sebastopol. Egypt and Palestine had it also. In Switzerland, which had been slightly touched before, Basle, Geneva, Zurich, and other places now suffered by it. The next year, 1856, still did not witness its withdrawal from Europe.

Since that period, until 1865, I have no means at hand for tracing the movements of epidemic cholera. Dr. Garvin Milroy says that the countries hitherto exempted have been as follows : Australia, New Zealand, and other islands in the Pacific ; the Cape of Good Hope and adjoining settlements ; the coast of Africa from the Cape as far northward as the Gambia, and including the islands of St. Helena and Ascension ; the Azores, Bermuda, Iceland, Faroe islands, and also the Orkney and Shetland ; the southern half of the eastern coast of South America, from the Rio Plata inclusive, Cape Horn, and the whole of the western coast of that continent, from the Cape and along the shores of Chili and Peru to Panama.

In 1865, every one was familiar with the accounts of cholera in Arabia and Egypt in the spring, at Constantinople in July,[1] and afterward in several parts of Europe, extending, though with but moderate violence, as far as England. While its vast migrations seem to be as capricious or incalculable as the flight of locusts, two local causes contributed at least to its *severity* in Mecca and on the Nile. These were the crowds of religious pilgrims at the former place, in the spring, and, in Egypt, the insalubrious circumstances attending the operations at the new Suez Canal. In both, "crowd-poison" was intensified to the greatest degree ; so that the pest-cause might well find there strength for the renewal of its flight onward to the northwest. In Paris, in 1865, 6383 deaths occurred during the late visitation.

I take from Dr. Brigham's treatise (published in 1832) the following table, of the deaths from cholera in 1832, and their proportion to population :—

---

[1] The first case occurred in that city on the 28th of June.

	Population.	Deaths.	Equal to
Moscow .	350,000	4,690	1 in 74
Petersburg	360,000	4,757	1 " 74
Vienna .	300,000	11,896	1 " 159
Berlin .	340,000	1,401	1 " 242
Hamburg	100,000	446	1 " 224
London .	1,500,000	1,223	1 " 1,228
Edinburgh	150,000	72	1 " 2,033
Glasgow .	180,000	395	1 " 455
Hungary	8,750,000	188,000	1 " 46
Paris .	800,000	20,000	1 " 40
Montreal	25,000	1,250	1 " 20
Quebec .	22,000	1,790	1 " 12
New York	200,000	2,000	1 " 100
Albany .	24,000	311	1 " 77

Supposing the population of Philadelphia to have been at that time 150,000, this, with a little over 600 deaths, would give a proportion for our city of 1 in 250 of the inhabitants. In 1849 the ratio was considerably less.[1]

It is an important fact in the history of cholera, that before, during, and after the epidemic has visited a place, many cases, greatly exceeding in number those of typical cholera, occur, of diarrhœa, sometimes also with vomiting, not violent, yielding easily to treatment. To these the name of *cholerine* is often given.[2]

Nature of Cholera.—Without discussing opinions at length, it may be asserted that cholera is not at all, like our ordinary cholera morbus, a disorder simply of the stomach and bowels. Being clearly an acute *systemic* affection, changes in the blood are *proved* to occur in it, and may well be believed to be primary; that is, that the morbid cause acts through the blood. But this is not all.

Cullen placed cholera, in his nosology, in the class *neuroses*, order *spasmi*. Many medical observers (Binaghi, Loder, Orton, Delpech, Lizars, Coste, Favell, C. W. Bell, Greenhow, G. Johnson, etc.) consider its principal effects to be referable to disturbed innervation, involving chiefly the ganglionic centres of organic life. Dr. Charles D. Meigs, years ago, graphically called the attack the " cholera squeeze." Velpeau, of Paris, lately repeats this, "le mal vous *tortille*." There, I think, is the pathology of cholera, in one word. As Dr. C. W. Bell says, it is not an adynamic, but a dynamic, or sthenic, collapse.

The heart, its left side at least, is, after death, contracted. The pulmonary artery and its branches are narrowed, making the lungs pale and anæmic. The gall-bladder is full of bile, but the duct is spasmodically closed, and detains it there. The urinary bladder is shrunken to half its size or less. The bloodvessels of the whole

---

[1] Moreau de Jonnes estimates the number *attacked* as, in France, 1 in 300 of the population; Russia, 1 in 20; Austria, 1 in 30; Prussia, 1 in 100; Poland, 1 in 32; Belgium, 1 in 120; Great Britain and Ireland, 1 in 131; Holland, 1 in 144; Germany, 1 in 700.

[2] The coincidence or anticipation of cholera by epidemic *influenza* and the potato blight, has been several times noticed. But there is, clearly, no uniformity in any such association.

alimentary canal press rigidly upon their contained fluid, and force its serum out into the stomach and bowels ; whence it is, by spasmodic ejections, thrown out. The very skin is, by its involuntary muscular fibres, as well as by vascular constriction everywhere, drawn tightly and closely upon the body. The voluntary muscles suffer with cramps. All is cramp, cramp, within and without. The brain is almost in anæsthesia during the collapse—no delirium, but apathy—as from cerebral anæmia. The blood, so compressed, grows thick as tar—it scarcely flows, is not aerated, and cyanosis follows ;—it is detained in the capillary and venous networks of the interior organs, in which congestion is found after death.

Cholera is, then, I say, a poison-spasm ; a *ganglionic tetanus.*

Causation.—As to this, all cannot yet be known. But it is clear that cholera must have a specific, material, migratory cause. I agree with Dr. G. B. Wood, Dr. Austin Flint, Dr. Snow of Providence, and some foreign authorities, for example, Dr. Southwood Smith, "the father of modern sanitary reform," in believing that cholera is *not personally contagious.*

My theory is as follows : That the cause of cholera is a (yet undiscovered) protozoon, or primal organism, of extreme individual minuteness ; which, on entering the human body, affects it as an organic poison. That the varying quantity or number of these organisms may in different cases account (along with individual predispositions and exposures) for the unequal violence of different epidemics ; as in the case of trichiniasis. Choleraic diarrhœa or cholerine, so frequent *before* as well as during and after the prevalence of cholera, may in some instances at least be explained by the action upon the alimentary canal only, of a minimum quantity of the cause. The dreadful fatality of some Indian seasons, is on the same view referred to an extreme *accumulation* of it.

A most important part of the theory is, further, that which concerns *promotive* causation. What conditions favor and maintain in life, multiplication, and migration, this *ens primalis?*

All the facts answer, as I believe, that *animal matter in a state of rapid and foul decomposition,* putrefaction, along with moderately high (not the highest) temperature, and ordinary moisture, will afford those conditions ; and that *nothing else* is required to explain the whole history of the propagation and extension of cholera. Nothing, I mean, but the admission of the existence of the "protozoon," which in ova or in maturity, or both, may fly "on the wings of the wind ;" or be conveyed to less distances by water ; and, with these the above-named conditions of its vital maintenance, as its food and "habitation."

It is, in my mind, *obvious* that this theory will explain all the facts. I believe, also, that some well-known facts can be explained by it *alone.* Such are the facts which account, by annual inundations, the mortality of the great fairs, the throwing of bodies into the river, and the inconceivable filth of the inhabitants, for the persistent residence of cholera in the Gangetic delta, while everywhere else it is only an occasional visitant.

In Europe and the United States, as well as in India, influences belonging to closely aggregated communities have always been observed to display a power to propagate cholera. It comes most often, stays longest, and is most destructive, in the densest and filth-iest cities, and in the worst quarters of those cities. Very important testimony exists as to the influence of the *drinking water* of localities. Dr. Snow, of England, asserted the theory that this was the almost universal medium of its propagation. This has been shown to be quite insufficient. But all such testimony is still available in regard to the propagating and extending power of *animal contamination.* Thus, Bethlehem Hospital, supplied by an artesian well, had, in 1849, among 400 inmates, no case of cholera. It was the *only* large lunatic hospital in London which escaped; as it was the only one supplied with spring water. In the districts of London supplied from the Thames above the entrance of the sewers, the mortality ranged from 8 to 33 in 10.000 of the inhabitants; in those supplied from below the entrance of the sewers, from 28 to 205 of the same number.

In this country, Dr. James Wynne's report[1] affords, upon almost every page, matter of exactly the same purport as the above. In St. Louis, Louisville, Buffalo, New York, Philadelphia, Boston, etc., similar facts were recorded. It is unnecessary to extract them, they are now so familiar and so commonly accepted.

The instances of apparent transmission of cholera by persons, which are quite exceptional, even if we admit a hundred or more authentic examples, are, as I believe, to be explained on the principle of *fomites;* of occasional, *very rare,* carrying of the material cause of the disease, the "germs" of it, in clothing, merchandise, or by the person of a human being; as one might carry skippers on a piece of cheese in his pocket, or a paper of flower seeds in his carpet bag.

All of Pettenkofer's and Thiersch's observations, in regard to sub-soil accumulation and transit, and fæcal fermentation after discharge, range themselves naturally under the one general fact which they exemplify, viz., that *animal decomposition is the one great promotive cause* of cholera; to which heat and moisture, etc., are merely adjuncts.

But, that which suggested first to me this opinion was. the singular history of the outbreak at Columbia, Lancaster County, Pennsylvania, in September, 1854. Cholera had never visited that town before. It is not large or populous, in a rural site, on the Susquehanna, not built densely enough to exclude malarial fevers. Why should it have cholera at all?

Visiting the town, with other physicians of our city, during the epidemic, I learned that an exceeding drought had reduced the channel of the river to an unusually low ebb, and that, in its bed, a short space above the town, a number of carcasses of sheep and other animals, thrown from the railroad trains, etc., were putrefying rankly in the sun. A reservoir which supplied many of the people with drinking water was filled from the river not far from that spot, and the wind blew from it directly over the town. The first subsidence in the disease, we were afterwards told, attended a decided change in the wind.

At Pittsburg, shortly after the above events, a similar epidemic oc-curred. A gentleman on a visit to that locality not many days before

---

[1] Presented to Parliament, and published in 1852.

the disease broke out, informed me that the same condition of the river existed there, with a like abundance of accumulated *putrefying animal matter*, exposed to the sun.

In Rhode Island, in the autumn of the same year, I was informed that the local existence of cholera in a few spots, otherwise very healthy, might be traced, in coincidence at least, with a practice not uncommon along the shore of the sea or bays, of dragging up fish in quantities by nets, and spreading them out to rot for manure.

Prevention.—Quarantine is urged by many, in this country as well as in Europe, to exclude cholera. Is it available? Will it do any good? I say, *no*. Theoretically, if the views advocated in the preceding pages are correct, it falls to the ground of course. But, more than that, it *never has succeeded*.

Dr. Alison, of Edinburgh, wrote thus in 1854:—

"It is a fact that cholera has made its way, not uniformly, but very generally, *in spite of cordons and quarantine regulations.*"

Dr. Gavin Milroy, one of the ablest and most industrious sanitarians of our time, published, about the same year, an essay with this title: "*The Cholera not to be Arrested by Quarantine.*"

Even Pettenkofer, the leading advocate of the hypothesis that cholera is diffused by the influence of the discharges from the bowels, has announced his conviction that local sanitary measures are much more reliable for its prevention than quarantine.

Quarantine, if sound in theory even, could not avail, never has availed in practice. Its infraction for smuggling and other inducements, is everywhere constant and notorious ; *this cannot be prevented.* Macaulay (History of England, vol. v. p. 52) states that when a contraband trade was, in the time of William III., carried on between France and England on the southeastern coast, "it was a common saying among the inhabitants, that if a gallows were set up every quarter of a mile along the coast, the trade would still go on briskly."

One might think the history of *blockade-running*, during the late rebellion in this country, might afford ample illustration and confirmation of this. Vain, indeed, would be the attempt to close our coast against the introduction of cholera, were it as contagious as smallpox, or as plague was once imagined to be.[1]

The *evils* of quarantine are great, almost incalculable. Sir John Bowring, speaking in the House of Commons in 1841, gave it as his belief that the losses from quarantine in the Mediterranean alone were not less than two or three millions sterling a year.

But what if, instead of preserving, quarantine actually involves, often, sacrifice of *life?* No doubt this has many times occurred. With yellow fever, the quarantine epidemic in New York harbor, a few years ago, exemplified this. In various quarters reports of travellers show the *miseries* and dangers of the lazaretto, and of the confinement on the vessel detained.

What more do we need to show this than the very recent instance of the steamer *England*, at Halifax? Forty passengers, one account says fifty (out of 1202), died on this vessel during the voyage. She

---

[1] The Governor of Eupatoria is said to have wished the British and French troops to undergo quarantine, at the opening of the Crimean war !

was prohibited from entering port; all were detained on board, and, by April 14th, 1866, 130 *more deaths* occurred! In all, 159 died while in quarantine. If the twelve hundred passengers had been *landed* and *scattered*, I, for one, doubt the occurrence of the disease in a dozen of their number; especially as it was reported as *altogether confined to the steerage.*

Were such measures sure to preserve from the epidemic the whole people of our continent, a *hecatomb* like this might find excuse. In the face of facts, I regard it as a barbarity. Pelissier, in Algiers, was, thought a monster, for suffocating a band of guerrillas in a cave; but what is this case of the *England* more like, except in motive? It is closing up hundreds of people for death; as though one might lock the doors, and bar the windows, against all escape of a thousand people from a burning church; such as that of which we read so harrowing an account, some time back, in South America.

But it will be said or asked, would you abolish *all* quarantine—abandon all *inspection of ships* whatever? No; I would not. But I would abandon altogether the whole *theory* of quarantine, as against cholera most particularly.

Ships should be inspected on approaching ports, because they may have unsanitary conditions intensified in them, on a scale sufficiently large to be important. This is, or should be, a part of sanitary police. Nor should it (and here is a great point of difference) include *any* restriction of *persons;* at the most, longer than enough for cleansing of the body and of the clothing. and purification of merchandise, by fresh air, and possibly by some disinfecting process in certain cases. I insist that SANITARY POLICE includes the sum total of available measures for the prevention of cholera in any place.

On this ground, the measures required are obvious, and familiar. The thorough and frequent cleansing of all streets, alleys, courts, wharves, and vessels, private and public buildings, and empty lots; the abatement of all nuisances; daily removal of offal; efficient sewerage; and *conservancy, i. e.,* the cleansing, ventilation, and disinfection of cess-pools and water-closets. Among all signs of danger of the location of cholera, none is more significant than the *privy odor.* Let it be everywhere annihilated. Lime, charcoal, dry earth, chloride of lime, Labarraque's chloride of soda, liquid coal tar, chloride of zinc, and sulphate of iron are the most available of disinfectants.

The fresh white-washing of cellars is useful; thorough ventilation and drying of them and of all parts of habitations, still more so. Chloride of lime may be placed, in a saucer, in any suspected room or other locality in a house. The same, in the solid form, or solution of green vitriol, may be thrown daily into a foul privy; and, during cholera time, especially in the case of patients with the disease, every water-closet and vessel used may and should be disinfected constantly, by a dilute solution of chloride of zinc, chloride of soda, permanganate of potassa, or carbolic acid. The immediate removal of all discharges from the sick-room, their disinfection and transportation to the safest possible place of elimination, ought to be imperatively maintained. All foul clothing must be promptly washed, or, if very bad, disinfected or burned.

These precautions have been proved to be capable of essentially limiting and mitigating the prevalence of epidemics.

Personal Prevention.—One principle will suffice here : *to keep the system at par;* neither above its level of excitement, nor below that of its due strength.

For this, regularity of life is required, in work, diet, mental movements, and all indulgences. The popular errors most common are, one, to suppose that living on rice or rice-water, avoiding fruits or vegetables, etc., will be preventive; another, to think constant alcoholic stimulation beneficial for that end, *Both are certainly wrong.*

In 1832 and 1849, the late Dr. Joseph Hartshorne, my father, then in very large practice, allowed in his family all its usual variety of food : boiled corn, peaches, watermelons, cantelopes, etc., everything but cucumbers; and no cholera resulted from the liberty. My own subsequent experience justifies the practice. Of course care is always needed as to *quality* and *quantity.*

Of all those most likely to die when attacked by this disease, the drunkard stands first, according to all records. Nor is he one whit less apt to be attacked than others. Temperance, *in all things,* is essential to safety during epidemics of every kind.

Treatment.—To *discuss*·all the modes of management proposed for cholera, would occupy too much space. I shall merely *enumerate* those which have attracted the most attention ; and then give my view as to what is so well sustained as to be worthy of further trial and some confidence.

1. **Bleeding.**—This was largely practised in India, in 1818-1825, by Corbyn, Scot, Annesley, and others. Without entering upon any argument about it, I will simply say, that (as Dr. Brigham's quotations show) as many positive facts have been asserted on behalf of the success of blood-letting as of any other remedy in cholera.[1] My father bled in several cases in 1832, and had confidence in the treatment, as "the most effectual *antispasmodic.*" In 1849 I bled in one case (a boy of twelve years of age), in incipient collapse. The blood at first was thick and black as tar; in a few minutes it flowed more freely, and the patient *recovered.* I confess that the only thing which makes it unlikely that I will ever try or advise the repetition of this practice is, the want of courage to stem the overwhelming tide of professional and popular opposition now existing against it. In this timidity I may be wrong ; if so, another generation may afford the demonstration of what is right, in such a way that no one can gainsay it.

2. **Calomel.**—This, too, was an old East Indian remedy. Suggested by the almost universal absence of bile in the discharges, which was thought to indicate the need of stimulation of the torpid liver, it has been more largely given than any other medicine in cholera.

Unhesitatingly, I hold the opinion that calomel is of no earthly use in cholera. The argument in its favor, from the absence of bile in the stools, is rebutted by the fact of its abundance in the gall-bladder ;

[1] In 1861, Surgeon G. R. Playfair, at Agra, India, found venesection to about 8 ounces, followed by stimulants (especially chloroform) the most successful treatment. He reports more than 77 per cent. cured, of genuine cholera. —*Edin. Med. Journal,* Sept. 1866, p. 275.

while the clinical experience quoted for its success is accounted for
by the addition to it, almost always, of opium, in the prescription.
Nor is the amount of success with it, even then, great. Such is Dr.
Gull's conclusion, based upon the examination of a great mass of evidence, given in his report.[1]

Dr. Ayre, a British practitioner of some note, gave prominence to
a modification of the old calomel treatment (in which twenty grains
were sometimes given at once), by prescribing a grain of calomel
every five minutes during the attack.

3. Saline Treatment.—Dr. Stevens, of Jamaica, proposed this,
upon the view that the main pathological element in cholera was the
loss of salts from the blood in the discharges. After the general failure of saline solutions (of common salt, carbonate and phosphate of
soda, etc.), given by the mouth, had been conceded, Dr. Mackintosh,
of Edinburgh, and others, tried the method of injection into the veins
(half an ounce of common salt, and four scruples of sesquicarbonate
of soda, dissolved in ten pints of water, at 105° to 120° Fahrenheit).
Under this plan, resorted to during collapse, of 156 patients in Dr.
Mackintosh's hands, only 25 recovered. Remarkable improvement,
almost like a resurrection, appeared in several, who afterwards fell
again into collapse, and died. The suggestion has been recently made,
that it may have been the *temperature* of the injected liquid which
produced the benefit, so promising and yet transient.

4. Eliminative Treatment.—Dr. George Johnson, of London, has
urged this with especial vigor. The castor-oil medication of cholera
owes its trial to him. Some recent lectures of his on the pathology
and treatment of the disorder give a full and very intelligent exposition of his views. A prominent idea with him is, that the general
collapse is due especially to anæmia of the lungs, owing to spasmodic
contraction of the pulmonary artery and its branches. I regard this
as only a *part* of the *universal* arterial (and other) involuntary muscular spasm, belonging to what I have called the *ganglionic tetanus* of
the collapse. But the essential feature of Dr. Johnson's pathology is
the opinion that, the disease being toxæmic, a morbid poison exists
which must be *eliminated* from the blood ; and that the discharges are
the media of this elimination. Therefore, the vomiting and diarrhœa
are salutary or relieving ; and ought to be rather encouraged than
checked. He goes even so far as to repudiate the commonly accepted
belief, that " premonitory diarrhœa" or " cholerine" ought to be
checked ; considering it a fallacy to assert that those who are relieved
of such symptoms by mild treatment were really, or would have been,
cases of cholera at all.

I am entirely unable, from observation or reflection, to assent to
these views. They have very few advocates or supporters, besides the
distinguished physician whose name and ability command for them at
present careful consideration. It is true that patients have died of
cholera without vomiting or purging. I saw in 1849 a woman in collapse (from which she recovered) for several hours without either ;
and many such cases are on record ; though, in some, after death, the
intestines have been found to be distended with the rice-water liquid.

---

[1] Report, etc., of Drs. Baly and Gull, already cited.

But the checking of the discharges is almost always the sign of improvement and recovery of the patient. And we cannot, on Dr. Johnson's dictum, set aside or quash all the accumulated evidence, in Europe and in this country,[1] which shows that it *is* desirable and important to *check all watery diarrhœas in cholera times*—such fluxes having been proved to be often premonitory of cholera attacks.

5. **Ice to the Spine.**—Dr. John Chapman's ice-bags threaten to become the "pathy" or therapy of the day, with those who are zealous and venturesome in experimental practice. Upon reasons of a physiological nature, not appropriate for discussion here, I disbelieve altogether in the theory of his therapeutics. In his pamphlet upon "Diarrhœa and Cholera," lately published, he gives but one case of the latter disease, and does not say whether the patient recovered or died.

As ice is so useful when internally given in cholera, it *may* be safe and beneficial when applied to the spine. Not having seen it tried, I am not prepared to deny the possibility. It is one of the experiments to consider, in so desperate a disease. But, if it *should* hereafter prove useful, I should explain that result quite otherwise than Dr. Chapman has done, in part at least.

6. **Sulphuric Acid.**—Dr. Cox, of England, afterward Mr. Buxton and Dr. Fuller, and recently Dr. Jules Worms, of Paris, have especially recommended dilute sulphuric acid in all stages of cholera. Many others especially report well of its action in the *premonitory diarrhœa*. Such an action would comport perfectly with the view I have taken of the *organic* nature of the poison of cholera; sulphuric acid being so potent a destroyer of everything organic, except such *mirabilia* as the Acarus Crossii.

Dr. Worms' treatment (based on the results in 238 cases of cholera, and 150 of cholerine, in 1865) is as follows : For prodromic diarrhœa, he makes a "mineral lemonade," of about half a drachm of concentrated sulphuric acid to a pint or more of sweetened decoction of salep (arrowroot would do as well). The patient is to take of this every hour a wineglassful, till relieved.

For confirmed cholera, the patient being kept in complete repose, there is administered every half hour a glass of a similar lemonade, of the strength (about) of a drachm to the pint ; ice and wine also being allowed *ad libitum*.

7. **Opium in large doses.**—This practice had once many advocates ; now they are few. Prof. Austin Flint, of New York, is one of them ; at least *morphia* is advised by him, in full dose, repeated if required. A great deal of evidence of the insufficiency of such a plan has been published ; although it is *not worse* than several other methods. Letting alone would probably be better. The *secondary fever* is apt to be more severe and more often fatal after treatment of the attack by large doses either of opiates or stimulants. Large quantities of brandy (I add, by the way) have been often used, with no good results.

Statistics are given, as follows, of the results of some of the most common modes of practice in cholera, by practitioners in Great

---

[1] See Lectures on Cholera, by Prof. A. Clark, of New York ; Report to the Royal College of Physicians, 1854 ; also, Madin, Briquet and Mignot, etc.

Britain, as reported to the "Treatment Committee of the Medical Council of the Board of Health," 1854-55.
Taking all grades of the disease, the deaths were—

	Per Cent.
With Eliminants	71.7
Stimulants	54.0
Calomel and Opium	36.2
Chalk and Opium	20.3

Of *collapsed* cases, the mortality was—

	Per Cent.
With Calomel and Opium	59.2
Larger doses of Calomel	60.9
Salines	62.9
Chalk and Opium	63.2
Calomel, small doses	73.9
Castor Oil	77.6
Sulphuric Acid	78.9

Much is uncertain, obviously, in such statistics, without farther account of dosage, circumstances, etc. But this seems to follow; that *neither treatment has much to boast of success.*

8. Treatment by antispasmodics and mild stimulants, in small doses at short intervals; with ice, and external frictions, etc.— In 1849, my first two cases of cholera were fatal; although assiduously watched, each for a day and a night. The third, I saw with the late Dr. Wm. E. Horner, Professor of Anatomy in the University of Pennsylvania. I left the treatment to him. He sat down by the bedside of the patient—a man, blue, cold, and with a scarcely perceptible pulse, copiously vomited and purged, with rice-water. Having ordered ice,[1] Dr. Horner took from his pocket a vial containing a mixture of chloroform, oil of camphor, and laudanum; which he gave in sweetened ice-water, in small doses, *every five minutes* by the watch. Each dose was followed by a piece of ice.

Soon the vomiting diminished, afterward the diarrhœa, and in an hour and a half the veins on the back of the hand began to fill up, and the blood to return in them more rapidly after pressure. Diminishing the frequency of the doses, we left him, an hour later, evidently convalescent. When I saw him after several hours again, he was sitting up in bed, at ease, and so changed, that I doubted at first his identity. No secondary fever followed; he was cured.

Naturally, I repeated this treatment in all my subsequent cases, some of which were of extreme severity; and with gratifying success. The memorandum book of the number of these cases has, to my present regret, been mislaid. After the treatment of Prof. Horner had been adopted, however, I saw no death, except in the instance of a drunkard, two or three hours in collapse before any medical treatment began.

Should I be attacked with cholera, such is the treatment I desire.

---

[1] Ice was used, and lauded, in cholera, by the celebrated Broussais, in 1832.

Conscientiously I believe, that nothing else will afford a better chance of recovery. I merely altered Prof. Horner's mixture to a tincture, for better preservation; adding some minor adjuvants. This recipe will be given directly. Frictions and sinapisms may also be added. The great merits of this plan are its antispasmodic nature, and the administering of small doses at *very short intervals*. This is eminently demanded in cholera. Phthisis may be a complaint of years; hooping-cough, of months; typhus, of weeks; pneumonia, of days; but cholera must be numbered by its hours, half hours, or even minutes.

Having reached, then, this conclusion, I may add, that a *rationale* for such a treatment is discernible. I only follow many good authorities in the opinion that cholera is, symptomatically and pathologically, a poison-spasm, or tetanus of the ganglionic system. Taken early, that condition may be *prevented*, by mild opiates and stimulants, in the *premonitory* stage. Later, while any medicines will act, these will do the most. What is needed in confirmation of this explanation, more, than is given by the action of quinine in preventing an anticipated chill, or, of the same, in full quininization, curing the paroxysmal disease (a toxæmic *neurosis*) of intermittent? An antagonistic influence against that which so perturbs innervation throughout the body; such is the whole definition that we can give of the remedial power shown in either case.

Let me be more specific in reference to treatment. Premonitory diarrhœa is very generally admitted to be present in a majority of cases of cholera.[1] In the East Indies, many writers, of different dates (Lawrie, 1832, Stewart Clark, 1864, etc.), assert such a stage to be an exception instead of the rule. But, in India, they have a premonitory or incipient stage of another kind; characterized by great languor or depression, with restlessness, and sometimes' ringing in the ears, occurring mostly in the night. Stewart Clark states[2] that, in this stage, a mild opiate ("with a little calomel or blue pill"), with a cup of warm tea or a small dose of a diffusible stimulant, as a few grains of carbonate of ammonia, or a little weak warm brandy and water, will arrest the attack in a great portion of cases otherwise to become serious.

Such symptoms, as well as diarrhœa, should be noticed here, during a cholera epidemic; and I believe the same treatment will meet either. Rest, warmth, and mild, composing, but gently stimulating draughts; paregoric, aromatic spirit of ammonia, tincture of ginger, lavender, etc., with a mustard-plaster over the abdomen, and a hot mustard foot-bath if coldness of the body increase, or vomiting begin; such are safe, and I believe will be efficient remedies. The above may be called the first or prodromic stage.[3]

The next has been well called, by Prof. A. Clark, the *rice-water* stage. For that, the treatment I have described as given to me by

---

[1] Barraut asserts fixed contraction of the pupil to be the first prodromic sign; M. Worms makes the same statement in regard to albuminuria.
[2] Hygiene of the Army in India, p. 12.
[3] The recently published experience of Dr. Hamlin, in Constantinople, confirms the importance of the above early treatment.

27

Prof. Horner is particularly adapted. My recipe, based upon his, is as follows :—

R.—Chloroform. et
   Tinct. Opii et
   Sp. Camph. et
   Sp. Ammon. Aromat. āā f3jss ;
   Creasot. gtt. iij ;
   Ol. Cinnamom. gtt. viij ;
   Sp. Vin. Gall. f3ij.—M.

Dissolve a teaspoonful of this in a wineglassful of ice-water; and give of that two teaspoonfuls *every five minutes;* followed each time by a lump of ice.[1] Iced water, or rice-water, to which common salt and carbonate of soda have been added, may be given, a little at a time, as a drink. I would also give a tablespoonful of brandy every hour or two.

Friction of the limbs with brandy and red pepper will be, along with large mustard-plasters on the back and pit of the stomach, useful to promote reaction.

The third stage is that of absolute collapse ; blue, pulseless, shrunken, voiceless. Should a case go on, in spite of the above-mentioned treatment, into this state, what else can be done ? All now seems to be desperate experimentation.[2] Let the ice-bags be *tried*, and judge them by the trial. I would also try belladonna internally, as an antagonist of vascular spasm. Leclerc, of Tours, introduced it in 1854 ; Barraut, of Mauritius, used it ($\frac{1}{4}$ grain every half hour), and reported success. He also employed *hypodermic injections of sulphate of atropia*. This should be tried again in bad cases. So might be, as was suggested by me in 1855, *warm baths of infusion of stramonium* (Jamestown weed) *leaves;* on the same indication. Also, the injection of hot liquids into the rectum ; the warm bath (hot baths cause *distress* in the collapse), with *carbonate of ammonia* added, as used sometimes in malignant scarlet fever (West) in children; or, the warm *mustard* bath. *Hot air*[3] bathing, if practicable, in the manner so praised of late by Erasmus Wilson and others, would be worth trying ; and so would the inhalation of nitrous oxide. Let us confess honestly, for it is wise to do so, our art is here very weak ; *fifty per cent. or more of collapsed cases die;* shall we not endeavor to *discover* new resources ? All honor to those who, at the risk of their own lives, contend yet, with so forlorn a hope, and so little glory to be won. There is room yet for, and possibility of obtaining, a final triumph.

---

[1] I take from Dr. Aitken's Practice the following recipe, much used and approved in India and England : R.—Ol. Anisi, Ol. Cajuput., Ol. Juniperi, āā 3ss ; Æther, 3ss; Liq. Acid. Halleri (*i. e.* one part concentrated sulphuric acid to three parts of rectified spirit), 3ss ; Tinct. Cinnam., 3ij.—M. Dose, 10 drops every ½ of an hour, in a tablespoonful of water.

[2] Duchaussoy and Vernois assert the non-absorption of medicines given by the stomach during the collapse ; but Magendie proved that a very slow absorption does occur.

[3] Dr. George Johnson states that he has seen the hot-air bath used without success.

Two words remain still to be said, with short comment: *house to house visitation*, and *houses of refuge*. These are measures of great consequence, shown to be of value during cholera epidemics. The latter, especially, is of notable importance ; that is, the establishment of houses of refuge in *salubrious places*, into which persons from tainted districts most liable to the disease may be received, on the occurrence there of the first cases.

That there *are* such tainted districts, has been amply proven. Thus, Dr. Laycock has shown that in York, England, the first death from cholera occurred in the spot where plague had been traditionally the worst, in a badly-drained district. In Edinburgh, the first case in 1848 occurred in the same house as did the first in 1832. In Holland, at the town of Groningen, in 1832 and 1848 but two houses in the better part of the town were attacked ; the same houses exactly in both epidemics.

Dr. Alison reports that in the first three months of the epidemic at Edinburgh, in 1832, 353 persons were taken in at Houses of Refuge, from 70 tainted districts, houses, and rooms in which decided cases or deaths had occurred. Of these, only 15 took the disease, and 7 died after removal. Of the 346 thus surviving brief exposure, it is very probable that more than half would have died had they remained in the midst of the infection. At Glasgow, in 1849, 401 persons were taken into Houses of Refuge from tainted districts ; only 19 of these took the disease and but 5 died. At Oxford, England, the same year, of 70 persons so taken in, none died. The London Board of Health, in its " General Report," gives the fact that of 1691 of whom the Board had accounts as taken into Houses of Refuge, but 33 were attacked, with only 10 deaths. These numbers would have been larger, but for the very common unwillingness of poor and ignorant people to leave their homes, chiefly from want of confidence in the greater safety of so doing. Could this be overcome, I have no doubt that an immense saving of life might be produced by Houses of Refuge, allowing also the places which are proved "foci of infection" to be thoroughly purified at once.

House to house visitation, by sanitary inspectors to abate nuisances, small and great, and by medical men to treat premonitory symptoms, might also have great preventive value. The establishment of cholera hospitals may be made necessary when the number of cases is great, especially as the greatest proportion always happens among the poor, who are ill provided for attendance at their homes.

RHEUMATISM.

Several affections are, in popular language. (partly sanctioned by medical usage), included under this term. 1. *Acute articular rheumatism*, or *rheumatic fever*. 2. "Chronic rheumatism," affecting the joints and sheaths of the muscles. 3. Syphilitic rheumatism, of the long and flat bones. 4. "Rheumatoid arthritis." 5. Myalgia. 6. "Gonorrhœal rheumatism."

Acute Rheumatism.—Only certain persons and families are liable to this affection, upon any exposure. It is characterized by high fever, with severe inflammation of several of the larger and smaller joints; which, mostly one after another, become swollen, red, hot, tender, and painful. The shoulders, wrists, knees, and ankles are most frequently so affected. Although with a full and rapid pulse, the skin, after the first week or so of the attack, is often bathed in perspiration. The duration of an attack under various modes of treatment has averaged nearly three weeks. Sometimes it extends over months ; and the *sequelæ*, or resulting *crippling* of the articulations, may remain for a lifetime.

The *danger* in rheumatic fever consists in the liability to endocarditis and pericarditis. A singular *complication* of it, occasionally met with at a late stage, is *chorea*. Rheumatism may undergo *metastasis* from the joints to the bronchial tubes (rheumatic bronchitis), or, much more rarely, to the membranes of the brain. In feeble persons, the bowels or the womb may occasionally be involved.

The *blood* in acute rheumatism is found to contain an excess of fibrin. *Lactic acid* has, upon some basis of observation and experiment (Richardson), been asserted to be in excess in the blood as the characteristic pathological element in rheumatism.

Apart from the cardiac affections possible in its course, rheumatic fever is not often dangerous to life ; but it is very painful and debilitating.

Treatment.—Many methods have been and still are in use. *Calomel and opium ; opium* alone, or with ipecac., as in Dover's powder; *lemon-juice ; quinine ; colchicum ;* and *alkalies ;* these are the most important. My conclusion upon the subject is, that the *alkaline* treatment is the best by far. Recoveries under it have, in my own practice as well as elsewhere. taken place several times *within a week*, where the symptoms indicated a probably long attack. Carbonate or bicarbonate of potassa, with the Rochelle salt or nitrate of potassa (in scruple doses of the carbonate, or half drachm of the bicarbonate, with about the same of either of the other salts), thrice daily, will answer [F. 37, 45, 46]. Opiates, especially Dover's powder, at night, may do great good. Local application of *laudanum* (detained by oiled silk) to the painful joints, gives great relief.

Lemon-juice has seemed to me a useful adjuvant (tablespoonful doses every three hours) in cardiac inflammations of rheumatic origin.

Quinine is sometimes very beneficial in enfeebled cases, with *free perspiration.* 10 or 15 grains may be given in a day.

Colchicum is of decided service only in the presence of the gouty diathesis.

Remarkable success has recently been reported in the treatment of rheumatism by "flying blisters ;" *i. e.*, the successive application, to different affected parts, of small blisters ; allowed to produce moderate vesication only. Several British physicians laud this practice.

Propylamin I believe to have been fully tried and found wanting in value.

Chronic Rheumatism.—Any one may have this affection, which is, however, most common in those advancing in age. It is a sort of slow inflammation of the fibrous tissues investing the joints and muscles, following exposure to cold and wet. The aching pains are apt to be worst at night.

Cold may produce pain, without any inflammation. Five minutes' exposure to a draught of damp air will often so affect different parts of the body;—relief being at once obtained on the application of warmth. This fact, of *cold directly producing pain*, especially in the muscles, ought not to be overlooked. It supports Inman's and Radcliffe's idea, that pain is always a sign of the local diminution of vitality.

The **treatment** of chronic rheumatism has been largely experimental. The medicines most given are iodide of potassium, guaiacum [F. 167], oil of turpentine, and cod-liver oil. Alkalies and colchicum do not signally affect it. Opium is seldom required unless locally. Local treatment generally does more for it than medicine. For this, various liniments are useful. I have found none better than those containing oil of turpentine, oil of sassafras, ammonia, and laudanum, diluted with soap liniment; or, where pain is considerable, chloroform or aconite liniment. Blisters may be applied in obstinate cases.

Dry cupping to the back, leaving a number of cups on for twenty or thirty minutes at a time, makes a more pervading favorable impression, sometimes, than might have been expected. For rigidity of the joints, and even for pain in them or in the muscles, *pouring hot water* continuously over the parts does great service. The *hot bath*, or *vapor bath*, or, as some prefer, the hot dry-air bath (130° to 200°) will be powerful for relief in many cases. Galvanism also will aid in hastening the restoration of use to the stiffened parts. Wrapping rheumatic joints in *cotton* is often very serviceable. Those subject to rheumatism should wear flannel the whole year.

Syphilitic Rheumatism.—As stated already, this affects the long and flat bones chiefly, and mostly *between* the joints, not at them. Generally there is *nodosity* upon the bones affected, or some degree of periosteal inflammation, at least.

The remedy for syphilitic rheumatism is iodide of potassium. I have never known it to fail to relieve the pains in a very few days. They may return in the course of months or weeks, when the same treatment should be renewed. (Ten to twenty grains of the iodide, thrice daily, will suffice) [F. 168].

Rheumatoid Arthritis.—This designation is applied by authors to a form of subacute or chronic inflammation of one or more large joints, of greater severity than ordinary chronic rheumatism. Effusion into the joint, with deformity and permanent or at least long continued lameness, may occur. I doubt the influence of the gouty diathesis in this affection ; while constitutional debility no doubt often promotes it.

Can rheumatism and gout ever actually be combined in the same patient, in a hybrid attack ? I am sure that they can, at least, be so far blended together, that inflammatory rheumatism, in a patient of gouty constitution, is more affected by the state of the digestive organs, and is more beneficially acted upon by colchicum, than in others. I will consider the diagnostic comparison between rheumatism and gout hereafter.

Myalgia.—Dr. Inman, of Liverpool, first gave this name to *muscular pain* without inflammation or other defined disease. It is more often met with in the *back* and *chest* than elsewhere. Debility and fatigue are its principal causes ; although, as I have said, muscular pain may follow from the direct impression of cold.

27*

Rest, warmth and tonics meet the general indications for the treatment of myalgia. Anodyne applications, as aconite liniment or tincture, or veratria ointment, will be required only in a few cases. The hot bath or douche will often give relief.

Gonorrhœal Rheumatism.—After Sir Astley Cooper, several English and French writers have described a peculiar inflammation of one or more joints, occasionally commencing in the course of gonorrhœa, or even of urethral inflammation from forced catheterism. The local affection may be severe, with suppuration in a few cases, and anchylosis of the joint in many. It appears to be an *ichorhœmic* affection; *i. e.*, the result of absorption into the blood of morbid matter effused into or formed in the membrane of the urethra.

Treatment.—Chambers and Brodhurst, on the ground of experience, recommend *active* treatment for this affection; by moderate bleeding, general in the robust, and local in others; followed by blisters, the hot air bath, muriate of ammonia, and opiates at night.

GOUT.

Synonyms.—*Podagra; Arthritis.*

Gout is a *diathesis*, or constitutional disorder, more or less persistent, with local affections, mostly inflammatory, occurring, in paroxysmal attacks.

Symptoms.—Premonition of a gouty spell is often witnessed for some days, with symptoms of indigestion; flatulence, acidity, constipation, palpitation of the heart. Then (or without such warning) a joint becomes very painful, swollen, red, and tender. In a majority of cases the *great toe* is affected. Other toes, the fingers, ankle, wrist, or knee, may be attacked; the large joints least often. Towards the end of the spell, *tophaceous* or chalk-like deposits (chiefly of urate of soda and lime, altering with time in part to carbonates) are thrown out about the joint, in some but not in all cases.

The suffering with the gouty inflammation is often very intense; but its duration is not commonly more than a few days at a time. Aptness to return, at intervals shortening with each attack, is an unpleasant feature. When the period of release is so short as to be almost absent, it is called *chronic* gout.

·**Retrocedent** or **Misplaced** gout is that in which, instead of the small joints, some internal organ is affected; as the stomach or heart. Such attacks are violent and threatening to life; but generally brief. Exposing an inflamed gouty foot to cold may thus "drive in" the disease, or produce a metastasis.

If the *stomach* be so involved, nausea, vomiting, and sense of spasm or cramp of the stomach are experienced, which, unless relieved in a short time, prostrate the patient very much. When the *heart* is the organ seized, its action is interfered with so as to cause distress in breathing, pallor, faintness, and debility.

The *urine*, during the attack of gout, is scanty, with its usual amount of urea, but a deficiency of uric acid, until near the close of the spell; when the latter is increased. The *perspiration* not unfrequently contains an excess of uric acid and the urates, particularly urate of soda.

Hereditary gout is sometimes genuine *Podagra*, or foot-gout, but more often is of the *wandering* kind. Neuralgia, indigestion, palpitation, and urticaria or eczema upon the skin, are its most common manifestations. In such a system, rheumatism and other affections are to a considerable degree modified by "the gouty tendency."

Morbid Anatomy.—Except the deposits of urates about the joints, and the proved excess of uric acid in the blood, the only peculiar alteration belonging to the anatomy of gout is, the shrinking and granular degeneration (with some deposit of urate of soda) of the kidney; the "gouty, contracted kidney" of Todd. The urate deposit is pathognomonic of gout.

Pathology.—Garrod has established the doctrine of the characteristic of gout being *excess of uric acid in the blood.* The *origin* of this excess is still doubtful. The view of Mialhe is plausible, that, urea being more highly oxidized than uric acid, deficiency of oxygenation of the blood may increase the amount of uric acid in it, unchanged.[1] Also, imperfect action of the kidneys may, by their not depurating the blood fully, induce the same accumulation.

Causation.—High living, with indolent habits, generates gout. Even excess of animal food, with scanty exercise, I have known to produce it. But strong wines and malt liquors much increase the tendency. Weak wines do not seem to have the same effect. In the Rhine region gout is rare. Nor do spirits produce it readily ; their effects, when abused, are different, though worse in the end. Hereditary transmission of the gouty constitution is very common.

Diagnosis.—Between gout and rheumatism there is great resemblance; and, as I have observed, they may be blended together. When clearly exemplified, the following differences exist :—
In gout, the small joints are chiefly affected ; in rheumatism, the larger joints. Repetition of attacks is much more frequent in gout; their duration is greater in rheumatism. In gout, the heart is seldom attacked, and *spasmodically ;* in rheumatism, the heart is often subject to *inflammation.* In gout, the stomach is sometimes spasmodically affected, with violent symptoms; in rheumatism, almost never, although the bowels may be.. In gout, and not in rheumatism, uric acid (or urate of soda) is in excess in the blood. In pure gout, colchicum generally does good; in pure rheumatism, hardly ever.

Treatment.—During the attack, colchicum and the alkalies are the remedies. Wine of the root (some prefer that of the seeds) of colchicum may be given in ten to twenty drop doses several times daily. The stomach and bowels are sometimes irritated by large doses ; but, for a few days, most patients will bear fifteen drops thrice daily. It should be stopped when relief has been obtained. Carbonate of potassa, ten to thirty grains at once, with half drachm doses of Rochelle salt, will be important in addition [F. 37, 45, 46]. Opiates or other anodynes may be craved by the patient during the extremity of his pain.

Shall any local application be made ? Not cold to reduce the inflammation. More than one death has occurred from this, by repulsion of the disorder to the heart, stomach, or brain. Laudanum may, I

---

[1] *Headland* and others advocate a quite different view.

believe, be safely applied to the part, as in rheumatism, by wetting a piece of linen or muslin with it, laying it on the painful joint, and covering it with oiled silk. Alkaline washes (not too cold) are sometimes used.

Gouty attacks affecting the stomach or heart spasmodically are usually sudden, violent, and prostrating; requiring prompt stimulation, as by brandy, laudanum, Hoffmann's anodyne, chloroform, or Warner's cordial (tinct. rhei et sennæ). Small or moderate doses of one or another of these should be given at *short intervals*. Mustard plasters to the epigastrium, or chest, and back, will be important; and the feet may be placed in hot mustard water for revulsion.

Breathing oxygen has been lately proposed as a remedy for the gouty state of the blood. Its utility has not yet been decided upon by sufficient trial.

The *prevention* of attacks, by the removal of the diathesis and predisposition, is often very difficult, even in the absence of hereditary taint. Regulation of the diet is of primary importance. But it should not be too low, especially when the patient's habits have been those of a free liver. Nourishment must be full, while the digestive power is economized, and positive stimulation avoided. Exercise, in proportion to strength, should be recommended. In some weak or old cases, tonics may be called for; vegetable bitters particularly. The state of the *skin* and that of the *bowels* are important.

Change of air, travelling, and mineral waters, are generally useful during the intervals between the paroxysms. Alkaline springs and baths (such as that of Vichy in France, or Ems in Germany) have an especial reputation as prophylactic against gout.

SCURVY.

Synonym.—*Scorbutus*. This affection was once very destructive to voyagers at sea, and explorers of barren regions, as well as, sometimes, to large armies. Captain Cook has the credit of proving the preventive value of vegetable food. Dr. Lind, his cotemporary, published a work on scurvy in 1757, advocating the anti-scorbutic use of oranges and lemons. Still, in their Arctic expeditions, Drs. Kane and Hayes were much incommoded by this disease. In the Crimean war, and during the late rebellion in this country, although uncomplicated scurvy was not very frequent, the *scorbutic diathesis* modified other diseases and increased mortality to a serious extent.

Symptoms.—Languor, debility, and lowness of spirits first occur. Then swelling, sponginess, and bleeding of the gums are observed; the teeth loosen, and the breath is offensive. Palpitation of the heart and dyspnœa may be present. Petechial spots (from subcutaneous extravasation of blood) appear on the limbs; sometimes the legs swell from fibrinous deposits, especially at the ham. Diarrhœa and dysentery often come on. Death may take place by gradual exhaustion, or by sudden syncope.

Diagnosis.—*Purpura hemorrhagica* is undoubtedly not identical with scurvy, although "purpuric" extravasations are common to both. Purpura does not depend, as scurvy does chiefly, upon a fault of diet; nor are the gums affected in purpura.

Causation and Pathology.—That the essential cause (*sine quâ non*) of scurvy is deprivation of fresh food, and, in almost all cases, of fresh *vegetable* food, is proved. Fresh meat will retard it, in the absence of vegetables; but neither this nor oranges and lemons will altogether prevent it, through long periods. Additional *promotive* causes are severe cold, fatigue and exposure, and mental anxiety or home-sickness.

Further than this, the pathology of scurvy has not been determined. The hypothesis that it depends upon *deficient alkalinity* of the blood is disproved by the failure, in many hands, of potassa and its compounds to hasten the cure, or insure prevention.

Treatment.—Medicine here is almost valueless. Fresh vegetables alone will restore what is wanting, though chemistry has not detected the nature of the need. Potatoes, tomatoes, oranges, lemonade are the three most generally available articles. If any medicine is useful as an adjuvant, it is the tincture of the chloride of iron, in moderate doses. Sometimes citric acid does good.

For the gums, a wash of solution of tannic acid or tincture of myrrh in diluted glycerin will be useful; or alum, brandy and water. Salt and whisky rubbing to the skin will aid in dissipating the petechiæ.

Prophylaxis.—Medical men in charge of expeditions to a distance from ordinary supplies should always insist on measures being taken to furnish enough fresh vegetables, or, next best, *desiccated* potatoes. After the latter, onions, tomatoes, turnips, &c., and oranges and lemons rank. *Wine* is also decidedly though not infallibly scorbutic. The leaves of the pokeberry plant (*phytolacca*) and of the *cactus opuntia*, are so. Raw meat is better, in the arctic regions, for the same end, than that which is cooked. The experience of the Army of the Potomac, during the late war, in the M'Clellan campaign, shows that neglect of the means of preventing this disease will sometimes cost far more than those means themselves, whatever difficulties they may seem to present.

SCORBUTIC DYSENTERY.

This term appears prominently in the sanitary and medical reports of the armies in the Crimea. In the peninsular campaign in our late war (just alluded to above), the Chickahominy region was the seat of a great amount of disease, partly febrile (typho-malarial fever) and partly scorbutic. While on duty in the summer of 1862 in two U. S. General Hospitals in this city, I met with many such cases. A record was especially kept of thirteen deaths in the Fourth Street Hospital, from what I then designated as "scorbutic marasmus."

These men were brought from the Chickahominy very much emaciated, pale, feeble, without appetite, almost without power of digestion, and with moderate diarrhœa. Vomiting occurred in some. Blue or nearly black purpura or petechial blotches appeared on their arms and legs; in the *fatal* cases, over the breast and abdomen also. But one of our men recovered in whom the extravasations occurred on the breast; a considerable number in whom only the limbs were so affected.

The diarrhœa was in none of them so great as of itself to threaten

life. Several improved under treatment for a while, and then relapsed
into a condition not unlike in aspect to the collapse of cholera; in
which they died. Two, after seemingly great improvement for a week
or more, died *suddenly*. It seemed that, in them, the blood or blood-
making power was hopelessly ruined.

Autopsy, in several of these, and in some patients in another ward
of the same hospital, under the care of Prof. S. D. Gross, exhibited
coincident lesions not often described together. These were, the
signs of extensive follicular colitis, and those of double pneumonia.
The latter invariably affected the posterior portions, only, of the
lungs. Suppuration had occurred in the lungs in two of our cases;
hepatization in three or four more.

The condition of the bowels in those instances was thus recorded
in my notes.

The large intestine, especially the rectum, was extensively inflamed;
with large, rugose tumefaction, the ridges covered thickly by an ash-
colored granular and diphtheritic deposit; the whole surface reddened
underneath this, and the bloodvessels generally enlarged. No pus
was found; and only slight, rare, and superficial spots of ulceration.
The ileum also was affected with marked hyperæmia and swelling of
the mucous membrane, without ulceration.

I give these facts and appearances as matters of medical and pathological history. The occasion of their occurrence, we may well trust,
will never occur again in this country.

SYPHILIS.

Few old subjects have been so completely re-opened lately as that
of syphilis. Twenty years ago, not many denied the unity of the
syphilitic poison (distinct from that of gonorrhœa), while all admitted
the multiplicity of its manifestations. Soft chancre, indurated chancre,
and phagedænic chancre were all recognized, as well as the specific
bubo, and secondary and tertiary syphilis. But now, prominent
authorities urge that at least two poisons exist, productive of venereal diseases, not mutually inoculable or convertible. This I am not
satisfied to pronounce proven. The topic is altogether rather surgical than medical; but, as the physician must often deal with it, I propose to state (perhaps, for brevity's sake, rather dogmatically) what
I conceive to be the most important practical points.

The "Hunterian" chancre is a sore on the male or female genitals,
slightly cup-shaped, upon an indurated base. From ten days to a
month or more elapse usually after infection before the chancre is
perceived. Then it begins as a red pimple, often unnoticed until it
ulcerates. Its secretion is moderate in amount, and scarcely purulent
except under irritation from without.

This is said not to be "auto-inoculable;" *i. e.*, matter from it will
not, if introduced anywhere on the patient's own body, produce a like
sore. The lymphatic glands may become affected, with enlargement
and hardening, not suppurating unless disturbed and inflamed. The
constitutional disease, called in its manifestations secondary and tertiary syphilis, results from infection by this sort of chancre.

The "simple, soft" chancre is described as having no period of incubation, and not commencing as a pimple or tubercle, but as an abrasion, which discharges pus. If a bubo follow it, it is the suppurating kind. The system is said, by recent authorities, not to be involved by this.

Phagedœnic chancre is characterized by unhealthy purulent discharge and a destructive tendency to erosion. Ulceration of the groin may follow its buboes. *Sloughing* chancre may be regarded as the extremest degree of this, observed under conditions of depressed vitality.

Now in the above discrimination between "infecting" and "noninfecting" chancre, the former being considered to be only that with indurated base and non-suppurating buboes, I follow late authors, not my own observation. A moderate, but not inconsiderable experience in the treatment of syphilis, in hospital and private practice, impresses me with different opinions ; viz., that either the hard-based or the simple soft chancre may have a suppurating bubo and a decided constitutional affection. I must assert that I have *seen* those results, repeatedly, follow *both*.

Treatment.—Without claiming opportunity to have put to the test all the different ideas, my conviction remains strong, that for all forms of primary syphilis except the sloughing or the extreme phagedænic variety, mercury is the specific antidote. I have not seen reason to believe in the efficacy, in primary syphilis, of any other medicine, internally administered.

Blue mass, calomel, iodide of mercury, &c., all have the effect. Enough must be given to produce the impression of mercury upon the system ; but decided salivation is not necessary. I never positively salivated more than two men ; one who had a bad chancre under a *phymosis*, and another peculiarly susceptible to ordinary doses. A grain of blue pill thrice daily, or half a grain of calomel as often, or from half to a grain of iodide of mercury [F. 169] twice a day, will do. The *earlier* it is begun with, the better.

Local treatment is also important. The caustic use of nitrate of silver (some prefer the stronger *potassa* caustic), used early, may *kill* the specific disease at the spot. To do so, it must burn out the whole substance of the chancre. After such application, astringent lotions, as lime-water, solution of sulphate of copper (gr. $\frac{1}{2}$ to gr. j in f$\mathfrak{z}$j), &c., may be applied, washing the part gently twice a day with castile soap and water. Many cases thus treated, early, will get well without taking any mercury. In obstinate venereal sores, however, sprinkling the part with powder of calomel is one of the most effectual remedies. The calomel air or vapor *bath* has lately been recommended.

Buboes, if they *inflame*, may be leeched and refrigerated with lead-water or soothed with poultices. When they suppurate, let them be freely opened with a bistoury. When, afterwards, they refuse to heal, the surgical treatment proper for indolent ulcers will be suitable for them ; besides the local use of powder of calomel.

CONSTITUTIONAL SYPHILIS.

Weeks or months after the primary disease, *secondary* syphilis may show itself. Once produced, although sometimes readily curable, it often impairs the constitution for life, and transmits the taint to off-spring.

The affections belonging to secondary syphilis are—peculiar copper-colored eruptions, rupia especially; warts about the genitals; ulcers of the throat; iritis; loss of hair (alopecia); affections of the testicle or uterus. These last, as well as *periostitis* and osseous tumors or *nodes*, cutaneous tubercles, and chronic degenerative inflammations of the brain, spinal marrow, liver, spleen, lungs, &c., are often called *tertiary* syphilis.

General experience and opinion have asserted that constitutional syphilis is not transmissible by inoculation. Some recent experiments have placed this question again " sub judice."

Treatment.—Mercury is available in the treatment of secondary as well as primary syphilis; but its power over it is less absolute. After moderate trial of its impression (especially of the iodide of mer-cury), iodide of potassium may be given; from ten to thirty grains thrice daily. It is an almost certain cure (I have never known it to fail) for syphilitic "*rheumatism*" or bone pains, with or without nodes. Over ulcers of the throat, also, it has great power. Such things, how-ever, often do not *stay* cured; they break out again, as may also the cutaneous eruptions; requiring the same treatment over and over.

Donovan's Solution,[1] internally, and mercurial ointment locally, are the only additional remedies among many proposed and often used, that I think it worth while to name in our brief consideration of the subject. Of course, enfeeblement of the constitution of the patient may require the employment of generous diet, salt bathing, change of air, iron, quinine, or cod-liver oil.

SYPHILIZATION.

Among the most remarkable curiosities of medical history is the attempt lately made to prevent, and even to cure syphilis by inocula-tion with the syphilitic virus. Auzias Turenne, Sperino of Turin, and Broeck of Christiania, have especially urged their assertions of suc-cess with this process. The immunity is said, like that of vaccination, to last for life. Out of place as it would be to discuss it here, it must be said only, that, after reading a good many pages of the evidence, pro and con, I do not find that, as yet, positive proof enough has been afforded to overcome the strong *a priori* improbability of it. Also, those who advocate it admit that it is a slow method of cure, as well as far from agreeable; and as to its *prophylactic* use, few physicians, at all events in this country, are likely to recommend it to their pa-tients instead of avoidance of the cause of the contamination.

GONORRHŒA.

Very few words must suffice us upon this topic. Gonorrhœa is a *specific urethritis;* in the female, also, vaginitis; produced by impure sexual congress. Its symptoms are, pain and soreness, redness and swelling, of the penis, with early and continued suppurative discharge.

[1] Liquor Hydrargyri et Arsenioi Iodidi. *Dose,* 3 to 5 drops.

Burning pain on passing water, and *chordee*, or painful rigidity of erection, are the principal causes of suffering, while the patient is at rest. Walking about aggravates very much the soreness and pain.

Urethritis, or balanitis (inflammation of the glans penis), may occasionally be brought on by contact with the matter of leucorrhœa, or the menstrual discharge. No perceptible difference exists in the symptoms, in this case, from gonorrhœa; but the latter is more obstinate, and is itself directly contagious. Such non-specific urethritis is, moreover, a very rare disorder.

The period of *incubation* of gonorrhœa is sometimes but a day; seldom many days. Its duration is ordinarily from ten days to three weeks. But a *gleet*, or chronic discharge, more or less muco-purulent, without active inflammation, may be left behind, of indefinite continuance.

Sympathetic non-suppurating *bubo* may attend gonorrhœa; so may also *orchitis*, or inflammation of the testicle. *Gonorrhœal rheumatism* is sometimes met with, ascribed to a metastasis or repulsion of the local affection to some of the joints.

Treatment.—At first, during the height of the inflammation, rest in bed, low diet, Epsom salts, and free draughts of flaxseed tea, comprise the best treatment. It is true, there is a period at the end of incubation, when the symptoms are just *commencing*, when *abortive* treatment may be practised; as by a strong injection of nitrate of silver (gr. vj to gr. x in f℥j) into the urethra. This is a *bold* and rather uncertain measure, however.

Bathing the penis frequently in warm water is very soothing to the pain and soreness. *Chordee* may be treated by that means, and by suppositories of opium and cocoa butter. A pill of camphor and belladonna (camphor five grains, ext. belladonna half a grain) at bedtime, will be useful in preventing chordee.

As soon as the activity of the urethritis has subsided, injections may be used; of nitrate of silver (gr. j to gr. iv in f℥j), acetate or subacetate of lead (subacetate, gr. xv in f℥j), sulphate of copper (gr. j in f℥j), sulphate of zinc (gr. ij in f℥j), or chloride of zinc (gr. j in f℥j). Glycerin may be added to the water for either of these solutions, with advantage.

Copaiba and cubebs are, time out of mind, medicines for gonorrhœa. Without any *specific* antidotal properties, they come in well, one after the other; first the copaiba, and then the cubebs (in half fluidrachm doses of the former, in mucilage, and ten to twenty grain doses of the latter), when the inflammation is subsiding [F. 174, 175].

For *gleet*, which is often very annoying, local treatment, with regulation of the diet (avoiding stimulants and condiments), must be depended on. Injections, of the same character as those above alluded to, may be repeated. Should they fail, a bougie smeared with an astringent ointment should be introduced every day or two, and left in the urethra for ten or twenty minutes. Ointment of nitrate of mercury; of carbonate of lead; spermaceti ointment; and ointment of nitrate of silver, are all recommended. A *flexible* bougie, of cacao (cocoa) butter will irritate the least. Very obstinate cases have sometimes been cured by the introduction of solid nitrate of silver by the *porte-caustique*.

28

Examination with the *endoscope* (recently introduced for specular examination of the urethra) may detect the exact spot which is the seat of the irritation and discharge. Blistering the perineum is practised by some, for gleet. Constitutional treatment, by tonics, may be called for when general relaxation maintains the complaint.

SCROFULA.

Prof. Aitken[1] defines (scrofulosis or) tuberculosis as follows:—
"A particular morbid condition of the system, attended by a persistent increase of temperature, followed by a continuous wasting of the body and the growth of a substance in various tissues and organs, especially the lungs, to which the name of tubercle or tuberculous matter has been applied. These phenomena are associated with peculiarities of outward appearance during life, and liability to certain diseases termed scrofulous, such as swellings of lymphatic glands and of joints, carious ulcerations of bones, frequent and chronic ulcerations of the cornea, ophthalmia, abscesses and cutaneous pustular eruptions, persistent swelling and catarrh of the mucous membrane of the nose, and characteristic thickening and swelling of the upper lip—lesions which, while they are distinguished by mildness of symptoms, are peculiarly persistent, and follow the application of exciting causes which would have no effect on a healthy person."

Scrofula is the term applied commonly to those of the above named local affections involving (most frequently in rather early life) the glands, bones, nose, ears, and eyes. The tubercular diathesis has already been sufficiently considered for our purpose and space. (See *General Pathology.*) A very few words of a practical bearing must be added.

The **causes** of scrofula are, chiefly, *hereditary transmission*, and *deprivation of pure air*. The former is well known to all. Baudelocque, McCormack, and Greenhow, among others, have proved the latter most thoroughly. All depression of the system by low living, such as insufficiency of food and warmth, &c., will promote it. It has been imagined, not proved, that the syphilitic taint of constitution may glide into it.

In **treatment** of scrofula, in any of its forms, but particularly in chronic enlargements, with or without cheesy softening, of the lymphatic glands (of the neck, armpit, or groin), iodine has general confidence [F. 176]. It is not, however, infallible. Iodide of ammonium (dose 3 grains) is now coming under trial. The external application of iodine to tumors, scrofulous or other, "to produce absorption," will very frequently disappoint. I am not sure that it has, locally, any effect but as a stimulant or irritant. That may sometimes be useful.

Cod-liver oil is also an anti-strumous remedy of great power; and one more readily taken by the young than by adults, generally. Iron may be serviceable in many debilitated scrofulous cases. Sea bathing and sea air are mostly the best of remedies. . Good diet is indispensable.

The *local* treatment of so-called scrofulous affections is to a great extent surgical. Slowly softening glands may sometimes be cut out.

---

[1] Science and Practice of Medicine, vol. ii. p. 188.

Scrofulous periostitis, threatening caries, I have seen arrested by free application of *cerate of carbonate of lead* over the affected bone. The leg is most frequently the seat of such disease; but it may attack any of the long bones. Removal of diseased or necrosed portions is to be recommended rarely, unless they are *loosened*. Extensive resections should be very exceptional.

RICKETS.

Synonym.— *Rachitis.* Infants upon learning to walk show the cachexia to which this name is given, by yielding of the bones, with muscular debility, and general failure of nutrition. The bones are brittle from imperfect development; the spine is apt to become curved, and the limbs crooked. The teeth are backward in coming, and fall out with early decay. Tenderness of the surface of the body, and irritability of the nervous system, also exist.

- **Treatment.**—*Hygienic* measures are of the first consequence. Well aired rooms, warm salt bathing, milk or beef-tea diet, cod-liver oil, iron, and phosphate or hypophosphite of lime, all have their value.

CARIES OF THE SPINE.

Synonym.—*Pott's Disease.*
In scrofulous children of either sex, between two and fifteen years of age, sometimes without, but oftener after a fall, blow, or other mechanical injury, caries of the body of one, or occasionally two or three of the vertebræ may occur. The *dorsal* region is most frequently attacked.

Symptoms.—Pallor. debility, pain in the abdomen,[1] in sudden and severe paroxysms; irritability of temper, stooping forward in walking, rigidity of muscles, a cautious, gliding gait, to avoid concussion of the spine; loss of appetite, swelling of the belly, uneasy sleep, hurried or impeded respiration; tenderness of the spine on pressure, an *angular deformity* or backward projection of a portion of the spine; paralysis in various degrees, abscesses of the back, discharging externally, or by the lungs, bowels, vagina; or, the pus entering the hip-joint.

Treatment.—Dr. Henry G. Davis,[2] of New York, claims, and I believe with reason, to have introduced an important improvement into the treatment of caries of the spine. Of the older methods, the best idea was *rest* to the back, with careful efforts at extension; and, especially in this country by the late Dr. John K. Mitchell, support (by means of corsets) dependent upon attachments quite outside of the body. Dr. Davis, reasoning upon the fact that the *bodies* of the vertebræ are the seats of the destructive process, aims at *separating* these, throwing all the weight upon the oblique processes. The spine is relieved then by *straightening* rather than extending it.
An apparatus of Dr. C. F. Taylor carries out this and other rational

[1] Dr. B. Lee (Angular Curvature of the Spine, 1867) speaks of "gastralgia" as an *initial symptom*.
[2] Conservative Surgery, 1867. Dr. Davis' first publication on the subject was in the Boston Medical and Surgical Journal, August, 1852.

principles of treatment very well. It is thus described:[1] "A broad band passes around the trunk low down, so low that in front it almost touches the thighs in sitting. It passes just above the pubes and entirely below the abdomen, so that the abdomen is sustained upward, instead of being, as in most instruments, pressed downward. There are two pieces or levers passing up the back, not over the spine, but each side of it, so that it is firmly held from lateral deviations. At the top is a cross-piece in the form of two T"s with the small ends united. The object of this arrangement is that the straps may pass directly forward and around the arms, and thus prevent a great loss of force by diagonal action; and also that they shall touch the person only where the pressure is needed, namely, on the forward part of the shoulders. At a part of the instrument opposite the seat of the disease, the point where we make our fulcrum, the pads are placed. These are made of chamois skin or Canton flannel, and are filled, not with cotton, which soon packs and becomes hard, but with long, elastic African or East Indian wool, which has no felting qualities. These pads are removable when they become compacted. The shoulder-straps and bands around the hips are likewise provided with removable pads to protect the skin against pressure and abrasion.

It will be seen that the instrument, like the spine itself, acts like a double lever with a common fulcrum at the curvature; this action is directly backward at the hips and shoulders, and directly forward at the middle of the back, or wherever the diseased part is located. . . . . . The instrument is provided with several hinges, *stop* hinges in *front*, but free to bend *backwards*, which allows the most unrestrained use of the muscles of the back . . . . . useful in causing the development of the spinal muscles instead of binding them up and causing their atrophy, as results from the use of instruments which prevent muscular action."

Constitutional treatment, by fresh air and sunshine, nourishing diet and cod-liver oil, iron, or other tonics, as well as purgatives (if required, as they are in most cases) must be added, of course, to mechanical means. Cures are thus sometimes effected in cases once thought hopeless.

Lateral curvature of the spine is very different—depending upon muscular weakness or inequality of development. Bad habitual positions often cause it. Training the subject of it to *use his muscles properly*, and thus develop and strengthen them, must be the leading idea in its treatment, apparatus here being quite secondary, though perhaps sometimes temporarily needful.

COXALGIA.

Synonyms.—*Morbus Coxarius; Hip-Disease.*

Though regarded, like spinal caries, as rather a "surgical" subject, a few words may not be out of place upon this theme also. Its *etiology* appears to be like that of disease of the spine; a constitutional tendency, tubercular or scrofulous, acted upon in many, though far from all cases, by a local injury. Inflammation of the hip-joint occurs, in

---

[1] Angular Curvature of the Spine, by Dr. B. Lee, p. 70.

some instances acute and violent, oftener active only at first and to a moderate degree; not rarely insidious in approach.

Symptoms of the most characteristic kind are, pain in the knee, without any other sign of disease about that part; and a limping gait, the knee being bent, the child treading only on the toe of the affected limb. Examining the hip-joint, it is found that pressing the head of the thigh-bone into it gives pain. Atrophy of the muscles over the hip may follow. General weakness and emaciation, with other symptoms of the scrofulous cachexia, usually attend. *Suppuration* in the joint, with chronic abscesses, ulceration of the cartilages, subluxation of the femur, and caries of the bones, with hectic fever and progressive debility, occur in severe cases.

Treatment.—Physick's celebrated treatment was, absolute rest of the joint by means of a carved splint, passive exercise in the open air in a carriage or, if a young child, in arms,—and systematic purgation with jalap and cream of tartar. To this, with less stress upon the not at all indispensable purging, Dr. Davis has added the use of *continued elastic extension* of the limb, so as to relieve the joint of the pressure of the head of the bone in its socket, caused by the contraction of the muscles. This continued elastic extension may be obtained in bed, by adhesive plaster strips, to which is suspended, by a cord and pulley, a *weight*, proportioned to the amount of power the muscles display, and tested by the comfort secured by it to the patient. Out of bed, a splint may be applied, maintaining elastic extension by a perineal band, best made of adhesive plaster spread (as proposed by Dr. Davis) upon *twilled* material, and kept for awhile before use so as to lose its unctuous property and remain more securely in place.

Simple inflammation of the hip-joint may, of course, follow an injury; and may find relief in a comparatively brief time, from rest, and local antiphlogistic measures, as cups, a blister, etc.

ANÆMIA.

Something has been said of this subject under *General Pathology*.

The **causes** of anæmia are, most often, either 1. Loss of blood, from disease or injury causing hemorrhage. 2. Excessive suckling in a mother or wet-nurse. 3. Severe or protracted diarrhœa, or (more rarely) leucorrhœa. 4. Typhoid or other form of fever. 5. The malarial influence, sustained for a considerable time. 6. Deficiency of food, light, warmth, or fresh air.

Anæmic **symptoms** are, pallor, slenderness of figure, debility, nervous excitability, cardiac palpitation. Anæmic *murmurs* in the heart and aorta have been mentioned under *Semeiology*.

In the **treatment** of anæmia, *good diet, pure air*, and *iron* or cod-liver oil are the essentials. Of the preparations of iron, numerous as they are, I have found the most satisfactory results from the tincture of the chloride, the pill of the carbonate (Valleix's mass), the iodide (syrupus ferri iodidi), the phosphate, and, in children, the citrate [F. 202, 203, 204, 205]. Dr. Aitken speaks very highly of the value of a combination designated as the "syrup of the phosphates of iron, quinine, and strychnia." This formula will be given at the end of the book [see F. 213].

28*

CHLOROSIS.

This not very common affection of girls, about the age of puberty, is by some regarded as simple anæmia; by others, as a pathologically distinct affection. Symptomatically, it is characterized by a peculiar waxy, yellowish, or greenish pallor of the face. The lips are also nearly colorless, and (as in common anæmia) the tongue often pale. Œdema of the feet or of the face may occur; or a dark circle may appear around each eye. Weakness, nervousness, and palpitations exist, with somewhat lowered temperature of the body. Ringing in the ears, lowness of spirits, and disturbed sleep are common. Digestion is impaired; and a *morbid appetite* is sometimes present, as for coal-ashes, slate-pencils, chalk, earth, or, in other cases, strong acids. Neuralgia, affecting especially the *abdominal* parietes, or myalgia, may occur. Menstruation is either absent (amenorrhœa), irregular, or painful (dysmenorrhœa).

The *blood* in chlorosis has been found deficient in corpuscles, and containing an excess of fibrin. One of the curiosities of medical history is the fact that crude theory led at one time to the employment of venesection in its treatment, to diminish the amount of fibrin, whose excess was supposed to constitute it an inflammatory disease !

The *duration* of chlorosis is variable. It may be protracted for years. It is perhaps never, alone, directly fatal.

In **treatment**, measures adapted to anæmia are suitable. Good diet, sea bathing, change of air, *light* gymnastics, iron, bitter tonics (sometimes even strychnia or nux vomica in small doses) will all have their place. Certain cases do not bear iron well, from tendency to fulness of the head. Some, even, chlorotic but not anæmic (though not on the *fibrin* theory), need to be relieved of that symptom by the application of a few leeches or cut cups to the back of the neck.

Special attention to the menstrual function will be demanded. Of this, a few words will be said in another place. (See *Amenorrhœa*.)

BERI-BERI.

This endemic disease of Ceylon and a part of Hindostan, being nowhere else met with, needs here to be only defined. This will be done in the words of Dr. Aitken:—

" A constitutional disease, expressed in the first instance by anæmia, culminating in acute œdema, and marked by stiffness of the limbs, numbness, and sometimes paralysis of the lower extremities; oppressed breathing; a swollen and bloated countenance. The urine is secreted in diminished quantity. The œdema is general, not only throughout the connective tissue of the muscles, but the connective tissue of solid and visceral organs in every cavity of the body is bathed in fluid. Effusion of serum into the serous cavities very generally precedes death."[1]

This disease may occur either in the acute or chronic form. Death may follow in a few hours, or be delayed for several weeks.

[1] Science and Practice of Medicine, vol. ii. p. 83.

Intemperance promotes it. But there must be some undiscovered element of *local* causation.

Tonics, stimulants, and generous diet would seem to be indicated in the treatment of beri-beri. Some native medicines have a reputation in India; but the management of the disease does not appear to have been satisfactory. Death is seldom averted, either from the first attack or after relapses.

LEUCOCYTHÆMIA.

We have defined this affection already. (See *General Pathology.*) The history of its discovery, which has been subject to controversy, appears to be, in brief, as follows. Dr. Craigie of Scotland reported (*Edinb. Med. & Surg. Journal*, vol. lxiv. 1845) a case of disease of the spleen, examined also by Dr. John Reid, in which a peculiar appearance of the blood occurred, supposed by them to be "purulent." Dr. Bennett of Edinburgh, in 1845, published an account of a similar case, describing it as "suppuration of the blood." A month latter, Virchow of Berlin described a case, presenting the same appearances under the microscope, as *leukæmia*, or white blood, asserting the view that excess of the colorless corpuscles, *not* suppuration, was the true nature of the affection. While, then, the first *facts* were Dr. Craigie's, the credit of discerning the pathology which explains them belongs to Prof. Virchow.[1] The first diagnosis of the disease during life was made by Dr. Fuller of London, in 1845.

The **causes** of leucocythæmia are, exposure to cold and wet, prostrating diseases such as typhus, typhoid, or puerperal fever, and affections of the lymphatic glands or of the spleen, often of undetermined origin.

Its **symptoms** are, debility, swelling of the abdomen, anasarca, often vomiting or diarrhœa, jaundice, and hemorrhages from the nose or gums. The spleen, and, sometimes, the liver, are enlarged. The lymphatic glands are often so, also. Cough may occur; and so may pustular eruptions. The tendency of the disease is towards death, and it is doubtful whether any case, well marked, has been cured. But it is slow, and may extend over many years.

Diagnosis of leucocythæmia is only possible by microscopic examination of the blood. A drop from a needle prick of a finger will suffice; placed under a microscope of 250 diameters or more. Instead of being but one to fifty of the red corpuscles, the white blood-cells may be one to six or four; perhaps even one to two or three. When a larger quantity of blood is drawn, it has, after heating, a whitish or milky look. Its coagulum is grayish-white on its surface, from excess of the colorless corpuscles. After death, coagula are found in the heart, consisting of such corpuscles almost alone.

We have said that the *cure* of leucocythæmia has not yet followed any of the many remedies tried for it. No doubt life may be prolonged under it, by hygienic management, and tonics. Nitric and nitro-

---

[1] Prof. Bennett's labored defence of his own claim to priority does not, I think, at all contravene the above view. See his "Clinical Lectures, &c." 2d ed. (N. Y.), p. 892.

muriatic acids are recommended; the latter by the bath as well as internally.

PYÆMIA.

"Absorption of *pus*," as such, through the walls of bloodvessels, being shown to be impossible on account of its cellular nature, the pus cells, moreover, being too large to pass through the capillaries, other views are now advanced. Under the name pyæmia, indeed, several affections are included. 1. *Septicæmia* or *ichorhæmia*; *i. e.*, blood-contamination from absorption, in a liquid state, of putrescent or otherwise morbific material; 2. Transfer by veins of actual pus, in cases of phlebitis, and its deposit in new localities; 3. *Thrombosis*, or coagulation in a vein during life, followed by *embolism*, or the con-veyance of a portion or portions of coagulum to different parts, causing irritation, or obstruction.

That inflammation of a vein (phlebitis) does not very unfrequently occur, there is no doubt. But the external coat and surrounding con-nective tissue are generally most involved; and suppuration of its internal lining is rare. Coagulation is much more frequent. Embol-ism. however, as well as thrombosis, may, and often does, take place, without any of those general symptoms to which the name of "puru-lent infection" is given. Most properly, I consider, the name pyæmia should be restricted to cases in which, to cite the words of J. Simon, "some diseased part (which need not be an external wound) so affects the blood circulating through it, that this blood afterwards excites destructive suppuration in parts to which the circulation carries it— namely, commonly first in the lungs, or (in certain cases) liver and lungs, and later, generally about the body." Putrid infection, septi-cæmia or ichorhæmia, may occur without local suppurations, but with symptoms otherwise similar. Clinical convenience may readily excuse the designation, common with many, of such cases, by the same term, pyæmia.

Symptoms of such an affection are, chills, low fever, rapidity and feebleness of the pulse, prostration, delirium, and swelling of the. joints. Death may occur in a few days, from devitalization of the blood; or, if purulent formations occur, by exhaustion caused by their presence and discharge.

In the **treatment** of pyæmia or septicæmia, support and depuration of the blood are the indications. *Pure air* is not only preventive, but positively curative of such affections. Of medicines, the attention of the profession is just now especially called to the sulphites and hypo-sulphites, of soda. lime and magnesia, proposed by Prof. Polli, of Milan, as antiseptic remedies. They are under trial. Several favora-ble cases of their use are reported; although, in the U. S army, during the late war, disappointment was experienced by a number of those who employed them. Sulphite of soda may be given safely to the extent of four or five drachms daily; the bisulphite (Wood), about half as much, or less. It is certainly proper to give these remedies a fair and prolonged trial.

EMBOLISM.

Cruveilhier, many years ago, proved that in inflamed veins a clot is formed, principally fibrinous. Gulliver ascertained that a granular degeneration of the central layers of such a coagulum may occur, giving a "puriform" character to their substance. Virchow then demonstrated that portions of such clots may be carried from their first seat in the circulation, and form *plugs* in the pulmonary or some other artery. Afterwards it was shown (Paget, Druitt, Kirkes, Good-fellow, &c.) that not only *thrombosis* in veins, with or without inflammation, but also inflammatory or degenerative deposits on the heart's valves, may give off *emboli* or floating masses, which may obstruct the arteries of the lungs, liver, brain, or other organs, causing atrophy, or irritation and inflammation. The septic degeneration of the debris of such clots may also contaminate the blood,—causing septicæmia or ichorhæmia.

Emboli are, apart from their origin, chiefly *arterial* or *venous* in their locality. The arteries most often so obstructed are, those at the base of the brain, the internal carotids, the femoral, brachial, splenic, renal, external carotid, and mesenteric arteries. One obstruction is apt to be the source of others. Cessation of the pulse of the arteries in a limb is an early positive sign. Gangrene is usually the last and fatal event if an extremity be involved.

When the right half of the heart has received an embolus, and the pulmonary artery is obstructed, collapse of the lungs, partial or entire, follows. Pleurisy, hemorrhage, or bronchitis may occur also, Or, the symptoms may be, great anxiety and dyspnœa, with reduction of the temperature of the body. A systolic murmur may be heard on auscultation; the rhythm of the heart becomes irregular; and pulsation of the jugular veins may be noticed. Giddiness may be present, with blueness and œdema of the hands, feet, or both. Death occurs in much the greater number of cases of embolism.

Where emboli have become broken up and decomposed, septicæmia results—commonly known as pyæmia; as before explained. The temperature in this disease is commonly high; from 106° to 107° in the evening exacerbation.

For **treatment** of embolism, without septicæmia, our only resources are *rest, support* by food and stimulants, and alleviation of nervous disturbance by opiates.

ANGEIOLEUCITIS.

Definition.—Inflammation of a lymphatic vessel.

Causation.—Any local irritation or injury may cause a neighboring lymphatic to inflame; but it is especially apt to follow a *poisoned* wound. Erysipelas may be attended by it. Dissecting wounds almost invariably produce it. In my own person this has happened several times; once, the absorbed matter so affected the whole trunk of the lymphatics proceeding from the right thumb, as to cause a large abscess in the axilla, with a severe illness. This experience has enabled me to arrive at a somewhat clear conclusion as to the nature and consequently proper *treatment* of "dissecting wounds" which, from want of care in prevention (by *sucking* and washing the part thoroughly at the moment of the injury) have been allowed to bring on local and lymphatic inflammation.

The pathognomonic *sign* of angeioleucitis is a distinct and somewhat elevated *red line* up the limb, or the part, with tenderness well-marked throughout its course.

That produced by a dissecting wound is, as I have proved, an inflammation, which may be quite *sthenic ;* not necessarily "typhoid," as some have imagined. I am sure that the free application of foreign leeches to the hand, and a large dose of Epsom salts, aborted one attack, which was threatening to be severe. Of course some cases may be asthenic or typhoid ; but of all that I have seen, with three examples in my own person, none have been so.

In ordinary angeioleucitis, the application of a light muslin or linen rag wet with lead-water and laudanum, allowed to evaporate, will be suitable. The part must, also, of course, be entirely at rest.

WHITLOW.

Synonym.—*Felon.*[1] The frequency with which this comes under every physician's notice makes it a proper topic for brief remark here. A felon or whitlow is a suppurating inflammation of one or more of the fingers. Velpeau's subdivision of its varieties is as good as any, into : 1. Sub-epidermic. 2. Subcutaneous. 3. Fibro-synovial. 4. Periosteal. The first is trifling, the second may be severe for several days, the third may cause great suffering for two or three weeks and lame the hand, the fourth threatens the loss of a phalanx or of the finger. Many practitioners always divide an inflamed finger down to the bone as soon as it is manifest that the inflammation is sure to progress. Velpeau advises early incision only in that form in which *periostitis* exists. I believe he is right. The only difficulty is in making sure of the diagnosis. But I would, upon experience, lean towards the doubt, and wait for suppuration unless satisfied of the deep-seated nature of the attack.

Leeches sometimes, water-dressing or irrigation, and poultices, comprise the rest of the treatment.

ONYCHIA.

Synonym.—*Paronychia.* Inflammation followed by suppuration or ulceration about the root of the nail. Injuries generally bring it on, but cachectic constitutions are most liable to it. The nail may become loosened, so as to be removable. Much more rarely, the last phalanx of the finger or toe suffers necrosis. Poultices, lime-water, solution of sulphite of soda, &c., with rest to the part, in bed if it be a toe, comprise the usual means of treatment.

ONYXIS.

This is commonly, but improperly, called *in-growing nail.* The great toe is its much most frequent seat. It is an inflammation of

---

[1] The term *p⟩ronychia* is best restricted to cases occurring near the nail.

the soft parts near the nail; their swelling pressing upon the latter; *not* the nail growing toward or into the flesh. The difference is important in reference to the treatment. For this, the patient must remain in bed, or at least avoid walking, until the inflammation of the toe subsides. Then the ulceration may be treated, if extensive, with lime-water, solution of sulphate of copper. &c.; and, if fungous protrusion of indolent granulations (proud flesh) exists, with touches of solid nitrate of silver every day or two. After this, or in milder cases from the first, a little strip of lint or cotton smeared with simple cerate or cold cream should be gently and carefully worked, with the back of a small knife blade, or the head of a large needle, in *between the flesh and the nail*, to be left there. Adhesive plaster may be put on so as to draw the flesh *away* from the nail, which is first trimmed closely and smoothly at its edge. Then *paint the parts thickly with collodion*. This makes an artificial cuticle; the cure will generally be rapid and complete, unless in malignant disease of the matrix of the nail itself. I do not think that the nail need ever be removed.

CARBUNCLE.

Synonym.—*Anthrax.* Though approaching or passing the bounds of surgery, the same reasons will excuse a word about this affection also. The **causes** of it, as well as of *furunculus* or boil, are undetermined. Boils and carbuncles are positive *opprobria medicinæ;* no one knows how to prevent them or to stop their continued recurrence. I have known ten or twenty boils or carbuncles to follow each other, in spite of purgatives, low diet, strong diet, tonics, refrigerants, alteratives, and even the sulphites, all tried in turn.

Carbuncle is a subcutaneous phlegmonous inflammation, more extensive than a boil, and attended by a larger sloughing of connective tissue under the skin; with much more pain and constitutional disturbance. It may even threaten life. The swelling is round, and flattened on its elevated surface. Redness may exist for some distance beyond it.

In the **treatment** of carbuncle, besides emollient poultices or warm water dressing under oiled silk, all surgeons agree that, at an early period, the tense skin must be divided or removed, to allow the extrusion of the slough and detained pus. Many make a *crucial* incision, quite across the tumor each way. Velpeau prefers a *radiated* incision; from the centre in several directions, extending a little beyond the circumference of the tumor. Probably no method is better than to congeal the part with Richardson's or some other spray-producer, with rhigolene or ether, and apply *caustic potassa* freely, until the whole top of the carbuncle is deeply blackened. Poultices, &c., will of course be afterwards required. Generally the patient requires to be supported by good diet, and, perhaps, tonics.

ADDISON'S DISEASE.

Synonym.—*Melasma Supra-renalis.* Hardly any clinical association of morbid changes is more obscure in its pathology than this. A bronze-like discoloration of the skin comes on gradually, preceded and accompanied by symptoms of anæmia and debility (muscular weakness, feebleness of the heart's impulse and pulse at the wrist, short breath upon exercise, impaired digestion, sometimes dimness of vision); after lasting from less than one year to four or five years, death occurs, and the only characteristic lesions are found to affect the supra-renal capsules. Dr. Wilks has been almost as prominent as the late Dr. Addison in the study of this disease. Dr. Greenbow has lately written an excellent monograph (Lectures) upon it.

It is manifestly a *cachexia.* Probably both the supra-renal capsular disease and the affection of the skin (olive-greenish darkening, mulatto-like, or like bronze without the gloss) depend upon the constitutional state. Perhaps caries of the vertebræ (scrofulous), which has been sometimes observed, may, by involving the *ganglia* in disease, thus produce the complex errors of nutrition, superficial, and general. Dr. Wilks describes the appearances of the supra-renal capsule as resembling those of scrofulous lymphatic glands; a lardaceous material being deposited, which afterwards softens into a putty-like mass (grayish translucent material with yellow cheesy nodules), or undergoes drying into a chalky concretion. The disease is fatal always, at last. Besides hygienic management, and perhaps iron or other tonics, little treatment is recommended for it. Dr. Greenbow asserts decided advantage to have followed the use of a combination of glycerin, in two drachm doses, combined with fifteen or twenty minims each of spirit of chloroform and tincture of chloride of iron.

GOITRE.

Synonym.—*Bronchocele.* In low and narrow valleys of the Alps, Andes, Himalayas, or other mountains, but especially often in Switzerland, whole families and village populations are affected with (congenital or early) enlargement of the thyroid gland, which sometimes becomes enormous. A stranger, upon a residence for a few months in one of the same localities, may be likewise affected; and, after leaving it for a high and open, salubrious country, may recover from it. Associated often, but not always, with this affection of the neck, is **cretinism**; a condition of bodily and mental weakness, stunting and deformity, most lamentable.

Occasionally, in any locality, a case of goitre or enlargement of the thyroid gland to a slight or moderate degree may be met with. I do not remember to have seen more than half a dozen cases of it in Philadelphia; none of them severe.

Causes of goitre, among the mountains, are believed to be, 1. Excess of magnesia and lime in the drinking water; 2. Dampness and deficiency of light; 3. Other unfavorable hygienic conditions; among them, frequent intermarriage of near relations in a stationary population.

In the **treatment** of goitre, *iodine* has had the reputation of a specific. It is not, however, infallible. Dr. Güggenbuhl many years since proved that the best management for goitrous and cretin children was to remove them from their valley and village homes to high, airy, and light situations, and there to give them good food, exercise, and other appliances of a health-producing regimen.

DISEASES OF THE SKIN.

As a clinical classification of cutaneous disorders, most convenient both for description and treatment, I prefer the following:—

Exanthemata.	Pustulæ.	Tuberculæ.
Papulæ.	Squamæ.	Hæmorrhagiæ.
Vesiculæ.	Maculæ.	Neuroses.
Bullæ.	Hypertrophiæ.	Parasiticæ.
	Syphilida.	

EXANTHEMATA.

In these, there is active congestion or hyperæmia of the "derma" or true skin. Besides scarlatina, measles, and erysipelas, already considered, this order contains *erythema, urticaria,* and *roseola.*

Erythema.—Superficial, circumscribed red patches, of variable shape and size, on the face, trunk, or limbs, not painful nor very sore, characterize this. Its causes are, all moderate but somewhat continued irritants to the skin. Its duration is generally but for a few days or a week or two. No fever attends it; nor is it either contagious or dangerous.

Varieties[1] of erythema are, *erythema fugax,* or fleeting; *erythema intertrigo,* from friction of two surfaces of the skin, as in not well cleaned children; *erythema rheumatica,* occurring now and then in rheumatic fever; *erythema pernio,* or unabraded chilblain; and *erythema nodosum,* on the legs, with rounded node-like prominent red patches, somewhat more inflamed than in the other forms.

Treatment of erythema must depend upon its cause more than upon its particular form. The stomach and bowels may need attention, with the use of antacids and laxatives; especially magnesia and rhubarb or Rochelle salts, or the citrate of magnesia.

Local applications may be, finely-powdered starch or arrowroot, dusted on, dry; cold cream (unguent. aq. ros.); lime-water and oil, equal parts (olive, or lard oil); ointment or glycerole of zinc [F. 177, 178, 179, 180]; glycero-cerate of lead; or glyceramyl [F. 148]. For erythema *pernio,* or frost-bite of mild degree, astringents are serviceable; as bathing the feet in tepid infusion or decoction of oak-bark, or solution of alum; or applying cerate or glycerole of carbonate of lead. Some recommend cabbage leaves.

Urticaria.—*Nettle-rash.* Elevated round or oval, red or white, patches or *wheals* characterize this. They may come and go in an hour, over the arms, trunk, or legs. Much burning, stinging, or itching attends them. The affection commonly lasts only a week or two; sometimes it is chronic and tedious.

Disorder of the stomach (as from unwholesome food) is rather more likely to cause nettle-rash than any other kind of eruption. Mild purgatives, especially salines or the antacid magnesia, with or without

---

[1] Here, as in other affections of the skin, only the *principal* varieties are named. Wilson makes sixteen varieties of erythema.

29

powdered charcoal, are commonly suitable for it, after a dose (two or three grains) of blue mass. Light diet is necessary. Vinegar and water, glycerin and rose-water, or the starch-powder, etc., mentioned for erythema, will answer for local applications. Much use of *cold* lotions should be avoided, lest the eruption be over hastily repelled, inducing gastric, hepatic, or other internal disturbance instead.

Roseola.—Bright, and yet generally *dark* red, damask rose-colored patches, irregular in shape and of various size, over any parts of the body, without much if any fever, belong to this affection. It is generally of but a few days' duration. Sometimes a certain amount of resemblance is presented by it to scarlet fever or measles; but the peculiar sore-throat of the former, and the catarrhal symptoms of the latter, are wanting.

Scarcely any treatment is called for in roseola; no local application, as the rash is but slightly irritating; and only such medicine as the general condition of the patient may indicate.

PAPULÆ.

These, *pimply* eruptions, involve *depositive* inflammation of the skin; which is raised in small, red, round, or conical points or minute tubercles, not very hard, and often, though not always, transitory. Papular affections are *Lichen* and *Strophulus*.

Lichen.—Pimples numerous, but of small size; red, and more or less heated and irritated. The principal forms of it are *lichen simplex*, common on the face, neck, etc., *lichen tropicus*, or *prickly heat*, and *lichen agrius*. The last named is the most inflamed and painful; sometimes quite severe. Lichen *simplex*, though mild, may be obstinate in its persistence; annoying ladies, sometimes, by remaining long on the face. In lichen *tropicus*, from which children, especially, often suffer in summer time, the eruption is not prominent, but the sense of irritation is very unpleasant.

Lichen *agrius* may become, in violent or neglected cases, a scabby confluent eruption, with cracks or fissures, and a serous, perhaps purulent discharge. This is not, however, very common.

Treatment.—Even for the simple form, and still more for l. agrius, constitutional alteratives are likely to be needed, doing more good than local applications. In l. tropicus, starch-powder, glycerin and rose-water, or glyceramyl, or weak lead-water will suffice, without any medicine. But in the other forms, rectification of any error of *balance* in the system must first be made. The plethoric must have low diet; the anæmic, lean meat, perhaps bitters, aromatic, sulphuric, or nitric acid, or iron. Costiveness must be overcome, as, by cream of tartar and sulphur, rhubarb and aloes, or other mild but decided laxatives. Blue mass may be given, a grain twice daily for two or three days. Then, arsenic may be prescribed; of Fowler's liq. potass. arsenit., three drops twice daily at first, increased every week one drop until ten twice daily have been taken; omitting the remedy if headache, nausea, diarrhœa, or puffiness of the face occur.

In lichen agrius, rest in bed may be required; with lime-water and oil dressing, or poultices of bread and milk, or flaxseed meal, or slippery elm bark powder, glyceramyl, etc.

Strophulus.—Red gum is a common name for this papular eruption of infancy. Indigestion, reflex irritation from dentition, and over thick clothing or living in hot rooms, produce it. The eruption is not severe, consisting of many small red pimples, close together, and often nearly all over the body. Attention to the stomach and bowels is necessary. *Lancing the gums is proper* (all authorities to the contrary notwithstanding) if they be swollen, tender, or so tense as evidently to distress the child. To the rash, only very soothing applications should be made, as starch-powder, ointment of oxide of zinc [F. 181], or glyceramyl. Care with the diet, if fed instead of being nursed, is of course also of great importance.

VESICULÆ.

These are *effusive* inflammations of the derma; characterized by numerous and small water blisters; the smallest are *sudamina;* the largest, *herpes;* eczema having vesicles of intermediate size, and scattered. *Sudamina* are met with in low fevers, consumption, &c., mostly when perspiration alternates with the febrile state in an enfeebled system.

Eczema.—This has been the subject of much disputation; as to whether it is a *disease per se*, going through stages not only of effusion, but also of incrustation, suppuration, desquamation, &c.; or, only a phase of cutaneous irritation and inflammation, called vesicular, whatever its cause, and eczematous to distinguish it from the herpetic eruptions. Unable to decide this question with positiveness, I am satisfied, nevertheless, that, while the eczematous vesicular eruption admits of very distinct description and recognition, it may come from or after a papular rash, and may in the same case be transformed (or progress) into a pustular or scabbing disease.

Eczema *simplex, rubrum, infantile*, and *impetiginodes* are its principal varieties. Besides others named in the books, there are also eczema *solare*, from heat, and eczema *mercurialis*, from the impression of mercury on the system. The *simple* form has but little inflammation; but there is always some soreness, and the vesicles may run together and break, oozing serum or lymph, or scabbing lightly. Eczema *rubrum* is more inflamed, with redness, heat, and some tumefaction. *Crusta lactea* or milk crust is a name often given to eczema *infantile* of the nursing time. It affects the face, sometimes very unpleasantly; scabbing, running and cracking all over it. E. impetiginodes appears to be an intermediate stage, or transition, between eczema and impetigo; water blisters appearing at first, and pustules afterwards.

Treatment.—An inflammatory state attends the eczematous eruption, nearly always; especially in *e. rubrum* and advanced *crusta lactea*. Saline laxatives, diuretics and diaphoretics (Rochelle salts, bitartrate of potassa, citrate of potassa, &c.) are often called for, perhaps to be repeated in moderate doses. Light diet is, in like case, proper. In children, small doses of calomel occasionally do good. Locally, weak lead-water when there is no scabbing; lime-water and oil when there is great irritation; decoction of bran; flaxseed infusion with bicarbonate of soda (∂j in f$\mathfrak{z}$iv); glyceramyl; glycerin with rose-water; carbonate of lead cerate; ointment of oxide of zinc; these are among the many applications used with advantage. The whole bath, tepid or slightly warm (never hot) two or three times a week, will be beneficial. In chronic eczema, the "Turkish" or dry, hot air bath (130° to 150°) is highly recommended by some.

Chronic eczema acquires alterative treatment internally. Arsenic is *the* alterative, par excellence. in obstinate cutaneous affections. Its peculiar action on the skin tends to displace the morbid process, and thus to restore, after its own transient influence is withdrawn, healthy nutrition and reparation. Five drops of Fowler's solution may be given at first, twice daily, increased gradually until the dose amounts to ten drops; sometimes even more. The medicine must be intermitted if the head, stomach, or bowels show its decided action. In case of its failure, particularly where syphilitic taint is possible, Donovan's solution (liq. arsenici et hydrargyri iodidi) may be given; three drops at first, cautiously increased. Scrofulous or otherwise feeble children may need cod-liver oil. In *crusta lactea*, or *eczema infantile*, the mother or nurse must be instructed not to burden the child with clothes, nor keep it in an overheated room. Daily bathing is particularly important to an infant suffering with such an eruption.

Herpes.—This has larger, more separated and less numerous vesicles than eczema; it is less apt to be chronic. **Varieties:** *herpes phlyctenodes, herpes zoster*, and *herpes circinatus*. The first is the most frequent; receiving also local names, according to its seat: as *h. labialis, præputialis*, &c. *Herpes labialis* is commonly called "fever blisters."

Herpes zoster is singular, but not very common. Half of the body, about the waist, is covered with vesicles, on an inflamed red surface. Sometimes neuralgic pains, quite severe, attend it. It generally affects the right side. Its duration is but for a week or two; unless in the feeble or old, in which it may be followed by ulcerations of a tedious, perhaps dangerous character.

Herpes circinatus is distributed in circular patches or rings. *Minute* vesicles appear around the circumference. By these, and the absence of the microscopic vegetation, and less disposition to chronicity, it is distinguished from *tinea tonsurans*, or true contagious ring-worm. *Herpes iris*, of writers, is an aborted *h. circinatus;* the rings being incomplete.

Herpes rarely appears in old persons; often in children and adolescents. All causes of irritation of the surface of the body may cause it; as febrile or catarrhal attacks, stimulating diet, violent exercise, &c.

For the **treatment** of herpes, the plan stated for eczema is, in principle, here also suitable. Cucumber ointment may be added to the applications recommended. Herpes *zoster* requires confinement to bed. The severe pains, in this, may call for anodynes. Herpes *labialis* is sometimes very annoying, especially to ladies. Pure *cologne-water* applied at the very start, may *abort* the vesicles. Magnesia powder is used by some to dust about the lips. Calomel ointment is recommended when the eruption is chronic, coming out in successive crops.

BULLÆ.

These are eruptions of *large* vesicles. *Pemphigus* and *Rupia* are the most distinct.

Pemphigus.—Bullæ of a circular or oval shape, from half an inch to two inches in diameter, and flattened. They may be distributed over any or all parts of the body. Fever, sometimes considerable, precedes and accompanies the eruption. I have seen it a very serious illness. After the vesicles mature, they burst, or dry away, leaving thin brown scabs. Ulceration may occur, but it is not deep or. obstinate, unless in a particularly unhealthy constitution. The duration of pemphigus is from one to three weeks, or more in bad cases. *Pompholyx* is the name given to a rare variety of pemphigus, in which the space continuously covered by bullæ is large, and there is little or no fever. A fly-blister causes artificial pompholyx.

Pemphigus is not usually considered to be contagious. One family came under my notice, however, in which five individuals were attacked by it, partly in succession, after travelling. It was difficult in that case not to suppose contagion.

In the **treatment** of pemphigus, gentle refrigerant laxatives at first, diuretics and diaphoretics next, and, often quite early, tonics and supporting regimen are called for. In one case I was obliged to stimulate quite freely ; the eruption being as confluent as in any case of small-pox, and prostrating, like an extensive burn. No local applications, other than the mildest lotions or unguents, will be suitable. The early puncture of each bullæ with a small needle is recommended ; but the raised cuticle must not be removed.

Rupia is probably but a modification of pemphigus ; with smaller blebs or bullæ. followed by thicker conical scabs, of dark color ; after whose removal ulcers are left, which may be weeks in healing. Rupia *simplex* is the variety in which the scabs are low and the ulcers slight ; rupia *prominens* in which they are elevated into irregular cones ; rupia *escharotica*, when the ulceration is deep and extended. *Syphilitic* rupia is quite common ; but every case of rupia is not, by authorities, admitted to be syphilitic. My observation goes to sustain this non-admission.

Treatment of rupia requires to be, generally, tonic and alterative. Quinine, cod-liver oil, and iodide of potassium, with good but simple diet, are apt to be wanted for it.

PUSTULÆ.

Suppurative inflammation of the skin (excluding smallpox, furuncle, and carbuncle, as well as the malignant pustule or *charbon* of the French, a rare affection said to be received from cattle) appears in the two forms *Ecthyma* and *Impetigo*.

Ecthyma.—Large, round, prominent pustules, upon any part of the body, not numerous ; ending in thick dark scabs, followed by slight (or in cachectic states, obstinate) ulcerations. Ointment of tartar emetic, or pure croton oil, or other strong cutaneous irritants, will produce it. Often, however, especially in syphilitic persons, or after acute fevers, &c., it occurs without local exciting cause. Sometimes it is chronic.

In **treatment** the causation is of great importance. If a local irritant produce it, local emollients, perhaps with general refrigerants, are to be used for its relief. Otherwise, diet, and *balancive* measures will be more in place : tonics for the feeble, purgatives and light regimen for the plethoric, &c.

As an eliminant and refrigerant in both ecthyma and impetigo (as well as rupia) I have found apparent benefit from the use of a prescription employed by Dr. Anderson of Scotland; equal parts of *wine of colchicum* and *wine of ipecac.*, say ten drops of each, thrice daily [F. 182]. Arsenic is called for in obstinate cases, as in other diseases of the skin ; Fowler's or Donovan's solution, in small doses carefully increased.

Impetigo.—Small and somewhat numerous pustules : varieties, *impetigo figurata* and *impetigo sparsa*. I. *figurata* is most common on the face, in circumscribed clusters of pustules, which may become confluent and scab. To this, in children, as well as to *eczema infantile*, the name of crusta lactea is given by authors. I. *sparsa* has the pustules scattered over more or less of the whole body.

Treatment.—When much irritation or inflammation exists, lead-water, glyceramyl, ointment of oxide of zinc, lime-water and olive oil, flaxseed tea and bicarbonate of soda, light poultices of flaxseed meal, slippery elm bark, or bread crumb, are to be applied. Daily use of castile soap and water is serviceable. ·Purgatives may be needed. Diet must be according to the general condition of the patient. Impetigo may affect the hairy scalp ; if so, the hair must be cut and kept very short. Colchicum and ipecac. may be given in acute cases ; arsenic in those which become chronic.

SQUAMÆ.

Scaly diseases are, *Lepra* (*Alphos* of Wilson), *Psoriasis, Leprosy of the Hebrews, Spedalsked* or Norwegian leprosy, *Pityriasis*, and *Ichthyosis*.

Lepra.—Always chronic, and very difficult to cure. Not regarded as contagious, though I have seen it occur successively in four persons in immediate contact (an infant at the breast, its wet-nurse, another infant suckled by her, and her husband). It is characterized by red desquamating patches, of various sizes, approximating to a circular shape, on any parts of the body ; especially on the arms and legs. Besides syphilitic lepra, its varieties are lepra *vulgaris*, with small patches and few thin scales, and lepra *inveterata* (*alphos diffusus* of Wilson) where they are large and desquamate extensively.

In both, the *margin* of the patch is the highest, reddest, and most squamous part.

Psoriasis.—Described under the names of ps. *vulgaris, gyrata*, and *inveterata*, psoriasis differs mainly from lepra in the irregular and varied forms of the desquamating patches ; and in the absence or less degree of depression near their centres. Wilson's view that psoriasis is only a kind of chronic eczema, does not seem to me to accord with the facts of its ordinary history. It is sometimes hereditary ; as is also lepra. No disease of the skin is so hard to eradicate, unless it be ichthyosis.

Treatment.—For lepra and psoriasis alike, all sorts of alterative agencies, local and systemic, are, if cautiously used, suitable for tentative practice. Our object is, to obtain the *making of a new skin*, unaffected by the morbid habitude of nutrition. Frequent bathing should be practised. Tar ointment, citrine ointment, ointment of sulphuret of potassium [F. 183], &c., may be applied. Arsenic, and the iodide of arsenic and mercury (Donovan's) should be given, carefully, but repeatedly, through long periods. Other medication must depend upon the conditions of each case.

Ichthyosis (Fish-skin disease).—This is rare; I have seen but one case of it. Hard, thick, dry scales form, continuously, over a part, or, sometimes, nearly the whole surface of the body; without much redness, soreness, or even itching. It is congenital and incurable. Frequent and thorough ablutions, and mild emollient applications, are palliative to it.

Pityriasis.—This is a chronic affection in which very numerous small white scales (dandriff) form upon the skin, particularly the scalp (p. capitis). Some redness, and often a good deal of itching, may attend it. It is difficult of cure in many cases. If it be upon the head, keeping the hair short, and washing daily with castile soap, followed by a spirituous lotion, or glycerin and rose-water, will do the best for it. Cleanliness and frequent bathing in tepid, cool, or, if the vigor of the system permit, cold water, are of essential importance in all cases.

The term *pityriasis versicolor* is sometimes applied to an *epiphytic* disease (*i. e.*, one connected with a vegetable parasitic growth), better called *chloasma versicolor*.

Spedalsked is a disorder only known in Norway and Sweden; especially among the fishermen. Accounts of it are given in medical journals and books;[1] but the mere reference to it will suffice here. (See *Elephantiasis Græcorum*.)

Leprosy of the Bible (Lepra Hebræorum) is of great historical interest.[2] It is still recognizable in the East, though not frequently met with. I saw a case of it in Alexandria, in 1859.

In the Book of Leviticus, three varieties of leprosy are described: dull or darkish white "freckled spots,"—dusky or shadowed,—and *bright white* (*bahereth lebhana*), the worst of all. *Tsorat* (whence *psora*, and sore) or malignant disease, was applied to the last two only. *Lepra* is an early Greek synonym of this term.

Mason Good thus describes the old leprosy: "A glossy, white, and spreading scale upon an elevated base; the elevation depressed into the middle, but without change of color; the black hair on the patches, which is the natural color of the hair in Palestine, participating in the whiteness, and the patches themselves perpetually widening their outline."

In favorable cases, after spreading over much of the person, though

[1] See Brit. and For. Medico-Chirurg. Rev., 1850, p. 71.
[2] See Neligan's Treatise on Diseases of the Skin, edited by Dr. Belcher (Philadelphia ed., 1866, p. 289).

without ulceration, the disease would die out; the scales would dry up and gradually disappear. In bad cases, ulceration would occur, with extensive sores, as well as desquamation. Then the leper was made an outcast, and treated as one dead; "unclean for life."

Not only the books of Moses, and others of the Bible, but also Hippocrates, Galen, and Celsus (under the names λεύχη, and λέπρα λευχή) speak of ancient leprosy as a *white scaly* disease. It thus differs decidedly from either kind of elephantiasis.

MACULÆ.

Ephelis, Vitiligo, and *Chloasma* may be included under this term; perhaps better, under that of *Decolorationes*.

Ephelis; *lentigo.* — Sunburn and freckles best correspond with these names; which, however, are by some authors extended further. Neither are of importance unless in regard to appearance. For the removal of freckles (which often disappear spontaneously with time) or the yellowish-brown spots called *chloasma*, or *melasma*, all applications may fail; dilute nitro-muriatic acid (fifteen to thirty drops in an ounce), left for some time in contact with the discolored spot, is more likely than anything else to take effect.

Vitiligo.—Literally, *veal-skin.* Unnatural whiteness from deficiency of coloring matter. When universal over the body (nearly always then congenital) it is *albinismus*. We see albinoes, sometimes families of them, occasionally, in all the races of mankind; as well as among the lower animals. *Leucopathia*, or white disease, is a name given by some writers to both the general and the local affection.

When local, vitiligo is seen mostly in rounded patches or spots, which slowly increase in size, though without regularity of shape. The head, chest, back, and thighs are the most frequent seats of them. The hairs on the parts involved become white; or fall out, causing baldness—*calvities*, or *alopecia*.

Treatment, for vitiligo, must be, first, general, for improvement of nutrition in the whole system,—and then local. Very hard it may be to cure the affection, although its importance is chiefly for appearance; no danger attends it. *Tannic acid* and *oil of turpentine* are the preferred local applications for it. Total *albinismus* is quite incurable.

Chloasma (pityriasis) *versicolor* will be spoken of under *Parasiticæ*. For *alopecia*, baldness, or premature loss of the hair, very many remedies are in vogue. Shaving the head repeatedly (*i. e.*, after an illness) may often save the hair. Stimulating applications sometimes help and sometimes hurt the case [F. 185, 186].

HYPERTROPHIÆ.

Morbid excesses of development of the skin or tissues connected with it, are thus named; *Nævus, Clavus, Verruca, Elephantiasis Arabum.*

Nævus.—(Mole, mother-mark.) This is always congenital. Discoloration and elevation of the part exists, with abnormal development of the capillaries and small veins of the skin; making a small, commonly flat, vascular enlargement. It is seldom more than an inch in diameter. Erectility sometimes belongs to the vessels of nævus.

Caustic, the ligature, the knife, and vaccination of the part, have all been employed for the removal of such formations. They may leave scars worse than the mole ; the operation ought to be exceptional. I have known it, performed early in infancy, to be quite successful.

Verruca.— *Wart.* A hypertrophy of the skin, with great development of the cuticle especially, upon a small surface ; such is a wart ; of which no one needs a further description. Some persons and families are especially liable to them ; why we cannot say.

Treatment.—Strong nitric acid ; chromic acid ; caustic potassa ; and in slight cases nitrate of silver, carefully applied only to the wart, after paring off nearly all the insensitive portion of it, will always, at least after repetitions, remove warts.

Clavus.— *Corn.* Most persons are well acquainted with this sort of localized hypertrophy of the skin of the foot, from irritating friction and intermittent pressure. Prevention is more easy, by far, than cure. Corns are either *hard* or *soft ;* the latter may become inflamed ; the former hurt only under decided pressure.

Pare a hard corn with a sharp knife or razor, closely, but *not* so as to hurt or draw blood. Soak the foot then in warm water for five or ten minutes, and pick out carefully the centre or "core." Two thicknesses of adhesive plaster, with the centre cut out (making a ring) should be put over the corn ; and a third piece, the centre not cut out, placed upon it and them.

Soft and *inflamed* corns require removal of all pressure for a while, and poulticing, &c., first ; then the above treatment.

Condylomata.—These are fleshy tumors or out-growths, more or less hard and wart-like sometimes, in other cases soft ; of syphilitic origin often, but not always. Especially apt are they to occur about the anus, prepuce, and vulva.

To remove such formations, if they be small and hard, nitric acid, pure, may be used, with care, to limit its contact to the part to be destroyed. When large, and soft, if troublesome enough to require destruction, the ligature is generally preferred. It may be, with a needle, passed through the centre of the mass, and then drawn and tied tightly about the base.

Elephantiasis Arabum.—*Bucnemia Tropica* of Wilson ; "Barbadoes Leg."

Enormous enlargement of the leg, scrotum, or neck, most often met with in warm countries, but occasionally anywhere, is thus called. Hard and nearly immovable, the parts become at last. The connective tissue as well as the dermoid texture proper is greatly hypertrophied. Impediment to the return of surplus material of nutrition by the lymphatics is the probable pathogenetic cause ; the nature of the impediment has seldom been discerned.

Ligature of a large artery is asserted to have arrested the growth of elephantiasis. No other treatment appears to be worth trying for it.

TUBERCULA.

Acne, Molluscum, Lupus, Elephantiasis Græcorum, Frambœsia, Keloid.

Acne.—Tuberculous elevations, from inflammation of the skin around sebaceous follicles, in which the secretion is detained, or is of a morbid character,—are called *acne*. Three varieties may include all those named by authors; viz., *acne simplex, acne pustulosa,* and *acne rosacea.*

Acne *simplex* or *punctata* has small and moderately red, rather hard tubercles, on the face principally. When very hard and chronic, it may be called acne indurata. Black points commonly mark the obstructed follicles. Acne *pustulosa* reaches a more mature suppuration, and is often painful, especially if upon the scalp.

Acne *rosacea* always affects the face; usually in adults, and most often in high livers. A good deal of soreness attends the eruption. First, the pimples are hard, red, and small; as they mature they grow somewhat larger; finally a little sanguinolent pus escapes, leaving a small scab. Rose redness around the pimples, or patches of them, has given rise to the name. It is generally a difficult disease to cure, and very unsightly. Not unfrequently it is hereditary.

Treatment.—Errors of digestion, brought on by gluttony or intemperance, or more moderate imprudence, often cause acne. They must be rectified for its cure. Attention to the state of the bowels, and to the action of the skin generally, is indispensable. Saline cathartics are useful in plethoric cases. Various mineral waters are recommended—saline and sulphurous especially. The pustules, when they mature, should be carefully punctured with a needle, avoiding irritating disturbances. Solution of carbonate or bicarbonate of soda (Ʒj in Oj) in water or flaxseed infusion, will be a good wash. Sulphuret of potassium, in lotion or ointment, is also advised; or ointment or glycerole of nitrate or amide of mercury (hydrarg. ammoniat.) [F. 187]. Obstinate cases justify more decided alterative treatment; as, the application, by a cotton tip upon a knitting needle, of a solution of corrosive sublimate, two to five grains to the ounce of water or alcohol, washing it off in a few moments; or, similarly, of pure Goulard's extract (liq. subacetate) of lead, followed by spermaceti ointment, cold cream, or glycerin and rose-water. Iodide of sulphur ointment (gr. xv to xxx in Ʒj of lard) is also much praised. In acne *indurata*, when very ugly, acid nitrate of mercury (mercury and nitric acid each an ounce) has been applied, and sometimes blistering the face with cantharidal collodion.

Molluscum.—**Acute** molluscum is a somewat *contagious* tuberculous eruption. The small tumors form without inflammation, increasing slowly, till they have almost the size and form of a currant, but without color, and nearly flat-based or sessile. They last from three to six months, either ulcerating finally and then shrinking away, or inflaming and sloughing off, leaving a pit or mark. Several crops of tubercles may succeed each other on the face and neck, in either adults or children, but especially in the latter.

Chronic molluscum is of still longer duration; is not contagious, and the tumors are *pedunculated, i. e.,* each has a stem, in many cases at least; they also become larger, and occur over different parts of the body. Neither form of molluscum is common. It is proper to add that some authorities do not admit the contagiousness of the acute variety. **Treatment** of acute molluscum seems not to be to any great extent available. In chronic molluscum the tumors may be cut off at the peduncle, the divided point being then touched with lunar caustic.

Lupus.—L. *exedens* and non-*exedens*, or l. *superficialis*, *serpiginosus*, and *devorans* (Neligan). Lupus *superficialis* is a rare disease, in which, most often on the cheek, a small, soft, sore, slow-gathering tubercle appears, which in time scabs, and ulcerates superficially, the scab and ulcer spreading for an indefinite time, and leaving behind them a permanent, whitish scam or scar. Irritation may make the tubercle very painful, and deepen the ulcer. It may last for years.

Lupus *serpiginosus* exhibits one or more livid, red, indolent tumors on the face, head, or elsewhere, sore, heated, and itching. In the course of months they become filled with pus, and suffer an undermining ulceration, which finally becomes an open, unhealthy-looking sore, forming upon it a hard, brown scab. Creeping from the edge of its original seat, in irregular rings, the disease extends, leaving behind it a depressed cicatrix. The same part may be again reached by its meandering progress. This is a very chronic affection, even of years' duration, without injuring the general health.

Lupus *exedens* or *devorans* (noli me tangere or *rodent ulcer*) is characterized by continuous destructive ulceration of the skin, subcutaneous connective tissue, muscles, and other parts, at length involving even bones; all following tubercles " rounded and dusky red," on the nose, cheek, eyelid, &c. An ichorous discharge belongs to it; cicatrization follows it, sometimes (as in the previous form) to be again attacked.

Young persons, from ten to thirty, are especially liable to lupus. Its progress is generally an affair of years, and it causes less suffering than its appearance would lead us to expect. Scrofula certainly, and probably syphilis, predisposes to it. It is very difficult to cure; sometimes, at least, incurable. The obvious alliance with cancer has induced some authorities to place lupus in a class of affections called *cancroid*. It differs from cancer, however, in not involving the glands, nor contaminating the general system. Lupus is a comparatively rare disease.

Iodine (as in Lugol's solution), cod-liver oil, and iron, internally, are commonly indicated in the treatment of lupus, especially the *exedens*. Fowler's or Donovan's solution may also, or each in its turn, be cautiously given. Chlorate of potassa has been suggested; I do not know of its trial. Sea-bathing is likely to assist in the treatment.

Locally, the animal oil of Dippel (made by dry distillation of hartshorn shavings) has a reputation in Europe for lupus superficialis as well as for l. devorans. So have dilute solutions of chloride of zinc, nitrate of silver, nitric acid, &c. In the superficial variety, *collodion*, softened -perhaps by adding $\frac{1}{16}$ of glycerin, may be painted lightly over the ulceration, every day or every few days.

Excision is sometimes practised for the exedent form, to prevent disfiguration; but the success of the operation is uncertain. So is that of strong caustics. Among these, nitrate of silver is preferred by most surgeons. Acetate of zinc, used solid for touching the ulcer, and applied every day or two, was much recommended by Neligan. He used also a lotion of the same salt, from three to five grains to an ounce of distilled water. Broadbent's new treatment for cancer, by injection of *acetic acid*, might be worth a fair trial in lupus. Its theory is very plausible.

Elephantiasis Græcorum.—Called by this name among the Greeks, probably because, as the elephant is a great and powerful animal, so is this a formidable disease. It was the leprosy of Europe in the middle ages; for whose treatment many hospitals were built, and an order of Christian knighthood (of St. Lazarus) was established.

It is characterized by many round tumors, from the size of a pea to that of an orange, livid, purple, yellowish or brownish, and soft; on the face and other parts of the body. The skin around them thickens irregularly, giving a repulsive aspect. Ulceration occurs, deepening even to the bones; all the organic functions suffer, and finally the mental faculties become enfeebled; diarrhœa, and perhaps tetanus, precede death.

This disease is probably identical with the *spedalsked* of Norway, already named. Allied to it are *radesyge* of Norway, the *morphie* of Brazil, *frambœsia* (raspberry disease), *Sibbens* of Scotland, and *Aleppo evil* (button of Aleppo; perhaps also the *Ngerengere* of New Zealand. *Pellagra*, of Lombardy, Spain, and France, is described by some as having a certain resemblance to it; but tumors do not belong to this disease; in which, with a general cachexia, the skin becomes discolored and somewhat thickened, with arrest of its normal functional action.

Treatment of elephantiasis and its allies must be upon the principles laid down for other serious cutaneous affections; *viz.*, to endeavor to *restore the balance of the general functions*, whatever may be wrong; whether that be by tonics, refrigerants, or purgatives, or other remedies acting upon the secretions; also improving the nutrition and repair of the skin, by local and general alteratives. I am not acquainted with any *specific* remedy for either of the forms of disease just named.

Keloid.—(*Kelis, Keloïs, Cheloïd, Sclerema.*) This is very rare. I saw one case of it, in a medical college *ambulatorium*, in 1860. Wilson, a few years since, stated that but twenty-four cases of it were upon record; more have been reported upon since. An irregular, cicatrix-like, smooth reddish and whitish, corrugated excrescence, painful, with a stinging sensation, sometimes, but not always; nearly in every case forming upon the front of the chest; slow in growth, not ulcerating, and not tender to the touch. It is not unfrequently spontaneously removed by absorption; but has not been shown to be amenable to treatment. Rayer advises constant firm compression.

HÆMORRHAGIÆ.

Purpura is the only affection of the skin belonging under this head. On parts, or often the whole, of the body, appear round red spots, which become gradually of a dark purple color; and then pass, as bruise-marks do, through green and yellow, till they disappear. They are extravasations of blood into or upon the true skin, from its capillary vessels. The duration of each spot is about a week or ten days. Feverishness may precede, and prostration may accompany purpura. In bad cases, hemorrhages may take place from the mucous membranes, as those of the mouth, stomach, bowels, bladder, vagina, &c.; producing, sometimes, even a fatal result.

Purpura is by some improperly confounded with scurvy. Although extravasation of blood occurs in scorbutus, it also may happen quite independently of it. Deficiency of fresh vegetable food is not at all necessary to engender purpura; the causation and pathology of which clinical experience and chemical investigation have both failed to show.

Treatment.—Although some assert plethora to be, as often as hydræmia (anæmia), antecedent to purpura, my own experience goes with the ordinary view, that rather a tonic than a depletory treatment is generally called for in it. Excessive stimulation, it is true, will aggravate its symptoms. Mineral acids, as elixir of vitriol, and Huxham's tincture of bark, or quinine, &c., are much given. Oil of turpentine is also recommended. Neligan prescribed it in large doses; even an *ounce* at once, with mucilage and an aromatic. This is beyond my degree of confidence in it; but it is said that it acts generally safely as a cathartic in such doses. Ammonio-ferric alum, tincture of chloride of iron, tannic and gallic acids, &c., are used as styptic medicines in some cases. Sponging the body with alum and brandy, or whisky, and water, at such temperature as is not chilling and yet is sedative to the circulation, will be the best local measure.

NEUROSES.

Under this head, of affections involving the innervation of the skin, I class *Prurigo*, *Anæsthesia*, and *Neuralgia cutis.*

Prurigo.—Often placed under *papulæ*, because sometimes minute pimples occur with it,—the essence of this disease really is, intense itching without eruption. It is commonly divided into prurigo *mitis*, *formicans*, and *senilis*. *Pruritus* is the technical name for itching as a symptom.

The difference between the first two varieties is one of degree. In the *mitis*, obstinacy rather than severity exists. In p. *formicans*, suffering may be extreme, pervading the body. Heat of a fire or of a bed, rubbing of the clothes, etc., may cause an irritation which drives the patient to rub and tear the skin, yet without relief. Sleep may thus be prevented, and the bodily as well as mental exhaustion so produced may be great. The complaint is occasionally intermittent. Very often it is confined to one or two portions of the body; as the scrotum, vulva, anus (pruritus *scroti*, *vulvæ*, *ani*, vel *podicis*), etc. Pruritus *ani* is often caused by worms; especially ascarides.

Prurigo senilis is so named because of its frequency in old people. Lice cause it not unfrequently. Papulæ attend it more often than the other forms.

Treatment.—This is sometimes a very hard disease to cure, or even relieve. We must consider and treat the general condition of the body; see that the bowels are regular, the digestion normal, the

30

skin kept clean and open by ablutions and proper change of clothing. Sometimes nervine tonics may be required ; as nux vomica, arsenic, or quinine, in small doses. Tincture of aconite is prescribed by some ; three or four drops at a time twice or thrice daily. Conium, belladonna, and other narcotics have been advised. The *hypodermic injection of morphia* may be resorted to to give rest in very distressing cases.

Locally, many things may, and should, be tried in succession, in the search for palliatives. Baths of flaxseed tea, with or without carbonate of soda or of potassa ; lathering with castile soap, with a shaving brush ; strong salt water, or whisky and salt ; dilute sulphuric, nitric, or acetic acid [F. 190, 191, 192, 193] ; mercurial ointment ; ointment of creasote [F. 224] ; cerate of white lead ; laudanum, sp. camphor, aconite, or chloroform, as lotion or in liniment ; spirituous solution of corrosive sublimate [F. 194] ; solution (dilute) of hydrocyanic acid [F. 195] ; glyceramyl ; pure glycerin ; tar ointment ; olive oil ; tobacco infusion ; the " Turkish," or hot air bath ; and the common hot water bath ; these are only a few of the measures which may be resorted to. The diet should be unstimulating. Advice should be given to the patient also to refrain as much as possible from violence in rubbing or scratching the parts affected ; and not to sleep in a very warm room or under too much cover.

Anæsthesia cutis is only a symptom of a larger affection—involving either the nervous system or the skin itself. It appears in one variety of elephantiasis Græcorum, called by some *lepra anæsthetica.* Vitiligo also is often attended by it, at the parts which undergo discoloration. Except stimulating frictions, when not contra-indicated by the other conditions of the case, and galvanism (faradization), under the same limitations, we have no special remedies to mention for loss of sensibility in the skin.

Neuralgia of the skin, temporarily, at least, limited to it, does undoubtedly occur, though seldom. I have experienced it in my own person. Its locality does not, however, so remove it from other forms of neuralgia as to require for it a special consideration.

PARASITICÆ.

Dermatologists are not all agreed upon the question, whether the *microphytes* or *epiphytes* (minute parasitic vegetations) discovered by aid of the microscope, in connection with certain skin diseases, are *essential* to these diseases, or accidental and secondary only. Wilson even denies their vegetative nature ; asserting them to be results of spontaneous granular degeneration of epithelium. Most authorities hold the opinion, which I fully believe to be correct (especially proved by the results of *treatment*), that the parasites are really the essential *causes* of the disorders they constantly attend ; that they may, under favorable circumstances, be *transplanted ;* and that, to cure those disorders, destruction of the parasitic forms is necessary. Again, Hebra, a high European authority, believes that all the epiphytes described are merely modifications of one and the same species, in different degrees of development. Tilbury Fox agrees with this opinion. E. Hallier[1] makes three series (Mucor, Achorion, Lepto-thryx) of forms, all capable of being educed from the same spores under different circumstances. Devergie believes in *spontaneous generation* of the epiphytes, although truly vegetable. Dr. McCall Anderson[2] gives proofs, by separate inoculation, of the non-identity of *three* vegetative parasites at least—*trichophyton, achorion,* and *microsporon.*

No doubt exists with the large majority of observers as to the cause of the animal parasitic eruption, *scabies* or itch.

Scabies.—Chiefly vesicular, this disease may be papular, scaly, or pustular in some instances. Ordinarily we see, especially between the fingers and on the back of the hand, next often on the arms, legs, and abdomen, occasionally on the scalp, hardly ever on the face,—a number of small red elevations with white or watery tops. Extreme itching is always present; often keeping the individual scratching night and day. King James I. is said to have described his experience of it as rather pleasurable ; but this is not the common account of it.

Closely looking at almost any of the vesicles, one may see a little red line or track, at the end of which may be found a slightly elevated point. In this is, generally, the animalcule—*Sarcoptes hominis* (Acarus scabiei) ; one of the *Arachnida,*—flat bellied, round backed, tortoise-shaped, eight-legged ; the female larger than the male, which is hard to find.

Treatment.—Sulphur is, not the only, but the most reliable and convenient parasiticide for itch. After thorough bathing, and wash-ing of the whole body with soap and water, strong sulphur ointment must be rubbed well into the parts affected. A few applications will usually suffice [F. 197]. The animalcule is killed, and the cure follows. There is evidence, however, that in some cases of long standing, recovery may follow but very slowly. The *habit* of the eruption has then become established in the skin ; this must be treated like eczema, or lichen, whichever it most resembles.

Oil of turpentine, kerosene or petroleum, ointment of sulphuric acid, and other powerful agents, may be also confidently relied upon to destroy the itch animalcule.

Army-Itch.—During and since the late war in this country, the inevitable filth of camp-life begot, among other evils, a very trouble-some contagious skin-disease, called by the above name. Itching, without any eruption except small papulæ, characterizes it. Outside of the army it has extended to a considerable number of persons. No better remedy for this affection, I believe, has been found than a lotion and ointment, composed of iodide of potassium and glycerin ; with water or rose-water for the lotion, and lard or cold cream for the ointment [F. 192, 199]. Mercurial ointment, and sulphuric acid ointment, are also efficacious for it.

The other parasitic affections of the skin depend upon the micro-phytes already alluded to. They are *Favus, Sycosis, Tinea Circina-tus. Tinea Decalvans, Chloasma versicolor,* and *Plica Polonica.*

Favus.—(*Porrigo, Tinea favosa.*) Generally appearing on the

---

[1] Archiv für Mikroscopische Anatomie, April, 1866.
[2] Brit. and For. Medico-Chirurg. Rev., July, 1866, p. 225.

scalp, this disease is peculiar in the formation of yellow cup-shaped crusts, in each of which one or two hairs grow. By joining together, these crusts may lose their regularity of shape, in a general scabbing; and a good deal of hair may fall out. A mealy powder is found in the crusts, which, on microscopic examination, is found to contain the formation called *achorion Schönleinii* by Remak. This presents minute tortuous branching tubes, straight or crooked not branching tubes, and sporules, free or united in bead-like strings. Granules and cellules of *mycelium*, the generative portion of the plant, are abundant. An offensive discharge occurs from the eruption in bad cases.

Favus is contagious, though seldom conveyed to cleanly persons. It is hard to cure, but not incurable. In its treatment, constitutional and local measures must be combined. Arsenic is as usual the most reliable alterative. Neligan has advised the iodide of arsenic, gr. $\frac{1}{16}$ thrice daily; intermitted if headache or dryness of the mouth come on.

For the local treatment, the hair must be *closely cut* with sharp scissors. Apply then a large flaxseed poultice for twelve hours or more,—perhaps repeatedly, to soften the crusts. Next, wash the head thoroughly, by means of a soft sponge, with solution of carbonate of potassa (one drachm to a pint of water); after which ointment of carbonate of potassa (potass. carb. ʒj, glycerin fʒj, adipis ʒj) may be applied spread thickly on lint, covered with oiled silk. This may be renewed daily; or, if there be much discharge, twice a day. The crusts will then come away in a few days. Ointment of iodide of lead may follow; washing the head night and morning, still with the carbonate of potassa lotion; and keeping the hair cropped short all the time. Three or four weeks will generally suffice for a cure. Cleanliness of person and regulated diet are at the same time, of course, essential.

For this and other parasitic affections of the skin, *tar ointment* is a far from contemptible remedy.

Sycosis (*Mentagra*).—This occurs on the bearded part of the face, chiefly the chin. It is contagious; sometimes being transmitted by uncleanly barbers in shaving. It presents slightly inflamed elevations about the roots of the hairs, covered by scurf; shaving decapitates these, inducing irritation and suppuration, as well as scabbing. The whole chin may become swollen and inflamed by it; and parts of the beard may be destroyed. The parasitic cause of this disease is the *trichophyton mentagrophytes* (*microsporon mentagrophytes* of Gruby). It is seen under the microscope to consist of minute stems, bifurcated at angles of from 40° to 80°, and granulated within.

Sycosis is not common. Acne, impetigo, and ecthyma of the bearded part of the face may be confounded with it. It is very hard to cure. In its treatment, keeping the beard constantly very short by close clipping (not shaving) is essential. Sponging twice daily with castile soap and water, or carbonate of potassa lotion, will be beneficial. Iodide of lead ointment, ointment of nitrate of mercury, and of calomel and camphor, &c., may be used in succession; besides the internal use of arsenic.

Tinea Circinatus (*Ring-worm. Scald-head*).—This is known by its circular form, occurring most often, though far from always, on the head or face. *Herpes circinatus* resembles it; but in that minute vesicles are unusual ; in *tinea*, rare and few. In *tinea* a thin powdery crust exists, whose examination will show the *trichophyton tonsurans*, closely allied to the parasite of sycosis.

Tinea decalvans is marked by a destruction of the hair in circular patches, making round spots of baldness. Its parasite is considered by many dermatologists as different from the *trichophyton*, and called *microsporon Audouinii*. Its sporules are rounder and smaller than those of trichophyton.

The **treatment** of both forms of tinea must be, besides cleansing, essentially *parasiticide*. Tar ointment ; "huile de cade ;" mercurial ointment ; solution of corrosive sublimate ; lotion and ointment of carbonate of potassa; lotion of sulphurous acid ; carbolic acid ; creasote ; cantharidal collodion, lightly applied ; these are among the many applications which may be used for the purpose, with generally successful result.

As has been observed, tinea is seldom transmitted to a cleanly person ; at least without very close and continued contact.

Chloasma Versicolor (*Pityriasis Versicolor*).—The parasite of this is *microsporon furfur*. The disease is recognized by the formation of dull, reddish-yellow spots of various size and shape, seldom numerous, on the front of the chest or abdomen. The same local applications may be used for it as for tinea ; besides the internal use of arsenic.

Plica Polonica.—This is an affection of the hairy scalp, endemic in Poland, Russia, and Tartary. The hair-follicles become diseased, and the hair is matted and glued together into felt-like masses. *Trichophyton tonsurans* and *trichophyton sporuloides* are the parasitic vegetations described as found connected with it. The disease has not been seen in this country.

SYPHILIDA.

Enough for our purpose and space has already been said of the general history of syphilis. Among its constitutional manifestations, cutaneous eruptions are very frequent. These are seldom vesicular, not very often papular ; most often squamous or scabbing. Lepra and rupia, particularly the latter, are prominent among syphilitic affections, though both may occur independently of syphilis. All eruptions in persons of this diathesis are marked by a *coppery color*, which remains long, even after their cure ; by a disposition to ulcerate, perhaps only superficially ; and by preference in locality for the face, shoulder, and back.

In the **treatment** of syphilitic eruptions, the diathesis must be met by our remedies. Iodide of mercury internally ; after that, iodide of potassium, and, in feeble persons, cod-liver oil, perhaps iodide of iron ; locally, mercurial ointment (besides palliatives, if required, as in other eruptions), or the calomel vapor bath, should be prescribed. Often, such affections will seem to be cured, but, after weeks or months, will return again ; then the treatment should be renewed, and discontinued when they disappear.

30*

POISON-VINE ERUPTION.

The common poison-vine (*Rhus toxicodendron*),[1] a species of sumach, and one or two other plants more rarely, cause, by contact, in some persons, an inflamed vesicular eruption of considerable severity. The hands and face are its most common localities; but it may come out on the lower limbs or about the anus and genitals. Its duration, when severe, may be from one to two weeks; but it is often quite limited and of shorter course.

In the **treatment** of this annoying but not dangerous attack I have had a good deal of experience in my own person as well as with others. I have found the most relief, and the greatest effect in shortening the course of the disease, by reducing the inflammation, from *lead-water, early*, freely and frequently applied, with a large camel's hair pencil. It should not be put upon the *opened* vesicles, which it irritates; but around them, upon the reddened skin. In the practice of my brother, Dr. E. Hartshorne, a very successful remedy has recently been the *fluid extract of serpentaria*, painted directly upon the eruption. It seems to kill it at once.

FROST-BITE; CHILBLAIN.

Gangrenous destruction of parts, especially of toes, not unfrequently follows actual congelation. Short of this, exposure to continued cold, especially when *suddenly* warmed again, may cause an erythematous inflammation, **erythema pernio**, already mentioned under that head. When the feet or other parts have been so chilled as to be almost frozen, *gradual* warming—for instance, at first, rubbing them with snow—is proper, as a preventive of frosting. In its **treatment** (as remarked before), cooling unguents, as Goulard's cerate, or lotions, as lead-water, may be first wanted, and then astringents, as alum water, infusion of oak bark, creasote ointment, &c. Cabbage-leaves are a popular domestic remedy for chilblains.

BURNS AND SCALDS.

If half of the body be so burned or scalded as to arrest the functions of the skin over that much of the surface, death will always result. *Collapse* comes on, from the terrible shock to the nervous system through the impression on the widely distributed cutaneous nerves. The pulse is then very low, the body cold, and, commonly, thirst is great. Even suffering is often, in a few hours, lost in apathy and prostration.

The **treatment** for this condition must be stimulant as well as anodyne. Brandy or whisky or wine should be given, as freely as in any other condition of positive debility or exhaustion.

For *local* treatment of burns, I believe nothing is better than *lime-water and oil*, equal parts (either linseed, olive, or lard oil) on cotton wadding, covered with oiled silk. Other remedies often used are dry cotton (which sticks too close in deep burns), glycerin, rye-meal, starch

---

[1] This resembles the Virginia Creeper, but is unlike it in being *three-leafed*.

powder, and molasses. To exclude the air seems to be the cardinal object.

UNCLASSIFIED AFFECTIONS.

AMENORRHŒA.

A few words seem appropriate here upon some of those affections of the sexual system which every practitioner must often meet with. Their full discussion belongs to books of a different kind.

Amenorrhœa, or suppression of the menstrual discharge in women, may be either an *interruption* of it, during its occurrence, or its habitual *non-appearance.* The former is commonly the result of cold and wet, or some nervous shock, to which the patient is exposed during the menstrual period.

Habitual amenorrhœa may occur with *plethora,* from disturbance of ovarian and uterine functions; or with *anæmia* and debility; or, as a secondary effect of chronic disease, *e. g.* phthisis. The greater number of cases is in anæmic females; but the opposite state is not uncommon. Vicarious hemorrhages, from the lungs, stomach, &c., sometimes accompany it.

As bearing upon the treatment of amenorrhœa, the question always comes up—is the suppression of the menstrual flow the *cause* of other symptoms or morbid effects, or, is the amenorrhœa itself the *effect* of a morbid condition, the removal of which will restore this arrested function? It is to be said in reply, that sometimes the one, and sometimes the other may be the case. In anæmia with *plethora,* generally the interruption of menstruation may be found to be a primary, though perhaps not the sole, cause of disturbance of the system. In *anæmic* amenorrhœa, most frequently the constitutional state is primary; and the restoration of general strength will be attended by the spontaneous return of the function.

Practically, then, we must, in any case, inquire into the general condition and history of the patient. If there is headache, increased by stooping, with a flushed face, and full, strong pulse, the patient having previously been vigorous in health, taking blood from the lumbar region by cups, or, in clear cases, from a vein in the arm by the lancet, is indicated. Also, purgatives; at first, in a sudden attack, senna, or. if much heat of the system exist, citrate or sulphate of magnesia; afterwards, when the amenorrhœa is obstinate, aloes. Hot mustard foot-baths, or warm hip-baths, and warm poultices to the breasts, every night, should be used in a case of sudden suppression of menstruation in the midst of its time. Tincture of aloes and myrrh is a favorite domestic emmenagogue; a teaspoonful twice or thrice daily, in hot water. Black hellebore, savin, seneka, &c., are also resorted to for similar action; but all emmenagogues are more uncertain, even, than diuretics.

In many cases of amenorrhœa, a delicate, and in some a difficult question, is, as to the possibility of the (physiologically) normal cause of pregnancy being present to account for it. Most of all may this difficulty present, of course, in young single women; who may, unfortunately, have reason for concealment. Apart from the very clear ethical principle, that a physician has no moral right to aid, in any way whatever, in producing an abortion, active emmenagogue treatment in the pregnant state is unsafe for the health of the subject of it herself. *Medicine* will fail to cause abortion, in eight or nine cases in ten, unless it be so used as to produce a serious, often dangerous effect upon the system of the patient.

When we *suspect* pregnancy, then, mild measures only are in place, —waiting for time to develop the nature of the case in full. *Anæmic* amenorrhœa requires tonics; above all, *iron*. Other medicinal and hygienic roborant agencies may also be called in. Aloes, in small doses, repeated daily [F. 201, 202], occasional or periodical hip-baths, foot-baths, and breast-poultices, especially near the time when the menstrual flow should occur,—may in many cases be super-added. *Strychnia*, in one-thirtieth of a grain doses, is a favorite tonic in amenorrhœa with some practitioners. *Galvanism*, or *statical electricity* (of the friction-machine) is much resorted to by others. The spinal and pelvic regions are the seats of the application.

DYSMENORRHŒA.

Painful menstruation is habitual with some women for years together. Pregnancy not unfrequently cures the habit. The affection seems to be of two kinds or origins; 1. functional or **physiological**, and 2. **mechanical** dysmenorrhœa. With the former, disorder of innervation and circulation occurs; even the ovaries may partake in this. I attended one woman in a number of attacks of monthly ovarian irritation (ovarian colic) of extreme violence and suffering, with fever. Ordinarily, before menstruation begins, the subject of functional dysmenorrhœa feels ill, with pain in the back, perhaps headache; followed by pains almost like labor-pains, of the first stage, in the womb. That organ becomes palpably swollen and heavy, its pain being somewhat assuaged by compression by the hand through the abdominal wall. When free discharge comes on, relief is obtained.

The symptoms · of mechanical dysmenorrhœa are not always strikingly different, but it is a more local affection. The direct cause of it is obstruction at the *os* or *cervix* uteri; the external or internal *os* usually, if constriction be the trouble; in the neck, when anteversion, retroversion, or lateral flexion produces it. On the indication of this causation, Dr. Simpson of Edinburgh some years since introduced the practice of *dilatation* of the os and cervix, for the cure not only of dysmenorrhœa, but of sterility also, dependent upon the same obstruction. A *sponge-tent* was used [F. 232]; sometimes, more lately, the *sea-tangle* (laminaria digitata) instead. Simpson and others, however, still more recently, prefer *incising* the neck of the uterus, with a hysterotome; asserting that this is more certain, and even less dangerous. Much discussion on this subject has transpired in late journals. I must refer upon it to works on special surgery; particularly the last work of Dr. Marion Sims. It is observable, however, that Dr. H. R. Storer, the distinguished obstetrician of Boston, adheres to careful dilatation, instead of incision. Drs. Tilt and H. Bennet of London also object to frequent hysterotomy.

Whatever the cause of dysmenorrhœa in any case, the subject of it should always avoid being much on her feet for a day or two before her monthly time; and should go to bed when the pain begins. Cloths wrung out of hot water, or spirits and water, may be placed upon the abdomen, and renewed as they cool. Internally, spirits of camphor, with compound spirits of lavender and hot water (sweetened to taste) may be given [F. 203]; or, if not relieved, paregoric, in teaspoonful doses. The advantage of avoiding much exercise or fatigue just before the time of the expected menses, ought to be impressed upon the patient. No medicine appears to have any important *prophylactic* effect.

MENORRHAGIA.

Excessive menstruation may be of two kinds: 1. its occurrence too often; 2. too great an amount or continuance of the discharge. Both very often occur together. Causes of menorrhagia are,—general relaxation of system; over-excitement of the genital apparatus; thinness of the blood, hemorrhagic diathesis; and over-fatigue, especially on the feet, promoting a descent of blood toward the pelvic organs about the time of menstruation. Ulceration, cancer, or tumors of the uterus, as well as abortion and *placenta prœvia*, cause uterine hemorrhage, not properly to be called menorrhagia.

This affection is much most common in the anæmic. Rest, iron, good diet, and astringents, internally and sometimes locally, are the remedies for it. Tincture of chloride of iron is, here, the favorite chalybeate. It may be given through the interval. During the attack, ammonio-ferric alum, in five grain doses, may be administered; or tannic or gallic acid, three to five grains several times daily. The patient must be kept still upon her back till the flow is controlled. Sometimes cold wet cloths (for a serious hemorrhage) have to be put upon the abdomen; or an ice-water sponge, or half a lemon, or a syringeful of tannin and solution, or of solution of tinct. ferri chlorid. (f3ss in f3viij) may be thrown into the vagina. Plugging, with a tampon of cotton, lint, or sponge,[1] in a few instances may have to be resorted to. In every case of severe or protracted menorrhagia, the practitioner must endeavor to be sure whether or not any malignant or other organic affection of the uterus is present.

LEUCORRHŒA.

Synonyms. *Fluor Albus; the Whites.*—This is a quite common trouble of women. The mucous discharge may be either from the vagina or from the uterine cavity. Irritation of the organs, followed by relaxation, is its general cause; but, often, relaxation alone seems capable of producing it. Procidentia or prolapsus uteri is a frequent source of it; the descended uterus pressing upon the vaginal walls, causing morbid increase of secretion and exhalation from one or both.

In **treatment** of leucorrhœa, tonics are often required; iron, bitters, &c. Also, *astringents*, by the mouth and locally; those mentioned

---

[1] Dr. T. E. Beesley has contrived a light *metallic* conical plug or cork for the vagina; to be kept in place by a bandage.

for menorrhagia will apply here also, but usually in less strength, for a longer time [F. 204, 205]. If prolapsus or procidentia exist, I believe that a well-adapted pessary (gutta percha or India rubber *ring*, or *double horse-shoe* of similar *light* material) will in a majority of cases do good service.

SPERMATORRHŒA.

Referring the reader for a full consideration of this subject to Bartholow,[1] Acton, or other authorities, the main facts only will be here stated. In continent men of full health, an involuntary seminal discharge during sleep once in two or three weeks is common ; and is then so innocent as to be regarded by many as physiological or normal. More frequent emissions are abnormal, in proportion to their frequency; and may cause much loss of strength. While hæmorrhoids, worms in the bowels, &c., may occasionally promote this, the cause of actually excessive spermatorrhœa in ninety-nine cases (at least) in a hundred must be believed to be self-abuse. The cure of this habit is, not always at once, but almost certainly in the end, the cure of the resulting spermatorrhœa. The disastrous effects so obvious in many cases are due first to the vicious habit, and, secondarily only, to the involuntary discharges.

In pathology, Lallemand has, for a long time, been allowed to impose upon the medical mind his opinion that irritation or inflammation of the prostatic portion of the urethra is the general or universal immediate cause of spermatorrhœa. As Bartholow more correctly states, this is quite exceptional. More largely by far, spermatorrhœa shows itself to be a cerebro-spinal *neurosis*. That is, the error is not in the local structure of the urethra, but in the morbid nervous excitability; which renews too often the sexual orgasm, somewhat after the manner of an eclampsia or convulsion, as a reflex act.

It is to be remembered that, in a relaxed state of the system, especially in those whose genital organs have been more or less abused, in natural or unnatural ways, sometimes a *mucous* discharge of small amount may occur from the urethra, like the leucorrhœa of the female. Only the presence of spermatozoa, visible with the microscope, *proves* seminal loss.

What are we to do, then, when consulted by a patient for spermatorrhœa? Ascertain the frequency of the discharges, the state of his general health, and, if possible, his habits. Relieve unwarranted alarm by stating the innocence of *bi-weekly spontaneous* evacuation of the seminal ducts; whose effect is quite different from that of the unnatural violence and mechanical irritation of self-indulgence. Impress upon him, whether the habit be acknowledged or not, that his danger lies in it, and that his cure depends upon entire and permanent abstinence.

To promote this, all *moral* impressions must be brought to bear upon his mind, as well as prudential considerations. Active muscular exercise in the open air (in proportion to strength) should be encour-

---

[1] On Spermatorrhœa, &c. By Roberts Bartholow, A. M., M. D., &c. New York, 1866.

aged, even to fatigue. He should eat very light suppers, sleep under light clothes, rise early, and bathe often in cool or cold water. The shower-bath will do very well. Iron is required in really weak cases, as a tonic. The diet should be nourishing, but not stimulant; avoiding high seasoning, and alcoholic beverages.

Of all drugs said to be anaphrodisiac (*i. e.*, capable of diminishing or quelling sexual appetite) I believe that none have any available power except lupulin and bromide of potassium. The dose of the former for this purpose is ten grains, at bedtime. Bromide of potassium is, however, the medicine of the day for *reducing excitability* of organs subject to reflex action. Twenty grains at bedtime, every night, will, according to my observation in practice, make a great difference with those who are troubled with frequent nocturnal discharges.

Lallemand's *porte-caustique* finds justification only upon his theory of urethral or prostatic disease as the cause of spermatorrhœa. Without feeling warranted in denying the occasional existence of such a lesion, and the possible benefit of limited cauterization in such an exceptional case, I am not prepared to believe in its frequency or great importance.

Acton,[1] however, has confidence in cauterization in a number of cases. He employs a *solution* of nitrate of silver, ten grains to the ounce of water; which he injects into the urethra by means of an instrument consisting of a glass syringe attached to a tube like a short catheter. The part to be acted upon is the irritable membranous portion of the urethra. Before using the caustic the patient should empty the bladder. The pain of the application is considerable. After the operation, Acton advises a copaiba capsule every eight hours, for two or three days; also, that the patient drink as little water as possible, and avoid passing urine as long-as he can. After once urinating, he is allowed to drink watery fluids as usual. The scalding and oozing of blood gradually disappear.

Mechanical means are sometimes employed to prevent nocturnal emissions; *e. g.*, a light metallic ring to surround the penis, having teeth projecting inwards; so that erection awakes the patient. In bad cases, where epilepsy, insanity, or extreme general exhaustion has followed a seemingly incurable habit of self-abuse, circumcision would really seem to be justifiable; more so, surely, than the more serious and dangerous operation of castration. Baker Brown's analogous operation to remove "peripheral irritation" as a cause of grave nervous maladies in the other sex, by excision of the clitoris, has met, recently, with decided opposition from no less an authority than Dr. Charles West, of London, as well as from others.

As signs of waste of substance and vigor by seminal losses, we find mentioned, pallor, with dark lines under the eyes, inability to look any one in the face, cold, moist hands, frequent flushing of the countenance, aversion to society. But these symptoms of general and nervous debility may all exist without being thus accounted for.

---

[1] On the Reproductive Organs, Phil. ed., p. 243.

WORMS.—ENTOZOA.

Helminthology, the study of worms, has assumed of late a very considerable importance in connection with medicine. About thirty entozoa inhabit different parts of the body of man. They have been generally classified as *Cœlelmintha* or hollow worms, and *Sterelmintha* or solid worms, *i. e.*, without any well-defined alimentary cavity. Broad or flat worms, *Platelmia*, and thread-like or cord-shaped worms, *Nematelmia*, constitute another arrangement. Of the flat worms, some are *Cestoid*, or riband-like ; others *Trematode*, or fluke-like. The most important ones are enumerated in the following table :—

Cestoid Worms :

Mature : Tænia solium ;	Tænia echinococcus ;
Tænia mediocauellata ;	Bothriocephalus latus ;
Immature : Cysticercus cellulosæ ;	Cysticercus t. mediocanellatæ.
Echinococcus hominis.	

Trematode Worms :

Distoma hepaticum (fasciola hepatica).	Bilharzia hæmatobia ;
Distoma ophthalmobium.	Tetrastoma renale.

Nematoid Worms :

Ascaris lumbricoides ;	Sclerostoma duodenale ;
Trichocephalus dispar ;	Filaria medinensis ;
Oxyuris (ascaris) vermicularis ;	Strongylus gigas.
Trichina spiralis.	

Most curious are the transformations some of these parasites undergo. Pallas, 1776, stated that all cystic worms were forms of a tape-worm. Steenstrup, in 1842, discovered the "alternation of generations" in some small aquatic worms, *cercariæ*. Kuchenmeister and Siebold proved by actual experiment that hydatid parasites are young or immature tape-worms. Humbert, of Geneva, in 1854, swallowed fourteen *cysticerci*, and in three months discharged fragments of *tæniæ*, which had developed within his body. The immature forms are non-sexual ; they may remain, as in hydatids, for a long time, in solid organs, without development. They only become mature and sexual, capable of reproduction, in parts of the body having some communication with the external air, as in the alimentary canal or lungs ; generally the former. Migration from one part of the body to another occurs with some ; as *trichina spiralis*.

Tænia solium and **tænia mediocanellata** look a good deal alike ; but the former is much the smaller. The immature *cysticercus* of the former is $\frac{1}{10}$ of an inch long : that of the latter, of the size of a pea. The *t. solium* has a circle of hooklets around a convexity of the head ; the *mediocanellata* is club-headed, with larger sucking disks than the *solium* has. One is designated as " armed " and the other " unarmed " tape-worm. The former is from the *cysticercus cellulosæ* of the hog ; the latter from the " *cysticercus bovis*"[1] (Cobbold) ; and is the most common. The unarmed is the easiest to drive out.

---

[1] Cobbold states that the *hydatid* of the cysticercus bovis has never yet been observed in man. Hydatid or "echinococcus" disease is especially frequent in Iceland.

The tape-worm is formed of flat segments, often several hundred in number, connected with the head by a slender neck. Each segment has male and female organs (hermaphrodite) ; as those at the tail mature, they are cast off. Some patients thus pass six or eight fragments from the bowels in a day. The whole length of the parasite is from ten to thirty feet.

The **symptoms** caused by tape-worm are not very determinate. They resemble those produced by other worms ; namely, uneasy sensations in the abdomen, and general nervous irritation ; bad sleep, attacks of *faintness*, and lowness of spirits, indigestion, irregularity of appetite and of the action of the bowels ; itching of the nose, and sometimes of the anus. Epilepsy and insanity are said to have sometimes been caused by it. The only *proof* of tape-worm is the finding of fragments of it in the stools. It is a common impression that it is never destroyed unless the head is discoverable ; but this is not exactly true. Conversely, if the head comes away, the parasite to which it belongs is no longer reproduced. More than one of them may, however, be present at once ; though this is rare.

The broad tape-worm, **bothriocephalus,** is known only in northern central Europe ; Russia, Sweden, Norway, Lapland, Finland, Poland, and Switzerland. Its head is elongated, compressed, obtuse ; its length from six to twenty or twenty five feet. It does not give off detached segments. Cobbold says it is indigenous to Ireland ; although he has never met with a patient born in that country who has been the subject of it.

Treatment of Tape-worm. *Oil of Turpentine,* in half ounce or ounce doses, will generally purge, and bring away the worm. It *intoxicates* some persons. In Egypt, *petroleum* is used for the same purpose, in doses of twenty or thirty drops. The *ethereal extract* (commonly called oil) *of male fern, extractum filicis liquidum,* U. S. Pharm., in the dose of a drachm and a half to two drachms, is esteemed highly by some practitioners. *Kousso,* the flower of the *Brayera anthelmintica* of Abyssinia, in half ounce doses, mixed with water, given on an empty stomach, is almost certain to destroy or remove the parasite. So is said to be *Kameela,* the *Rottlera tinctoria* of botanists. Pumpkin seeds, plentifully taken on an empty stomach, are quite effectual.

Prevention of Tape-worm.—As immature tape-worms find residence in the bodies of animals used for food, and thus get the opportunity to enter the human alimentary canal, the *avoidance of raw or under-cooked meat* is the precept of prophylaxis suggested, and confirmed by experience. This applies not only to the prevention of tape-worm, but, also, to that of other parasites, especially *trichinæ.* Tape-worms are derivable from infested beef, even oftener (Cobbold) than from pork. Mutton has been found occasionally to contain cysticerci.

TREMATODE WORMS.

These are the *Distomata, Bilharzia hæmotobia, Tetrastoma renale,* and others. They are of a flattened oval shape, soft and smooth. They have a bifurcating alimentary canal, with a mouth, but no anus.

Both sexes are upon one individual. They exist in two conditions, mature and encysted, and immature and free. Their methods of reproduction are very curious, but of greater importance in zoological than in pathological science.

Distoma hepaticum, found sometimes in the liver and its ducts, measures about an inch in length when mature, and rather less than half an inch in width.

Distoma ophthalmobium has been found in the eye of a child having congenital cataract. It is about half a line ($\frac{1}{24}$ in.) in length.

Bilharzia hæmatobia is found in great abundance in Egypt; where it inhabits the *veins* of the *abdominal organs* of the inhabitants, in the proportion of nearly one-third of the population. Hemorrhage from the kidney, and the symptoms of dysentery, may follow from its presence. It is not more than three or four lines ($\frac{1}{4}$ to $\frac{1}{3}$ in.) in length. The sexes are on different individuals.

Tetrastoma renale is occasionally found in the substance of the kidney. It is nearly half an inch long.

NEMATOID, OR ROUND WORMS.

Ascaris lumbricoides is the commonest of entozoa. It inhabits mostly the small intestines; but may get into the stomach, and, of course, the large intestines. I have repeatedly known them to be vomited from the stomach. This round worm is from five to fifteen inches in length, light-brown in color, tapering to a point at each end. A considerable number of them may exist together; it is only then that their presence in the bowels is likely to do much harm, unless in very susceptible children. Their escape into the stomach may cause nausea, vomiting, and indigestion, sometimes difficult to account for until the throwing up of the worm explains the cause. I have known this to happen in an adult, in whom the symptoms of gastric irritation continued for two or three weeks. These worms probably enter the body chiefly in the drinking water of shallow wells, muddy streams, etc.

Treatment; Diagnosis.—Two things are wanted: to expel the worms present, and to prevent their re-accumulation. As to the evidence of the existence of lumbricoid worms in the bowels, it is always doubtful unless some of them pass out with the evacuations. Signs of gastro-intestinal and nervous irritation attend them, especially in infants and young children. So, grinding the teeth during sleep, itching of the nose and anus, bad or irregular appetite, and tumidity of the abdomen, are regarded commonly as signs of worms. But other sources of indigestion and disturbance may be thus made known. Convulsions may undoubtedly be caused by worms in children; and so may laryngismus stridulus, and spasmodic croup.

When there is good reason to believe that they do exist in the bowels, anthelmintics may be given, with purgatives, in safe doses, watching their effects. Besides the *vermicides* mentioned in connection with tape-worm, many other drugs have more or less such effect; as santonin (most certain of all), calomel, pink-root (spigelia), bark of pomegranate root, azedarach. chenopodium, cowhage (mucuna), powder of tin, etc. [F. 206, 207, 208].

Infusion of senna and spigelia, half an ounce of each to a pint; for an adult, a winglassful every morning before breakfast; this is very popular in this country. Instead, may be given *fluid extract of spigelia and senna*, a teaspoonful for a dose. As above said, santonin is the most effectual of the vermicides or vermifuges. It requires care in its use, however; producing serious vomiting, prostration, and nervous symptoms in over-dose. A child should not take more than half a grain of santonin once or twice daily; an adult, from three to six grains.

Trichocephalus dispar.—This worm inhabits the large intestine. It has a length of an inch and a half to two inches. The head is attenuated or hair-like; whence its name. The sexes are on different individuals. The trichocephalus is much less common than the lumbri-coid worm.

Oxyuris vermicularis (*Ascaris vermicularis*)— *White seat-worm*. Of this the male is about a line ($\frac{1}{12}$ inch) and a half long; the female, five or six lines. It is found in the rectum, generally of children; sometimes in considerable numbers. They cause a great deal of itching; occasionally, other nervous irritation. Females may have them find their way into the vagina; more rarely, they get into the urethra.

For the **treatment** of seat-worms, I know of nothing equal to *suppositories of santonin;* made with cacao butter, three grains of the drug in each; one to be introduced into the rectum every night [F. 209]. Other common remedies are, injections of lime-water, infusion of aloes, mercurial ointment, etc.

Trichina spiralis.—Since 1822, when Tiedeman discovered it (Hilton, 1832), and especially since it was described by Owen in 1835, the dissecting room has furnished observers with specimens of this parasite, long supposed to be harmless. Zenker of Dresden first showed that, although a few *trichinæ* may be innocent, they sometimes abound to such an extent as to cause serious disease, and even to destroy life. Such an affection is called trichinous disease, *trichiniasis* or *trichinosis*. It has occurred particularly often in Germany, where it has been recognized since 1860. In 1863, in a Prussian town, of 103 persons in good health who dined together on a festive occasion, nearly all became ill from eating sausage made of the meat of an ill-conditioned pig; and quite a number died. Another outbreak, at Hedersleben, in 1865, caused 40 deaths in 300 cases. The first cases in America were reported by Dr. Schoetter of New York.[1] At Marion, Iowa, in 1866, nine cases occurred in one family, under the care of Dr. J. H. Wilson; five died. In the same county, eating raw ham containing trichinæ (as proved afterwards by examination) caused the disease in six children at the same time; reported by Dr. Ristine. An examination of pork in Chicago by a committee of the Academy of Sciences of that city proved the existence of trichinæ in 1 in 50 of the hogs inspected; some of their muscles containing from 10,000 to 18,000 in a cubic inch. Such animals are not themselves nearly always out of health. Cattle, also, are, to a less degree, subject to the same parasite. The meat of those so infected should of course

---

[1] Clymer, in Phil. ed. Aitken's Practice of Medicine, vol. i. p. 858.

not be used for food. In some German cities the butchers have microscopic examination made of the flesh of their animals.

To the naked eye, the muscles of a trichinous animal present whitish dots, which a lens will show to be the capsules or cysts of immature trichinæ. Those not encysted are invisible without a microscope. The capsule is hard and transparent; the worm is coiled spirally within it. Under the tongue is the preferred place to search for the trichinæ in the living animal; a delicate *harpoon* being used.

The trichina is a minute bi-sexual worm, reproducing in the intestinal canal of animals or men; the offspring then finding their way out through the walls of the intestines to become finally encysted in the muscles. The disease produced by them has two distinct stages: 1. that of the presence of the worms in the alimentary canal, and their multiplication there; 2. that of their migration to and location in the muscles. Of the first period, *malaise*, vomiting, and diarrhœa are the leading symptoms. Of the second, fever, resembling typhoid, severe pains, with stiffness, in the muscles, and prostration. As the muscles of the larynx are often attacked, hoarseness is a common symptom. The complication of pneumonia is not infrequent. The first stage above mentioned lasts about a week or less; the second may terminate fatally within six days, but usually has a duration of from two to four or five weeks.

We are not informed of any success with the **treatment** of trichiniasis. Its **prevention** is always possible. Besides proper inspection of animals, every piece of meat which may be suspected must be *well cooked*. Reliance cannot be had upon salting and smoking; at least unless they be very thoroughly done.

Sclerostoma duodenale is common in Egypt and in parts of Europe. It exists in the small intestines, and causes a chlorosis-like anæmia. The worm is from a third to half an inch long. Its *vermicide* is said to be turpentine.

Strongylus gigas (*Eustrongylus gigas*) inhabits the kidney. It is rare in man.

Filaria medinensis (*Dracunculus*) or Guinea-worm lives in the subcutaneous tissue. It is common in the tropical regions of the old world. The female worm it is, that enters the skin of a human being, and develops, with its contained young, in a whipcord-like shape, to a length varying from six inches to four, five, or six feet, and a width of about one twelfth of an inch. A dozen or more of the worms may exist upon the same person. The lower limbs are especially invaded by them; but they can migrate almost all over the body. They evidently get into the legs and feet of those who bathe in shallow streams or ponds, or walk barefooted in damp and muddy places. An incubation of a year or more is required for the development of the worm to a perceptible size.

A characteristic vesicle appears, generally upon the lower part of the leg, when the worm matures. This bursts, emitting the young filariæ; a good deal of itching and irritation ensues, and sometimes ulceration. The natives often rid themselves of the worm by letting a stream of water run or pour for a time upon the leg. When it creeps partly out, they draw upon it until it is dislodged.

EPIZOA.

Parasitic animals living upon the *surface* of the body are (besides the *sarcoptes*), chiefly **lice, fleas,** and **ticks.** The former are the **head** *louse* (pediculus capitis), **body** *louse* (p. corporis), and **crab** *louse* (p. pubis). These are true insects, without wings. The *preventive* of them is cleanliness, with avoidance of contact with unclean persons. Their *destruction* must be accomplished either by assiduous search and slaughter, or by parasiticide lotions, ointments, or powders; as, corrosive sublimate, cinnabar, pyrethrum, cocculus indicus, sulphur, carbolic acid, staphisagria, sabadilla, alcohol, essential oils [F. 210, 211, 212].

Two or three grains of corrosive sublimate dissolved in an ounce of water with a drachm or so of alcohol, will be as effectual as any of these. Powder or ointment of cocculus indicus is a good deal used. The *flea-powder* of the East (quite useful in stupefying fleas in a bed, if sprinkled before lying down) is probably *pyrethrum.*

Ticks belong with the *arachnida* of naturalists. People living in the country often have them to enter the skin from other animals or from plants; *e. g.*, the harvest-tick (leptus autumnalis). The irritation is moderate and local only.

Fleas in most parts of the world produce only annoying *bites*, larger and somewhat more inflamed than mosquito-bites. In Brazil and other parts of South America, the chigoe or jigger (*pulex penetrans*), makes a more permanent lodgment, and causes a very considerable irritation.

POISONS.

A few *memoranda* upon the more common poisons may be convenient here. Toxic agents are: 1. Irritant; 2. Narcotic; 3. Unclassified.

Irritant or Corrosive Poisons.—1. **Acids;** *e. g., sulphuric, nitric, chlorohydric, oxalic.* For all but the last, any alkaline substance (carbonate of soda, potassa, magnesia, or lime; or magnesia or lime alone) dissolved in or mixed with water, will be suitable as an antidote. *Oxalic acid* should have lime-water freely used as its antidote. 2. **Alkalies;** *e. g.,* caustic potassa, soda, strong solution of ammonia; earths, baryta, lime. For these vinegar or lemon-juice will act antidotally, making neutral salts. Olive or castor oil will *saponify* the alkaline material, and thus render it innocent.

3. **Corrosive Sublimate.**—For this, whites of eggs, or wheat flour mixed with water will be the best. 4. **Arsenic.**—Hydrated peroxide of iron has the reputation of being an almost certain antidote for the common arsenical compound, arsenious acid or white oxide of arsenic. It may be made fresh by adding *aqua ammoniæ* to *liquor ferri persulphatis;* or *aqua ammoniæ* to tincture of the chloride of iron. It is well for every physician to have a pint of each of the two first-named articles always within reach. The precipitated hydrate should be given promptly and freely. Some toxicologists regard *magnesia* as an almost equally reliable antidote for arsenic. 5. **Sulphate of Copper; Salts of Tin.**—Whites of eggs, milk, or flour mixed with

water may be given freely. 6. **Tartar Emetic.**—Infusion of galls or oak bark, or tannic acid in solution, may be administered copiously. Afterwards, opiates, as paregoric, will help to compose the stomach and bowels. 7. **Acetate of Lead.**—Sulphate of magnesia is antidotal for this; making the insoluble and inert sulphate of lead. 8. **Sulphate of Iron** (green vitriol); **Sulphate of Zinc** (white vitriol).—Carbonate of soda is recommended for these; copiously diluted. Flaxseed tea is a good diluent for any corrosive poison. 9. **Nitrate of Silver.**—Common salt is its antidote; making chloride of silver, inert. 10. **Phosphorus.**—A mustard emetic may be the first thing. In *any* poisoning, not accompanied by vomiting as an effect, this will be proper. Magnesia and mucilaginous infusions may then be given, quickly and largely. 11. **Iodine.**—Starch neutralizes iodine; but it will not neutralize iodide of potassium; for which no strict chemical antidote is known. 12. **Creasote.**—Whites of eggs, or milk, or flour and water, will combine with it. But, while waiting for these, free draughts of water should be given.

Narcotic Poisons.—Opium. When this or any other such poison has been taken, if the patient can swallow, an emetic should be given; ten grains of sulphate of copper, twenty grains of sulphate of zinc, half a teaspoonful of powder of ipecacuanha, or a tablespoonful of mustard; each dissolved or mixed in a large draught of warm water. Vomiting must be insured by repeated doses. If swallowing be impossible, the *stomach pump* must be used; introducing the flexible tube through the pharynx into the stomach, and washing it out by gently injecting, and then withdrawing, half a pint of water at a time by a syringe.

If any antidote for opium or its alkaloids has given reason for confidence, it is belladonna. Facts fully warrant its administration. Twenty minims of tincture of belladonna may, in opiate narcotism, be given every hour. Strong coffee is an older remedy, upon a similar indication. To counteract the comatose tendency, also, cold water may be dashed or poured at intervals over the head and face; strong sinapisms may be applied to the back, epigastrium, and limbs; or the patient, if able, may be made to walk about; or flagellation, with the hand or a wet towel, may be used for the same end.

In the extremest cases, *faradization* may be used; the interrupted current being applied to the spine and chest. *Galvano-puncture* is justifiable if other means fail; the fine needle being made to penetrate so as to reach the diaphragm, for the immediate stimulation of its muscular power. The needle for such a purpose should be of soft-tempered steel, platinized; fine and sound, well polished, three or four inches long, with a lance-shaped point. *Artificial respiration* is resorted to in some cases. (See *Asphyxia*.)

Belladonna or *atropia* may be antagonized by opium, upon the same kind of evidence as that just alluded to. The antidotal action is not chemical, but physiological. **Stramonium** (Jamestown weed) must also stand in the same relation to opium; and so must **hyoscyamus**, perhaps in a less positive degree.

Unclassified Poisons.—Prussic Acid. For this no certain antidote exists; and the arrest of life is so sudden as scarcely to allow its use, if we had one. T. and T. C. Smith, English chemists, assert that they have proved the following recipe to be antidotal for it:—

"Take of liquor of perchloride of iron 57 minims; protosulphate of iron in crystals, pure, 25 grains; as much water as will make a solution of a proto-sesqui-salt of iron, measuring about half an ounce. Dissolve, on the other hand, 77 grains of crystallized carbonate of soda in about half an ounce of water. These quantities destroy the poisonous action of between 100 and 200 drops of prussic acid, officinal strength, by giving first the one liquid, and then the other. For *cyanide of potassium* the antidote is the same except that the solution of proto-sesqui-salt of iron is to be used without the soda solution; the hydrocyanic acid having been already combined with an alkaline substance. The use of the soda would, however, not be injurious. The quantities given, as above, would decompose 35 grains of cyanide of potassium."

Cold affusion, chlorine water, and *ammonia* are the older measures advised for prussic acid poisoning.

Aconite, digitalis, hemlock, ergot, tobacco, lobelia, veratrum viride, aniline, strychnia, poisonous fungi, &c., have no known antidotes. **Emetics** should be promptly given when any of them is known to have been taken. Castor oil is also recommended, especially for those least depressing in their action. Aconite, lobelia and tobacco are the most powerfully sedative. **Animal charcoal** is advised, to absorb and render innoxious organic poisons in the stomach; teaspoonful doses should be given, repeatedly. For the spasms caused by *strychnia* or *nux vomica,* inhalation of chloroform is thought to be beneficial. For *tobacco, lobelia, aconite, digitalis,* or *veratrun viride,* taken poisonously, brandy or whisky as a stimulant would seem to be indicated.

BITES OF SERPENTS.

When a person is bitten by a venomous serpent, or by a rabid dog or other animal, the part should be, if accessible, at once *sucked* strongly with the mouth, to avoid loss of time. Wash it then thoroughly with *hot* water. Apply a cupping-glass for some minutes. Cauterize it with caustic potassa; or, if practicable, *excise* the part bitten. *Aqua ammoniæ* is useful also as a local antidote[1] for snake-poison, and for that of venomous insects; and so is fluid extract of *serpentaria.*

Should symptoms of poisoning have already followed the bite of a rattlesnake, experience seems to countenance the antagonistic and supporting use of whisky. Cures are said to have occurred, in several instances, where the person bitten drank large amounts of this; intoxication not being produced, on account of the counteracting impression of the poison upon the system.

ASPHYXIA.

Whether from drowning, breathing coal-gas in an unventilated apartment, or excessive inhalation of chloroform, &c., the treatment for suspended animation must be essentially the same in principle.

---

[1] Bibron's antidote consists of bromine ℨijss; iodide of potassium, gr. ij; corrosive sublimate gr. j; dilute alcohol fℨxxx. Dose, fℨj, in wine or brandy, p. r. n.

First, loosen everything about the neck. Draw the tongue forward and clear the mouth. Laying the patient upon the back, let both arms be raised (Sylvester) as far as possible above the head, and then brought down again; this should be repeated at least *twenty* times a minute; that is, faster than the ordinary rate of breathing. Blowing into the mouth or nostrils, with or without a trachea-tube, is available sometimes, especially in a child ; at the same time, when oxygen gas can be obtained (as, of course, it very rarely can), a jet of it may be used.

Passing a vial of solution of ammonia at intervals under the nostrils will aid to excite the nerve-centres. Rubbing the limbs and trunk, vigorously, and chiefly *towards* the heart, to hurry the venous circulation, is useful. So, also, is the application of mustard, or friction with red pepper and brandy or whisky. Hot bottles may be applied to the feet and legs. Excessive heat will not be expedient before respiration is established; but moderate warmth always promotes vitality. The expedient of applying a red hot iron, momentarily, to the epigastrium or the back of the neck, for intensely stimulant effect, is not unreasonable in idea. Galvanism may be employed in any case of suspended animation.

FORMULÆ.

Every physician should acquire such knowledge of the remedies he employs as to prescribe and combine them according to the indications of particular cases; not by the routine of names of diseases; and still less by fixed recipes or formulæ. While this is obvious, all *routine* being, as such, bad practice, a beginner may yet find advantage, and a practitioner of experience may occasionally save time, by having some *exemplars* of prescriptions at hand for reference. A selection of such examples is therefore given. Many of the recipes are original, and all are carefully made; the number being very much less extended than it might easily have been, in accordance with the principle above laid down. Those first given will for convenience, follow mainly the order of the diseases for which they are most likely to be required, as those diseases are treated of in Part II. of this volume.

The doses in these prescriptions, unless otherwise stated, are intended for adults. To reduce the dose of any drug (except narcotics, and, perhaps, mercurials) according to the age of a child, the rule suffices, to divide the dose for an adult, in proportion to the number of years of the child's age, increased by 12. Thus, for a child of two years, the dose will be $\frac{2}{14}$ (2 divided by 2 + 12) or $\frac{1}{7}$th of that for an adult : for a child of three years (3 divided by 3 + 12 = 15), $\frac{3}{15}$ or $\frac{1}{5}$th, &c. Opium and other narcotics act more powerfully, in proportion, upon children; so that their dose should be reduced in a greater degree. Calomel and other mercurials do not so readily affect the glands, at least, in children, as in adults.

Simplicity is made an especial aim in the following formulæ; considering, in this, the advantage rather of the *tyro* than of the practitioner of experience.

MEDICINES REFERRED TO IN PART II.

1. *Solution of Tartar Emetic.*

℞.—Tartrate of Antimony and Potassa, two grains ; Water, four fluid-ounces ; dissolve. Take one or two teaspoonfuls every two, three, or four hours.

In active pneumonia, pleurisy, severe bronchitis, pericarditis, etc.

2. *Quinine Solution.*

℞.—Sulphate of Quinine, half a drachm ; Aromatic Sulphuric Acid (elixir of vitriol), a fluidrachm and a half ; Oil of Cloves, four drops ; Mucilage of Gum Arabic, a fluidounce ; Peppermint Water, enough to make in all four fluidounces ; mix. Take a teaspoonful or two every three or four hours, in *asthenic pneumonia, low fevers,* etc., as a supporting remedy ; larger doses, or the same at shorter intervals, for *intermittent fever,* etc.

3. *Ammonia Mixture.*

℞.—Carbonate of Ammonia, one drachm ; Mucilage of Gum Arabic, four fluidounces ; Orange-flower Water, or Peppermint Water, two fluid-ounces ; mix. Dose, a dessert-spoonful, or tablespoonful, every hour.

In cases of general prostration ; typhoid pneumonia, influenza of old people, etc.

4. *Nitrate of Potassa.*

℞.—Nitrate of Potassa, two drachms ; Powder of Gum Arabic, or White Sugar, two drachms ; divide into twelve papers. Take one every two or three hours.

In mild pneumonia, bronchitis, etc.

5. *Wine of Ipecacuanha.*

℞.—Wine of Ipecacuanha, half a fluidounce. Take twenty drops every two or three hours, in a tablespoonful of water.

In tonsillitis, erysipelas, etc.

6. *Calomel, Ipecacuanha, and Nitre.*

℞.—Calomel and Ipecacuanha Powder, each six grains ; Nitrate of Potassa, half a drachm, or a drachm ; mix, and divide into twelve pow-ders. Take one powder every three hours.

In pneumonia, pleurisy, etc.

7. *Solution of Acetate of Ammonia.*

Dissolve two scruples of Carbonate of Ammonia in four fluidounces of Water, and add pure Vinegar, slowly, until it ceases to effervesce. This will substitute the "liquor ammoniæ acetatis" or spiritus Mindereri. Dose, a dessert-spoonful, or a tablespoonful, with as much of water, every two or three hours ; in any *febrile affection* where purging is not desirable, as a *diaphoretic.*

8. *Acetate of Potassa.*

℞.—Acetate of Potassa, five drachms and a half; Sweet Spirits of Nitre, two fluidrachms ; Water, enough to make eight fluidounces; dissolve. Take a tablespoonful every three or four hours.

In feeble cases of pneumonia, instead of tartar emetic ; also, as diuretic, in pleuritic effusion, etc.

9. *Calomel, Opium, and Tartar Emetic.*

℞.—Calomel, six grains ; Opium, three to six grains ; Tartar Emetic, a grain and a half; mix, and divide into twelve powders. Take one every three or four hours, in water.

In acute pleurisy.

10. *Squills and Digitalis.*

℞.—Powder of Squills, half a drachm ; Powder of Digitalis, eight to sixteen grains; mix, and divide into sixteen pills. Take one thrice daily.

In pleuritic effusion.

11. *Compound Spirits of Juniper.*

• ℞.—Compound Spirit of Juniper, two fluidounces. Take one or two teaspoonfuls thrice daily, in a wineglassful of water.

As diuretic, in pleuritic effusion, etc. ; especially in feeble cases.

12. *Juniper Infusion and Cream of Tartar.*

℞.—Bruised Juniper Berries, one ounce ; infuse for two hours in a pint of hot water ; pour off, and add a tablespoonful or two of Bitartrate of Potassa. Stir and drink in portions through the day.

In dropsical effusion of any kind.

·13. *Squills, Nitre, and Digitalis.*

℞.—Nitrate of Potassa, two drachms ; Oxymel of Squills, a fluid-ounce ; Tincture of Digitalis, half a fluidrachm ; Vinegar, a tablespoonful; Sugar and Gum Arabic, each two drachms ; Water enough to make in all six fluidounces ; mix. Take a tablespoonful every three hours.

In acute bronchitis, influenza, etc.

14. *Squills and Tartar Emetic.*

℞.—Tartar Emetic, one grain ; Syrup of Squills, four ounces ; mix. Take a teaspoonful every three or four hours.

In bronchitis, with dry cough.

15. *Squills and Paregoric.*

℞.—Syrup of Squills, three fluidounces ; Paregoric (Camphorated Tincture of Opium), one fluidounce ; mix. Take a teaspoonful three or four times daily ; or two teaspoonfuls at night.

In bronchitis or influenza, after loosening the cough.

16. *Muriate of Ammonia.*

℞.—Muriate of Ammonia, three drachms ; Mucilage of Gum Arabic, four fluidounces ; mix. Take a tablespoonful four times daily.
In chronic bronchitis.

17. *Copaiba Mixture.*

℞.—Balsam of Copaiba, three fluidrachms ; Compound Spirit of Lavender, two fluidrachms ; White Sugar and Gum Arabic, each two drachms ; Water, enough to make six fluidounces ; mix. Take a table-spoonful thrice daily.

18. *Lobelia and Ipecacuanha.*

℞.—Tincture of Lobelia, and Wine of Ipecacuanha, each half a fluid-ounce ; mix. Take one half teaspoonful every half hour until expectoration or nausea occurs.
In asthma.

19. *Musk Mixture.*

℞.—Musk, two scruples ; Syrup of Orange, one fluidounce ; Mucilage of Gum Arabic, three fluidounces ; mix. Take a tablespoonful every two or three hours.
In spasmodic cough of any kind ; or other spasmodic affections.

20. *Hydrocyanic Acid.*

℞.—Dilute Hydrocyanic Acid, sixteen drops ; Syrup of Wild Cherry, and Camphor Water, each one fluidounce ; mix. Dose, a teaspoonful every two or three hours.
In violent, troublesome cough.

21. *Nitromuriatic Acid.*

℞.—Nitromuriatic Acid, half a fluidounce (or, Nitric Acid, one flui-drachm and a half ; Muriatic Acid, two and a half fluidrachms). Take three or four drops twice or thrice daily, with water, *in a glass.*
In general or gastric debility, chronic or subacute jaundice, etc.

22. *Bromide of Potassium.*

℞.—Bromide of Potassium, half an ounce ; Peppermint or Cinnamon Water, or Pure Water, six fluidounces. Dose, from a dessert-spoonful to a tablespoonful.
In insomnia, hysteria, spermatorrhœa, etc.

23. *Citrate of Iron.*

℞.—Citrate of Iron, two drachms ; Orange-flower Water, five ounces and a half ; Simple Syrup, half an ounce. Take from a teaspoonful to a tablespoonful thrice daily, before or after meals.
For anæmic children.

24. *Ipecacuanha and Alum.*

℞.—Powder of Ipecac. and Powder of Alum, each half a teaspoonful ; mix with water. Repeat in ten minutes if it does not vomit.
In threatening croup.

25. Calomel and Nitrate of Potassa.

℞.—Calomel, six to twelve grains; Nitrate of Potassa, one drachm; Sugar, one scruple; mix, and divide into twelve powders. Take one every three hours.
In inflammatory croup.

26. Nitrate of Silver Solution.

℞.—Nitrate of Silver, five to ten grains; Rose-water, or Distilled Water, half a fluidounce; dissolve. Apply with camel's hair pencil to the throat, *in membranous croup, or scarlet fever.*

27. Tincture of Aconite Root.

℞.—Saturated Tincture of Aconite Root, one teaspoonful. To be rubbed gently into the skin, in *neuralgia.*

28. Chloroform Liniment.

℞.—Chloroform, three fluidounces; Olive Oil, four fluidounces; mix. *Pure chloroform*, prevented from evaporating by oiled silk, or a watch glass, acts as a strong rubefacient; burning like mustard.

29. Ointment of Veratria.

℞.—Veratria, ten to twenty grains; Pure Lard, one ounce; mix. *In severe neuralgia;* applied to the part.

30. Cod-Liver Oil.

℞.—Cod-Liver Oil, Syrup of Ginger, and Mucilage of Gum Arabic, each two fluidounces; Oil of Cloves, six drops; mix. Take a table-spoonful three or four times daily.
In wasting diseases.

31. Cod-Liver Oil and Glycerin.

℞.—Cod-Liver Oil and Glycerin, each two fluidounces; Gum Arabic, two drachms; Oil of Bitter Almonds, two drops; Oil of Cloves, twelve drops. *Take a tablespoonful thrice daily.*

32. Cod-Liver Oil, Glycerin, Iron, and Quinine.

℞.—Take of Citrate of Ammonia, Iron, and Quinine, ten grains; Cod-Liver Oil and Glycerin, each two fluidounces; mix.
Dose, a tablespoonful.

33. Iodide of Iron.

℞.—Liquor of the Iodide of Iron, half a fluidounce. Take twelve to twenty drops, in water, thrice daily.
In anæmia, scrofula, etc.

34. Tincture of Nux Vomica.

℞.—Tincture of Nux Vomica, half a fluidounce. Take from ten to thirty drops, thrice daily.
In nervous debility, aggravated dyspepsia, etc.

35. *Wild Cherry and Lactucarium.*

℞.—Syrup of Wild Cherry, and Syrup of Lactucarium, each two fluid-ounces; mix. Take a dessert-spoonful or two, at night, or one or two teaspoonfuls in the daytime.

In frequent and troublesome cough ; as in phthisis.

36. *Hoffmann's Anodyne, Squills, and Morphia.*

℞.—Syrup of Squills, a fluidounce and a half ; Hoffmann's Anodyne (Compound Spirit of Ether) and Solution of Morphia (one grain in the ounce) each a fluidounce ; Camphor Water and Mucilage Gum Arabic, of each a fluidounce and a quarter ; mix. Dose, from a teaspoonful to a tablespoonful.

In troublesome coughs.

37. *Carbonate of Potassa and Nitre.*

℞.—Carbonate of Potassa and Nitrate of Potassa, each two drachms and a half ; Water, eight fluidounces ; dissolve. Take a tablespoonful thrice daily.

In gouty attacks.

38. *Digitalis, Squills, &c.*

℞.—Citrate of Potassa, two hundred grains ; Tincture of Squills, two fluidrachms ; Wine of Colchicum Root, one fluidrachm ; Liquor of Acetate of Ammonia, two fluidrachms ; Infusion of Digitalis, two fluidounces ; Peppermint Water, enough to make eight fluidounces ; mix. Take half a wineglassful thrice daily.

In dropsical effusions.

39. *Cream of Tartar and Dandelion.*

℞.—Take of Bitartrate of Potassa, an ounce ; Extract of Taraxacum, half a drachm ; Decoction of Taraxacum, eight fluidounces ; mix. Take half a wineglassful two or three times daily.

In dropsy or jaundice.

40. *Cider Mixture.*

℞.—Bruised Juniper Berries, Mustard Seed, and Ginger, each half an ounce ; Bruised Horseradish and Parsley Root, each an ounce ; sound old Cider, a quart ; infuse. Dose, a wineglassful thrice daily.

In dropsy.

41. *Acetate of Lead Pills.*

℞.—Acetate of Lead, half a drachm ; Opium, five grains ; Conserve of Roses, or Crumb of Bread, a sufficient quantity ; mix, and divide into twenty pills. Take one thrice daily.

In hypertrophy of the heart.

42. *Digitalis.*

℞.—Powder of Digitalis, twelve grains ; divide into twelve pills. Take one thrice daily.

In cases of over-rapid action of the heart.

32

43. *Digitalis.*

℞.—Tincture of Digitalis, half a fluidounce.   Take ten drops thrice daily, in water.
As above.

44. *Veratrum Viride.*

℞.—Norwood's Tincture of Veratrum Viride, half a fluidounce. Take from two to five drops every three or four hours.   If nausea or prostration follow, withdraw it or diminish the dose.
In hypertrophy of the heart and inflammatory fever.

45. *Colchicum and Magnesia.*

℞.—Wine of Colchicum Root, one fluidrachm ; Husband's Magnesia, one drachm ; Peppermint Water, four fluidounces ; mix.   Take a tablespoonful thrice daily.
In gout and gouty rheumatism.

46. *Colchicum and Alkalies.*

℞.—Wine of Colchicum Root, one fluidrachm ; Bicarbonate of Potassa and Rochelle Salts, each two drachms and a half; Peppermint Water, four fluidounces ; mix.   Take a tablespoonful thrice daily.
In gout and gouty rheumatism.

47. *Hoffmann's Anodyne, Ammonia, and Soda.*

℞.—Bicarbonate of Soda, four scruples; Aromatic Spirit of Ammonia, one fluidrachm ; Compound Spirit of Ether, one fluidounce ; Compound Tincture of Cardamom, three fluidrachms ; Camphor Water and Mucilage of Gum Arabic, each a fluidounce and a quarter ; mix.   Take a dessert-spoonful or tablespoonful at once.
In angina pectoris, or gout of the stomach or heart.

48. *Warner's Cordial and Laudanum.*

℞.—Tincture of Rhubarb and Senna, a fluidounce and a half; Syrup of Ginger, three fluidrachms ; Laudanum, one fluidrachm ; mix.   Take a teaspoonful at once, in hot water.
In angina pectoris, or spasmodic gout.

49. *Chloroform, Hoffmann's Anodyne, &c.*

℞.—Chloroform and Aromatic Spirit of Ammonia, each two fluidrachms ; Hoffmann's Anodyne and Paregoric, each half an ounce ; Mucilage of Gum Arabic, half an ounce ; mix.   Take a teaspoonful at once.
In angina pectoris, retrocedent gout, &c.

50. *Glycerin and Rose-Water.*

℞.—Glycerin, one part ; Rose-water, five parts ; mix.   Use as a lotion for the skin, or a mouth-wash.

51. *Prepared Chalk and Gum Arabic.*

℞.—Equal parts of finely powdered prepared chalk, and powder of Gum Arabic ; mix.
Apply to ulcerated places in the mouth.

52. *Borax, Myrrh, &c.*

R.—Biborate of Soda, two drachms; Powdered Myrrh, one drachm; Water, six fluidounces; mix.
Use as mouth-wash.

53. *Sulphate of Zinc and Rose-Water.*

R.—Sulphate of Zinc, from two to ten or twenty grains; Rose-water, a fluidounce; dissolve.
Use as mouth-wash, with care, in severe cases.

54. *Chlorate of Potassa.*

R.—Chlorate of Potassa, half an ounce; Water, six fluidounces; dissolve. Take a tablespoonful every three or four hours.
In ulceration of the mouth or throat, diphtheria, etc.

55. *Muriatic Acid and Honey.*

R.—One part of Muriatic Acid and two parts of Honey; mix. To be applied to the throat in diphtheria, with a soft sponge, firmly fastened to a (probang) piece of whalebone.
In diphtheria.

56. *Tincture of Chloride of Iron.*

R.—Tincture of Chloride of Iron, half a fluidounce. Take from ten to thirty drops thrice daily, in water.
In anæmia, diphtheria, menorrhagia, leucorrhœa, asthenic erysipelas, etc.

57. *Chlorinated Soda and Glycerin.*

R.—Labarraque's Solution of Chlorinated Soda, one fluidrachm Bower's Glycerin, and Water, each two fluidounces; mix.
Use as mouth-wash, in gangræna oris.

58. *Creasote and Glycerin.*

R.—Creasote, two or three drops; Bower's Glycerin, and Water, each half a fluidounce; mix.
Use as mouth-wash, in cancrum or gangræna oris, or severe aphthæ or thrush.

59. *Alum, Brandy, and Water.*

R.—Alum one drachm, dissolve in six fluidounces of water; add two fluidounces of brandy.
To wash the mouth in salivation.

60. *Tannic Acid Solution.*

R.—Tannin, ten to thirty grains; Water, a fluidounce; dissolve.
To be applied with a hair-pencil, to enlarged tonsils, etc.

61. *Iodide of Potassium.*

R.—Iodide of Potassium, one to two drachms; Cinnamon or Peppermint Water, six fluidounces; dissolve. Take a tablespoonful thrice daily.
As alterative in syphilitic rheumatism, and in many other affections.

62. *Nitrate of Silver Pills.*

℞.—Nitrate of Silver, five grains ; Opium, two grains and a half ; mix, and divide into twenty pills. Take one thrice daily.
In chronic gastritis.

63. *Subnitrate of Bismuth.*

℞.—Subnitrate of Bismuth, one to three drachms ; divide into twelve powders. Take one three or four times daily, in water.
In gastric or intestinal irritation.

64. *Lime-water and Milk.*

Mix together equal parts of clear Lime-water and good Milk. Take a dessert-spoonful or tablespoonful of the mixture at once.
To check vomiting, or give nourishment when the stomach is irritable.

65. *Effervescing Draught.*

Dissolve two drachms and a half of Bicarbonate of Potassa in four fluidounces of Water. Pour out, for administration, a tablespoonful of this solution, and add to it a tablespoonful of water. Then add a tablespoonful of fresh Lemon-juice ; or of a solution containing two drachms of Citric Acid in four fluidounces of Water.
In fever, with irritability of stomach; also, in sea-sickness.

66. *Cardamom and Potassa Mixture.*

℞.—Bicarbonate of Potassa, one drachm ; Compound Tincture of Cardamom, a fluidounce ; Syrup of Ginger, two fluidrachms ; Orange-flower Water, enough to make four fluidounces ; mix. Take a dessert-spoonful at once.
To relieve nausea and vomiting.

67. *Ammonia, Soda, and Morphia.*

℞.—Bicarbonate of Soda, four scruples ; Aromatic Spirit of Ammonia, one fluidrachm ; Solution of Morphia, two fluidrachms ; Cinnamon Water, enough to make four fluidounces. Take one or two teaspoonfuls at once.
For vomiting.

68. *Creasote, Soda, and Morphia.*

℞.—Creasote, eight drops ; Bichlorate of Soda, one drachm ; Solution of Morphia, a fluidrachm and a half ; Peppermint Water, enough to make four fluidounces ; mix. Take one or two teaspoonfuls at once.
For vomiting.

69. *Calomel Powders.*

℞.—Calomel, two grains ; divide into eight powders. Take one every two hours.
For vomiting.

70. *Spice Poultice.*

℞.—Of Powdered Cloves, Ginger, and Cinnamon, each one or two teaspoonfuls ; Wheat Flour, a tablesponful ; Brandy, enough to make a

mass moist enough to spread upon thin, soft flannel. Double the flannel over it, and apply it to the abdomen.
In obstinate vomiting, etc.

71. *Nux Vomica, Iron, and Quinine.*

℞.—Pill of Carbonate of Iron (Valleix's Mass), two scruples (or, Quevenne's Metallic Iron, per hydrogen, one scruple); Sulphate of Quinia, one scruple; Alcoholic Extract of Nux Vomica, five grains; mix, and divide into twenty pills. Take one thrice daily.
In prolonged atonic dyspepsia, general debility, or ganglionic cachexia.

72. *Tincture of Gentian and Rhubarb.*

℞.—Compound Tincture of Gentian, and Tincture of Rhubarb, each two fluidounces; mix. Take two teaspoonfuls before each meal.
In dyspepsia.

73. *Gentian and Rhubarb Pills.*

℞.—Extract of Gentian, and Powder of Rhubarb Root, each half a drachm; mix and divide into twenty pills. Take one or two thrice daily.
In dyspepsia, flatulence, or tendency to colic.

74. *Gentian, Rhubarb, and Blue Mass.*

℞.—Extract of Gentian, and Powder of Rhubarb, each half a drachm; Blue Mass, four grains; mix and divide into twenty pills. Take one three or four times daily, for a few days.
To prevent recurring bilious colic or sick headache.

75. *Rhubarb Pills.*

℞.—Rhubarb Root, and Castile Soap, each half a drachm; Oil of Anise, four drops; mix, and divide into twenty pills. Take one or two as required.
For slight constipation.

76. *Rhubarb and Colocynth.*

℞.—Rhubarb, and Compound Extract of Colocynth, each half a drachm; mix, and divide into twenty pills. Take one or two as required.
For constipation.

77. *Rhubarb and Aloes, &c.*

℞.—Rhubarb, two scruples; Aloes, one scruple; Extract of Nux Vomica, four grains; mix, and divide into twenty pills. Take one as required.
For obstinate constipation.

78. *Carminative Mixture.*

℞.—Bicarbonate of Soda, one drachm; Compound Tincture of Cardamom, one fluidounce; Spirit of Camphor, one fluidrachm (or, Paregoric, half a fluidounce); Spiced Syrup of Rhubarb, half a fluidounce; Peppermint Water, enough to make four fluidounces. Take a teaspoonful at once.

32*

79. *Oil of Cajuput.*

℞.—Oil of Cajuput, half a fluidrachm ; Compound Spirit of Lavender, half a fluidounce ; Syrup of Ginger, two fluidrachms ; Mucilage of Gum Arabic, enough to make two fluidounces. Take a dessert-spoonful at once.
For flatulent pain in the bowels.

80. *Ammonio-ferric Alum.*

℞.—Ammonio-ferric Alum, two scruples ; Cinnamon Water, four fluid ounces ; dissolve. Take a tablespoonful every two or three hours.
An excellent tonic astringent.

81. *Creasote Pills.*

℞.—Creasote, twenty drops ; Conserve of Roses (or Extract of Gentian), one drachm ; mix, and divide into twenty pills. Take one every two, three, or four hours.
As astringent, in hæmatemesis, ulcer of stomach, etc.

82. *Podophyllum, etc.*

℞.—Resin of Podophyllum, two grains ; Fluid Extract of Rhubarb and Fluid Extract of Senna, each a fluidounce ; Oil of Cloves, four drops ; Syrup of Ginger, half a fluidounce ; Mucilage of Gum Arabic, enough to make four fluidounces. Dose for an adult, a tablespoonful.
For constipation.

83. *Suppository of Soap.*

Cut a piece of good Yellow Soap to the shape, and rather less than the size, of the last joint of the little finger. Dip it in Castor Oil, Olive Oil, or Lard, and introduce it within the rectum.
To act upon the bowels, instead of an enema.

84. *Nux Vomica, Colocynth, and Soap.*

℞.—Compound Extract of Colocynth and White Soap, each half a drachm ; Extract of Nux Vomica, five grains ; mix, and divide into twenty pills. Take one night and morning.
For torpor of the bowels.

85. *Aloes, Rhubarb, and Belladonna.*

℞.—Rhubarb and Aloes, each half a drachm ; Extract of Belladonna, three grains ; Oil of Cloves, three drops ; mix, and divide into twenty pills. Take one twice daily.
For habitual constipation.

86. *Calomel and Opium Pills.*

℞.—Calomel and Opium, each six grains ; mix, and divide into twelve pills. Take one every two, three, or four hours.
In peritonitis, bilious colic, etc.

87. *Pills of Opium and Ipecacuanha.*

℞.—Powder of Opium and Powder of Ipecacuanha, each six grains ; mix, and divide into twelve pills. Take one every three hours.
In typhlitis.

88. Cerate of Carbonate of Lead.

℞.—Carbonate of Lead, two drachms; Simple Cerate, one ounce; mix. For external use, in chronic ophthalmia, periostitis, hæmorrhoids, etc.

89. Aromatics, etc., for Colic.

℞.—Aromatic Spirit of Ammonia and Spirit of Camphor, each a fluidrachm; Tincture of Ginger, two fluidrachms; Bicarbonate of Soda, four scruples; Peppermint Water, enough to make four fluidounces. Dose, a tablespoonful.

90. Carminative Anodyne.

℞.—Spiced Syrup of Rhubarb, Compound Tincture of Cardamom, Paregoric, and Cinnamon Water, each a fluidounce; mix. Dose, from a dessert-spoonful to a tablespoonful.
For crapulent colic.

91. Chloroform Mixture.

R.—Chloroform, a fluidounce; Camphor Water, Peppermint Water, and Mucilage of Gum Arabic, each a fluidounce; mix. Dose, from a teaspoonful to a tablespoonful, repeated cautiously.
For colic, etc.

92. Chloroform Paregoric—No. 1.

℞.—Chloroform, Laudanum, Spirit of Camphor, and Aromatic Spirit of Ammonia, each a fluidrachm and a half; Creasote, three drops; Oil of Cinnamon, eight drops; Alcohol, two fluidrachms; mix. Dose, from ten drops to half a teaspoonful, in water.
In cholera.

93. Chloroform Paregoric—No. 2.

℞.—Chloroform, two fluidrachms; Spirit of Camphor, a fluidrachm and a half; Laudanum, a fluidrachm; Oil of Cinnamon, five drops; Alcohol, three and a half fluidrachms; mix. Dose, ten drops to half a teaspoonful, in water.

94. Carminative for Infants.

℞.—Bicarbonate of Soda, half a drachm; Aromatic Spirit of Ammonia, half a fluidrachm; Solution of Morphia, half a fluidrachm; Syrup of Ginger, half a fluidounce; Camphor Water, enough to make two fluidounces; mix. Dose, a teaspoonful, repeated if necessary.
In colic.

95. Podophyllum, Rhubarb, etc.

℞.—Resin of Podophyllum, one grain; Simple Syrup of Rhubarb, a fluidounce; Oil of Fennel, one drop; mix. Dose, ten drops to a teaspoonful.
For constipation in infants.

96. Castor Oil and Spiced Syrup of Rhubarb.

Mix one tablespoonful of Castor Oil thoroughly with two tablespoonfuls of Spiced Syrup of Rhubarb; and administer it immediately after mixture. This is the least disagreeable way of taking castor oil.

97. *Castor Oil and Laudanum.*

To the above prescription, add ten, twenty, or thirty drops of Laudanum. *Useful in incipient acute dysentery.*

98. *Assafœtida Mixture.*

℞.—Rub one drachm of Assafœtida gradually with four ounces of Water, until thoroughly mixed. Then add two fluidounces of Syrup of Ginger.
Dose for a child, a teaspoonful.

99. *Magnesia and Ammonia Mixture.*

℞.—Best Magnesia (Husband's or Ellis'), a drachm ; Aromatic Spirit of Ammonia, a fluidrachm ; Peppermint Water, four fluidounces; mix. To be shaken before administration. Take a teaspoonful every half hour.
In common summer cholera morbus.
Half a fluidounce of Paregoric may be added to the above, if there is much purging. ·

100. *Chloroform and Camphor.*

℞.—Chloroform, half a troyounce ; Camphor, one drachm ; the yolk of one Egg ; Water, six fluidounces. Rub the yolk in a mortar, first by itself, then with the Camphor, previously dissolved in the Chloroform, and lastly, with the Water, gradually added. This is the " Mixture of Chloroform" of the United States Pharmacopœia.
Dose, from a teaspoonful to a tablespoonful.

101. *Spiced Rhubarb and Magnesia.*

℞.—Spiced Syrup of Rhubarb, half a fluidounce ; Magnesia (Husband's), fifteen grains ; Cinnamon Water and Camphor Water, each two fluidrachms ; mix. Take in two doses, three hours apart.
As a corrective in slight diarrhœa.

102. *Chalk Mixture.*

℞.—Prepared Chalk, two drachms ; White Sugar and Gum Arabic, each a drachm and a half; Tincture of Kino, two fluidrachms and a half ; Laudanum, twenty to forty drops ; Peppermint Water, enough to make six fluidounces; mix. *Dose, a tablespoonful.*
In diarrhœa.

103. *Camphor Mixture.*

℞.—Compound Spirits of Lavender, a fluidounce ; Spirit of Camphor, a fluidrachm ; Laudanum, half a fluidrachm ; Sugar and Gum Arabic, each a drachm and a half ; Cinnamon Water, enough to make six fluidounces ; mix. Dose, a tablespoonful once in three hours.
In diarrhœa.

104. *Lead and Morphia Mixture.*

℞.—Acetate of Lead, eight to sixteen grains ; Acetate of Morphia, one grain ; Gum Arabic, two drachms; Cinnamon Water, enough to make eight fluidounces ; mix. Take a teaspoonful every three or four hours.
In obstinate diarrhœa.

105. *Catechu and Paregoric.*

℞.—Tincture of Catechu and Paregoric, each half a fluidounce ; mix. Take a teaspoonful every three or four hours.
In severe diarrhœa.

106. *Tannic Acid and Opium.*

℞.—Tannic Acid, thirty-six grains ; Powder of Opium, three to four grains ; mix, and divide into twelve pills. Take one every three or four hours.
To check diarrhœa.

107. *Calomel, Soda, and Ginger.*

℞.—Calomel, two grains ; Bicarbonate of Soda, one scruple ; Powder of Ginger, twelve grains ; mix, and divide into twelve powders. Give one three or four times daily.
In incipient cholera infantum.

108. *Mercury with Chalk, and Cinnamon.*

℞.—Mercury with Chalk, and Powder of Cinnamon, each twelve grains ; mix, and divide into twelve powders. Give one thrice daily.
In the early stage of cholera infantum.

109. *Ammonia, Rhubarb, and Paregoric.*

℞.—Aromatic Spirit of Ammonia, twenty-five drops ; Paregoric, half a fluidrachm to a fluidrachm ; Spiced Syrup of Rhubarb, a fluidounce ; Peppermint Water, enough to make two fluidounces; mix. Give a teaspoonful every three hours.
In cholera infantum.

110. *Rhatany and Paregoric.*

℞.—Tincture of Krameria and Paregoric, each a fluidrachm ; Sugar and Gum Arabic, each half a drachm ; Cinnamon Water, diluted, enough to make two fluidounces; mix. Give a teaspoonful every two, three, or four hours.
To check the diarrhœa of cholera infantum.

111. *Blue Mass and Ipecacuanha.*

℞.—Blue Mass, twelve grains; Powder of Ipecacuanha, six to twelve grains; mix, and divide into twelve pills. Take one every three hours.
In incipient dysentery.

112. *Blue Mass, Ipecacuanha, and Camphor.*

℞.—Blue Mass, eight grains; Ipecacuanha, six grains ; Camphor, twelve grains ; mix, and divide into twelve pills. Take one every three hours.
In the early stage of dysentery.

113. *Camphor, Ipecac., and Opium.*

℞.—Camphor, eighteen grains; Ipecacuanha, six grains; Opium, three to six grains; mix, and divide into twelve pills. Take one every three or four hours.

In dysentery.

114. *Acetate of Lead and Opium Pills.*

℞.—Acetate of Lead, twelve to twenty-four grains; Opium, three to twelve grains; mix, and divide into twelve pills. Take one every three or four hours.

In dysentery.

115. *Enema of Laudanum and Starch.*

Prepare half an ounce of Starch, thin enough to be drawn into a small syringe; add from twenty to sixty or more drops of Laudanum, according to the case; mix, and inject into the bowel.

In severe dysentery, retention of urine, very painful hæmorrhoids, etc.

116. *Enema of Sulphate of Zinc and Laudanum.*

To four fluidounces of Flaxseed Tea, made without boiling, add forty drops of Laudanum, and from four to ten grains of Sulphate of Zinc; mix, and inject into the rectum.

In obstinate chronic dysentery.

117. *Quinine, Ipecac., Camphor, and Opium.*

℞.—Quinine, one scruple; Camphor, two scruples; Ipecacuanha, five grains; Opium, ten grains; mix, and divide into twenty powders (or pills). Take one every three or four hours.

In asthenic, malarious dysentery.

118. *Ointment of Galls and Opium.*

℞.—Powder of Galls, two drachms; *Opium, ten grains;* Lard, one ounce; mix. Apply as ointment.

For piles.

119. *Spermaceti Ointment and Opium.*

℞.—Ointment of Spermaceti, Ointment of Rose Water (Cold Cream) or Glyceramyl, an ounce; Opium, ten grains; mix. To be used as ointment.

For piles.

120. *Belladonna Ointment.*

℞.—Extract of Belladonna, a drachm; Spermaceti Ointment, an ounce; mix. Use as ointment.

For painful piles.

121. *Tannic Acid Wash.*

℞.—Tannic Acid, twenty to thirty grains; Water, six fluidounces; dissolve. To be injected into the rectum (cooled with ice) *for bleeding piles.*

122. Soda and Sweet Spirits of Nitre.

R.—Bicarbonate of Soda, three drachms; Sweet Spirit of Nitre, six fluidrachms; Peppermint Water, enough to make six fluidounces; dissolve. Take a tablespoonful three or four times daily.
In uric acid gravel.

123. Benzoic Acid and Soda.

R.—Bicarbonate of Soda, two drachms; Phosphate of Soda, half an ounce; Benzoic Acid and Gum Arabic, each two drachms; Sweet Spirit of Nitre, half a fluidounce; Peppermint Water, enough to make six fluidounces; mix. Take from a teaspoonful to a tablespoonful, occasionally.
In gravel.

124. Opium Suppositories.

R.—Opium, one or two grains; Cacao (Cocoa) Butter, a sufficient quantity; mix, and introduce into the rectum.
For painful hæmorrhoids, dysmenorrhœa, irritation of the bladder, etc.

125. ·Belladonna Suppositories.

R.—Extract of Belladonna, one to four grains; Cacao Butter, a sufficient quantity; mix, and introduce into the bowel.
For painful hæmorrhoids, etc.

126. Benzoic Acid.

R.—Benzoic Acid, two drachms; Cinnamon Water, six fluidounces; dissolve. Take a tablespoonful thrice daily.
In irritation of the bladder, incontinence of urine, etc.

127. Croton Oil.

R.—Croton Oil, four drops; Crumb of Bread or Conserve of Roses, a sufficient quantity to make four pills; mix, and divide. Take one every four hours, until they operate.
As a powerful cathartic, in rare cases.

128. Lead-water for the Eyelids.

To a fluidounce of pure River or Spring Water, add one drop of Goulard's Extract of Subacetate of Lead. *Apply this with a camel's hair pencil, to the outside of the lids, frequently.*

129. Alum Eye-water.

R.—Two to four grains of Alum; Water, one fluidounce; dissolve. Drop into the eye from a quill, or a hair pencil, once or twice daily.

130. Collyrium of Nitrate of Silver.

R.—Nitrate of Silver, two to four grains; Distilled Water, one fluidounce; dissolve. Apply to the inside of the lids with a hair pencil, or drop between the lids.

131. *Atropia Solution for the Eye.*

℞.—Sulphate of Atropia, two grains; Water, one fluidounce; dissolve. Drop into the eye once or twice daily.
To dilate the pupil; as in iritis.

132. *Lotion for the Ear.*

℞.—Glycerin and Warm Water, each half a teaspoonful ; mix. Pour into the ear from a teaspoon (in preference to a syringe) twice daily.
For otalgia, or irritation of the ear.

133. *Olive Oil and Laudanum.*

Mix half a teaspoonful of warm Olive Oil with ten drops of Laudanum; pour it into the ear.
For earache.

134. *Bromide of Potassium.*

℞.—Bromide of Potassium, half an ounce; Cinnamon Water, six fluidounces; dissolve. Take a tablespoonful at night, or thrice daily.
In hysteria, spermatorrhœa, etc.

135. *Resin of Jalap.*

℞.—Resin of Jalap, one scruple ; divide into three parts. Give one every four hours till they operate.
An active cathartic.

136. *Podophyllum Pills.*

℞.—Resin of Podophyllum, two grains ; Turkey Rhubarb, eight grains ; Oil of Anise, two drops ; divide into eight pills. Take one or two at once.
Cathartic and cholagogue.

137. *Strychnia.*

℞.—Strychnia, half a grain ; Conserve of Roses, sixteen grains ; mix, and divide into sixteen pills. Take one or two, thrice daily.
Cautiously, in some cases of paralysis, etc.

138. *Stimulating Liniment.*

℞.—Oil of Turpentine, Spirit of Camphor, Water of Ammonia, and Olive Oil, each two tablespoonfuls; mix well together, for external application.
In chronic rheumatism, bruises, sprains, etc.

139. *Sassafras Liniment.*

℞.—Oil of Sassafras, two fluidrachms ; Water of Ammonia, a tablespoonful ; Camphorated Soap Liniment, three fluidounces ; mix.
For swellings of joints, etc.

140. *Iodide of Potassium.*

℞.—Iodide of Potassium, a drachm ; Cinnamon Water, six fluidounces ; dissolve. Take a tablespoonful thrice daily.
In syphilitic rheumatism, etc.

141. *Enema of Castor Oil, Soap, and Molasses.*

Mix together a tablespoonful of Oil, and the same of Molasses, with a pint of warm Water, in which a little Castile or good yellow Soap has been dissolved. Inject into the rectum with a syringe.
To empty the bowel promptly.

142. *Phosphate of Iron.*

℞.—Phosphate of Iron, one drachm; divide into twelve powders. Take one thrice daily, in water.
A good chalybeate tonic.

143. *Assafœtida Pills.*

℞.—Assafœtida, one drachm; divide into twenty pills. Take one every two or three hours.
For hysterical nervousness.

144. *Solution of Morphia with Valerian.*

℞.—Solution of Sulphate of Morphia, and Fluid Extract of Valerian, each one fluidounce; mix. Take one or two teaspoonfuls as required.
In delirium tremens.

145. *Pills of Opium, Camphor, and Hyoscyamus.*

℞.—Opium, four grains; Camphor, twelve grains; Extract of Hyoscyamus, a scruple; mix, and divide into twelve pills. Take one every three or four hours; or, one or two at night.
A good calmative.

146. *Gallic Acid.*

℞.—Gallic Acid, two drachms and a half; Syrup of Cinnamon, four fluidounces; mix. Take a dessert-spoonful every two, three, or four hours.
An astringent, in hemorrhages, diabetes, etc.

147. *Oil of Turpentine Mixture.*

R.—Oil of Turpentine, two to four fluidrachms; Compound Spirit of Lavender, half a fluidounce; Laudanum, twenty minims; Sugar and Gum Arabic, each two drachms; Water, enough to make six fluidounces; mix. Take a tablespoonful at once.
In hemorrhages, typhoid fever, etc.

148. *Glyceramyl.*

R.—Mix together two drachms of Starch and two fluidounces of Bower's or Price's Glycerin, cold; heat gradually to about 240° Fahr., stirring all the time; then let it cool.
A very soothing local emollient.

149. *Neutral Mixture.*

R.—Citrate of Potassa, two drachms; Lemon Syrup, half a fluidounce; Water, three and a half fluidounces; mix. Dose, a tablespoonful every two or three hours, with one of water, *in fever.* The lemon syrup may be omitted without disadvantage.

33

150. *Spiritus Mindereri with Nitre.*

℞.—Liquor of Acetate of Ammonia, three fluidounces and a half; Sweet Spirit of Nitre, half a fluidounce; mix. Take a tablespoonful every two or three hours, with a little water.
In typhoid fever.

151. *Jalap and Squills.*

℞.—Resin of Jalap, half a drachm to a drachm; Squills, twelve grains to a scruple; mix, and divide into twelve powders. Take one at once.
As cathartic, in cerebral congestion, hydrocephalus, etc.

152. *Sulphite of Soda.*

℞.—Sulphite of Soda, two to four drachms; Mucilage of Gum Arabic, six fluidounces; mix. Take a tablespoonful every three or four hours.
In zymotic diseases, as glanders, etc.

153. *Assafœtida and Expectorants.*

℞.—Syrup of Ipecac., two fluidrachms and a half; Syrup of Squills, three or four fluidrachms; Mixture of Assafœtida, enough to make two fluidounces; mix. Give one or two teaspoonfuls at once.
In hooping-cough.

154. *Belladonna Mixture.*

℞.—Extract of Belladonna, one grain; Mucilage of Gum Arabic, two fluidounces; mix. Give one or two teaspoonfuls thrice daily.
In severe hooping-cough.

155. *Fluid Extract of Hyoscyamus.*

℞.—Fluid Extract of Hyoscyamus, half a fluidrachm; Orange-flower Water, or Camphor Water, four fluidounces; mix. Give from a teaspoonful to a tablespoonful, every three or four hours.
In severe hooping-cough.

156. *Chlorate of Potassa.*

℞.—Chlorate of Potassa, two drachms and a half; Peppermint Water, four fluidounces; dissolve. Take a tablespoonful every three hours.
In diphtheria, ulcerated sore mouth, etc.

157. *Chlorate of Potassa and Chloride of Iron.*

℞.—Chlorate of Potassa, two drachms; Tincture of Chloride of Iron, one fluidrachm: Simple Syrup and Peppermint Water, each two fluidounces; mix. Take a tablespoonful every three hours.
In diphtheria.

158. *Creasote in Glycerin.*

Dissolve four to eight drops of Creasote in two fluidounces of Glycerin, diluted with an equal bulk of Water.
Use as gargle.

159. *Quinine Pills.*

℞.—Divide twenty grains of Sulphate of Quinine into ten or twenty pills. Take one as required.
For intermittent fever, etc.

160. *Sulphate of Cinchonia Solution.*

℞.—Sulphate of Cinchonia, forty-eight grains ; Aromatic Sulphuric Acid (Elixir of Vitriol), a fluidrachm and a half; Compound Tincture of Cardamom, half a fluidounce ; Peppermint Water, enough to make four fluidounces. Take a teaspoonful or two as required.
For intermittent, etc.

161. *Sulphate of Cinchonia Pills.*

℞.—Sulphate of Cinchonia, two scruples ; divide into twenty pills. Take one as required.
As tonic, or for intermittent.

162. *Quinine and Iron Pills.*

℞.—Sulphate of Quinine, a scruple ; Pill of Carbonate of Iron (Valleix's mass) a drachm ; mix, and divide into twenty pills. Take one three or four times daily.
An admirable tonic, after intermittent, etc.

163. *Capsicum Pills.*

℞.—Powdered Capsicum, a drachm ; divide into twelve pills. Take one every hour or two.
In the chill of pernicious fever.

164. *Calomel, Quinine, Camphor, and Opium.*

℞.—Calomel, Quinine, and Camphor, each eight grains ; Opium, two grains ; divide into eight pills. Take one every half hour, hour, or two hours.
In pernicious fever.

165. *Nitromuriatic Acid, Nitre, and Camphor Water.*

℞.—Nitromuriatic Acid, half a fluidrachm ; Sweet Spirit of Nitre, half a fluidounce ; Camphor Water, five and a half fluidounces ; dissolve. Take a tablespoonful every two or three hours.
In low fever.

166. *Nitric Acid.*

℞.—Nitric Acid, forty drops ; Water, eight fluidounces ; dissolve. Take one or two tablespoonfuls every three hours.
In typhus fever.

167. *Guaiacum.*

℞.—Tincture of Guaiacum, two fluidounces. Take a teaspoonful thrice daily, in water.
For chronic rheumatism.

168. *Iodide of Potassium Solution.*

℞.—Iodide of Potassium, two drachms ; Peppermint Water, six fluid-ounces ; dissolve.   Take a tablespoonful thrice daily.
For syphilitic rheumatism.

169. *Iodide of Mercury.*

℞.—Green Iodide (Protiodide) of Mercury, twelve grains ; Conserve of Roses, a scruple ; divide into twelve or twenty-four pills.   Take one twice daily.
In syphilis.

170. *Donovan's Solution.*

℞.—Liquor of the Iodide of Mercury and Arsenic, half a fluidounce. Take from three to five drops twice or thrice daily.
In secondary syphilis, obstinate cutaneous eruptions, etc.

171. *Injection for Gonorrhœa.*

℞.—Sulphate of Zinc, four grains ; Water, two fluidounces ; dissolve.
Use once daily.

172. *Injection for Gonorrhœa.*

℞.—Solution of Subacetate of Lead (Goulard's) half a drachm to a drachm ; Water, four fluidounces.
Use once daily.

173. *Injection for Gonorrhœa.*

℞.—Chloride of Zinc, two grains ; Glycerin and Water, each a fluid-ounce ; dissolve.
Use once daily.

174. *Copaiba Mixture.*

℞.—Copaiba, half a fluidounce ; Compound Spirit of Lavender, two fluidrachms ; Sugar and Gum Arabic, each two drachms ; Peppermint Water, enough to make six fluidounces.   Take a tablespoonful thrice daily.
In gonorrhœa.

175. *Cubebs Mixture.*

℞.—Oil of Cubebs, two drachms ; Sweet Spirit of Nitre, half a fluid-ounce ; Sugar and Gum Arabic, each two drachms ; Peppermint Water, enough to make six fluidounces ; mix.   Take a tablespoonful thrice daily.
In gonorrhœa.

176. *Lugol's Solution.*

℞.—Iodine, six drachms ; Iodide of Potassium, a troyounce and a half ; Distilled Water, a pint ; dissolve.   Dose, five or six drops, twice daily, in water.
In scrofulous affections.

177. *Glycerole of Zinc.*

℞.—Oxide of Zinc, half a drachm; Bower's or Price's Glycerin, four fluidounces; mix. Apply externally, as an emollient. Shake before using it.

178. *Cold Cream with Zinc.*

℞.—Acetate of Zinc, two grains, dissolved in one fluidrachm of Rose Water; mix with one ounce of Ointment of Rose Water (Cold Cream). *Apply externally, for erythema.*

179. *Lead Ointment.*

℞.—Carbonate of Lead, four grains; Glycerin, a fluidrachm; Simple Cerate, one ounce; mix.
For chronic erythema, etc.

180. *Glycerole of Lead.*

℞.—Carbonate of Lead, one drachm; Glycerin, four fluidounces; mix. As a local application, for *ophthalmia (to the outside of the lids with a hair pencil), inflamed hæmorrhoids, periostitis,* etc.

181. *Ointment of Oxide of Zinc.*

℞.—Oxide of Zinc, two drachms; Lard, one ounce; mix. Apply locally, for *eruptions on the face,* etc.

182. *Colchicum and Ipecacuanha.*

℞.—Wine of Colchicum Root, and Wine of Ipecac., each, two fluidrachms; mix. Take twenty drops, in water, thrice daily.
In pustular diseases of the skin.

183. *Ointment of Sulphuret of Potassium.*

℞.—Sulphuret of Potassium, and Carbonate of Soda, each two drachms; Lard, two ounces; mix.
For tinea capitis, etc.

184. *Sulphite of Soda and Glycerin.*

℞.—Sulphite of Soda, two ounces; Glycerin, four fluidounces; Water, enough to make a pint; mix.
Use as lotion, in chronic diseases of the skin.

185. *Stimulating Embrocation.*

℞.—Aromatic Spirit of Ammonia, Spirit of Rosemary, and Glycerin, each a fluidounce; Tincture of Cantharides, three fluidrachms; Rosewater, enough to make eight fluidounces; mix.
For the scalp, in premature baldness.

186. *Cantharides and Castor Oil Pomade.*

℞.—Balsam of Tolu, two drachms; Oil of Rosemary, twenty minims; Tincture of Cantharides, two fluidrachms; Castor Oil, four fluidrachms; Lard, an ounce and a half; mix.
For premature baldness.

33*

187. *Ointment of White Precipitate.*

℞.—Ammoniated Mercury, one scruple ; Glycerin, a fluidrachm ; Oil of Bitter Almonds, three drops ; Lard, or Simple Ointment, an ounce ; mix.
For acne rosacea, etc.

188. *Iodide of Sulphur Ointment.*

℞.—Iodide of Sulphur, one scruple ; Lard, one ounce ; mix.
For army itch, etc.

189. *Astringent Powder.*

℞.—Powder of Krameria, half an ounce ; Prepared Chalk, two drachms ; Dry Starch, an ounce and two drachms ; mix.
To be dusted on the skin in eczema, lichen agrius, etc.

190. *Juniper Tar Soap.*

℞.—Oil of Juniper (Huile de Cade), Soft Soap, and Alcohol, each a fluidounce ; mix.
Apply as local alterative, in obstinate skin diseases.

191. *Anti-pruriginous Lotion.*

℞.—Oil of Juniper and Alcohol, each a fluidounce ; Water, six fluidounces ; mix.
For itching of the skin, in prurigo senilis, etc.

192. *Acid Lotion.*

℞.—Muriatic Acid, twenty drops ; Water, four fluidounces ; dissolve.
For obstinate itching.

193. *Lotion of Blue Vitriol.*

℞.—Sulphate of Copper, six grains ; Elderflower Water, two fluidounces ; dissolve.
Use as lotion, for chronic erythema, etc.

194. *Lotion of Corrosive Sublimate.*

℞.—Bichloride of Mercury, four grains ; Alcohol and Distilled Water, each a fluidounce ; dissolve.
In favus, etc., as paraciticide.

195. *Astringent and Sedative Lotion.*

℞.—Creasote, eight drops ; Tincture of Krameria, two fluidrachms ; Hydrocyanic Acid, eight drops ; Distilled Water, four fluidounces ; mix.
In irritative and obstinate skin diseases.

196. *Sedative Lotion.*

℞.—Cyanide of Potassium, fifteen grains ; Water, eight ounces ; dissolve. Apply externally. It should be kept in a dark place.
For lichen or prurigo.

197. Sulphuro-alkaline Ointment.

R.—Two drachms of Sulphur; one drachm of Carbonate of Potassa; and one ounce of Lard; mix.
For itch.

198. Iodide of Potassium and Glycerin.

R.—Iodide of Potassium, half an ounce; Glycerin, two fluidounces; mix. Use as lotion.
For scabies, army itch, etc.

199. Iodide of Potassium and Iodide of Sulphur.

R.—Iodide of Potassium, half an ounce; Iodide of Sulphur, a drachm; Glycerin and Rose-water, each two fluidounces; Oil of Bitter Almonds, three drops; mix.
For itch, etc.

200. Ointment of Carbonate of Potassa.

R.—Carbonate of Potassa, one drachm; Glycerin, one fluidrachm; Lard, an ounce; mix.
For eczema, herpes, etc.

201. Aloes Pills.

R.—Powder of Aloes, one to two scruples; Oil of Cloves, four drops; mix, and divide into twenty pills. Take one twice or thrice daily.
For amenorrhœa.

202. Aloes and Iron.

R.—Aloes, twenty grains; Quevenne's Metallic Iron (per hydrogen), half a drachm; Oil of Cloves, three drops; mix, and divide into twenty pills. Take one thrice daily.
For amenorrhœa.

203. Camphor, Lavender, Paregoric, and Ginger.

R.—Spirits of Camphor, one fluidrachm; Paregoric, two fluidrachms; Tincture of Ginger, half a fluidrachm; Compound Spirit of Lavender, half a fluidounce; Water, enough to make two fluidounces; mix. Take a dessert-spoonful every hour or two.
In dysmenorrhœa.

204. Alum Lotion.

R.—Alum, two drachms; Water, eight fluidounces; dissolve. Inject into the vagina, once or twice daily.
For leucorrhœa.

205. Tannin Lotion.

R.—Tannic Acid, two drachms; Glycerin, a fluidounce; Water, seven fluidounces; mix. Inject into the vagina daily.
For leucorrhœa.

206. Santonin.

R.—Santonin, half a drachm; divide into twelve pills. Take one twice daily.
An excellent vermifuge. For a child, the dose should be reduced.

207. *Senna and Pink-Root Infusion.*

℞.—Leaves of Senna, and Root of Spigelia, each half an ounce; Boiling Water, a pint and a quarter; infuse, covered, for two hours. Take a wineglassful morning and night.
A good vermifuge.

208. *Fluid Extract of Senna and Spigelia.*

℞.—Fluid Extract of Senna and Spigelia, a dessert-spoonful; take it in the morning, on an empty stomach.
For worms.

209. *Suppository of Santonin.*

℞.—Santonin, twelve grains; Cocoa Butter, a sufficient quantity to make four suppositories; mix, and divide. Introduce one into the bowel at bedtime.
For seat-worms; an infallible remedy.

210. *Corrosive Sublimate Lotion.*

℞.—Bichloride of Mercury, a scruple; Water, four fluidounces; dissolve. Use as wash.
To destroy lice.

211. *Cocculus Indicus.*

℞.—Seeds of Cocculus Indicus, eighty grains; Prepared Lard, an ounce. Bruise the seeds well in a mortar, and mix with the lard.
To destroy lice.

212. *Carbolic Acid and Glycerin.*

℞.—Carbolic Acid, one or two drachms; Glycerin, a fluidounce; Water, enough to make eight fluidounces; mix. Use as lotion.
To destroy lice, or relieve pruritus.

MISCELLANEOUS PRESCRIPTIONS.

213. *Syrup of Iron, Quinine, Strychnia.*

℞.—Ferri Sulphatis, ʒv; Sodæ Phosphatis, ʒvj—ʒj; Quiniæ Sulphatis, gr. cxcij; Acidi Sulphurici Diluti, quantum sufficit; Aquæ Ammoniæ, quantum sufficit; Strychniæ, gr. vj; Acidi Phosphorici Diluti, fℨxiv; Sacchari Albi, ℥xiv. Dissolve the sulphate of iron in one ounce of boiling water, and the phosphate of soda in two ounces of boiling water. Mix the solutions, and wash the precipitated phosphate of iron until the washings are tasteless. With sufficient diluted sulphuric acid, dissolve the sulphate of quinia in two ounces of water. Precipitate the quinia with ammonia water, and carefully wash it. Dissolve the phosphate of iron and the quinia thus obtained, and also the strychnia, in the diluted phosphoric acid. Then add the sugar, dissolve the whole, and mix without heat.

Dose, a teaspoonful thrice daily, in *anæmia, chlorosis, leucocythæmia,* &c. This is a favorite prescription with Dr. Aitken.

214. - *Chlorodyne.*

℞.—Chloroformi, fʒss; Ætheris sulphurici, ℳxc; Olei Menthæ Piperitæ, gtt. viij; Resinæ Cannabis Indicæ, gr. vj; Capsici, gr. ij. Mix, shake occasionally, and allow it to stand for a few days. Also, Morphiæ Muriatis, gr. xvj; dissolved by the aid of heat in fʒij of Water; to which, when cold, add of Scheele's Hydrocyanic Acid, ℳlxv; Perchloric Acid, fʒj; of Simple Syrup (or treacle), fʒij. Add this gradually to the first mixture, and add enough water to make four fluidounces in all. Dose, thirty minims.
A powerful narcotic, whose pretensions, however, have been exaggerated.

215. *Dr. Hammond's Alterative for Syphilis.*

℞.—Potassii Iodidi, ʒj; Hydrargyri Bichloridi, gr. vj; Aquæ, fʒxij; misce.
Take a teaspoonful thrice daily.
In syphilitic rupia etc.

216. *Cinchonated Syrup of Iron.*

℞.—Ferri Phosphatis, ʒj;
    Aquæ, fʒjss; solve.
    Cinchoniæ Sulphatis, Əj;
    Acidi Sulphurici Diluti, gtt. xx;
    Aquæ, fʒjss; solve.
Misce, et adde Syrupi Aurantii quantum sufficit ut fiat mistura, fʒiv.
Dose, a dessert-spoonful, as a tonic.

217. *Antidote for Arsenic.*

℞.—Liquoris Ferri Tersulphatis (U. S. P.), et
    Aquæ Ammoniæ, āā fʒiv;
    Aquæ, Oj; misce.
Pour this mixture into a small muslin bag, strain and wash it; then dilute the precipitate with half a pint of water. Give a tablespoonful every five minutes.

218. *Compound Rhubarb Pills* (U. S. P.).

℞.—Pulveris Radicis Rhei, gr. xxiv; Aloës, gr. xviij; Myrrhæ, gr. xij; Olei Menthæ Piperitæ, ℳ ij; Aquæ, q. s.; misce bene, et divide in pil. no. xij.
A good, moderately active cathartic.

219. *Compound Cathartic Pills* (U. S. P.).

℞.—Extracti Colocynthidis Compositi, gr. xvj; Extracti Jalapæ (pulv.), et Hydrargyri Chloridi Mitis, āā gr. xij; Gambogiæ, gr. ijss; Aquæ, q. s.; misce, et divide in pil. no. xij.
A decidedly active cholagogue cathartic.

220. *Effervescing Solution of Citrate of Magnesia* (U. S. P.).

℞.—Magnesiæ Carbonatis, ʒv; Acidi Citrici, ʒviiss; Syrupi Limonis (vel Syrupi Acidi Citrici), fʒij; Aquæ, q. s. Dissolve the citric acid in four fluidounces of water, and add four drachms of the carbonate of magnesia, previously rubbed with three fluidounces of water. When the reaction has ceased, filter into a strong glass bottle holding twelve fluidounces; into which the syrup has been introduced. Rub the remaining carbonate of magnesia with two fluidounces of water, pour into the bottle, cork quickly, and secure the cork with twine. Dose, from half a tumblerful to a whole bottle.

This is the least disagreeable of all cathartic medicines.

221. *Hope's Mixture.*

℞.—Aquæ Camphoræ, f℥iv; Acidi Nitrosi, ℳ xxx; Tincturæ Opii, ℳ xx; misce.

Dose, a tablespoonful, every two or three hours.

In diarrhœa and asthenic dysentery.

222. *Quinine and Chloride of Iron.*

℞.—Quiniæ Sulphatis, ℈j; Tincturæ Ferri Chloridi, f℥ijss. Fiat solutio.

Dose, fifteen drops, in solution.

In diphtheria, asthenic erysipelas, etc.

223. *Quinine for Children.*

℞.—Quiniæ Sulphatis, ℨss; Pulveris Gummi Acaciæ, ℨss; Syrupi Zingiberis, f℥iv; misce.

Dose, a teaspoonful (one grain of quinine) or less, *as required, in intermittent,* etc.

224. *Effervescing Fever Powders.*

℞.—Acidi Citrici, ℨv; divide in parts xij. Wrap each of these in a white paper.

℞.—Potassæ Bicarbonatis, ℨvjss; divide in parts xij. Wrap each of these in blue paper.

For use, dissolve the acid powder in four tablespoonfuls of cold water in a tumbler, and add, with stirring, the other powder. One dose every two or three hours will be suitable *in inflammatory or remittent fever,* etc.

225. *Liquid Substitute for Dover's Powders.*

℞.—Vini Ipecacuanhæ, ℳxvj; Tincturæ Opii, ℳxiij; Spiritûs Ætheris Nitrici, fℨj; misce. To be taken with water, at bedtime.

For influenza, etc.

226. *Soda Powders.*

℞.—Soda Bicarbonatis, gr. xxiij.

℞.—Acidi Tartarici, gr. xx.

Dissolve each in four tablespoonfuls of water, separately; then pour the solutions together, and drink while effervescing. Ginger syrup may be added if desirable.

227. *Scudamore's Gout Mixture, Modified.*

℞.—Magnesiæ Sulphatis, ℥j; Magnesiæ Optimæ, ℨij; Vini Colchici Radicis, f℥j; Aquæ Menthæ Piperitæ, f℥x. Misce.

Dose, a tablespoonful every hour until it operates.

228. *Black Wash.*

℞.—Hydrargyri Chloridi Mitis, ʒj ; Liquoris Calcis, fʒiv ; misce. Apply on lint.
A popular lotion for chancre.

229. *Yellow Wash.*

℞.—Hydrargyri Chloridi Corrosivi, gr. xvj ; Liquoris Calcis, fʒviij ; misce. Apply on lint.
For chancre.

230. *Volatile Liniment.*

℞.—Aquæ Ammoniæ et Olei Olivæ, āā fʒss. Misce.
To bathe an inflamed throat, etc.

231. *Iodine Ointment.*

℞.—Iodinii, ℈j ; Potassii Iodidi, gr. iv ; Aquæ, ℳ vj ; Adipis, ʒj·
Misce.
For tumors, chronic inflammation of joints, etc.

232. *Tar Ointment.*

℞.—Picis Liquidæ, et Sevi (vel Adipis) ʒij. Mix with the aid of heat, and strain through muslin.
For tinea capitis, etc.

233. *Glycerin Ointment.*

℞.—Cetacei, ʒss ; Ceræ Albæ, ʒj ; Olei Amygdalæ Dulcis, fʒij ; Glycerinæ, fʒj.
Melt the wax and spermaceti with the oil of almonds at a moderate heat ; put these into a mortar, add the glycerin and triturate until cold.
For chapped hands, etc.

234. *Calomel and Camphor Ointment.*

℞.—Hydrargyri Chloridi Mitis, gr. viij ; Camphoræ, ℈j ; Glycerinæ, fʒj ; Cetacei, ʒss ; Adipis, ʒjss. Misce.
For lichen or herpes of the face, etc.

235. *Sulphur Ointment.*

℞.—Sulphuris, ʒj ; Adipis, ʒij ; misce.
For itch, etc.

236. *Lozenges for Hoarseness.*

℞.—Pulveris Cubebæ, ʒss ; Ammoniæ Muriatis, ʒj ; Olei Sassafras, fʒj ; Pulveris Glycyrrhizæ, Sacchari Albi, et Gummi Acaciæ, āā ʒiij ; Syrupi Tolutani, q. s.
Rub the powders thoroughly together, then add the oil, lastly the syrup. Divide the mass into lozenges of ten grains each.

237. *Liquorice and Opium Lozenges* (U. S. P.).

℞.—Opii Pulveris, ʒss ; Glycyrrhizæ, Gummi Acaciæ, et Sacchari Albi, āā ʒx ; Olei Anisi, fʒj. Rub the powders thoroughly together,

then add the oil of anise, and lastly add enough water to form a mass.
Divide into lozenges, each of ten grains.

Like "Wistar's Cough Lozenges," these are very soothing to cough,
when taken at night.

238. Rhubarb, Magnesia, and Charcoal.

Take of Powder of Rhubarb Root, Husband's Magnesia, and Prepared
Charooal Powder, each a teaspoonful; Powdered Ginger, half a teaspoonful; mix, and divide into three parts. Take one on rising in the
morning.

For "biliousness," etc.

239. Iodide of Lead Ointment.

℞.—Plumbi Iodidi, ℨj ; Adipis, ℥j ; misce.
For scrofulous and other tumors.

240. Stevens' Saline Draught.

℞.—Sodii Chloridi, ℈iv; Potassæ Chloratis, gr. xxviij ; Sodæ Carbonatis, ℥ij ; Aquæ, f℥vj ; dissolve. Take two or three tablespoonfuls every
half hour, as the "saline treatment" of *cholera.*

241. Radcliffe's Phosphorus Pills.

Take of Phosphorus, six grains ; Suet, six hundred grains. Melt the
suet in a stoppered bottle capable of holding twice the quantity indicated. Put in the phosphorus, and, when liquid, agitate the mixture
until it becomes solid. Roll into three-grain pills, and cover with gelatin.
Each pill will contain one thirty-third of a grain of phosphorus.

242. Trousseau's Syrup of Lime.

Saturate Simple Syrup with unslaked Lime. Or, mix two ounces of
Lime and eight ounces of Sugar in a mortar, and pour over them a pint
of boiling water. Take half a teaspoonful two or three times daily, in
milk.

For rheumatism.

243. To make a Sponge-Tent.

Cut a small elongated and conical piece of sponge, dip it in water, and
bind it tightly, with fine strong twine or cord, around a central wire ;
then dry it, remove the cord, coat it with a mixture of equal parts of
wax, lard, and glycerin, and fasten a piece of tape four inches long to
the larger end.

For uterine dilatation, etc. Great care is necessary in its use.

The dried stem of the sea-tangle (laminaria digitata) is preferred for
the same purpose by some practitioners.

ALIMENTARY PREPARATIONS.

Barley Water.

Take of pearl barley, two ounces ; boiling water, two quarts. Before adding the water let the barley be well washed. Then boil to one half, and strain the liquor. A little lemon-juice and sugar may be added, if desirable.

Rice Water.

Take of rice, two ounces ; water, two quarts. Boil it for an hour and a half, and then add sugar and nutmeg to taste. Some prefer *salt*. *An excellent drink in diarrhœa, dysentery, etc.*

Toast Water.

Cut a slice of stale bread half an inch thick, and toast it brown, without scorching. Pour over it a pint of boiling water ; cover closely till it cools ; then pour off and strain it.

Oat-Meal Gruel.

Boil a pint of water in a saucepan ; when boiling, mix with it two tablespoonfuls of oat meal, half a pint of milk, and a little salt. Let it then simmer for half an hour ; strain it through a hair-sieve, sweeten, and add a little nutmeg. A few raisins may be added before the boiling.

Vegetable Soup.

Put two potatoes, one onion, and a piece of bread, into a quart of water ; boil it down to a pint. Then throw in a little chopped celery or parsley, and salt. Cover, remove from the fire, and allow it to cool.

Bread and Butter Broth.

Spread a slice of well-baked bread with good fresh butter ; sprinkle it moderately with salt and black pepper. Pour a pint of boiling water over it, cover, and let it stand to cool.

Lime-water and Milk.

Take of clear saturated lime-water, and fresh milk, each a wineglassful ; mix. Let a tablespoonful, or less, be taken at once. This will sometimes remain upon an irritable stomach which will retain nothing else.

Chicken Broth.

Clean half a chicken ; pour on it a quart of cold water, and salt to taste ; add a tablespoonful of rice, and boil slowly for two hours ; skim well, and add a little parsley.

Panada.

Cut two slices of stale bread, without crust ; toast them brown, cut them up into squares of about two inches, lay them in a bowl and sprinkle with salt and a little nutmeg. Pour on a pint of boiling water, and stand to cool.

34

Arrow-Root.

Mix a tablespoonful, or a tablespoonful and a half, with a little cold water, till it makes a paste. Boil a pint of water, stir in the arrow-root, and boil it a few minutes. Sweeten with white sugar. Brandy or wine may be added if necessary; and half or all milk may be used instead of water. A little lemon or orange-peel added before boiling will improve the flavor.

Tapioca.

Cover two tablespoonfuls of tapioca with a teacupful or more of cold water, and soak for two or three hours, or over night. Put it then into a pint of boiling water, and boil it until it is clear and of the desired consistence. Sugar, nutmeg, or wine, etc., may be added as required.

Sago Jelly.

Mix well together four tablespoonfuls of sago, the juice and rind of one lemon, and a quart of water. Sweeten to taste, let it stand half an hour, and boil it, stirring constantly, until clear. Then add a wineglassful of wine; currant wine will do.

Beef-Tea.

Chop a pound of lean beef into very small pieces, pour over it a pint of cold water, cover, and let it stand two hours by the side of the fire. Then put it on the fire and boil it for half an hour. Remove the scum, skim off all the oil drops, and salt to taste. *Do not filter or strain it.* Good beef-tea should have a rich brown appearance when stirred.

Essence of Beef.

Cut up a pound of lean beef into small pieces; put it into a pint bottle, without water, cork it loosely, and immerse the bottle to its neck in cold water in a stewpan. Bring the water to a boil, and let it boil for two hours. Then pour off the essence.

Liebig's Broth.

Chop half a pound of beef, mix it well with one drachm of table salt, four drops (ten would be better) of muriatic acid, and eighteen ounces of distilled water. Macerate for an hour, and strain through a fine hair sieve. Dose, a teacupful. This contains the *soluble* constituents of the meat; but not all its nutritive elements.

Liebig's Food for Infants.

Mix together half an ounce of wheat flour, the same of malt flour, seven and a quarter grains of bicarbonate of potassa, and an ounce of water. Add five ounces of fresh milk, and put the whole upon a gentle fire. When it begins to thicken, take it from the fire, stir it for five minutes, heat and stir again until it becomes quite fluid; finally boil it for a short time. Filter it through a sieve to separate the bran; it is then ready for use. It will keep for twenty-four hours. Its effect is slightly aperient.

Camplin's Bran Loaf, for Diabetes.

Take two or three quarts of wheat bran, boil it in two successive waters for ten minutes, each time straining it through a sieve, then wash it well with cold water (on the sieve), until the water runs off perfectly clear; squeeze the bran in a cloth as dry as you can, then spread it thinly on a dish, and place it in a slow oven.—If put in at night let it remain until the morning, when, if perfectly dry and crisp, it will be fit for grinding. The bran thus prepared must be ground in a fine mill, and sifted through a wire sieve of sufficient fineness to require the use of a brush to pass it through : that which does not pass at first ought to be ground and sifted again, until the whole is soft and fine.

Take of this bran-powder three troyounces ; three fresh eggs; an ounce and a half of butter, and rather less than half a pint of milk. Mix the eggs with part of the milk, and warm the butter with the other portion ; then stir the whole well together, adding a little nutmeg and ginger, or other spice. Just before putting into the oven, stir in, first, thirty-five grains of bicarbonate of soda, and then three drachms of dilute hydrochloric acid. Bake the loaf in a basin (well buttered) for an hour or rather more.

Wine Whey.

Boil half a pint of milk, and, while boiling, add a wineglassful of Madeira or Sherry wine. Separate the curd, by straining through muslin or a sieve. Sweeten the whey to taste, and grate upon it a little nutmeg.

Egg and Wine, or Brandy.

Beat up a raw fresh egg, and stir with it two tablespoonfuls of wine, or one of brandy. Sweeten or not, according to taste.

Caudle.

Beat up a raw egg with a wineglassful of Sherry, and add to it half a pint of hot gruel. Flavor with lemon-peel, nutmeg, and sugar.

Milk Punch.

Into a tumblerful of milk put one or two tablespoonfuls of brandy, whisky, or Jamaica rum. Sweeten well, and grate nutmeg on top.

Ferruginous Chocolate.

Mix sixteen ounces of chocolate with half an ounce of carbonate of iron. Divide the mass into cakes of one ounce each. One may be dissolved in half a pint of hot milk, to be taken night and morning.

DISINFECTANTS.

The best preventives of infection are **ventilation** and **cleanliness**. No agencies can be made to take the place of these. The following are the most available temporary aids in purification of insalubrions places.

For disinfection of **privies**: *sulphate of iron*, a pound dissolved in a gallon of water; or the same amount of *chloride of lime* may be thoroughly mixed in water.

Burnet's Liquid consists of solution of *chloride of zinc*, twenty-five grains in each fluidrachm of water. Of this a pint may be put into a gallon of water for use.

For **water closets, bed-pans**, etc., Labarraque's solution of *chlorinated soda* may be employed,—a fluidounce to a quart of water; or *permanganate of potassa*, from one to five grains to the ounce of water; or carbolic acid, twenty grains to the pint or quart. This last disinfectant, much valued in England, is not yet made cheaply here. *Coal tar* possesses its virtues in a dilute form.

Articles of clothing, contaminated by discharges, etc., from patients, if very bad, should be *burned*. Otherwise, they should be *boiled* thoroughly; or, at least, plunged into boiling water. Woollens, and all clothing which cannot be washed, as well as bedding, should be exposed for several hours to a *dry heat* of from 200° to 250° Fahrenheit.

Occupied rooms and houses may be disinfected (besides ventilation) by diffusing in spray through the air Ledoyen's liquid (solution of *nitrate of lead*, made by dissolving one pound of litharge in seven ounces of nitric acid and two gallons of water). Or, by placing in shallow vessels the solid *chloride of lime* (bleaching salt). Fresh whitewashing is beneficial to the air of a cellar. *Charcoal* and *quicklime* are absorbent (especially the former) of gases, and thus aid in purifying the air.

Hospital wards may be disinfected (besides ventilation and cleansing) by Ledoyen's liquid, or chloride of lime, or *bromine* left exposed to the air in shallow vessels, or *iodine*, heated moderately.

Heaps of filth, solid or semi-liquid, may be covered with charcoal, two or three inches deep, or with *dry earth*. **Drains, ditches**, and **sewers** may be disinfected with sulphate of iron, coal tar, chloride of lime, &c. A pound of good chloride of lime will suffice for a thousand gallons of running sewage.

On the subject of *ozone* as a disinfectant, I refer to works on chemistry and hygiene.

INDEX OF FORMULÆ.

34*

GENERAL INDEX.

AN ANALYTICAL COMPENDIUM OF THE VARIOUS

BRANCHES OF MEDICAL SCIENCE; for the Use and Examination of Students. By JOHN NEILL, M. D., and FRANCIS G. SMITH, M. D., Prof. of the Institutes of Medicine in the Univ. of Penn'a. A new edition, revised and improved. In one very large and handsomely printed royal 12mo. volume of about 1000 pages, with 374 wood-cuts; extra cloth, $4; strongly bound in leather, with raised bands, $4 75.

The compend of Drs. Neill and Smith is incomparably the most valuable work of its class ever published in this country. Attempts have been made in various quarters to squeeze Anatomy, Physiology, Surgery, the Practice of Medicine, Obstetrics, Materia Medica, and Chemistry into a single manual; but the operation has signally failed in the hands of all up to the advent of "Neill and Smith's" volume, which is quite a miracle of success. The outlines of the whole are admirably drawn and illustrated, and the authors are eminently entitled to the grateful consideration of the student of every class. —*N. O. Med. and Surg. Journal.*

This popular favorite with the student is so well known that it requires no more at the hands of a medical editor than the annunciation of a new and improved edition. There is no sort of comparison between this work and any other on a similar plan, and for a similar object.—*Nashville Journal of Medicine.*

There are but few students or practitioners of medicine unacquainted with the former editions of this unassuming though highly instructive work. The whole science of medicine appears to have been sifted, as the gold-bearing sands of El Dorado, and the precious facts treasured up in this little volume. A complete portable library so condensed that the student may make it his constant pocket companion.—*Western Lancet.*

To compress the whole science of medicine in less than 1000 pages is an impossibility, but we think that the book before us approaches as near to it as is possible. Altogether, it is the best of its class, and has met with a deserved success. As an elementary text-book for students, it has been useful, and will continue to be employed in the examination of private classes, whilst it will often be referred to by the country practitioner.—*Virginia Medical Journal.*

As a handbook for students it is invaluable, containing in the most condensed form the established facts and principles of medicine and its collateral sciences.—*N. H. Journal of Medicine.*

Having made free use of this volume in our examinations of pupils, we can speak from experience in recommending it as an admirable compend for students, and especially useful to preceptors who examine their pupils. It will save the teacher much labor by enabling him readily to recall all of the points upon which his pupils should be examined. A work of this sort should be in the hands of every one who takes pupils into his office with a view of examining them; and this is unquestionably the best of its class. Let every practitioner who has pupils provide himself with it, and he will find the labor of refreshing his knowledge so much facilitated that he will be able to do justice to his pupils at very little cost of time or trouble to himself.—*Transylvania Medical Journal.*

LUDLOW'S MANUAL OF EXAMINATIONS.

A MANUAL OF EXAMINATIONS upon Anatomy, Physiology, Surgery, Practice of Medicine, Obstetrics, Materia Medica, Chemistry, Pharmacy, and Therapeutics. To which is added a Medical Formulary. By J. L. LUDLOW, M. D. Third edition, thoroughly revised and greatly extended and enlarged. With 370 illustrations. In one handsome royal 12mo. volume of 816 large pages; extra cloth, $3 25; leather, $3 75.

The arrangement of this volume in the form of question and answer renders it especially suitable for the office examination of students, and for those preparing for graduation.

We know of no better companion for the student during the hours spent in the lecture-room, or to refresh, at a glance, his memory of the various topics crammed into his head by the various professors to whom he is compelled to listen. — *Western Lancet.*

As it embraces the whole range of medical studies it is necessarily voluminous, containing 816 large duodecimo pages. After a somewhat careful examination of its contents, we have formed a much more favorable opinion of it than we are wont to regard such works. Although well adapted to meet the wants of the student in preparing for his final examination, it might be profitably consulted by the practitioner also, who is most apt to become rusty in the very kind of details here given, and who, amid the hurry of his daily routine, is but too prone to neglect the study of more elaborate works. The possession of a volume of this kind might serve as an inducement for him to seize the moment of excited curiosity to inform himself on any subject, and which is otherwise too often allowed to pass unimproved.—*St. Louis Med. and Surg. Journal.*

HENRY C. LEA—Philadelphia.

DUNGLISON'S MEDICAL DICTIONARY.

MEDICAL LEXICON; A DICTIONARY OF MEDICAL SCIENCE: Containing a concise explanation of the various Subjects and Terms of Anatomy, Physiology, Pathology, Hygiene, Therapeutics, Pharmacology, Pharmacy, Surgery, Obstetrics, Medical Jurisprudence, and Dentistry. Notices of Climate and of Mineral Waters; Formulæ for Officinal, Empirical, and Dietetic Preparations; with the Accentuation and Etymology of the Terms, and the French and other Synonymes; so as to constitute a French as well as English Medical Lexicon. By ROBLEY DUNGLISON, M. D., Professor of Institutes of Medicine in Jefferson Medical College, Philadelphia. Thoroughly Revised, and very greatly Modified and Augmented. In one very large and handsome royal octavo volume of 1048 double-columned pages, in small type; strongly done up in extra cloth, $6; leather, raised bands, $6 75.

The object of the author from the outset has not been to make the work a mere lexicon or dictionary of terms, but to afford, under each, a condensed view of its various medical relations, and thus to render the work an epitome of the existing condition of medical science. Starting with this view, the immense demand which has existed for the work has enabled him, in repeated revisions, to augment its completeness and usefulness, until at length it has attained the position of a recognized and standard authority wherever the language is spoken. The mechanical execution of this edition will be found greatly superior to that of previous impressions. By enlarging the size of the volume to a royal octavo, and by the employment of a small but clear type, on extra fine paper, the additions have been incorporated without materially increasing the bulk of the volume, and the matter of two or three ordinary octavos has been compressed into the space of one not unhandy for consultation and reference.

It would be a work of supererogation to bestow a word of praise upon this Lexicon. We can only wonder at the labor expended, for whenever we refer to its pages for information we are seldom disappointed in finding all we desire, whether it be in accentuation, etymology, or definition of terms.—*New York Medical Journal*, November, 1865.

It would be mere waste of words in us to express our admiration of a work which is so universally and deservedly appreciated. The most admirable work of its kind in the English language. As a book of reference it is invaluable to the medical practitioner, and in every instance that we have turned over its pages for information we have been charmed by the clearness of language and the accuracy of detail with which each abounds. We can most cordially and confidently commend it to our readers.—*Glasgow Medical Journal*, January, 1866.

A work to which there is no equal in the English language.—*Edinburgh Med. Journ.*

It is something more than a dictionary, and something less than an encyclopædia. This edition of the well-known work is a great improvement on its predecessors. The book is one of the very few of which it may be said with truth that every medical man should possess it.—*London Medical Times*, Aug. 26, 1865.

Few works of the class exhibit a grander monument of patient research and of scientific lore. The extent of the sale of this lexicon is sufficient to testify to its usefulness, and to the great service conferred by Dr. Robley Dunglison on the profession, and indeed on others, by its issue.—*London Lancet*, May 13, 1865.

It is as necessary a work to every enlightened physician as Worcester's English Dictionary is to every one who would keep up his knowledge of the English tongue to the standard of the present day. It is, to our mind, the most complete work of the kind with which we are acquainted.—*Boston Med. and Surg Journal*, June 22, 1865.

We are free to confess that we know of no medical dictionary more complete; no one better, if so well adapted for the use of the student; no one that may be consulted with more satisfaction by the medical practitioner.—*Am. Journ. Med. Sciences*, April, 1865.

A DICTIONARY OF THE TERMS USED IN MEDICINE
AND THE COLLATERAL SCIENCES. By RICHARD D. HOBLYN, M. D. A new American edition, revised, with numerous additions, by ISAAC HAYS, M. D., Editor of the "American Journal of the Medical Sciences." In one large royal 12mo. volume of over 500 double-columned pages; extra cloth, $1 50; leather, $2.

It is the best book of definitions we have, and ought always to be upon the student's table.—*Southern Med. and Surg. Journal*.

HENRY C. LEA—Philadelphia.

CATALOGUE OF BOOKS

PUBLISHED BY

HENRY C. LEA,

(LATE LEA & BLANCHARD.)

The books in the annexed list will be sent by mail, post-paid, to any Post Office in the United States, on receipt of the printed prices. No risks of the mail, however, are assumed. Gentlemen will therefore, in most cases, find it more convenient to deal with the nearest bookseller. Detailed catalogues furnished or sent free by mail on application. An illustrated catalogue of 64 octavo pages, handsomely printed, mailed on receipt of 10 cents. Address,

HENRY C. LEA,

Nos. 706 and 708 Sansom Street, Philadelphia.

AMERICAN JOURNAL OF THE MEDICAL SCIENCES. Edited by Isaac Hays, M.D., published quarterly, about 1100 large 8vo. pages per annum, **MEDICAL NEWS AND LIBRARY**, monthly, 384 large 8vo. pages per annum, } For five Dollars per annum in advance.

OR,

AMERICAN JOURNAL OF THE MEDICAL SCIENCES, Quarterly, **MEDICAL NEWS AND LIBRARY**, monthly, **RANKING'S HALF-YEARLY ABSTRACT OF THE MEDICAL SCIENCES.** 2 vols. a year, of about 300 pages each. In all, over 2000 large 8vo. pages per annum, } For six Dollars per annum in advance.

ABSTRACT, RANKING'S HALF-YEARLY, per volume, $1 50; per annum, $2 50.

ALLEN (J. M.) THE PRACTICAL ANATOMIST; or, THE STUDENT'S GUIDE IN THE DISSECTING ROOM. With 266 illustrations. 1 vol. royal 12mo., over 600 pages, cloth, $2.

ASHTON (T. J.) ON THE DISEASES, INJURIES, AND MALFOR-MATIONS OF THE RECTUM AND ANUS. With remarks on Habitual Constipation. Second American from the fourth London edition, with illustrations. 1 vol. 8vo. of about 300 pp., cloth, $3 25.

ABEL AND BLOXAM'S HANDBOOK OF CHEMISTRY, THEORE-TICAL, PRACTICAL, AND TECHNICAL. With illustrations. 1 vol. 8vo. of 662 pages, cloth, $4 50.

ARNOTT (NEIL). ELEMENTS OF PHYSICS; or, NATURAL PHILO-SOPHY, GENERAL AND MEDICAL. 1 vol. 8vo., with illustrations, cloth, $2 25.

ASHWELL (SAMUEL). A PRACTICAL TREATISE ON THE DIS-EASES OF WOMEN. Third American from the third London edi-tion. In one 8vo. vol. of 528 pages, cloth, $3 50.

BRINTON (WILLIAM). LECTURES ON THE DISEASES OF THE STOMACH; with an introduction on its Anatomy and Physiology. From the second London edition, with illustrations. 1 vol. 8vo. of about 300 pages, cloth, $3 25.

BRANDE (WM. T.), AND ALFRED S. TAYLOR. CHEMISTRY. Second American edition, thoroughly revised by Dr. Taylor. In one large and handsome octavo volume, extra cloth, $5; leather, $6.

BUMSTEAD (F. J.) THE PATHOLOGY AND TREATMENT OF VENEREAL DISEASES. Including the results of recent investigations upon the subject. A new and revised edition, with illustrations. 1 vol. 8vo., of 640 pages, cloth, $5.

—— AND CULLERIER'S ATLAS OF VENEREAL. See "CULLERIER."

BARWELL (RICHARD). A TREATISE ON DISEASES OF THE JOINTS. Illustrated with engravings on wood. 1 vol. 8vo., of about 500 pages, cloth, $3.

BUCKNILL (J. C.) AND DANIEL M. TUKE. A MANUAL OF PSYCHOLOGICAL MEDICINE. Containing the History, Nosology, Description, Statistics, Diagnosis, Pathology, and Treatment of Insanity. With a Plate. 1 vol. 8vo., of 536 pages, cloth, $4 25.

BARCLAY (A. W.) A MANUAL OF MEDICAL DIAGNOSIS; being an Analysis of the Signs and Symptoms of Disease. Third American from the second revised London edition. 1 vol. 8vo., of 451 pages, cloth, $3 50.

BENNET (HENRY). A PRACTICAL TREATISE ON INFLAMMATION OF THE UTERUS, ITS CERVIX AND APPENDAGES, AND ON ITS CONNECTION WITH UTERINE DISEASE. Sixth American, from the fourth and revised English edition. 1 vol. 8vo., of about 500 pages, cloth, $3 75.

—— A REVIEW OF THE PRESENT STATE OF UTERINE PATHOLOGY. 1 small vol. 8vo., cloth, 50 cents.

BARLOW (GEORGE H.) A MANUAL OF THE PRACTICE OF MEDICINE. With additions by D. F. Condie, M.D. 1 vol. 8vo., of over 600 pages, cloth, $2 50.

BROWN (ISAAC BAKER). ON SOME DISEASES OF WOMEN ADMITTING OF SURGICAL TREATMENT. With illustrations. 1 vol. 8vo., of 276 pages, cloth, $1 60.

BROWNE (R. W.) A HISTORY OF GREEK CLASSICAL LITERATURE. Second American, from a revised English edition. 1 vol. crown 8vo., of about 500 pages, cloth, $1 90.

—— A HISTORY OF ROMAN CLASSICAL LITERATURE. Second American, from a revised English edition. 1 vol. crown 8vo., of about 500 pages, cloth, $1 90.

BAIRD (ROBERT). IMPRESSIONS AND EXPERIENCES OF THE WEST INDIES AND UNITED STATES. 1 vol. royal 12mo., cloth, 75 cents.

BUDD (GEORGE). ON DISEASES OF THE LIVER. Third American, from the third and enlarged London edition. With four colored plates and numerous wood-cuts. 1 vol. 8vo., of 500 pages, cloth, $4.

BUCKLER (THOMAS H.) ON FIBRO-BRONCHITIS AND RHEUMATIC PNEUMONIA. 1 vol. 8vo., of 150 pages, cloth, $1 25.

BLAKISTON (PEYTON). ON CERTAIN DISEASES OF THE CHEST. 1 vol. 8vo., cloth, $1 25.

BOWMAN (JOHN E.) A PRACTICAL HAND-BOOK OF MEDICAL CHEMISTRY. Edited by C. L. Bloxam. Fourth American, from the fourth and revised London edition. With numerous illustrations. 1 vol. royal 12mo. of 350 pages, cloth, $2 25.

—— INTRODUCTION TO PRACTICAL CHEMISTRY, INCLUDING ANALYSIS. Edited by C. L. Bloxam. Fourth American, from the fifth and revised London edition, with numerous illustrations. 1 vol. royal 12mo. of 350 pages, cloth, $2 25.

BRODIE (SIR BENJAMIN). CLINICAL LECTURES ON SURGERY. 1 vol. 8vo., of 350 pages, cloth, $1 25.

CHAMBERS (T. K.) THE INDIGESTIONS; OR, DISEASES OF THE DIGESTIVE ORGANS FUNCTIONALLY TREATED. Second American, from the second and enlarged London edition. 1 vol. 8vo., of over 300 pages, cloth, $3 00.

COLUMBAT DE L'ISERE. THE DISEASES OF FEMALES. Translated by Charles D. Meigs, M.D. Second edition, with numerous illustrations. 1 vol. 8vo., of 720 pages, cloth, $3 75.

CARPENTER (WM. B.) PRINCIPLES OF HUMAN PHYSIOLOGY, WITH THEIR CHIEF APPLICATIONS TO PSYCHOLOGY, PATHOLOGY, THERAPEUTICS, HYGIENE, AND FORENSIC MEDICINE. A new American edition edited by Francis G. Smith, M.D. With nearly 300 illustrations. In one large vol. 8vo., of nearly 900 closely printed pages, cloth, $5 50; leather, raised bands, $6 50.

—— PRINCIPLES OF COMPARATIVE PHYSIOLOGY. New American, from the fourth and revised London edition. With over 300 beautiful illustrations. 1 vol. 8vo., of 752 pages, cloth, $5 00.

—— THE MICROSCOPE AND ITS REVELATIONS. With an Appendix containing the applications of the Microscope to Clinical Medicine, by Francis G. Smith, M.D. With 434 handsome illustrations. 1 vol. 8vo., of 724 pages, cloth, $5 25.

—— PRIZE ESSAY ON THE USE OF ALCOHOLIC LIQUORS IN HEALTH AND DISEASE. New edition, with a Preface by D. F. Condie, M.D. 1 vol. 12mo. of 178 pages, cloth, 60 cents.

CARSON (JOSEPH). A SYNOPSIS OF THE COURSE OF LECTURES ON MATERIA MEDICA AND PHARMACY, delivered in the University of Pennsylvania. Fourth and revised edition. 1 vol. 8vo., extra cloth, $3 00. (Now ready.)

CHRISTISON (ROBERT.) DISPENSATORY OR COMMENTARY ON THE PHARMACOPŒIAS OF GREAT BRITAIN AND THE UNITED STATES. With a Supplement by R. E. Griffith. In one 8vo. vol. of over 1000 pages, containing 213 illustrations, extra cloth, $4 00.

CHURCHILL (FLEETWOOD). ON THE THEORY AND PRACTICE OF MIDWIFERY. A new American from the fourth revised London edition. With notes and additions by D. Francis Condie, M.D. With about 200 illustrations. In one handsome 8vo. vol. of nearly 700 pages, extra cloth, $4 00; leather, $5 00.

—— ON THE DISEASES OF WOMEN: INCLUDING THOSE OF PREGNANCY AND CHILDBED. A new American edition revised by the author. With notes and additions by D. Francis Condie, M.D. In one large and handsome 8vo. vol. of 768 pages, with numerous illustrations, extra cloth, $4 00; leather, $5 00.

—— ESSAYS ON THE PUERPERAL FEVER, AND OTHER DISEASES PECULIAR TO WOMEN. In one neat octavo vol. of about 450 pages, extra cloth, $2 50.

CLYMER ON FEVERS. In one 8vo. vol. of 600 pages, leather, $1 75.

CONDIE (D. FRANCIS). A PRACTICAL TREATISE ON THE DISEASES OF CHILDREN. Sixth edition, revised and enlarged. In one large octavo volume of nearly 800 pages, extra cloth, $5 25; leather, $6 25.

COOPER (B. B.) LECTURES ON THE PRINCIPLES AND PRACTICE OF SURGERY. In one large 8vo. vol. of 750 pages, extra cloth, $2 00.

CURLING (T. B.) A PRACTICAL TREATISE ON DISEASES OF THE TESTIS, SPERMATIC CORD, AND SCROTUM. 1 vol. 8vo. of 420 pages, extra cloth, $2 00.

CULLERIER (A.) AN ATLAS OF VENEREAL DISEASES. Translated and edited by FREEMAN J. BUMSTEAD, M.D. A large imperial quarto volume, with 26 plates containing about 150 figures, beautifully colored, many of them the size of life. To be issued in five parts, at $3 each. (Parts I., II., and III. now ready.)

CYCLOPEDIA OF PRACTICAL MEDICINE. By Dunglison, Forbes, Tweedie, and Conolly. In four large super royal octavo volumes, of 3254 double-columned pages, leather, raised bands, $15; extra cloth, $11.

CAMPBELL'S LIVES OF LORDS KENYON, ELLENBOROUGH, AND TENTERDEN. Being the third volume of "Campbell's Lives of the Chief Justices of England." In one crown octavo vol., cloth, $2.

DALTON (J. C.) A TREATISE ON HUMAN PHYSIOLOGY. Fourth edition, revised, with nearly 300 illustrations on wood. In one very handsome octavo volume of about 700 pages, extra cloth, $5 25; leather, $6 25.

DE JONGH, ON THE THREE KINDS OF COD-LIVER OIL. 1 small 12mo. vol., 75 cents.

DEWEES (W. P.) A TREATISE ON THE DISEASES OF FEMALES. With illustrations. In one 8vo. vol. of 536 pages, extra cloth, $3.

—— A COMPREHENSIVE SYSTEM OF MIDWIFERY. In one octavo volume of 600 pages, with plates, extra cloth, $3 50.

—— A TREATISE ON THE PHYSICAL AND MEDICAL TREATMENT OF CHILDREN. In one octavo volume of 548 pages, extra cloth, $2 80.

DICKSON (S. H.) ELEMENTS OF MEDICINE. Second edition, revised. 1 vol. 8vo., of 750 pages, extra cloth, $4.

DRUITT (ROBERT). THE PRINCIPLES AND PRACTICE OF MO- DERN SURGERY. A revised American, from the eighth London edition. Illustrated with 432 wood engravings. In one handsome 8vo. vol. of nearly 700 large and closely printed pages, extra cloth, $4; leather, $5.

DUNGLISON (ROBLEY). MEDICAL LEXICON; a Dictionary of Medical Science. Containing a concise explanation of the various subjects and terms of Anatomy, Physiology, Pathology, Hygiene, Therapeutics, Pharmacology, Pharmacy, Surgery, Obstetrics, Medical Jurisprudence, and Dentistry. Notices of Climate and of Mineral Waters; Formulæ for Officinal, Empirical, and Dietetic Preparations, with the accentuation and Etymology of the Terms, and the French and other Synonymes; so as to constitute a French as well as English Medical Lexicon. In one very large royal 8vo. vol. of 1048 double columned pages, in small type; strongly bound in cloth, $6; leather, raised bands, $6 75.

—— HUMAN PHYSIOLOGY. Eighth edition, thoroughly revised. In two large 8vo. vols. of about 1500 pages, with 532 illustrations, extra cloth, $7.

—— GENERAL THERAPEUTICS AND MATERIA MEDICA. Sixth edition, revised and improved. With 193 illustrations. In two large 8vo. vols. of about 1100 pages, extra cloth, $6 50.

—— NEW REMEDIES, WITH FORMULÆ FOR THEIR PREPARA- TION AND ADMINISTRATION. Seventh edition. In one very large 8vo. vol. of 770 pages, extra cloth, $4.

DE LA BACHE'S GEOLOGICAL OBSERVER. In one large 8vo. vol. of 700 pages, with 300 illustrations, cloth, $4.

DON QUIXOTE DE LA MANCHA. Translated by Chas. Jarvis, Esq., with illustrations by Tony Johannot. In two handsome vols. crown 8vo., fancy cloth, $3; plain cloth, $2 50; library sheep, $3 20; half morocco, $3 70.

DANA (JAMES D.) ON ZOOPHYTES. In one imperial quarto vol. with numerous illustrations on wood, cloth. Also an Atlas of 60 colored plates, imperial folio, half morocco. Price for the whole, $45. A few copies of the Atlas can be had separate; price, $30.

—— THE STRUCTURE AND CLASSIFICATION OF ZOOPHYTES. With illustrations on wood. In one imperial 4to. vol., cloth, $4.

ELLIS (BENJAMIN). THE MEDICAL FORMULARY. Being a collection of prescriptions derived from the writings and practice of the most eminent physicians of America and Europe. Twelfth edition, carefully revised. In one 8vo. vol. (Nearly Ready.)

ERICHSEN (JOHN). THE SCIENCE AND ART OF SURGERY. A new and improved American, from the second enlarged and revised London edition. Illustrated with over 400 engravings on wood. In one large 8vo. vol. of 1000 closely printed pages, extra cloth, $6; leather, raised bands, $7.

—— ON RAILWAY AND OTHER INJURIES OF THE NERVOUS SYSTEM. In one small 8vo. vol., extra cloth, $1. (Just Issued.)

ENCYCLOPÆDIA AMERICANA. Complete in 14 large 8vo. vols. Containing nearly 9000 double columned pages, cloth, $22.

ENCYCLOPÆDIA OF GEOGRAPHY. In three large 8vo. vols. Illustrated with 83 maps and about 1100 wood-cuts, cloth, $5.

FISKE FUND PRIZE ESSAYS ON TUBERCULOUS DISEASE. In one small 8vo. vol., cloth, $1.

FLINT (AUSTIN). A TREATISE ON THE PRINCIPLES AND PRACTICE OF MEDICINE. Second edition, revised and enlarged. In one large 8vo. volume of nearly 1000 pages, extra cloth, $6 50; leather, raised bands, $7 50. (Just Issued.)

—— A PRACTICAL TREATISE, ON THE PHYSICAL EXPLORATION OF THE CHEST, AND THE DIAGNOSIS OF DISEASES AFFECTING THE RESPIRATORY ORGANS. Second and revised edition. One 8vo. vol. of 595 pages, cloth, $4 50. (Just issued.)

—— A PRACTICAL TREATISE ON THE DIAGNOSIS AND TREATMENT OF DISEASES OF THE HEART. In one neat 8vo. vol. of nearly 500 pages, extra cloth, $3 50.

FOWNE (GEORGE). A MANUAL OF ELEMENTARY CHEMISTRY. With 197 illustrations. In one royal 12mo. vol. of 600 pages, extra cloth, $2; leather, $2 50.

FULLER (HENRY). ON DISEASES OF THE LUNGS AND AIR PASSAGES. Their Pathology, Physical Diagnosis, Symptoms and Treatment. From the second English edition. In one 8vo. vol. of about 500 pages, extra cloth, $3 50. (Just issued.)

FLETCHER'S NOTES FROM NINEVEH, AND TRAVELS IN MESOPOTAMIA, ASSYRIA, AND SYRIA. In one 12mo. vol., cloth, 75 cts.

GARDNER'S MEDICAL CHEMISTRY. In one 12mo. vol. of 396 pages, cloth, $1.

GLUGE (GOTTLIEB). ATLAS OF PATHOLOGICAL HISTOLOGY. Translated by Joseph Leidy, M.D., Professor of Anatomy in the University of Pennsylvania, &c. In one vol. imperial quarto, with 320 copper plate figures, plain and colored, extra cloth, $4.

GRAHAM (THOMAS). THE ELEMENTS OF INORGANIC CHEMISTRY, INCLUDING THE APPLICATION OF THE SCIENCE IN THE ARTS. A new and enlarged edition by H. Watts and Robert Bridges, M.D. In one 8vo. vol., of over 800 pages, with 232 wood-cuts, extra cloth, $5 50.

GIBSON'S INSTITUTES AND PRACTICE OF SURGERY. In two 8vo. vols. of about 1000 pages, leather, $6 50.

*

GRAY (HENRY). ANATOMY, DESCRIPTIVE AND SURGICAL. Second American, from the second revised London edition. In one large imperial 8vo. vol. of over 800 pages, with 388 large and elaborate engravings on wood. Extra cloth, $6 ; leather, raised bands, $7.

GRIFFITH (ROBERT E.) A UNIVERSAL FORMULARY, CONTAINING THE METHODS OF PREPARING AND ADMINISTERING OFFICINAL AND OTHER MEDICINES. In one large 8vo. vol. of 650 pages, double columns, extra cloth, $4 ; leather, $5.

GROSS (SAMUEL D.) A SYSTEM OF SURGERY, PATHOLOGICAL, DIAGNOSTIC, THERAPEUTIC, AND OPERATIVE. Illustrated by over 1300 engravings. Fourth edition, revised and improved. In two large royal 8vo. vols. of 2200 pages, strongly bound in leather, raised bands, $15.

—— A PRACTICAL TREATISE ON THE DISEASES, INJURIES, AND MALFORMATIONS OF THE URINARY BLADDER, THE PROSTATE GLAND, AND THE URETHRA. Second edition, with 184 illustrations. One large 8vo. vol. of over 900 pages, extra cloth, $4.

—— A PRACTICAL TREATISE ON FOREIGN BODIES IN THE AIR PASSAGES. In one 8vo. vol. of 468 pages. Extra cloth, $2 75.

—— ELEMENTS OF PATHOLOGICAL ANATOMY. Third edition. In one large 8vo. vol. of nearly 800 pages, with about 350 illustrations, extra cloth, $4.

GUIZOT'S HISTORY OF OLIVER CROMWELL. In two royal 12mo. vols. Containing 900 pages, cloth, $2.

HARTSHORNE (HENRY). ESSENTIALS OF THE PRINCIPLES AND PRACTICE OF MEDICINE. In one 12mo. vol. of about 350 pages, cloth, $2 38 ; half bound, $2 63. (Just issued.)

HABERSHON (S. O.) PATHOLOGICAL AND PRACTICAL OBSERVATIONS ON DISEASES OF THE ALIMENTARY CANAL, ŒSOPHAGUS, STOMACH, CÆCUM, AND INTESTINES. In one 8vo. vol. of 312 pages, extra cloth, $2 50.

HUDSON (A.) LECTURES ON THE STUDY OF FEVER. 1 vol. 8vo. (Publishing in the Med. News and Library for 1867 and 1868.)

HAMILTON (FRANK H.) A PRACTICAL TREATISE ON FRACTURES AND DISLOCATIONS. Third edition, revised. In one handsome 8vo. vol. of 777 pages, with 294 illustrations, extra cloth, $5 75.

HARRISON'S ESSAY TOWARD A CORRECT THEORY OF THE NERVOUS SYSTEM. In one vol. 8vo. of 292 pages, cloth, $1 50.

HOBLYN (RICHARD D.) A DICTIONARY OF THE TERMS USED IN MEDICINE AND THE COLLATERAL SCIENCES. In one 12mo. vol. of over 500 double columned pages, cloth, $1 50 ; leather, $2.

HODGE (HUGH L.) ON DISEASES PECULIAR TO WOMEN, INCLUDING DISPLACEMENTS OF THE UTERUS. In one 8vo. vol. (A new edition preparing.)

—— THE PRINCIPLES AND PRACTICE OF OBSTETRICS. Illustrated with large lithographic plates containing 159 figures from original photographs, and with numerous wood-cuts. In one large quarto vol. of 550 double-columned pages. Strongly bound in extra cloth, $14.

HOLLAND (SIR HENRY). MEDICAL NOTES AND REFLECTIONS. From the third English edition. In one 8vo. vol. of about 500 pages, extra cloth, $3 50.

HODGES (RICHARD M.) PRACTICAL DISSECTIONS. Second edition. In one neat royal 12mo. vol., half bound, $2.

HORNER (WILLIAM E.). SPECIAL ANATOMY AND HISTOLOGY. Eighth edition, revised and modified. In two large 8vo. vols. of over 1000 pages, containing 300 wood-cuts, extra cloth, $6.

HUGHES (H. M.) A CLINICAL INTRODUCTION TO AUSCULTA-TION AND OTHER MODES OF PHYSICAL DIAGNOSIS. Second edition. In one 12mo. vol. of 304 pages, cloth, $1 25.

HILLIER (THOMAS). HAND BOOK OF SKIN DISEASES. In one neat 12mo. vol. of about 300 pages, with two plates, extra cloth, $2 25.

HUGHES' SCRIPTURE GEOGRAPHY AND HISTORY, with 12 colored maps. In 1 vol. 12mo., cloth, $1.

HALL (MRS. M.) LIVES OF THE QUEENS OF ENGLAND BEFORE THE NORMAN CONQUEST. In one handsome 8vo. vol., cloth, $2 25; crimson cloth, $2 50; half morocco, $3.

HUMBOLDT'S ASPECTS OF NATURE. In one large 12mo. vol., extra cloth, $1.

JONES (T. WHARTON). THE PRINCIPLES AND PRACTICE OF OPHTHALMIC MEDICINE AND SURGERY, with 117 illustra-tions. Third American, with additions from the second London edi-tion. In one 8vo. vol. of 455 pages, extra cloth, $3 25.

JONES (C. HANDFIELD), AND SIEVEKING (E. D. H.) A MANUAL OF PATHOLOGICAL ANATOMY. In one large 8vo. vol. of nearly 750 pages, with 397 illustrations, extra cloth, $3 50.

JONES (C. HANDFIELD). CLINICAL OBSERVATIONS ON FUNC-TIONAL NERVOUS DISORDERS. Second American Edition. In one 8vo. vol. of 348 pages, extra cloth, $3 25. (Just Issued.)

KIRKES (WILLIAM SENHOUSE). A MANUAL OF PHYSIOLOGY. From the third London edition, with 200 illustrations. In one large 12mo. vol. of 586 pages, cloth, $2 25; leather, $2 75.

KNAPP (F.) TECHNOLOGY; OR CHEMISTRY APPLIED TO THE ARTS AND TO MANUFACTURES, with American additions, by Prof. Walter R. Johnson. In two 8vo. vols., with 500 illustrations, cloth, $6.

KENNEDY'S MEMOIRS OF THE LIFE OF WILLIAM WIRT. In two vols. 12mo., cloth, $2.

LEA (HENRY C.) SUPERSTITION AND FORCE; ESSAYS ON THE WAGER OF LAW, THE WAGER OF BATTLE, THE ORDEAL, AND TORTURE. In one handsome royal 12mo. vol. of 406 pages, extra cloth, $2 50.

LALLEMAND (M.) AND WILSON (MARRIS). A PRACTICAL TREATISE ON THE CAUSES, SYMPTOMS, AND TREATMENT OF SPERMATORRHŒA. Translated and edited by Henry J. McDougall. Fifth American edition. To which is added——ON DISEASES OF THE VESICULÆ SEMINALES. With special re-ference to the Morbid Secretions of the Prostatic and Urethral Mucous Membrane. By Marris Wilson, M. D. In one neat octavo volume, of about 400 pages, extra cloth, $2 75.

LA ROCHE (R.) YELLOW FEVER IN ITS HISTORICAL, PATHO-LOGICAL, ETIOLOGICAL, AND THERAPEUTICAL RELA-TIONS. In two 8vo. vols. of nearly 1500 pages, extra cloth, $7.

—— PNEUMONIA, ITS SUPPOSED CONNECTION, PATHOLO-GICAL AND ETIOLOGICAL, WITH AUTUMNAL FEVERS. In one 8vo. vol. of 500 pages, extra cloth, $3.

LAURENCE (J. Z.) AND MOON (ROBERT C.) A HANDY BOOK OF OPHTHALMIC SURGERY. With numerous illustrations. In one 8vo. vol., extra cloth, $2 50. (Just Issued.)

LAWSON (GEORGE). INJURIES OF THE EYE, ORBIT, AND EYE-LIDS, with about 100 illustrations. From the last English edition. In one handsome 8vo. vol., extra cloth, $3 50. (Just issued.)

LAYCOCK (THOMAS). LECTURES ON THE PRINCIPLES AND METHODS OF MEDICAL OBSERVATION AND RESEARCH. In one 12mo. vol., extra cloth, $1.

LEHMANN (C. G.) PHYSIOLOGICAL CHEMISTRY. Translated by George F. Day, M. D., and edited by R. E. Rogers, M. D., Prof. of Chemistry, in the University of Pennsylvania. With plates, and nearly 200 illustrations. In two large 8vo. vols., containing 1200 pages, extra cloth, $6.

—— A MANUAL OF CHEMICAL PHYSIOLOGY. Translated with notes and additions, by J. Cheston Morris, M. D. With an Introductory Essay on Vital Force, by Prof. Samuel Jackson. In one very handsome 8vo. vol. of 336 pages, extra cloth, $2 25.

LUDLOW (J. L.) A MANUAL OF EXAMINATIONS UPON ANATOMY, PHYSIOLOGY, SURGERY, PRACTICE OF MEDICINE, OBSTETRICS, MATERIA MEDICA, CHEMISTRY, PHARMACY, AND THERAPEUTICS. To which is added a Medical Formulary. Third edition. In one royal 12mo. vol. of over 800 pages, extra cloth, $3 25 ; leather, $3 75.

LYONS (ROBERT D.) A TREATISE ON FEVER. In one neat 8vo. vol. of 362 pages, extra cloth, $2 25.

LYNCH (W. F.) A NARRATIVE OF THE UNITED STATES EX-PEDITION TO THE DEAD SEA AND RIVER JORDAN. In one large and handsome octavo vol., with 28 beautiful plates and two maps, cloth, $3.

MACLISE (JOSEPH). SURGICAL ANATOMY. In one large imperial quarto vol., with 68 splendid plates, beautifully colored ; containing 190 figures, many of them life size, extra cloth, $14.

MALGAIGNE'S OPERATIVE SURGERY. With numerous illustrations. In one 8vo. vol. of nearly 600 pages, cloth, $2 50.

MANUALS OF BLOOD AND URINE. By Griffith, Reese, and Markwick. 1 vol. 12mo. of 460 pages, extra cloth, $1 25.

MAYNE'S DISPENSATORY AND THERAPEUTICAL REMEMBRANCER. · Edited by R. E. Griffith, M.D. In one 12mo. vol. of about 300 pages, extra cloth, 75 cents.

MACKENZIE (W.) A PRACTICAL TREATISE ON DISEASES AND INJURIES OF THE EYE. In one handsome 8vo. vol. of 1027 pages, with plates and numerous wood-cuts, extra cloth, $6 50.

MEIGS (CHAS. D.) OBSTETRICS, THE SCIENCE AND THE ART. Fifth edition, revised, with 130 illustrations. In one beautifully printed 8vo. vol. of 760 pages, extra cloth, $5 50 ; leather, $6 50. (Just issued.)

—— WOMAN : HER DISEASES AND THEIR REMEDIES. Fourth and improved edition. In one large 8vo. vol. of over 700 pages, extra cloth, $5 ; leather, $6.

—— ON THE NATURE, SIGNS, AND TREATMENT OF CHILD-BED FEVER. In one 8vo. vol. of 365 pages, extra cloth, $2.

MILLER (HENRY). PRINCIPLES AND PRACTICE OF OBSTETRICS, &c. In one very handsome 8vo. vol. of over 600 pages, extra cloth, $3 75.

MILLER (JAMES). PRINCIPLES OF SURGERY. Fourth American, from the third Edinburgh edition. In one large 8vo. vol. of 700 pages, with 240 illustrations, extra cloth, $3 75.

—— THE PRACTICE OF SURGERY. Fourth American, from the last Edinburgh edition. In one large 8vo. vol. of 700 pages, with 364 illustrations, extra cloth, $3 75.

MONTGOMERY (W. F.) AN EXPOSITION OF THE SIGNS AND SYMPTOMS OF PREGNANCY. From the second English edition. In one handsome 8vo. vol. of nearly 600 pages, extra cloth, $3 75.

MORLAND (W. W.) DISEASES OF THE URINARY ORGANS. With illustrations. In one handsome 8vo. vol. of about 600 pages, extra cloth, $3 50.

MORLAND (W. W.) ON THE RETENTION IN THE BLOOD OF THE ELEMENTS OF THE URINARY SECRETION. In one vol. 8vo., extra cloth, 75 cents.

MILLWRIGHT'S GUIDE. By Oliver Evans. Fourteenth edition. In one vol. 8vo. with numerous plates, extra cloth, $2 50.

MULLER (J.) PRINCIPLES OF PHYSICS AND METEOROLOGY. In one large 8vo. vol. with 550 wood-cuts, and two colored plates, cloth, $4 50.

MIRABEAU; A LIFE HISTORY. In one royal 12mo. vol., cloth, 75 cents.

MACFARLAND'S TURKEY AND ITS DESTINY. In 2 vols. royal 12mo., cloth, $2.

MARSH (MRS.) A HISTORY OF THE PROTESTANT REFORMA-TION IN FRANCE. In 2 vols. royal 12mo., extra cloth, $2.

NEILL (JOHN) AND SMITH (FRANCIS G.) COMPENDIUM OF THE VARIOUS BRANCHES OF MEDICAL SCIENCE. In one handsome 12mo. vol. of about 1000 pages, with 374 wood-cuts, extra cloth, $4; leather, raised bands, $4 75.

NELIGAN (J. MOORE). A PRACTICAL TREATISE ON DISEASES OF THE SKIN. Fifth American, from the second Dublin edition. In one neat royal 12mo. vol. of 462 pages, extra cloth, $2 25.

—— AN ATLAS OF CUTANEOUS DISEASES. In one handsome quarto vol. with beautifully colored plates, &c., extra cloth, $5 50.

NIEBUHR (B. G.) LECTURES ON ANCIENT HISTORY; comprising the history of the Asiatic Nations, the Egyptians, Greeks, Macedonians, and Carthagenians. Translated by Dr. L. Schmitz. In three neat volumes, crown octavo, cloth, $5 00.

PARRISH (EDWARD). A TREATISE ON PHARMACY. With many Formulæ and Prescriptions. Third edition. In one handsome 8vo. vol. of 850 pages, with several hundred illustrations, extra cloth, $5.

PEASLEE (E. R.) HUMAN HISTOLOGY IN ITS RELATIONS TO ANATOMY, PHYSIOLOGY, AND PATHOLOGY. With 434 illustrations. In one 8vo. vol. of 600 pages, extra cloth, $3 75.

PIRRIE (WILLIAM). THE PRINCIPLES AND PRACTICE OF SURGERY. In one handsome octavo volume of 780 pages, with 316 illustrations, extra cloth, $3 75.

PEREIRA (JONATHAN). MATERIA MEDICA AND THERAPEU-TICS. An abridged edition of the late Dr. Pereira's "Elements of Materia Medica." With numerous additions and references to the United States Pharmacopœia. In one large octavo volume, of 1040 pages, with 236 illustrations, extra cloth $7; leather, raised bands, $8. (Just issued.)

PULSZKY'S MEMOIRS OF AN HUNGARIAN LADY. In one neat royal 12mo. vol., extra cloth, $1.

PAGET'S HUNGARY AND TRANSYLVANIA. In two royal 12mo. vols., cloth, $2.

RAMSBOTHAM (FRANCIS H.) THE PRINCIPLES AND PRAC-TICE OF OBSTETRIC MEDICINE AND SURGERY. In one imperial 8vo. vol. of 650 pages, with 64 plates, besides numerous wood-cuts in the text. Strongly bound in leather $7.

ROBERTS (WILLIAM). A PRACTICAL TREATISE ON URINARY
AND RENAL DISEASES. With numerous illustrations. In one
very handsome 8vo. vol. of 516 pages, extra cloth, $4 50.

RIGBY (EDWARD). THE CONSTITUTIONAL TREATMENT OF
FEMALE DISEASES. In one neat royal 12mo. vol. of about 250
pp., extra cloth, $1.

—— A SYSTEM OF MIDWIFERY. Second American edition. In
one handsome 8vo. vol. of 422 pages, extra cloth, $2 50.

ROKITANSKY (CARL). A MANUAL OF PATHOLOGICAL ANA-
TOMY. Translated by W. E. Swaine, Edward Sieveking, C. H.
Moore, and G. E. Day. Four vols. 8vo., bound in two. About 1200
pages, extra cloth, $7 50.

ROYLE (J. FORBES). MATERIA MEDICA AND THERAPEUTICS.
Edited by Jos. Carson, M. D. In one large 8vo. vol. of about 700
pages, with 98 illustrations, extra cloth, $3.

RANKE'S HISTORY OF THE TURKISH AND SPANISH EMPIRES
in the 16th and beginning of 17th Century. In one 8vo. volume,
paper, 25 cts.

—— HISTORY OF THE REFORMATION IN GERMANY. Parts I.
II. III. In one vol., extra cloth, $1.

SARGENT (F. W.) ON BANDAGING AND OTHER OPERATIONS
OF MINOR SURGERY. New edition, with an additional chapter
on Military Surgery. In one handsome royal 12mo. vol. of nearly
400 pages, with 184 wood-cuts, extra cloth, $1 75.

SHARPEY (WILLIAM) AND QUAIN (JONES AND RICHARD).
HUMAN ANATOMY. With notes and additions by Jos. Leidy,
M. D., Prof. of Anatomy in the University of Pennsylvania. In two
large 8vo. vols. of about 1300 pages, with 511 illustrations, extra
cloth, $6.

SIMPSON (SIR JAMES Y.) CLINICAL LECTURES ON THE DIS-
EASES OF WOMEN. With numerous illustrations. In one 8vo.
vol. of 500 pages, extra cloth. (A new edition preparing.)

SIMON'S GENERAL PATHOLOGY. In one 8vo. vol. of 212 pages,
extra cloth, $1 25.

SKEY (FREDERIC C.) OPERATIVE SURGERY. In one 8vo. vol.
of over 650 pages, with about 100 wood-cuts, cloth, $3 25.

SLADE (D. D.) DIPHTHERIA; ITS NATURE AND TREATMENT.
Second edition. In one neat royal 12mo. vol., extra cloth, $1 25.

SMITH (HENRY H.) AND HORNER (WILLIAM E.) ANATOMICAL
ATLAS. Illustrative of the structure of the Human Body. In one
large imperial 8vo. vol., with about 650 beautiful figures, extra
cloth, $4 50.

SMITH (EDWARD). CONSUMPTION; ITS EARLY AND REME-
DIABLE STAGES. In one 8vo. vol. of 254 pp., extra cloth, $2 25.

SOLLY (SAMUEL). THE HUMAN BRAIN; ITS STRUCTURE,
PHYSIOLOGY, AND DISEASES. In one neat 8vo. vol. of 500 pp.
with 120 wood-cuts, extra cloth, $2 50.

STILLE (ALFRED). THERAPEUTICS AND MATERIA MEDICA.
Third edition, revised and enlarged. In two large and handsome
8vo. vols., extra cloth, $10; leather, $12. (Now Ready.)

SALTER (H. H.) ASTHMA; ITS PATHOLOGY, CAUSES, CONSE-
QUENCES, AND TREATMENT. In one volume 8vo., extra cloth,
$2 50.

SMALL BOOKS ON GREAT SUBJECTS. Twelve works; each one 15
cents, sewed, forming a neat and cheap series; or done up in 3 vols.,
extra cloth, $1 50.

SCHOEDLER (FREDERICK) AND MEDLOCK (HENRY). WONDERS OF NATURE. An elementary introduction to the Sciences of Physics, Astronomy, Chemistry, Mineralogy, Geology, Botany, Zoology, and Physiology. Translated from the German by H. Medlock. In one neat 8vo. vol., with 679 illustrations, extra cloth, $3.

STRICKLAND (AGNES). LIVES OF THE QUEENS OF HENRY THE VIII. AND OF HIS MOTHER. In one crown octavo vol., extra cloth, $1; black cloth, 90 cents.

——MEMOIRS OF ELIZABETH, SECOND QUEEN REGNANT OF ENGLAND AND IRELAND. In one crown octavo vol., extra cloth, $1 40; black cloth, $1 30.

SCHMITZ AND ZUMPT'S CLASSICAL SERIES. In royal 18mo. CORNELII NEPOTIS LIBER DE EXCELLENTIBUS DUCIBUS EXTERARUM GENTIUM, CUM VITIS CATONIS ET ATTICI. With notes, &c. Price in extra cloth, 60 cents; half bound, 70 cts.

C. I. CÆSARIS COMMENTARII DE BELLO GALLICO. With notes. map, and other illustrations. Price in extra cloth, 60 cents; half bound, 70 cents.

C. C. SALLUSTII DE BELLO CATILINARIO ET JUGURTHINO. With notes, map, &c. Price in extra cloth, 60 cents; half bound, 70 cents.

Q. CURTII RUFII DE GESTIS ALEXANDRI MAGNI LIBRI VIII. With notes, map, &c. Price in extra cloth, 80 cents; half bound, 90 cents.

P. VIRGILII MARONIS CARMINA OMNIA. Price in extra cloth, 85 cents; half bound, $1.

M. T. CICERONIS ORATIONES SELECTÆ XII. With notes, &c. Price in extra cloth, 70 cents; half bound, 80 cents.

ECLOGÆ EX Q. HORATII FLACCI POEMATIBUS. With notes, &c. Price in extra cloth, 70 cents; half bound, 80 cents.

SCHMITZ'S ADVANCED GRAMMAR OF THE LATIN LANGUAGE. Half bound, price 80 cents.

ADVANCED LATIN EXERCISES, WITH SELECTIONS FOR READING. Revised, with additions. Extra cloth, price 60 cents; half bound, 70 cents.

TANNER (THOMAS HAWKES). A MANUAL OF CLINICAL MEDICINE AND PHYSICAL DIAGNOSIS. Third American from the second revised English edition. In one handsome 12mo. vol. (Preparing for early publication.)

—— ON THE SIGNS AND DISEASES OF PREGNANCY. First American from the second English edition. With four colored plates and numerous illustrations on wood. In one vol. 8vo. of about 500 pages, extra cloth, $4 25. (Now ready.)

TAYLOR (ALFRED S.) MEDICAL JURISPRUDENCE. Sixth American from the eighth London edition. With notes and references to American Decisions, by C. B. Penrose of the Philadelphia Bar. In one large 8vo. vol. of 776 pages, extra cloth, $4 50; leather, $5 50.

THOMAS (T. GAILLARD). A COMPLETE PRACTICAL TREATISE ON THE DISEASES OF FEMALES. In one large and handsome octavo volume of 625 pages, with 219 illustrations, extra cloth, $5; leather, $6. (Now ready.)

TODD (ROBERT B.) AND BOWMAN (W.) PHYSIOLOGICAL ANATOMY AND PHYSIOLOGY OF MAN. In one large 8vo. vol. of about 950 pages, with 300 illustrations on wood, extra cloth, $4 75.

TODD (ROBERT BENTLEY). CLINICAL LECTURES ON CERTAIN ACUTE DISEASES. In one vol. 8vo. of 320 pp., extra cloth, $2 50.

TOYNBEE (JOSEPH). THE DISEASES OF THE EAR: Their nature, Diagnosis, and Treatment. Second American edition. In one handsome 8vo. vol. of 440 pp., with 100 illustrations, extra cloth, $4.

WALES (PHILIP S.) MECHANICAL THERAPEUTICS: A Practical Treatise on Surgical Apparatus, Appliances, and Elementary Operations ; embracing Minor Surgery, Bandaging, Orthopraxy, and Treatment of Fractures and Dislocations. In one large 8vo. vol. of about 700 pages, with 642 illustrations on wood, extra cloth, $5 75 ; leather, $6 75. (Just issued.)

WALSHE (W. H.) PRACTICAL TREATISE ON THE DISEASES OF THE HEART AND GREAT VESSELS. Third American from the third revised London edition. In one 8vo. vol. of 420 pages, extra cloth, $3.

WHAT TO OBSERVE AT THE BEDSIDE AND AFTER DEATH IN MEDICAL CASES. In one royal 12mo. vol., extra cloth, $1.

WATSON (THOMAS). LECTURES ON THE PRINCIPLES AND PRACTICE OF PHYSIC. A new American from the last revised English edition, with additions by D. Francis Condie. With 185 illustrations on wood. In one very large volume imperial 8vo. of over 1200 pages, in small type, extra cloth, $6 50 ; strongly bound in leather, raised bands, $7 50.

WEST (CHARLES). LECTURES ON THE DISEASES PECULIAR TO WOMEN. Third American from the Third English edition. In one octavo volume of 550 pages, extra cloth, $3 75 ; leather, $4 75. (Now ready.)

——— LECTURES ON THE DISEASES OF INFANCY AND CHILD-HOOD. Fourth American from the fifth revised English edition. In one large 8vo. vol. of 656 closely printed pages, extra cloth, $4 50 ; leather, $5 50.

——— AN ENQUIRY INTO THE PATHOLOGICAL IMPORTANCE OF ULCERATION OF THE OS UTERI. In one vol. 8vo., extra cloth, $1 25.

WILLIAMS (CHARLES J. B.) PRINCIPLES OF MEDICINE. A new American from the third revised London edition. In one 8vo. vol. of about 500 pages, extra cloth, $3 50.

WILSON (ERASMUS). A SYSTEM OF HUMAN ANATOMY. A new and revised American from the last English edition. Illustrated with 397 engravings on wood. In one handsome 8vo. vol. of over 600 pages, extra cloth, $4 ; leather, $5.

——— THE DISSECTOR'S MANUAL. Third American from the last revised London edition. In one large 12mo. vol. of 582 pages, with 154 illustrations, extra cloth, $2.

——— ON DISEASES OF THE SKIN. The seventh American from the last English edition. In one large 8vo. vol. of over 800 pages, extra cloth, $5. (Just ready.)

Also A SERIES OF PLATES, illustrating " Wilson on Diseases of the Skin," consisting of 20 plates, thirteen of which are beautifully colored, representing about one hundred varieties of Disease. $5 50.

Also, the TEXT AND PLATES, bound in one volume, extra cloth, $10.

——— THE STUDENT'S BOOK OF CUTANEOUS MEDICINE. In one handsome royal 12mo. vol., extra cloth, $3 50.

——— ON SKIN AND HAIR IN RELATION TO HEALTH. In one vol. 12mo. of 291 pages, cloth, $1.

WINSLOW (FORBES). ON OBSCURE DISEASES OF THE BRAIN AND DISORDERS OF THE MIND. In one handsome 8vo. vol. of nearly 600 pages, extra cloth, $4 25.

www.ingramcontent.com/pod-product-compliance
Lightning Source LLC
Chambersburg PA
CBHW032305280326
41932CB00009B/699